SUPERSYMMETRY

An Introduction with Conceptual and Calculational Details

SUPERSYMMETRY

An Introduction with Conceptual and Calculational Details

H. J. W. Müller-Kirsten
A. Wiedemann

Department of Physics, University of Kaiserslautern
Kaiserslautern, West Germany

World Scientific
Singapore • New Jersey • Hong Kong

Published by

World Scientific Publishing Co. Pte. Ltd.
P.O. Box 128, Farrer Road, Singapore 9128

U.S.A. office: World Scientific Publishing Co., Inc.
687 Hartwell Street, Teaneck NJ 07666, USA

Library of Congress Cataloging-in-Publication data is available.

SUPERSYMMETRY — AN INTRODUCTION WITH CONCEPTUAL AND CALCULATIONAL DETAILS

ISBN 9971-50-354-9
9971-50-355-7 pbk

Printed in Singapore by Kim Hup Lee Printing Co. Pte. Ltd.

PREFACE

This text is a detailed version of material presented by both of us in seminars and lectures in the theory group of this department. Except for parts of Chapters 9 and 10 the material has also been covered in a series of seminars by one of us (M.-K.) in the Department of Physics of the University of Adelaide, Adelaide, Australia, in August and September 1985 and in the Department of Physics of Shanxi University, Taiyuan, China, in March and April 1987. The interest and criticism of the audiences at these departments and, in particular, the support and enthusiasm of Professor A.W. Thomas (Adelaide) and Professor Zhang Jianzu (Taiyuan) are gratefully acknowledged.

The text was compiled with the belief that the majority of potential readers is more interested in actually using or applying supersymmetry in some model theory than in painstakingly rediscovering the results of others for themselves. It seemed plausible, therefore, to revise various relevant concepts and in particular, to include the proof or verification of almost every formula. In this way the reader can select the problems he wants to tackle himself, compare his solutions with the calculations given here, and thus gain the confidence in his own calculations which he needs for his discussions of supersymmetry in other contexts. It has been our experience that (except for the last two chapters) the material presented here can be covered in a one-semester course for graduate or post graduate students with some knowledge of field theory.

In compiling this text we have, of course, used previous reviews. The choice of our sequence of topics was motivated by the lecture notes of Legovini[1]. Standard texts which we have consulted are the monograph by Wess and Bagger[2] and the review by Fayet and Ferrara[3]. For the detailed treatment of the on-shell Wess-Zumino model we consulted the lecture notes of de Roo[4]. In the text we do not discuss any experimental signatures of supersymmetry. For an introduction into this topic we refer to articles by Haber and Kane[5,6]; further details can be found in the Proceedings of the Thirteenth SLAC Summer Institute on Particle Physics[7] and in the reviews by Nilles[8] and Dragon, Ellwanger and Schmidt[9]. As further general references we refer to the nontechnical review by Wess[10], to a very brief review of topics covered here by Campbell and Fogleman[11] and to the lectures of Wess[12], Ferrara[13] and Witten[14]. A more advanced text is the book by Gates, Grisaru, Rocek and Siegel[15]. The very readable review by Sohnius[16] appeared after completion of the first draft of our text. Meanwhile several other texts have been published, each, however, with its emphasis in a different direction. We refer here to the books by West[17], Srivastava[18] and Freund[19]. For more specific topics we refer to the article by Salam and Strathdee[20] to the Proceedings of the 28th Scottish Universities' Summer School in Physics[21], and to the Proceedings of the NATO Advanced Study Institute on Supersymmetry[22]. All considerations of this text refer to a four-dimensional Minkowski space. For the basic technicalities in the context of supersymmetric quantum mechanics we refer to the work by Cooper and Freedman[23], whereas those of two-dimensional field theories can be found in reference 24.

CONTENTS

INTRODUCTION

Symmetries are of fundamental importance in the description of physical phenomena. In the realm of particle physics symmetries are believed to permit ultimately a classification of all observed particles. A fundamental symmetry of particle physics, which has been firmly established both theoretically and experimentally is that of the Poincaré group, i.e. of rotations and translations in four-dimensional Minkowski space. Besides this fundamental symmetry there are other socalled internal symmetries (such as the symmetry of the SU(3) flavour group) which have also been firmly established over the last few decades, although their manifestation in Nature is not exact. As is well known, the consistent search for more fundamental symmetries led to the development of nonabelian gauge theories and the spectacular experimental confirmation of several predictions of the latter in recent years.

In the course of time several attempts have been made to unify the space-time symmetry of the Poincaré group with the symmetry of some internal group. Such attempts have, however, been shown to be futile if the theory, which necessarily has to be a quantum field theory, is expected to satisfy certain basic requirements. In fact, the socalled "no-go" theorem of Coleman and Mandula [25] shows that if one makes the plausible assumptions of locality, causality, positivity of energy and finiteness of the number of particles (and one more technical assumption) the invariance group of the theory can at best be the direct product of the Poincaré group and a compact (internal) group, and this therefore does not offer a genuine unification of one group with the other.

Now the generators of the Poincaré group satisfy well known commutation relations, and Noether's theorem relates these to conserved currents. In their turn the conserved currents are functions of relativistic fields. The commutation relations of the field operators which quantize these fields are therefore directly related to those of the generators. It was realized by Wess and Zumino[26,27] that if one allows also anticommutation relations of generators of supersymmetry transformations which transform bosons into fermions and vice versa, then the unification of the space-time symmetries of the Poincaré group with this internal symmetry can be achieved. The formal proof of this discovery, i.e. the proof that anticommuting generators which respect the other assumptions of the theorem of Coleman and Mandula[25] do exist, was established by Haag, Lopuszanski and Sohnius[28].

Supersymmetry thus arises as a symmetry which combines bosons and fermions in the same representation or multiplet of the enlarged group which encompasses both the transformations of the Poincaré group and the appropriate supersymmetry transformations. Thus every bosonic particle must have a fermionic partner and vice versa. In view of the fact that such a spectrum of particles is not compatible with observation, supersymmetry must be badly broken at the level of presently available energies. Clearly only experimental observation can decide whether supersymmetry is indeed inherent in Nature. It can be argued that one of the most immediate ways to observe evidence of supersymmetry is to see if there is a missing energy and momentum in the final e^+e^- spectrum of the reaction

$$e^+ + e^- \longrightarrow \gamma \longrightarrow \tilde{e}^+ + \tilde{e}^- \longrightarrow e^+ + e^- + \tilde{\gamma} + \tilde{\gamma}$$

where $\tilde{e}^+$, $\tilde{e}^-$ and $\tilde{\gamma}$ are the supersymmetry partners of e^+, e^- and γ respectively. If there is such a missing energy and momentum it could be that carried away by the neutral photinos $\tilde{\gamma}$ (charged supersymmetry particles at energies presently available would have been detected long ago). Since supersymmetry must be broken, the photinos $\tilde{\gamma}$ would not be massless.

However, supersymmetry does not only open the possibility of a much more complex spectrum of particles than heretofore envisaged; supersymmetry also has some intriguing theoretical consequences which could make it a desirable symmetry. It is well known that a realistic quantum field theory in the traditional sense is plagued by the problem of ultraviolet divergences and the consequent necessity of renormalization. Supersymmetry, however, provides a mechanism for the cancellation of such divergences in view of the same number of bosonic and fermionic degrees of freedom in each particle multiplet. Clearly such a built-in cancellation of divergent terms is a highly desirable feature of a quantum field theory.

In Chapter 1 we begin with a recapitulation of basic aspects of the Lorentz group, including a discussion of Casimir operators and the classification of representations in terms of their eigenvalues. We then consider the group SL(2,C) and its basic representations, i.e. the self-representation and the complex conjugate self-representation. The elements of the appropriate representations spaces are the undotted and dotted Weyl spinors. In view of the importance of Weyl spinors throughout the entire text, we consider these here in more detail than is generally done in the literature. We then introduce the concept of Grassmann number and perform some basic manipulations involving Weyl spinors, thereby deriving a number of useful formulae. In the subsequent section the connection between SL(2,C) and the proper orthochonous Lorentz group is established. It is then natural to discuss four-component Dirac spinors and the Weyl representation. The connection with two-component Weyl spinors is obtained by introducing four-component Majorana spinors. Then again various formulae are derived which are useful in later calculations.

Chapter 2 begins with a discussion of the "no-go" theorems of Coleman and Mandula[25] and Haag, Lopuszanski and Sohnius[28]. The latter leads to a consideration of graded Lie algebras which we approach in successive steps by defining first the characteristics of a Lie algebra, then those of a graded algebra and finally those of a graded Lie algebra, i.e. the properties of grading, supersymmetrization and generalized Jacobi identities. As an example we calculate the graded Lie algebra of the algebra su(2,C). The final section of Chapter 2 deals with graded matrices and their properties.

Chapter 3 deals with the grading, i.e. supersymmetric extension of the Poincaré algebra. We demonstrate explicitly that for the grading chosen all possible Jacobi identities are satisfied. Having established the algebra of the Super-Poincaré group with the fermionic generators in the Dirac four-component form, we then decompose it into the appropriate relations of the Weyl two-component method.

In Chapter 4 we use the method of Casimir operators to classify the irreducible representations of the Super-Poincaré algebra, and it is shown that supersymmetry implies an equal number of bosonic and fermionic degrees of freedom.

Chapter 5 deals with the most immediate field theoretical realization of the Super-Poincaré algebra, the Wess-Zumino model, which is a field theory involving a scalar field, a pseudoscalar field and one spinor field, all with the same mass. We demonstrate by explicit calculation that the spinor charges of the theory, considered as linear operators in Fock space, satisfy the commutation and anti-commutation relations of the Super-Poincaré algebra.

In Chapter 6 we introduce the concepts of superspace and superfields, and define differentiation with respect to Grassmann numbers. Then three different but related operators are constructed which describe three different but equivalent actions of the supersymmetry group on functions in superspace. These operators define three different types of superfields. By considering infinitesimal supersymmetry transformations we obtain the corresponding three differential operator representations of the fermionic generators of the Super-Poincaré group. Then covariant derivatives are introduced as a prerequisite for the construction of manifestly supersymmetric action integrals. These covariant derivatives also permit the definition of projection operators. The search for irreducible representations of the Super-Poincaré algebra then becomes a search for solutions of constraint equations expressed in terms of these projection operators. The final section of Chapter 6 is devoted to the derivation of the explicit supersymmetry transformations of the component fields of the supermultiplet. In this context it is seen that the highest order component field always transforms into a total Minkowski derivative and thus is a candidate for a supersymmetric Lagrangian density.

In Chapter 7 we begin with an investigation of the constraint equations which define left-handed and right-handed chiral superfields (also known as scalar superfields). Then vector superfields are defined by an appropriate constraint equation, and the supersymmetric generalization of the abelian gauge transformation is discussed. Finally left-handed and right-handed spinor superfields are discussed which represent the components of the supersymmetric field strength for an arbitrary vector superfield.

Chapter 8 deals with the construction of supersymmetric action integrals. We begin with the definition of integration over Grassmann numbers. Then Lagrangians are constructed from scalar superfields and from vector superfields (i.e. the supersymmetric field strength). The case of the former is shown to contain the Wess-Zumino model as a special case, whereas the case of the latter yields the supersymmetric generalization of the pure Maxwell theory (i.e. with no interaction with matter fields) which contains in addition to the massless vector field also the massless spinor field of the photino.

Chapter 9 deals with the spontaneous breaking of supersymmetry. For the convenience of discussions the concept of superpotential is introduced. In view of the necessity of evaluating action integrals over superspace an equivalent and convenient Grassmann projection technique is developed. Some general aspects of spontaneous symmetry breaking are then discussed and, in particular, the Goldstone theorem is established for the general case of the breaking of supersymmetry or some other symmetry. Finally the O'Raifeartaigh model, which is a specific theory involving three scalar superfields, is considered and the spectrum resulting from the spontaneous breaking of supersymmetry is investigated. In this case supersymmetry breaking results from the nonvanishing vacuum expectation value of some auxiliary field of a superfield.

Finally, in Chapter 10, we consider supersymmetric gauge theories. Introducing first global and local U(1) gauge transformations of scalar superfields and the corresponding supersymmetric version of minimal coupling, we consider super quantum electrodynamics. We then investigate the Fayet-Iliopoulos mechanism of spontaneous breaking of supersymmetry in which the latter results from the nonvanishing vacuum expectation value of the highest order component field of a vector superfield. The last section contains a brief introduction to nonabelian gauge transformations for superfields with the appropriate tensorial transformation properties.

CHAPTER 1

LORENTZ GROUP, POINCARÉ GROUP, SL(2,C), DIRAC AND MAJORANA SPINORS

1.1 The Lorentz Group[a]

A point in the space-time manifold is denoted by $(x^\mu) = (x^0, x^1, x^2, x^3)$ where $x^0 = t$ and x^1, x^2, x^3 are the space components of the four-vector x^μ. The laws of physics are invariant under the Lorentz group. Transformations of this group are linear transformations acting on four-vectors

$$x'^\mu = \Lambda^\mu{}_\nu x^\nu \tag{1.1}$$

leaving the quadratic form

$$\begin{aligned} x^2 &= x^\mu x_\mu \\ &= \eta_{\mu\nu} x^\mu x^\nu \\ &= (x^0)^2 - (\vec{x})^2 \end{aligned} \tag{1.2}$$

invariant, i.e.

a Sections 1.1 and 1.2 serve mainly the purpose of completeness, to define notation and to recollect some formulae which will be needed later in the text. The reader familiar with Sections 1.1 and 1.2 could start immediately with Section 1.3 .

$$x'^2 = x'^\mu x'_\mu = \eta_{\mu\nu} x'^\mu x'^\nu$$
$$= \eta_{\mu\nu} \Lambda^\mu{}_\rho x^\rho \Lambda^\nu{}_\tau x^\tau$$
$$\stackrel{!}{=} \eta_{\rho\tau} x^\rho x^\tau$$

Hence

$$\eta_{\mu\nu} \Lambda^\mu{}_\rho \Lambda^\nu{}_\tau = \eta_{\rho\tau} \tag{1.3}$$

Here $\eta_{\mu\nu}$ = diag (1, -1, -1, -1) is the metric tensor; it lowers indices and its inverse $\eta^{\mu\nu}$ raises indices.

<u>Proposition</u>: The constraints

$$\det \Lambda = \pm 1, \quad |\Lambda^0{}_0| \geq 1 \tag{1.4}$$

define four disconnected pieces in the parameter space.

<u>Proof</u>: The determinant of a product of matrices is the product of the determinants. Hence, taking the determinant of (1.3) yields

$$[\det \Lambda]^2 = 1 \quad \text{or} \quad \det \Lambda = \pm 1$$

Taking the oo-component of (1.3) we obtain

$$\eta_{00} = \eta_{\mu\nu} \Lambda^\mu{}_0 \Lambda^\nu{}_0$$
$$= \eta_{00} \Lambda^0{}_0 \Lambda^0{}_0 + \eta_{ii} \Lambda^i{}_0 \Lambda^i{}_0$$
$$= (\Lambda^0{}_0)^2 - (\Lambda^k{}_0)^2$$
$$= 1$$

or $$(\Lambda^0{}_0)^2 = 1 + (\Lambda^k{}_0)^2$$

Hence

$$(\Lambda^0{}_0)^2 \geq 1$$

The second of constraints (1.4) distinguishes socalled orthochronous Lorentz transformations with $\Lambda^0{}_0 \geq 1$ and nonorthochronous Lorentz transformations with $\Lambda^0{}_0 \leq -1$.

<u>Proposition</u>: The matrices $(\Lambda^\mu{}_\nu)$ form a noncompact Lie

group, the Lorentz group

$$L := O(1,3) = \{\Lambda \in GL(4,R) \mid \Lambda^T \eta \Lambda = \eta\}$$

with Lie algebra

$$o(1,3) := \{\mathfrak{a} \in M_{4\times 4}(R) \mid \mathfrak{a}^T = -\eta \mathfrak{a} \eta\}$$

where GL(4,R) denotes the set of all invertible 4x4-matri - ces with real components, and $M_{4x4}(R)$ is the set of all 4x4-matrices with real elements.

Proof: From Lie algebra theory we know that each $\Lambda \in O(1,3)$ can be written in the form

$$\Lambda(t) = \exp(t\mathfrak{a})$$

where t is a real parameter and $\mathfrak{a} \in o(1,3)$ is an element of the Lie algebra. Matrices of O(1,3) are subject to the condition

$$\Lambda^T(t)\, \eta\, \Lambda(t) = \eta$$

Inserting the above expression we obtain

$$[\exp(t\mathfrak{a})]^T \eta [\exp(t\mathfrak{a})] = \eta .$$

and considering

$$\frac{d}{dt}\left\{[\exp(t\mathfrak{a})]^T \eta [\exp(t\mathfrak{a})]\right\}\Big|_{t=0} = 0$$

we obtain the condition for Lie algebra elements, since the Lie algebra of any Lie group is isomorphic to the tangent space at the identity of the group. It follows that

$$\frac{d}{dt}[\exp(t\mathfrak{a})]^T \eta \exp(t\mathfrak{a})\Big|_{t=0} + [\exp(t\mathfrak{a})]^T \eta \frac{d}{dt}\exp(t\mathfrak{a})\Big|_{t=0} = 0$$

which leads to

$$\mathfrak{a}^T \eta + \eta \mathfrak{a} = 0$$

or

$$\mathfrak{a}^T = -\eta \mathfrak{a} \eta \qquad \forall \mathfrak{a} \in o(1,3)$$

In summary we have the following classification: Let Λ be any invertible 4x4-matrix with real elements, i.e. $\Lambda \in$ GL(4,R), then

(i) The full Lorentz group is

$$L := O(1,3) = \{\Lambda \in GL(4,R) \mid \Lambda^T \eta \Lambda = \eta\}$$

(ii) Proper Lorentz transformations are

$$L_+ := SO(1,3) = \{\Lambda \in O(1,3) \mid \det \Lambda = +1\}$$

L_+ being a subgroup of L.

(iii) Improper Lorentz transformations are

$$L_- := \{\Lambda \in O(1,3) \mid \det \Lambda = -1\}$$

L_- is not a subgroup of L, since the identity element is not an element of L_-. Note, however, that discrete transformations such as time- or space-reflection are elements of L_-.

(iv) Orthochronous Lorentz transformations are

$$L^{\uparrow} := \{\Lambda \in O(1,3) \mid \Lambda^0{}_0 \geq +1\}$$

$L^{\uparrow}$ is a subgroup of L.

(v) Nonorthochronous Lorentz transformations are

$$L^{\downarrow} := \{\Lambda \in O(1,3) \mid \Lambda^0{}_0 \leq -1\}$$

(vi) The restricted Lorentz group is

$$\begin{aligned} L^{\uparrow}_+ &= L^{\uparrow} \cap L_+ \\ &= \{\Lambda \in O(1,3) \mid \det \Lambda = +1, \Lambda^0{}_0 \geq +1\} \end{aligned}$$

This subgroup of L is also called the proper orthochronous Lorentz group; it does not contain time- or space-reflections.

<u>Remark</u>: Lorentz transformations which are orthochronous map the forward light-cone onto itself and $\Lambda \in L^{\downarrow}$ maps the backward light-cone onto itself.

<u>Generators of the Lorentz group</u>

In the neighbourhood of the identity $1_{SO(1,3)}$, a Lorentz transformation $\Lambda \in L^{\uparrow}_+$ can be written

$$\Lambda = 1_{4\times 4} + \omega \tag{1.6}$$

Inserting this expression into (1.3) we obtain

$$\eta_{\mu\nu}\Lambda^{\mu}{}_{\rho}\Lambda^{\nu}{}_{\sigma}$$
$$= \eta_{\mu\nu}\{\delta^{\mu}{}_{\rho}+\omega^{\mu}{}_{\rho}\}\{\delta^{\nu}{}_{\sigma}+\omega^{\nu}{}_{\sigma}\}$$
$$= \eta_{\mu\nu}\delta^{\mu}{}_{\rho}\delta^{\nu}{}_{\sigma}+\eta_{\mu\nu}\delta^{\mu}{}_{\rho}\omega^{\nu}{}_{\sigma}+\eta_{\mu\nu}\omega^{\mu}{}_{\rho}\delta^{\nu}{}_{\sigma}$$
$$= \eta_{\rho\sigma}+\omega_{\rho\sigma}+\omega_{\sigma\rho}$$
$$\overset{!}{=} \eta_{\rho\sigma}$$

Hence

$$\omega_{\rho\sigma} = -\omega_{\sigma\rho} \tag{1.7}$$

i.e. the infinitesimal parameters ω are antisymmetric in ρ and σ. Now

$$x'^{\mu} = \Lambda^{\mu}{}_{\nu}x^{\nu} \quad \text{(with (1.1))}$$
$$= (1+\omega)^{\mu}{}_{\nu}x^{\nu} \quad \text{(with (1.6))}$$
$$= x^{\mu}+\omega^{\mu}{}_{\nu}x^{\nu}$$

and the variation of x^{μ} due to infinitesimal transformations is

$$\delta x^{\mu} := x'^{\mu}-x^{\mu}$$
$$= \omega^{\mu}{}_{\nu}x^{\nu}$$

with antisymmetric infinitesimal parameters as demonstrated in (1.7). On the other hand we may consider the vector representation of the restricted Lorentz group $L_+^{\uparrow}$, and we write $\Lambda\in L_+^{\uparrow}$ in the form

$$\Lambda^{\mu}{}_{\nu} = [\exp(-\tfrac{i}{2}\,\omega^{\rho\sigma}M_{\rho\sigma})]^{\mu}{}_{\nu} \tag{1.8}$$

where the 4x4-matrices $(M_{\rho\sigma})$ constitute a basis of the Lie algebra o(1,3), to be verified later. $(M_{\rho\sigma})$ are antisymmetric in ρ and σ and the factor i is chosen in such a way that the $(M_{\rho\sigma})$ are hermitian. For infinitesimal transformations, i.e. $\omega^{\rho\sigma}$ infinitesimal, we consider

$$\begin{aligned}\delta x^\mu &= x'^\mu - x^\mu \\ &= \Lambda^\mu{}_\nu x^\nu - x^\mu \quad \text{(now use (1.8))} \\ &= [\exp(-\tfrac{i}{2}\omega^{\rho\sigma}M_{\rho\sigma})]^\mu{}_\nu x^\nu - x^\mu \\ &= \{1_{4\times4} - \tfrac{i}{2}\omega^{\rho\sigma}M_{\rho\sigma}\}^\mu{}_\nu x^\nu - x^\mu \\ &= -\tfrac{i}{2}\omega^{\rho\sigma}(M_{\rho\sigma})^\mu{}_\nu x^\nu \end{aligned} \tag{1.9}$$

and we conclude

$$\omega^\mu{}_\nu = -\frac{i}{2}\omega^{\rho\sigma}(M_{\rho\sigma})^\mu{}_\nu \tag{1.10}$$

Therefore the generators $M_{\rho\sigma}$ of the Lorentz group have the matrix form

$$(M_{\rho\sigma})^\mu{}_\nu = i(\eta_{\sigma\nu}\delta_\rho{}^\mu - \eta_{\rho\nu}\delta_\sigma{}^\mu) \tag{1.11}$$

Checking:

$$\begin{aligned}-\frac{i}{2}\omega^{\rho\sigma}(M_{\rho\sigma})^\mu{}_\nu &= \frac{1}{2}\omega^{\rho\sigma}(\eta_{\sigma\nu}\delta_\rho{}^\mu - \eta_{\rho\nu}\delta_\sigma{}^\mu) \\ &= \frac{1}{2}(\omega^\mu{}_\nu - \omega_\nu{}^\mu) = \omega^\mu{}_\nu \quad \text{(with (1.7))}\end{aligned}$$

We now derive (1.11) explicitly.
According to the definition of the Lie algebra o(1,3), any $\mathfrak{a} \in$ o(1,3) satisfies the relation

$$\mathfrak{a}^T = -\eta\,\mathfrak{a}\,\eta$$

Explicitly this is

$$\begin{pmatrix} a_{00} & a_{01} & a_{02} & a_{03} \\ a_{10} & a_{11} & a_{12} & a_{13} \\ a_{20} & a_{21} & a_{22} & a_{23} \\ a_{30} & a_{31} & a_{32} & a_{33} \end{pmatrix}^T$$

$$= -\begin{pmatrix} 1 & & & \\ & -1 & 0 & \\ 0 & & -1 & \\ & & & -1 \end{pmatrix} \begin{pmatrix} a_{00} & a_{01} & a_{02} & a_{03} \\ a_{10} & a_{11} & a_{12} & a_{13} \\ a_{20} & a_{21} & a_{22} & a_{23} \\ a_{30} & a_{31} & a_{32} & a_{33} \end{pmatrix} \begin{pmatrix} 1 & & & \\ & -1 & 0 & \\ 0 & & -1 & \\ & & & -1 \end{pmatrix}$$

which leads to the set of equations

$$a_{00} = a_{11} = a_{22} = a_{33} = 0$$
$$a_{10} = a_{01}, \quad a_{20} = a_{02}, \quad a_{30} = a_{03}$$
$$a_{21} = -a_{12}, \quad a_{31} = -a_{13}, \quad a_{32} = -a_{23}$$

Therefore a typical matrix $\mathfrak{A} \in$ o(1,3) has the form

$$\mathfrak{A} = \begin{pmatrix} 0 & a_{01} & a_{02} & a_{03} \\ a_{01} & 0 & -a_{12} & a_{13} \\ a_{02} & a_{12} & 0 & -a_{23} \\ a_{03} & -a_{13} & a_{23} & 0 \end{pmatrix}$$

We choose a basis of o(1,3) of the following form

$$M_1 := \begin{pmatrix} 0 & 0 & 0 & 0 \\ 0 & 0 & 0 & 0 \\ 0 & 0 & 0 & -1 \\ 0 & 0 & 1 & 0 \end{pmatrix}, \quad M_2 := \begin{pmatrix} 0 & 0 & 0 & 0 \\ 0 & 0 & 0 & 1 \\ 0 & 0 & 0 & 0 \\ 0 & -1 & 0 & 0 \end{pmatrix}$$

$$M_3 := \begin{pmatrix} 0 & 0 & 0 & 0 \\ 0 & 0 & -1 & 0 \\ 0 & 1 & 0 & 0 \\ 0 & 0 & 0 & 0 \end{pmatrix}, \quad N_1 := \begin{pmatrix} 0 & 1 & 0 & 0 \\ 1 & 0 & 0 & 0 \\ 0 & 0 & 0 & 0 \\ 0 & 0 & 0 & 0 \end{pmatrix}$$

$$N_2 := \begin{pmatrix} 0 & 0 & 1 & 0 \\ 0 & 0 & 0 & 0 \\ 1 & 0 & 0 & 0 \\ 0 & 0 & 0 & 0 \end{pmatrix}, \quad N_3 := \begin{pmatrix} 0 & 0 & 0 & 1 \\ 0 & 0 & 0 & 0 \\ 0 & 0 & 0 & 0 \\ 1 & 0 & 0 & 0 \end{pmatrix}$$

The three matrices M_i generate the SO(3) subgroup of SO(1,3), the three matrices N_i, $i = 1,2,3$ generate Lorentz boosts. As can be shown by explicit calculation, these generators obey the following commutation relations:

$$[M_i, M_j] = \varepsilon_{ijk} M_k$$

$$[N_i, M_j] = -\varepsilon_{ijk} N_k$$

$$[N_i, N_j] = -\varepsilon_{ijk} M_k$$

where ε_{ijk} is totally antisymmetric in i,j,k. The matrices M_i, $i = 1,2,3$, are antisymmetric, i.e.

$$M_i^T = -M_i$$

whereas the Lorentz boost generators N_i are symmetric:

$$N_i^T = N_i$$

We now construct hermitian matrices

$$J_\ell := iM_\ell, \quad \ell = 1,2,3$$

with

$$J_\ell^+ = (iM_\ell)^+ = -iM_\ell^T = iM_\ell = J_\ell$$

and antihermitian matrices

$$K_\ell := iN_\ell$$

such that

$$K_\ell^+ = (iN_\ell)^+ = -iN_\ell^T = -iN_\ell = -K_\ell$$

These matrices obey the following commutation relations:

$$[J_i, J_k] = i\varepsilon_{ikl} J_l$$

$$[J_i, K_l] = i\varepsilon_{ilm} K_m$$

$$[K_i, K_j] = -i\varepsilon_{ijk} J_k$$

The first commutation relation shows that J_i, i= 1,2,3, generate the rotation subgroup of $L_+^\uparrow$. Usually physicists take the matrices K_i and J_i as generators of Lorentz boosts and rotations respectively.

Starting with matrices K_i and J_i one can construct a covariant formalism, defining an antisymmetric 4x4-matrix $M_{\mu\nu}$, μ,ν = 0,1,2,3:

$$\varepsilon_{kij} M_{ij} := J_k \quad \text{(i,j,k = 1,2,3 in cyclic order)}$$

$$M_{0i} := -K_i \quad \text{(i = 1,2,3)}$$

Explicitly:

$$(M_{\mu\nu}) := \begin{pmatrix} 0 & -K_1 & -K_2 & -K_3 \\ K_1 & 0 & J_3 & -J_2 \\ K_2 & -J_3 & 0 & J_1 \\ K_3 & J_2 & -J_1 & 0 \end{pmatrix} \qquad (1.12)$$

In covariant notation the matrix $M_{\mu\nu}$ is given by (1.11), i.e.

$$(M_{\rho\sigma})^\mu{}_\nu = i(\eta_{\sigma\nu}\delta_\rho{}^\mu - \eta_{\rho\nu}\delta_\sigma{}^\mu)$$

In order to prove this we consider separately:

i) $\rho = 0, \sigma = 1$. Using (1.11) we have

$$(M_{01})^\mu{}_\nu = i(\eta_{1\nu}\delta_0{}^\mu - \eta_{0\nu}\delta_1{}^\mu)$$

and the nonvanishing elements of the matrix M_{01} are

$$(M_{01})^0{}_1 = -i \ , \quad (M_{01})^1{}_0 = -i$$

On the other hand using (1.12),

$$(M_{01})^\mu{}_\nu = -(K_1)^\mu{}_\nu = -i(N_1)^\mu{}_\nu$$

and the explicit form of N_1 gives the only nonvanishing elements:

$$(N_1)^0{}_1 = 1 = (N_1)^1{}_0$$

Hence

$$(M_{01})^0{}_1 = -i = (M_{01})^1{}_0$$

ii) $\rho = 2, \sigma = 3$. Using (1.11) we have

$$(M_{23})^\mu{}_\nu = i(\eta_{3\nu}\delta_2{}^\mu - \eta_{2\nu}\delta_3{}^\mu)$$

and the nonvanishing elements of M_{23} are

$$(M_{23})^2{}_3 = -i \ , \quad (M_{23})^3{}_2 = +i$$

On the other hand using (1.12),

$$(M_{23})^\mu{}_\nu = (J_1)^\mu{}_\nu = i(M_1)^\mu{}_\nu$$

and from the explicit form of M_1 we obtain

$$(M_{23})^2{}_3 = i\,(M_1)^2{}_3 = -i$$

$$(M_{23})^3{}_2 = i\,(M_1)^3{}_2 = +i$$

The other matrices can be checked analogously. Q.e.d.

We now define

$$\begin{aligned}(M_{\rho\sigma})_{\mu\nu} &= \eta_{\mu\tau}(M_{\rho\sigma})^\tau{}_\nu \\ &= i\eta_{\mu\tau}(\eta_{\sigma\nu}\delta_\rho{}^\tau - \eta_{\rho\nu}\delta_\sigma{}^\tau) \\ &= i(\eta_{\rho\mu}\eta_{\sigma\nu} - \eta_{\rho\nu}\eta_{\sigma\mu})\end{aligned} \qquad (1.11')$$

where we used (1.11).

Proposition: The matrices

$$[(M_{\rho\sigma})_{\mu\nu}]$$

are hermitian, i.e.

$$[(M_{\rho\sigma})_{\mu\nu}]^{+} = [(M_{\rho\sigma})_{\mu\nu}]$$

Proof: Consider

$$[(M_{\rho\sigma})_{\mu\nu}]^{+} = [i(\eta_{\rho\mu}\eta_{\sigma\nu} - \eta_{\rho\nu}\eta_{\sigma\mu})]^{+}$$
$$= -i(\eta_{\rho\mu}\eta_{\sigma\nu} - \eta_{\rho\nu}\eta_{\sigma\mu})^{T}$$

We observe that transposition means interchanging the matrix indices μ and ν, i.e.

$$[(M_{\rho\sigma})_{\mu\nu}]^{+} = -i(\eta_{\rho\nu}\eta_{\sigma\mu} - \eta_{\rho\mu}\eta_{\sigma\nu})$$
$$= i(\eta_{\rho\mu}\eta_{\sigma\nu} - \eta_{\rho\nu}\eta_{\sigma\mu}) = [(M_{\rho\sigma})_{\mu\nu}]$$

(using (1.11')).

This implies for the generators K_i and J_i

$$[(J_i)_{\mu\nu}]^{+} = [(J_i)_{\mu\nu}]$$
$$[(K_i)_{\mu\nu}]^{+} = [(K_i)_{\mu\nu}]$$

Explicitly we have

$$(J_1)_{\mu\nu} = (\eta_{\mu\rho})(J_1)^{\rho}{}_{\nu}$$
$$(J_1)_{23} = \eta_{2\rho}(J_1)^{\rho}{}_{3} = \eta_{22}(J_1)^{2}{}_{3} = -i(M_1)^{2}{}_{3}$$
$$= i$$
$$(J_1)_{32} = \eta_{3\rho}(J_1)^{\rho}{}_{2} = \eta_{33}(J_1)^{3}{}_{2} = -i(M_1)^{3}{}_{2} = -i$$

All other elements of this matrix are zero. Hence

$$[(J_1)_{\mu\nu}] = \begin{pmatrix} 0 & 0 & 0 & 0 \\ 0 & 0 & 0 & 0 \\ 0 & 0 & 0 & i \\ 0 & 0 & -i & 0 \end{pmatrix}$$

Similarly one shows that

$$[(J_2)_{\mu\nu}] = \begin{pmatrix} 0 & 0 & 0 & 0 \\ 0 & 0 & 0 & -i \\ 0 & 0 & 0 & 0 \\ 0 & i & 0 & 0 \end{pmatrix}, \quad [(J_3)_{\mu\nu}] = \begin{pmatrix} 0 & 0 & 0 & 0 \\ 0 & 0 & i & 0 \\ 0 & -i & 0 & 0 \\ 0 & 0 & 0 & 0 \end{pmatrix}$$

For the boost generators we have

$$(K_1)_{\mu\nu} = \eta_{\mu\rho}(K_1)^{\rho}{}_{\nu}$$

$$(K_1)_{01} = \eta_{0\rho}(K_1)^{\rho}{}_{1} = \eta_{00}(K_1)^{0}{}_{1} = i(N_1)^{0}{}_{1} = i$$

$$(K_1)_{10} = \eta_{1\rho}(K_1)^{\rho}{}_{0} = \eta_{11}(K_1)^{1}{}_{0} = -i(N_1)^{1}{}_{0} = -i$$

and all other elements of this matrix are zero. Therefore the explicit form of $[(K_1)_{\mu\nu}]$ is

$$[(K_1)_{\mu\nu}] = \begin{pmatrix} 0 & i & 0 & 0 \\ -i & 0 & 0 & 0 \\ 0 & 0 & 0 & 0 \\ 0 & 0 & 0 & 0 \end{pmatrix}$$

K_2 and K_3 are given by

$$[(K_2)_{\mu\nu}] = \begin{pmatrix} 0 & 0 & i & 0 \\ 0 & 0 & 0 & 0 \\ -i & 0 & 0 & 0 \\ 0 & 0 & 0 & 0 \end{pmatrix}, [(K_3)_{\mu\nu}] = \begin{pmatrix} 0 & 0 & 0 & i \\ 0 & 0 & 0 & 0 \\ 0 & 0 & 0 & 0 \\ -i & 0 & 0 & 0 \end{pmatrix}$$

Hence in this form the six matrices

$$\{K_i\}_{i=1,2,3}, \qquad \{J_i\}_{i=1,2,3}$$

form a set of hermitian generators of the Lorentz group. The Lie algebra o(1,3) describes the Lorentz group locally and is determined by the commutator of the basis elements.

<u>Proposition</u>: The generators $M_{\mu\nu}$ of the Lorentz group obey the following commutation relation

$$[M_{\mu\nu}, M_{\rho\sigma}] = -i(\eta_{\mu\rho}M_{\nu\sigma} - \eta_{\mu\sigma}M_{\nu\rho} - \eta_{\nu\rho}M_{\mu\sigma} + \eta_{\mu\sigma}M_{\mu\rho}) \tag{1.13}$$

<u>Proof</u>: Using (1.11) and (1.11') we have

$$[M_{\mu\nu}, M_{\rho\sigma}]_{\alpha\beta}$$
$$= (M_{\mu\nu})_{\alpha\gamma}(M_{\rho\sigma})^{\gamma}{}_{\beta} - (M_{\rho\sigma})_{\alpha\gamma}(M_{\mu\nu})^{\gamma}{}_{\beta}$$
$$= i(\eta_{\mu\alpha}\eta_{\nu\gamma} - \eta_{\mu\gamma}\eta_{\nu\alpha})\,i(\delta_{\rho}^{\gamma}\eta_{\sigma\beta} - \eta_{\rho\beta}\delta_{\sigma}^{\gamma})$$
$$- i(\eta_{\rho\alpha}\eta_{\sigma\gamma} - \eta_{\rho\gamma}\eta_{\sigma\alpha})\,i(\delta_{\mu}^{\gamma}\eta_{\nu\beta} - \eta_{\mu\beta}\delta_{\nu}^{\gamma})$$
$$= -(\eta_{\mu\alpha}\eta_{\nu\rho}\eta_{\sigma\beta} - \eta_{\mu\alpha}\eta_{\nu\sigma}\eta_{\rho\beta} - \eta_{\mu\rho}\eta_{\nu\alpha}\eta_{\sigma\beta} + \eta_{\mu\sigma}\eta_{\nu\alpha}\eta_{\rho\beta})$$
$$+ (\eta_{\rho\alpha}\eta_{\sigma\mu}\eta_{\nu\beta} - \eta_{\rho\alpha}\eta_{\sigma\nu}\eta_{\mu\beta} - \eta_{\rho\mu}\eta_{\sigma\alpha}\eta_{\nu\beta} + \eta_{\rho\nu}\eta_{\sigma\alpha}\eta_{\mu\beta})$$

$$= -[\eta_{\mu\alpha}(\eta_{\nu\rho}\eta_{\sigma\beta} - \eta_{\nu\sigma}\eta_{\rho\beta})$$
$$- \eta_{\nu\alpha}(\eta_{\mu\rho}\eta_{\sigma\beta} - \eta_{\mu\sigma}\eta_{\rho\beta})$$
$$- \eta_{\rho\alpha}(\eta_{\sigma\mu}\eta_{\nu\beta} - \eta_{\sigma\nu}\eta_{\mu\beta})$$
$$+ \eta_{\sigma\alpha}(\eta_{\rho\mu}\eta_{\nu\beta} - \eta_{\rho\nu}\eta_{\mu\beta})]$$
$$= -i\eta_{\mu\rho}(i)(\eta_{\nu\alpha}\eta_{\sigma\beta} - \eta_{\nu\beta}\eta_{\sigma\alpha})$$
$$+ i\eta_{\mu\sigma}(i)(\eta_{\nu\alpha}\eta_{\rho\beta} - \eta_{\rho\alpha}\eta_{\nu\beta})$$
$$+ i\eta_{\nu\rho}(i)(\eta_{\mu\alpha}\eta_{\sigma\beta} - \eta_{\sigma\alpha}\eta_{\mu\beta})$$
$$- i\eta_{\nu\sigma}(i)(\eta_{\mu\alpha}\eta_{\rho\beta} - \eta_{\rho\alpha}\eta_{\mu\beta})$$
$$= -i\eta_{\mu\rho}(M_{\nu\sigma})_{\alpha\beta} + i\eta_{\mu\sigma}(M_{\nu\rho})_{\alpha\beta}$$
$$+ i\eta_{\nu\rho}(M_{\mu\sigma})_{\alpha\beta} - i\eta_{\nu\sigma}(M_{\mu\rho})_{\alpha\beta}$$

Dropping the matrix indices α, β, we obtain (1.13).

Next we want to rederive (as a consistency check) the commutation relations of the generators K_i, J_j starting from the commutator (1.13). We had

$$M_{mn} = \varepsilon_{mni} J_i \tag{1.14}$$

$$M_{oi} = -K_i \tag{1.15}$$

Equation (1.13) reads for $\mu = \rho = 0$, $\nu = i$, $\sigma = j$; i,j = 1,2,3

$$[M_{oi}, M_{oj}] = [K_i, K_j] \quad \text{(with (1.15))}$$
$$= -i(\eta_{oo}M_{ij} - \eta_{oj}M_{io} - \eta_{io}M_{oj} + \eta_{ij}M_{oo})$$
$$= -i\eta_{oo}M_{ij} \quad \text{(using (1.13))}$$
$$= -i\varepsilon_{ijk}J_k$$

Hence

$$[K_i, K_j] = -i\,\varepsilon_{ijk} J_k \tag{1.16}$$

For $\mu = 0$, $\nu = i$, $\rho = h$, $\sigma = l$; i,j,k = 1,2,3 we have

$$[M_{oi}, M_{kl}] = -[K_i, \varepsilon_{klm} J_m] \quad \text{(with (1.14),(1.15))}$$

$$= -i(\eta_{ok} M_{il} - \eta_{ol} M_{ik} - \eta_{ik} M_{ol} + \eta_{il} M_{ok})$$

$$= -i\,\eta_{ik} K_l + i\,\eta_{il} K_k \quad \text{(with (1.13))}$$

$$= i(\delta_{ik} K_l - \delta_{il} K_k) \qquad (\eta_{ik} = -\delta_{ik})$$

Hence

$$\varepsilon_{klm} [K_i, J_m] = -i(\delta_{ik} K_l - \delta_{il} K_k) \tag{1.16'}$$

Then, using

$$\varepsilon_{kln} \varepsilon_{klm} = 2\,\delta_{nm} \tag{1.17}$$

we obtain (with (1.17) and (1.16'))

$$\varepsilon_{kln} \varepsilon_{klm} [K_i, J_m] = 2\delta_{nm} [K_i, J_m]$$

$$= 2[K_i, J_n] = -i\,\varepsilon_{kln} (\delta_{ik} K_l - \delta_{il} K_k)$$

$$= -i\,\varepsilon_{iln} K_l + i\,\varepsilon_{kin} K_k$$

$$= 2i\,\varepsilon_{inl} K_l$$

Hence

$$[K_i, J_k] = i\,\varepsilon_{ikl} K_l \tag{1.18}$$

and finally we have to evaluate (1.13) for the case $\mu = i$, $\nu = j$, $\rho = k$, $\sigma = l$: Using (1.13) we have

$$[M_{ij}, M_{kl}] = [\varepsilon_{ijm} J_m, \varepsilon_{kln} J_n]$$

$$= \varepsilon_{ijm} \varepsilon_{kln} [J_m, J_n]$$

$$= -i(\eta_{ik} M_{jl} - \eta_{il} M_{jk} - \eta_{jk} M_{il} + \eta_{jl} M_{ik})$$

$$= -i(\eta_{ik}\varepsilon_{jla} - \eta_{il}\varepsilon_{jka} - \eta_{jk}\varepsilon_{ila} + \eta_{jl}\varepsilon_{ika})J_a$$

Then (with (1.17))

$$\frac{1}{4}\varepsilon_{ijf}\varepsilon_{klg}\varepsilon_{ijm}\varepsilon_{kln}[J_m, J_n]$$

$$= \delta_{fm}\delta_{gn}[J_m, J_n] = [J_f, J_g]$$

$$= -\frac{i}{4}\varepsilon_{ijf}\varepsilon_{klg}(\eta_{ik}\varepsilon_{jla} - \eta_{il}\varepsilon_{jka} - \eta_{jk}\varepsilon_{ila} + \eta_{jl}\varepsilon_{ika})J_a$$

$$= -\frac{i}{4}\{-\varepsilon_{ijf}\varepsilon_{ilg}\varepsilon_{jla} + \varepsilon_{ijf}\varepsilon_{kig}\varepsilon_{jka} + \varepsilon_{ijf}\varepsilon_{jlg}\varepsilon_{ila} - \varepsilon_{ijf}\varepsilon_{kjg}\varepsilon_{ika}\}J_a$$

$$= \frac{i}{4}\{(\delta_{jl}\delta_{fg} - \delta_{jg}\delta_{fl})\varepsilon_{jla} + (\delta_{jk}\delta_{fg} - \delta_{jg}\delta_{fk})\varepsilon_{jka} + (\delta_{il}\delta_{fg} - \delta_{ig}\delta_{lf})\varepsilon_{ila} + (\delta_{ik}\delta_{fg} - \delta_{ig}\delta_{kf})\varepsilon_{ika}\}J_a$$

$$= -\frac{i}{4}\{\varepsilon_{gfa} + \varepsilon_{gfa} + \varepsilon_{gfa} + \varepsilon_{gfa}\}J_a$$

$$= i\varepsilon_{fga}J_a$$

where we made use of $\eta_{ij} = -\delta_{ij}$ and

$$\varepsilon_{ijk}\varepsilon_{ilm} = \delta_{jl}\delta_{km} - \delta_{jm}\delta_{kl}$$

Hence

$$[J_i, J_j] = i\varepsilon_{ijk}J_k \qquad (1.19)$$

Equation (1.19) defines the rotation group SO(3) as a subgroup of $L_+^\uparrow$, whereas (1.18) states that $\vec{K}$ is a vector under the Lorentz group. The minus-sign in (1.16) is significant; it expresses the difference between the non-compact group SO(1,3) and its compact form SO(4) or between SL(2,C) and SU(2)xSU(2), since locally homomorphic Lie groups (SO(1,3) and SL(2,C) are locally homomorphic as we shall see explicitly in Section 1.3.3, as well as SO(4) and SU(2)xSU(2)) have homomorphic Lie algebras.

In order to be able to classify the irreducible, finite-dimensional, non-unitary[29] representations of the restricted Lorentz group $L_+^\uparrow$, we have to change the basis K_i, J_i of the Lie algebra so(1,3) by introducing the complex linear combinations

$$S_i := \frac{1}{2}(J_i + iK_i) \tag{1.20}$$

$$T_i := \frac{1}{2}(J_i - iK_i) \tag{1.21}$$

It is easy to verify, using (1.16), (1.18) and (1.19), that these nonhermitian generators decouple the commutation relations of K_i and J_i so that

$$\begin{aligned} [S_i, S_j] &= i\varepsilon_{ijk} S_k \\ [T_i, T_j] &= i\varepsilon_{ijk} T_k \\ [T_i, S_j] &= 0 \end{aligned} \tag{1.22}$$

This means that the generators S_i and T_j obey the commutation relations of the Lie algebra of SU(2). In addition, the commutation relations (1.22) show that the Lie algebra so(1,3) decomposes into the direct sum of two su(2) Lie algebras. However, this decomposition holds only for the complexified Lie algebra $so(1,3)^C$, i.e. considering the set of real 4x4-matrices $\mathfrak{a}$ satisfying

$$\mathfrak{a}^T = -\eta\,\mathfrak{a}\,\eta$$

as a complex vector space, this allows complex linear combinations of the form S_i and T_i. Hence the decomposition

$$so(1,3)^C \quad = \quad su(2) \times su(2)$$

is valid only for the complexified Lie algebra of the Lorentz group. However, we can use the classification of irreducible representations of the complex Lie algebra $so(1,3)^C$ to find the irreducible representations of the real Lie algebra $so(1,3) \cong sl(2,C)$, since there is a one-to-one correspondence between representations of a complex Lie algebra and representations of any of its real forms[30,31].

But the classification of finite-dimensional irreducible representations of su(2)xsu(2) is well known. According to the theorem of Racah[32], there is one Casimir operator for every su(2) subalgebra, i.e.

$$\sum_{i=1}^{3} S_i S_i \qquad \text{and} \qquad \sum_{i=1}^{3} T_i T_i$$

are Casimir operators, commuting with any element of the algebra, with eigenvalues $n(n+1)$ and $m(m+1)$ respectively, where $n, m = 0, 1/2, 1, 3/2, \ldots$. n and m are eigenvalues of S_3 and T_3 respectively. Therefore we can label representations of so(1,3) by the pair (n,m), and since

$$J_3 = S_3 + T_3$$

we can identify the spin of the representation with n+m. The dimension of the representation space is $(2n+1).(2m+1)$ for an (n,m)-representation of the Lorentz group.

It is important to note that the two su(2) subalgebras are not independent since they can be interchanged by the operation of parity. Parity acts as follows on rotation and boost generators:

$$J_i \longrightarrow J_i \ , \qquad K_i \longrightarrow - K_i$$

and (1.20), (1.21) show that parity transforms S_i into T_i and T_i into S_i. In addition the operation of hermitian conjugation also interchanges S_i and T_i since, as demonstrated above, J_i and K_i can be chosen to be hermitian:

$$S_i^+ = \frac{1}{2}(J_i + iK_i)^+ = \frac{1}{2}(J_i - iK_i) = T_i$$
$$T_i^+ = \frac{1}{2}(J_i - iK_i)^+ = \frac{1}{2}(J_i + iK_i) = S_i$$

Hence the parity operation is equivalent to hermitian conjugation.

As examples we consider the following representations:

(a) $(0,0)$ with total spin zero is the scalar representation; the dimension of the representation space is one.

(b) $(\frac{1}{2},0)$ with total spin 1/2 is called the left-handed spinor representation; the dimension of the representation space is two.

(c) $(0,\frac{1}{2})$ with total spin 1/2 is called the right-handed spinor representation; the dimension of the representation space is again two.

The handedness is a convention. In subsequent sections we shall discuss these two spinor representations in great detail. Since parity switches S_i to T_i and T_i to S_i, representations of the Lorentz group in general are not parity eigenstates. In particular, the left-handed spinor representation (1/2,0) transforms under parity into the (0, 1/2) representation and vice versa. Therefore to obtain a representation such that parity acts as a linear transformation, one has to consider the direct sum of the spinor representations (b) and (c), $(1/2,0) \oplus (0, 1/2)$, which yields a Dirac-spinor representation. This representation of the Lorentz group will be considered in detail in Section 1.4 .

The importance of the representations (b) and (c) is due to the fact, that any other representation of the Lorentz group can be generated from these spinor representations. In Section 1.3 we shall consider a few examples explicitly. For instance, the Kronecker product of (b) and (c)

$$\left(\tfrac{1}{2},0\right)\otimes\left(0,\tfrac{1}{2}\right)=\left(\tfrac{1}{2},\tfrac{1}{2}\right)$$

gives a spin 1 representation (see (1.129 a)) with four components, and the Kronecker product of two left-handed spinor representations decomposes into a scalar representation and a spin 1 representation, given by an antisymmetric, selfdual second rank tensor (see (1.129 b)), i.e. $\left(\tfrac{1}{2},0\right)\otimes\left(\tfrac{1}{2},0\right)=(0,0)\oplus(1,0)$

1.2 The Poincaré Group

As stated above, the Lorentz group leaves the interval $(x - y)^2$ in Minkowski space invariant. On the other hand the translations

$$x^\mu \rightarrow x'^\mu = x^\mu + a^\mu$$

where a^μ is a constant four-vector, also leave the length squared $(x - y)^2$ invariant. This leads to the definition of the Poincaré group P as the group of all real transformations in Minkowski space

$$x^\mu \rightarrow x'^\mu = \Lambda^\mu{}_\nu x^\nu + a^\mu \qquad (1.23)$$

which leave the length squared $(x - y)^2$ invariant. Definition (1.23) leads to the following composition law for the elements of P: Let

$$\begin{aligned} x' &= \Lambda_1 x + a_1 \\ x'' &= \Lambda_2 x' + a_2 \\ &= \Lambda_2\{\Lambda_1 x + a_1\} + a_2 \\ &= \Lambda_2\Lambda_1 x + \Lambda_2 a_1 + a_2 \end{aligned}$$

Writing (Λ, a) for an element of P, we have

$$(\Lambda_2, a_2)\cdot(\Lambda_1, a_1) = (\Lambda_2\Lambda_1, \Lambda_2 a_1 + a_2) \qquad (1.24)$$

This demonstrates that P is the semidirect product $L \overset{s}{\otimes} T_4$ of the Lorentz group L and the translation group

T_4 in Minkowski space. Similarly, as for the Lorentz group, P decomposes into four pieces identified by det Λ and Λ^0_0; i.e.

$$P^{\uparrow}_{+}\,,\quad P^{\downarrow}_{+}\,,\quad P^{\uparrow}_{-}\,,\quad P^{\downarrow}_{-}$$

The identity element of P is $(1_{4x4}, 0)$ and the inverse of $(\Lambda, a) \in$ P is $(\Lambda^{-1}, -\Lambda^{-1}a)$ such that (with (1.24))

$$\begin{aligned}(\Lambda, a)\circ(\Lambda^{-1}, -\Lambda^{-1}a) &= (\Lambda\Lambda^{-1}, -\Lambda\Lambda^{-1}a + a) \\ &= (1_{4\times4}, 0) \\ (\Lambda^{-1}, -\Lambda^{-1}a)\circ(\Lambda, a) &= (\Lambda^{-1}\Lambda, \Lambda^{-1}a - \Lambda^{-1}a) \\ &= (1_{4\times4}, 0)\end{aligned}$$

The Lie algebra of $P^{\uparrow}_{+}$ is determined by the commutation relation (1.13) of the Lorentz group $L^{\uparrow}_{+}$, the trivial commutation relation of the translation group (observe that T_4 is abelian) and the commutator of translations and Lorentz transformations still to be determined. In order to obtain this commutator, we consider a faithful representation of $P^{\uparrow}_{+}$

$$(\Lambda, a) \longrightarrow g(\Lambda, a) \tag{1.25}$$

in a vector space V such that

$$\begin{aligned} g(\Lambda_2, a_2)\, g(\Lambda_1, a_1) &= g(\Lambda_2\Lambda_1, \Lambda_2 a_1 + a_2) \\ g^{-1}(\Lambda, a) &= g(\Lambda^{-1}, -\Lambda^{-1}a) \end{aligned} \tag{1.26}$$

Infinitesimally we can write (id_V = identity in V)

$$g(\Lambda, a) = id_V - \frac{i}{2}\omega_{\rho\sigma}M^{\rho\sigma} + i a_\mu P^\mu \tag{1.27}$$

where $\omega_{\rho\sigma} = -\omega_{\sigma\rho}$ are six infinitesimal parameters leading to an infinitesimal Lorentz transformation $\Lambda \in L^{\uparrow}_{+}$ and a_μ denotes four infinitesimal parameters, leading to an infinitesimal translation.

Now $M^{\rho\sigma} = -M^{\sigma\rho}$ and P^μ are generators of Lorentz transformations and translations respectively in the corresponding representation. Consider (using (1.26))

$$g^{-1}(\Lambda,0)\,g(\Lambda',a')\,g(\Lambda,0) = g^{-1}(\Lambda,0)\,g(\Lambda'\Lambda,a')$$
$$= g(\Lambda^{-1},0)\,g(\Lambda'\Lambda,a')$$
$$= g(\Lambda^{-1}\Lambda'\Lambda,\,\Lambda^{-1}a') \tag{1.27'}$$

For infinitesimal $(\Lambda',a') \in P_+^\uparrow$ the left hand side gives

$$g^{-1}(\Lambda,0)\,g(\Lambda',a')\,g(\Lambda,0) \qquad \text{(now use (1.27))}$$
$$= g^{-1}(\Lambda,0)\{id_V - \tfrac{i}{2}\omega'_{\mu\nu}M^{\mu\nu} + ia'_\mu P^\mu\}\,g(\Lambda,0)$$
$$= id_V - \tfrac{i}{2}\omega'_{\mu\nu}\,g^{-1}(\Lambda,0)M^{\mu\nu}g(\Lambda,0)$$
$$+ ia'_\mu\, g^{-1}(\Lambda,0)P^\mu g(\Lambda,0)$$

The right hand side of (1.27') may be expanded as

$$g(\Lambda^{-1}\Lambda'\Lambda,\Lambda^{-1}a') = id_V - \tfrac{i}{2}(\Lambda^{-1}\omega'\Lambda)_{\rho\sigma}M^{\rho\sigma} + i(\Lambda^{-1}a')_\rho P^\rho$$
$$= id_V - \tfrac{i}{2}(\Lambda^{-1})_\rho{}^\mu\,\omega'_{\mu\nu}\Lambda^\nu{}_\sigma M^{\rho\sigma} + i(\Lambda^{-1})_\rho{}^\mu a'_\mu P^\rho$$
$$= id_V - \tfrac{i}{2}\omega'_{\mu\nu}\Lambda^\mu{}_\rho\Lambda^\nu{}_\sigma M^{\rho\sigma} + ia'_\mu\Lambda^\mu{}_\rho P^\rho$$

Hence we obtain

$$g^{-1}(\Lambda,0)M^{\mu\nu}g(\Lambda,0) = \Lambda^\mu{}_\rho\Lambda^\nu{}_\sigma M^{\rho\sigma}$$
$$g^{-1}(\Lambda,0)P^\mu g(\Lambda,0) = \Lambda^\mu{}_\rho P^\rho \tag{1.28}$$

The first equation states that $M^{\mu\nu}$ is an antisymmetric tensor operator under $L_+^\uparrow$; the second relation shows that P^μ is a vector operator under $L_+^\uparrow$. Now consider infinitesimal $\Lambda \in L_+^\uparrow$; for the first of equations (1.28) we obtain

$$g^{-1}(\Lambda,0)M^{\mu\nu}g(\Lambda,0) = g(\Lambda^{-1},0)M^{\mu\nu}g(\Lambda,0)$$

$$= (id_V + \frac{i}{2}\omega_{\rho\sigma} M^{\rho\sigma}) M^{\mu\nu} (id_V - \frac{i}{2}\omega_{\rho\sigma} M^{\rho\sigma})$$
$$= M^{\mu\nu} + \frac{i}{2}\omega_{\rho\sigma} [M^{\rho\sigma}, M^{\mu\nu}]$$

On the other hand

$$\Lambda^\mu{}_\rho \Lambda^\nu{}_\sigma M^{\rho\sigma} = (\delta^\mu_\rho + \omega^\mu{}_\rho)(\delta^\nu_\sigma + \omega^\nu{}_\sigma) M^{\rho\sigma}$$
$$= M^{\mu\nu} + \omega^\mu{}_\rho \delta^\nu_\sigma M^{\rho\sigma} + \delta^\mu_\rho \omega^\nu{}_\sigma M^{\rho\sigma}$$
$$= M^{\mu\nu} + \omega^\mu{}_\rho M^{\rho\nu} + \omega^\nu{}_\sigma M^{\mu\sigma}$$
$$= M^{\mu\nu} + \eta^{\mu\sigma}\omega_{\sigma\rho} M^{\rho\nu} + \eta^{\nu\rho}\omega_{\rho\sigma} M^{\mu\sigma}$$
$$= M^{\mu\nu} + \frac{1}{2}\{\eta^{\mu\sigma}\omega_{\sigma\rho} M^{\rho\nu} + \eta^{\mu\rho}\omega_{\rho\sigma} M^{\sigma\nu} + \eta^{\nu\rho}\omega_{\rho\sigma} M^{\mu\sigma} + \eta^{\nu\sigma}\omega_{\sigma\rho} M^{\mu\rho}\}$$
$$= M^{\mu\nu} + \frac{1}{2}\omega_{\rho\sigma}\{\eta^{\mu\rho} M^{\sigma\nu} - \eta^{\mu\sigma} M^{\rho\nu} - \eta^{\nu\sigma} M^{\mu\rho} + \eta^{\nu\rho} M^{\mu\sigma}\}$$

Hence

$$[M^{\rho\sigma}, M^{\mu\nu}] = -i\begin{pmatrix}\eta^{\mu\rho} M^{\sigma\nu} - \eta^{\mu\sigma} M^{\rho\nu} \\ -\eta^{\nu\sigma} M^{\mu\rho} + \eta^{\nu\rho} M^{\mu\sigma}\end{pmatrix}$$

or

$$[M^{\mu\nu}, M^{\rho\sigma}] = -i\begin{pmatrix}\eta^{\mu\rho} M^{\nu\sigma} - \eta^{\mu\sigma} M^{\nu\rho} \\ -\eta^{\nu\rho} M^{\mu\sigma} + \eta^{\nu\sigma} M^{\mu\rho}\end{pmatrix}$$

Thus we have rederived the commutation relation of so(1,3), i.e. equation (1.13).

The second of equations (1.28) leads to

$$g^{-1}(\Lambda, 0) P^\rho g(\Lambda, 0)$$
$$= (id_V + \frac{i}{2}\omega_{\mu\nu} M^{\mu\nu}) P^\rho (id_V - \frac{i}{2}\omega_{\mu\nu} M^{\mu\nu})$$

$$= P^\rho + \frac{i}{2}\omega_{\mu\nu}[M^{\mu\nu}, P^\rho]$$

$$= \Lambda^\rho{}_\nu P^\nu \quad \text{(using (1.28))}$$

$$= (\delta^\rho{}_\nu + \omega^\rho{}_\nu)P^\nu$$

$$= P^\rho + \omega^\rho{}_\nu P^\nu$$

$$= P^\rho + \eta^{\rho\mu}\omega_{\mu\nu}P^\nu$$

$$= P^\rho + \frac{1}{2}\{\eta^{\rho\mu}\omega_{\mu\nu}P^\nu + \eta^{\rho\nu}\omega_{\nu\mu}P^\mu\}$$

$$= P^\rho + \frac{1}{2}\omega_{\mu\nu}\{\eta^{\rho\mu}P^\nu - \eta^{\rho\nu}P^\mu\}$$

and we obtain

$$[M^{\mu\nu}, P^\rho] = -i(\eta^{\mu\rho}P^\nu - \eta^{\nu\rho}P^\mu) \tag{1.29}$$

In summary we see that the Poincaré algebra is given by the following set of commutators

$$[M_{\mu\nu}, M_{\rho\sigma}] = -i(\eta_{\mu\rho}M_{\nu\sigma} - \eta_{\mu\sigma}M_{\nu\rho} - \eta_{\nu\rho}M_{\mu\sigma} + \eta_{\nu\sigma}M_{\mu\rho})$$

$$[M_{\mu\nu}, P_\rho] = i(\eta_{\nu\rho}P_\mu - \eta_{\mu\rho}P_\nu)$$

$$[P_\mu, P_\nu] = 0 \tag{1.30}$$

<u>Proposition</u>: $P^2 = P_\mu P^\mu$ is a Casimir operator of the Poincaré algebra (1.30), i.e.

$$\left.\begin{aligned}[M_{\mu\nu}, P^2] &= 0\\ [P_\mu, P^2] &= 0\end{aligned}\right\} \tag{1.31}$$

<u>Proof</u>: The second of relations (1.31) is trivial. Hence consider the first. Using (1.29) we have

$$[M_{\mu\nu}, P^2] = [M_{\mu\nu}, P^\rho P_\rho]$$
$$= [M_{\mu\nu}, P^\rho]P_\rho + P^\rho[M_{\mu\nu}, P_\rho]$$
$$= [M_{\mu\nu}, \eta^{\rho\sigma}P_\sigma]P_\rho + P^\rho[M_{\mu\nu}, P_\rho]$$
$$= -\eta^{\rho\sigma} i(\eta_{\mu\sigma}P_\nu - \eta_{\nu\sigma}P_\mu)P_\rho$$
$$- P^\rho i(\eta_{\mu\rho}P_\nu - \eta_{\nu\rho}P_\mu)$$
$$= -i\eta^{\rho\sigma}\eta_{\mu\sigma}P_\nu P_\rho + i\eta^{\rho\sigma}\eta_{\nu\sigma}P_\mu P_\rho$$
$$- iP^\rho\eta_{\mu\rho}P_\nu + iP^\rho\eta_{\nu\rho}P_\mu$$
$$= -i\delta^\rho{}_\mu P_\nu P_\rho + i\delta^\rho{}_\nu P_\mu P_\rho$$
$$- iP_\mu P_\nu + iP_\nu P_\mu$$
$$= -i[P_\nu, P_\mu] - i[P_\mu, P_\nu]$$
$$= 0$$

The second Casimir operator of the Poincaré algebra (1.30) can be constructed from the Pauli-Ljubanski polarization vector defined by

$$W_\mu := \frac{1}{2}\varepsilon_{\mu\nu\rho\sigma}P^\nu M^{\rho\sigma} \tag{1.32}$$

where $\varepsilon_{\mu\nu\rho\sigma}$ is the Levi-Civita tensor with $\varepsilon_{0123} = +1$.

Proposition: The Pauli-Ljubanski vector is invariant under translations, i.e.

(i) $$[P_\mu, W_\nu] = 0 \tag{1.33a}$$

It can be written

(ii) $$W^\mu = [I, P^\mu] \tag{1.33b}$$

where

$$I := \frac{i}{8}\varepsilon_{\mu\nu\rho\sigma} M^{\mu\nu} M^{\rho\sigma}$$

and is a vector under $L_+^\uparrow$.

(iii)

$$[M_{\mu\nu}, W_\rho] = -i(\eta_{\mu\rho} W_\nu - \eta_{\nu\rho} W_\mu) \qquad (1.34)$$

<u>Proof</u>:

(i) Proof of (1.33a): Using (1.32) and (1.29) we have

$$\begin{aligned}
[P_\mu, W_\nu] &= \tfrac{1}{2}[P_\mu, \varepsilon_{\nu\rho\sigma\tau} P^\rho M^{\sigma\tau}] \\
&= \tfrac{1}{2}\varepsilon_{\nu\rho\sigma\tau}[P_\mu, P^\rho M^{\sigma\tau}] \\
&= \tfrac{1}{2}\varepsilon_{\nu\rho\sigma\tau}\eta_{\mu\gamma}\{[P^\gamma, P^\rho]M^{\sigma\tau} \\
&\qquad + P^\rho[P^\gamma, M^{\sigma\tau}]\} \\
&= \frac{i}{2}\varepsilon_{\nu\rho\sigma\tau}\eta_{\mu\gamma}P^\rho\{\eta^{\sigma\gamma}P^\tau - \eta^{\tau\gamma}P^\sigma\} \\
&= \frac{i}{2}\{\varepsilon_{\nu\rho\mu\tau}P^\rho P^\tau - \varepsilon_{\nu\rho\sigma\mu}P^\rho P^\sigma\} \\
&= \frac{i}{4}\{\varepsilon_{\nu\rho\mu\tau}P^\rho P^\tau + \varepsilon_{\nu\tau\mu\rho}P^\tau P^\rho \\
&\qquad - \varepsilon_{\nu\rho\sigma\mu}P^\rho P^\sigma - \varepsilon_{\nu\sigma\rho\mu}P^\sigma P^\rho\} \\
&= 0
\end{aligned}$$

(ii) Proof of (1.33b):

$$\begin{aligned}
[I, P^\mu] &= [\tfrac{i}{8}\varepsilon_{\alpha\beta\gamma\delta}M^{\alpha\beta}M^{\gamma\delta}, P^\mu] \\
&= \frac{i}{8}\varepsilon_{\alpha\beta\gamma\delta}[M^{\alpha\beta}M^{\gamma\delta}, P^\mu] \\
&= \frac{i}{8}\varepsilon_{\alpha\beta\gamma\delta}\{M^{\alpha\beta}[M^{\gamma\delta}, P^\mu] + [M^{\alpha\beta}, P^\mu]M^{\gamma\delta}\} \\
&= \frac{i}{8}\varepsilon_{\alpha\beta\gamma\delta}\{M^{\alpha\beta}(-i)(\eta^{\gamma\mu}P^\delta - \eta^{\delta\mu}P^\gamma) \\
&\qquad - i(\eta^{\alpha\mu}P^\beta - \eta^{\beta\mu}P^\alpha)M^{\gamma\delta}\} \\
&= \frac{1}{8}\{\varepsilon_{\alpha\beta\gamma\delta}\eta^{\gamma\mu}M^{\alpha\beta}P^\delta
\end{aligned}$$

$$- \varepsilon_{\alpha\beta\gamma\delta}\eta^{\delta\mu}M^{\alpha\beta}P^{\gamma} + \varepsilon^{\mu}{}_{\beta\gamma\delta}P^{\beta}M^{\gamma\delta}$$
$$- \varepsilon_{\alpha}{}^{\mu}{}_{\gamma\delta}P^{\alpha}M^{\gamma\delta}\}$$
$$= \frac{1}{4}\varepsilon^{\mu}{}_{\alpha\beta\delta}M^{\alpha\beta}P^{\delta} + \frac{1}{4}\varepsilon^{\mu}{}_{\alpha\gamma\delta}P^{\alpha}M^{\gamma\delta}$$

Now

$$M^{\alpha\beta}P^{\delta} = P^{\delta}M^{\alpha\beta} - i(\eta^{\alpha\delta}P^{\beta} - \eta^{\beta\delta}P^{\alpha})$$

and contracting with $\varepsilon^{\mu}{}_{\alpha\beta\delta}$

$$\frac{1}{4}\varepsilon^{\mu}{}_{\alpha\beta\delta}M^{\alpha\beta}P^{\delta} = \frac{1}{4}\varepsilon^{\mu}{}_{\alpha\beta\delta}P^{\delta}M^{\alpha\beta}$$
$$= \frac{1}{4}\varepsilon^{\mu}{}_{\delta\alpha\beta}P^{\delta}M^{\alpha\beta}$$

so that

$$[I, P^{\mu}] = \frac{1}{2}\varepsilon^{\mu}{}_{\alpha\gamma\delta}P^{\alpha}M^{\gamma\delta} = W^{\mu}$$

(iii) In order to prove (1.34) we consider the following commutator and use the Jacobi identity

$$[M^{\mu\nu}, W^{\rho}] = [M^{\mu\nu}, [I, P^{\rho}]]$$
$$= -[I, [P^{\rho}, M^{\mu\nu}]] - [P^{\rho}, [M^{\mu\nu}, I]]$$
$$= -[I, i(\eta^{\mu\rho}P^{\nu} - \eta^{\nu\rho}P^{\mu})] - 0$$
$$= -i(\eta^{\mu\rho}[I, P^{\nu}] - \eta^{\nu\rho}[I, P^{\mu}])$$
$$= -i(\eta^{\mu\rho}W^{\nu} - \eta^{\nu\rho}W^{\mu})$$

where $[M^{\mu\nu}, I] = 0$

since I is invariant under Lorentz transformations.

Proposition: The square of the Pauli-Ljubanski vector is

$$W^2 = W_{\mu}W^{\mu} = -\frac{1}{2}M_{\mu\nu}M^{\mu\nu}P^2 + M^{\rho\sigma}M_{\nu\sigma}P_{\rho}P^{\nu} \qquad (1.35)$$

<u>Proof</u>: $W^2 = W_\mu W^\mu = \eta^{\mu\nu} W_\mu W_\nu$

$$= \eta^{\mu\nu}\{\tfrac{1}{2}\varepsilon_{\mu\alpha\beta\gamma}P^\alpha M^{\beta\gamma}\}\{\tfrac{1}{2}\varepsilon_{\nu\rho\sigma\tau}P^\rho M^{\sigma\tau}\}$$

$$= \tfrac{1}{4}\varepsilon^\mu{}_{\alpha\beta\gamma}\varepsilon_{\mu\rho\sigma\tau}P^\alpha M^{\beta\gamma}P^\rho M^{\sigma\tau}$$

$$\begin{aligned} = -\tfrac{1}{4}\{ &\eta_{\alpha\rho}(\eta_{\beta\sigma}\eta_{\gamma\tau} - \eta_{\beta\tau}\eta_{\gamma\sigma}) \\ &- \eta_{\alpha\sigma}(\eta_{\beta\rho}\eta_{\gamma\tau} - \eta_{\beta\tau}\eta_{\gamma\rho}) \\ &+ \eta_{\alpha\tau}(\eta_{\beta\rho}\eta_{\gamma\sigma} - \eta_{\beta\sigma}\eta_{\gamma\rho})\}\cdot \\ &\cdot P^\alpha M^{\beta\gamma}P^\rho M^{\sigma\tau} \end{aligned}$$

$$\begin{aligned} = -\tfrac{1}{4}\{ &P_\rho M_{\sigma\tau}P^\rho M^{\sigma\tau} - P_\rho M_{\tau\sigma}P^\rho M^{\sigma\tau} \\ &- P_\sigma M_{\rho\tau}P^\rho M^{\sigma\tau} + P_\sigma M_{\tau\rho}P^\rho M^{\sigma\tau} \\ &+ P_\tau M_{\rho\sigma}P^\rho M^{\sigma\tau} - P_\tau M_{\sigma\rho}P^\rho M^{\sigma\tau}\} \end{aligned}$$

Now (using (1.30))

$$\begin{aligned} &P_\rho M_{\sigma\tau}P^\rho M^{\sigma\tau} \\ &= \{M_{\sigma\tau}P_\rho - i(\eta_{\tau\rho}P_\sigma - \eta_{\sigma\rho}P_\tau)\}P^\rho M^{\sigma\tau} \\ &= M_{\sigma\tau}P_\rho P^\rho M^{\sigma\tau} - i(P_\sigma P_\tau - P_\tau P_\sigma)M^{\sigma\tau} \\ &= M_{\sigma\tau}P^2 M^{\sigma\tau} - i[P_\sigma, P_\tau]M^{\sigma\tau} \\ &= M_{\sigma\tau}M^{\sigma\tau}P^2 \qquad \text{(with (1.30), (1.31))} \end{aligned}$$

and using the antisymmetry of $M^{\sigma\tau}$

$$\begin{aligned} &-P_\sigma M_{\rho\tau}P^\rho M^{\sigma\tau} + P_\sigma M_{\tau\rho}P^\rho M^{\sigma\tau} \\ &+ P_\tau M_{\rho\sigma}P^\rho M^{\sigma\tau} - P_\tau M_{\sigma\rho}P^\rho M^{\sigma\tau} \end{aligned}$$

$$= -2P_\sigma M_{\rho\tau} P^\rho M^{\sigma\tau} + 2P_\tau M_{\rho\sigma} P^\rho M^{\sigma\tau}$$
$$= -4P_\sigma M_{\rho\tau} P^\rho M^{\sigma\tau}$$
$$= -4[M_{\rho\tau} P_\sigma - i(\eta_{\tau\sigma} P_\rho - \eta_{\rho\sigma} P_\tau)] P^\rho M^{\sigma\tau}$$
$$= -4M_{\rho\tau} P_\sigma P^\rho M^{\sigma\tau} + 4iP^2 M_\tau{}^\tau$$
$$- 4iP_\tau P_\sigma M^{\sigma\tau}$$
$$= -4M_{\rho\tau} P_\sigma P^\rho M^{\sigma\tau}$$

since $M_\tau{}^\tau = 0$ and because the antisymmetry of $M^{\sigma\tau}$ implies that $P_\tau P_\sigma M^{\sigma\tau}$ vanishes.

In the same way one can show that

$$-4M_{\rho\tau} P_\sigma P^\rho M^{\sigma\tau} = -4M_{\rho\tau} M^{\sigma\tau} P_\sigma P^\rho$$

and therefore

$$W^2 = -\tfrac{1}{2} M_{\mu\nu} M^{\mu\nu} P^2 + M^{\mu\rho} M_{\nu\rho} P_\mu P^\nu$$

The importance of the Pauli-Ljubanski vector is due to the fact, that its square W^2 is the second Casimir operator of the Poincaré algebra.

<u>Proposition</u>: W^2 commutes with the generators P_μ , $M_{\mu\nu}$, i.e.

$$[M_{\mu\nu}, W^2] = 0, \ [P_\mu, W^2] = 0 \qquad (1.36)$$

<u>Proof</u>: Using (1.34) we have

$$[M_{\mu\nu}, W^2] = [M_{\mu\nu}, W^\rho W_\rho]$$
$$= [M_{\mu\nu}, W_\rho] W^\rho + W^\rho [M_{\mu\nu}, W_\rho]$$
$$= -i(\eta_{\mu\rho} W_\nu - \eta_{\nu\rho} W_\mu) W^\rho - iW^\rho(\eta_{\mu\rho} W_\nu - \eta_{\nu\rho} W_\mu)$$
$$= -i(W_\nu W_\mu - W_\mu W_\nu + W_\mu W_\nu - W_\nu W_\mu)$$
$$= 0$$

With (1.33) the second of relations (1.36) is trivial, i.e.

$$[P_\mu, W^2] = [P_\mu, W_\rho]W^\rho + W^\rho[P_\mu, W_\rho] = 0 \quad \text{(with (1.33))}$$

Proposition: The Pauli-Ljubanski polarization vector has the additional property

$$W_\mu P^\mu = 0 \tag{1.37}$$

Proof:

$$W_\mu P^\mu = \frac{1}{2}\varepsilon_{\mu\nu\rho\sigma}P^\nu M^{\rho\sigma}P^\mu \quad \text{(with (1.32))}$$

$$= \frac{1}{2}\varepsilon_{\mu\nu\rho\sigma}P^\nu P^\mu M^{\rho\sigma} + \frac{i}{2}\varepsilon_{\mu\nu\rho\sigma}P^\nu \eta^{\sigma\mu}P^\rho$$

$$- \frac{i}{2}\varepsilon_{\mu\nu\rho\sigma}P^\nu \eta^{\rho\mu}P^\sigma \quad \text{(with (1.30))}$$

$$= \frac{1}{4}\varepsilon_{\mu\nu\rho\sigma}P^\nu P^\mu M^{\rho\sigma} + \frac{1}{4}\varepsilon_{\nu\mu\rho\sigma}P^\mu P^\nu M^{\rho\sigma}$$

$$+ \frac{i}{2}\varepsilon_{\mu\nu\rho}{}^{\mu}P^\nu P^\rho - \frac{i}{2}\varepsilon_{\mu\nu}{}^{\mu}{}_{\sigma}P^\nu P^\sigma$$

$$= \frac{1}{4}\varepsilon_{\mu\nu\rho\sigma}[P^\nu, P^\mu]M^{\rho\sigma}$$

$$= 0 \quad \text{(with (1.30))}$$

The representation theory of the Poincaré group has been discussed extensively in the literature using the formalism of induced representations and the concept of little groups. We do not go into any detail here, therefore, and simply quote the following results[33,34]:

The unitary (infinite-dimensional) representations of the Poincaré group can be split into three main classes. These are:

i) $P^2 = P_\mu P^\mu = m^2 > 0 \; , \quad W^2 = -m^2 s(s+1)$ (1.38)

The eigenvalue of the second Casimir operator W^2 is $-m^2 s(s+1)$ where s denotes the spin which assumes discrete values s = 0, 1/2, 1, ... From (1.37) one deduces that in the rest frame ($P^\mu = (m, \vec{0})$) the zero component of the Pauli-Ljubanski vector must vanish, and the space components in the rest frame are given by

$$W_i = \frac{1}{2} \varepsilon_{i0jk} P^0 S^{jk}$$

such that

$$W^2 = -\vec{W}^2 = -m^2 \vec{S}^2 \tag{1.39}$$

where

$$S^i = \frac{1}{2} \varepsilon^{ijk} S_{jk} \tag{1.40}$$

is the spin operator. This representation is specified in terms of the mass m and spin s. Physically a state in a representation (m,s) corresponds to a particle of rest mass m and spin s; moreover, since the spin projection s_3 can take on any value from -s to +s, massive particles fall into (2s+1)-dimensional multiplets.

<u>ii) $P^2 = 0$, $W^2 = 0$</u> (1.41)

In this case W and P are linearly dependent:

$$W_\mu = \lambda P_\mu \tag{1.41a}$$

The constant of proportionality is called the helicity and is equal to $\pm s$ where s = 0, 1/2, 1, ... is the spin of the representation. The time component of W^μ is

$$W^0 = \frac{1}{2} \varepsilon^{0ijk} P_i M_{jk}$$

$$= \vec{P} \cdot \vec{J} \quad \text{(using (1.12))} \tag{1.42}$$

so that (1.41a) implies

$$\lambda = \frac{\vec{P} \cdot \vec{J}}{P_0} \tag{1.43}$$

which is the definition of the helicity of a massless particle. Examples of particles which fall into this category are the photon with spin 1 and helicity states ± 1,and the neutrino with spin 1/2 and helicity states $\pm 1/2$.

<u>iii) $P^2 = 0$, $W^2 = -\rho^2$</u> (1.44)

This type of representation describes a particle of rest mass zero with an infinite number of polarization states labeled by the continuous variable ρ . These representations do not seem to be realized in nature.

Remark: For this case the calculations of case i) do not work since we cannot make a transformation to the rest frame. However, we can always transform to a system where

$$P_\mu = (P_0, 0, 0, P_0)$$

If ω_μ is the eigenvalue of W_μ, (1.37) implies

$$0 = \omega_\mu P^\mu = \omega^0 p^0 - \omega^3 p^0$$

i.e.

$$\omega^0 = \omega^3$$

and

$$\omega^2 = \eta^{\mu\nu}\omega_\mu\omega_\nu = \omega_0^2 - \vec{\omega}^2$$
$$= -(\omega_1^2 + \omega_2^2)$$
$$= -\rho^2$$

Thus in this case the eigenvalues of the Casimir operator W^2 can assume any value.

1.3 SL(2,C), Dotted and Undotted Indices

1.3.1 Spinor Algebra

We consider the special linear Lie group in 2 dimensions with complex parameters

$$SL(2,C) := \{ M \in GL(2,C) \mid \det M = +1 \} \tag{1.45}$$

A linear representation of this group is a map from SL(2,C) into the automorphism group of a certain vector space F [b]. This means

$$M \in SL(2,C) \rightarrow D(M) \tag{1.46}$$

b The reader who wants to refresh his memory of definitions of mathematical terms without delving into mathematical texts is advised to consult the article by A.S. Sciarrino and P. Sorba [35] which is a "Junior Dictionary" of group theory concepts commonly met in particle physics.

the automorphism group being defined as the set of linear bijective maps from F to F, the group multiplication being the composition of maps. As usual we demand the representation properties

$$D(1_{SL(2,C)}) = 1_F \tag{1.47}$$
$$D(M_1)D(M_2) = D(M_1 \cdot M_2), \forall M_1, M_2 \in SL(2,C)$$

where $1_{SL(2,C)}$ is the unit element of SL(2,C) and 1_F the identity map in the vector space F.

Let ψ be any element of F and $\{\hat{e}_i\}$ the canonical basis in F. Then

$$\psi = \sum_{n=1}^{\dim F} \psi_n \hat{e}_n \tag{1.48}$$

and representation matrices act on ψ in the following way:

$$D(M)\psi = \sum_{n=1}^{\dim F} \psi_n' \hat{e}_n \tag{1.49}$$

where

$$\psi_n' = \sum_{n=1}^{\dim F} D_n{}^i(M)\psi_i \tag{1.50}$$

The $\left(D_n{}^m(M)\right)$ are dim F x dim F - dimensional matrices called representation matrices; dim F is called the dimension of the representation.

Two representations $D^{(1)}, D^{(2)}$ are called equivalent if an invertible dim F x dim F - matrix U can be found, such that

$$D^{(1)}(M) = U D^{(2)}(M) U^{-1}, \quad U \in GL(F) \tag{1.51}$$

SL(2,C) admits two inequivalent spinor representations, as we shall see.

<u>a) The self-representation</u>

The self-representation is defined by

$$D(M) := M \quad , \forall M \in SL(2,C) \tag{1.52}$$

The dimension of the representation space is therefore two. According to (1.50) elements of the representation space $\psi \in F$ transform under the self-representation as

$$\psi_A' = M_A{}^B \psi_B \ , \quad A,B = 1,2 \tag{1.53}$$

In the literature the elements $\psi \in F$ transforming according to (1.53) under SL(2,C) are called left-handed Weyl spinors, and this representation is denoted by (1/2, 0)[36,37].

b) The complex conjugate self-representation

The socalled complex conjugate self-representation is defined by

$$D(M) := M^*, \ \forall M \in SL(2,C) \tag{1.54}$$

and as in case a) the dimension of the representation space is two. We call the corresponding representation space $\dot{F}$. Elements $\bar{\psi}$ of this representation space transform under SL(2,C) according to

$$\bar{\psi}_{\dot{A}}' = (M^*)_{\dot{A}}{}^{\dot{B}} \bar{\psi}_{\dot{B}} \ , \quad \dot{A},\dot{B} = \dot{1},\dot{2} \tag{1.55}$$

where $\bar{\psi}_{\dot{B}} \in \dot{F}$. The elements $\bar{\psi}_{\dot{A}} \in \dot{F}$ transforming according to (1.55) under SL(2,C) are called right-handed Weyl spinors, and the representation is denoted by (0, 1/2).

Left- and right-handed Weyl spinors are related by complex conjugation; i.e. taking $\psi_A \in F$ then[1]

$$(\psi_A)^* =: \bar{\psi}^{\dot{A}} \in \dot{F} \qquad \text{(see also (1.177))} \tag{1.56}$$

It should be observed that this equation does not exhibit the same index structure on both sides. As long as we deal only with Weyl spinors this does not give rise to difficulties. A relation consistent with (1.56) which does exhibit the same index structure on both sides will be given in Section 1.4.4 . The self-representation and its complex conjugate representation are inequivalent, i.e. it is not possible to find a 2x2-matrix C such that

$$M = C M^* C^{-1}$$

Proposition: The representation

$$\mathcal{D}(M) =: M^{-1T} \tag{1.57}$$

is equivalent to the self-representation (1.52).

Proof: We have to show the existence of a matrix $\mathcal{E}$ such that

$$\varepsilon M \varepsilon^{-1} = M^{-1T}, \qquad \varepsilon \in \mathrm{GL}(2,C) \tag{1.58}$$

Let

$$\varepsilon = (\varepsilon_{AB}) = \begin{pmatrix} \varepsilon_{11} & \varepsilon_{12} \\ \varepsilon_{21} & \varepsilon_{22} \end{pmatrix}$$

and

$$\varepsilon^{-1} = (\varepsilon_{AB})^{-1} = \frac{1}{\det \varepsilon} \begin{pmatrix} \varepsilon_{22} & -\varepsilon_{12} \\ -\varepsilon_{21} & \varepsilon_{11} \end{pmatrix}$$

and

$$M := \begin{pmatrix} M_{11} & M_{12} \\ M_{21} & M_{22} \end{pmatrix}$$

Then

$$M^{-1T} = \frac{1}{\det M} \begin{pmatrix} M_{22} & -M_{12} \\ -M_{21} & M_{11} \end{pmatrix}^T = \begin{pmatrix} M_{22} & -M_{21} \\ -M_{12} & M_{11} \end{pmatrix}$$

since $M \in \mathrm{SL}(2,C)$ and so $\det M = +1$. Then we have to show that

$$\frac{1}{\det \varepsilon} \begin{pmatrix} \varepsilon_{11} & \varepsilon_{12} \\ \varepsilon_{21} & \varepsilon_{22} \end{pmatrix} \begin{pmatrix} M_{11} & M_{12} \\ M_{21} & M_{22} \end{pmatrix} \begin{pmatrix} \varepsilon_{22} & -\varepsilon_{12} \\ -\varepsilon_{21} & \varepsilon_{22} \end{pmatrix}$$

$$\stackrel{!}{=} \begin{pmatrix} M_{22} & -M_{21} \\ -M_{12} & M_{11} \end{pmatrix}$$

Evaluating the product of the three matrices we have

$$(\det \varepsilon)^{-1} \begin{pmatrix} M_{11}\varepsilon_{11}\varepsilon_{22} - M_{12}\varepsilon_{11}\varepsilon_{21} + M_{21}\varepsilon_{12}\varepsilon_{22} - M_{22}\varepsilon_{12}\varepsilon_{21} & -M_{11}\varepsilon_{11}\varepsilon_{12} + M_{12}\varepsilon_{11}\varepsilon_{22} - M_{21}\varepsilon_{12}^2 + M_{22}\varepsilon_{12}\varepsilon_{22} \\ M_{11}\varepsilon_{21}\varepsilon_{22} - M_{12}\varepsilon_{21}^2 + M_{21}\varepsilon_{22}^2 - M_{22}\varepsilon_{22}\varepsilon_{21} & -M_{11}\varepsilon_{12}\varepsilon_{21} + M_{12}\varepsilon_{21}\varepsilon_{22} - M_{21}\varepsilon_{12}\varepsilon_{22} + M_{22}\varepsilon_{22}^2 \end{pmatrix}$$

$$\overset{!}{=} \begin{pmatrix} M_{22} & -M_{21} \\ -M_{12} & M_{11} \end{pmatrix}$$

which leads to the following set of equations

$$M_{11}\varepsilon_{11}\varepsilon_{22} - M_{12}\varepsilon_{11}\varepsilon_{21} + M_{21}\varepsilon_{12}\varepsilon_{22} - M_{22}\varepsilon_{12}\varepsilon_{21} = M_{22}\det\varepsilon$$

$$M_{11}\varepsilon_{21}\varepsilon_{22} - M_{12}\varepsilon_{21}^2 + M_{21}\varepsilon_{22}^2 - M_{22}\varepsilon_{22}\varepsilon_{21} = -M_{12}\det\varepsilon$$

$$-M_{11}\varepsilon_{11}\varepsilon_{12} + M_{12}\varepsilon_{11}\varepsilon_{22} - M_{21}\varepsilon_{12}^2 + M_{22}\varepsilon_{12}\varepsilon_{22} = -M_{21}\det\varepsilon$$

$$-M_{11}\varepsilon_{12}\varepsilon_{21} + M_{12}\varepsilon_{21}\varepsilon_{22} - M_{21}\varepsilon_{12}\varepsilon_{22} + M_{22}\varepsilon_{22}^2 = M_{11}\det\varepsilon$$

From the first equation we deduce by equating coefficients of M_{ab} on both sides

$$\varepsilon_{11} = \varepsilon_{22} = 0 \qquad \text{and} \qquad -\varepsilon_{21}\varepsilon_{21} = \det \varepsilon$$

From the second

$$\varepsilon_{21}^2 = \det \varepsilon$$

and from the third

$$\varepsilon_{12}^2 = \det \varepsilon$$

Then

$$-\varepsilon_{12}\varepsilon_{21} = \det \varepsilon = \varepsilon_{12}^2$$

i.e.

$$-\varepsilon_{12} = \varepsilon_{21}$$

We choose $\varepsilon_{12} = -1$ so that $\varepsilon_{21} = +1$ and we have

$$(\varepsilon_{AB}) = \begin{pmatrix} 0 & -1 \\ 1 & 0 \end{pmatrix} =: (\varepsilon^{AB})^{-1} \tag{1.59}$$

where the right hand side defines a matrix with upper indices. Then the matrix ε with upper indices is

$$(\varepsilon^{AB}) = \begin{pmatrix} 0 & 1 \\ -1 & 0 \end{pmatrix} = (\varepsilon_{AB})^T \tag{1.60}$$

such that

$$(\varepsilon^{AB})(\varepsilon^{AB})^{-1} \equiv \varepsilon . \varepsilon^{-1} = 1_{2\times 2}$$

implies

$$\begin{pmatrix} 0 & 1 \\ -1 & 0 \end{pmatrix}\begin{pmatrix} 0 & -1 \\ 1 & 0 \end{pmatrix} = \begin{pmatrix} 1 & 0 \\ 0 & 1 \end{pmatrix}$$

or in mixed form

$$\varepsilon^{AB}\varepsilon_{BC} = \delta^A{}_C \, , \quad \varepsilon_{AB}\varepsilon^{BC} = \delta_A{}^C$$

$$\varepsilon_{AB}^T \varepsilon^{BC} = -\delta_A{}^C \tag{1.61}$$

With this index convention, i.e. writing ε with upper and ε^{-1} with lower indices, so that ε plays the role of a metric, we can write (1.58)

$$\varepsilon^{AB} M_B{}^C \varepsilon_{CD} = (M^{-1T})^A{}_D \tag{1.62}$$

Thus M^{-1T} is equivalent to M. Multiplying (1.62) by ε_{EA} from the left and by ε^{DF} from the right we obtain

$$\varepsilon_{EA}\varepsilon^{AB} M_B{}^C \varepsilon_{CD}\varepsilon^{DF} = \varepsilon_{EA}(M^{-1T})^A{}_D \varepsilon^{DF}$$

i.e.

$$\delta_E{}^B M_B{}^C \delta_C{}^F = \varepsilon_{EA}(M^{-1T})^A{}_D \varepsilon^{DF}$$

i.e.

$$M_E{}^F = \varepsilon_{EA}(M^{-1T})^A{}_D \varepsilon^{DF} \tag{1.63}$$

where we used (1.61).

Now, taking a $\psi_A \in \mathcal{F}$ we know from (1.53) that

$$\psi'_A = M_A{}^B \psi_B$$

$$= \varepsilon_{AC}(M^{-1T})^C{}_D \varepsilon^{DB}\psi_B \quad \text{(using (1.63))}$$

Multiplying this equation by ε^{EA} from the left and again using (1.61) we arrive at

$$\varepsilon^{EA}\psi_A{}' = (M^{-1T})^E{}_D \varepsilon^{DB}\psi_B \tag{1.64}$$

We now define a Weyl spinor with contravariant spinor index by

$$\psi^A := \varepsilon^{AB}\psi_B \tag{1.65}$$

Then (1.64) reads

$$\psi'^A = (M^{-1T})^A{}_B \psi^B \tag{1.66}$$

Proposition: The representation

$$\mathcal{D}(M) =: M^{*-1T} \tag{1.67}$$

is equivalent to the complex conjugate representation (1.54).

Proof: As in (1.58) we have to search for a 2x2-matrix $\bar{\varepsilon}$ such that

$$\bar{\varepsilon} M^* \bar{\varepsilon}^{-1} = M^{*-1T} \tag{1.68}$$

In the same manner as for (1.58) one can show that

$$\bar{\varepsilon} = \begin{pmatrix} 0 & 1 \\ -1 & 0 \end{pmatrix} =: \varepsilon^{\dot{A}\dot{B}}$$

and (1.69)

$$\bar{\varepsilon}^{-1} = \begin{pmatrix} 0 & -1 \\ 1 & 0 \end{pmatrix} =: \varepsilon_{\dot{A}\dot{B}}$$

In index notation (1.68) is written

$$\varepsilon^{\dot{A}\dot{B}} M^*{}_{\dot{B}}{}^{\dot{C}} \varepsilon_{\dot{C}\dot{D}} = (M^{*-1T})^{\dot{A}}{}_{\dot{D}} \tag{1.70}$$

Multiplying from the left by $\varepsilon_{\dot{E}\dot{A}}$ and from the right by $\varepsilon^{\dot{D}\dot{F}}$ and using

$$\varepsilon^{\dot{A}\dot{B}} \varepsilon_{\dot{B}\dot{C}} = \delta^{\dot{A}}{}_{\dot{C}}$$

$$\varepsilon_{\dot{A}\dot{B}} \varepsilon^{\dot{B}\dot{C}} = \delta_{\dot{A}}{}^{\dot{C}} \tag{1.71}$$

we obtain

$$(M^*)_{\dot{A}}{}^{\dot{B}} = \varepsilon_{\dot{A}\dot{C}} (M^{*-1T})^{\dot{C}}{}_{\dot{D}} \varepsilon^{\dot{D}\dot{B}} \tag{1.72}$$

Using (1.55) we have the transformation property of dotted Weyl spinors

$$\begin{aligned} \bar{\psi}'_{\dot{A}} &= (M^*)_{\dot{A}}{}^{\dot{B}} \bar{\psi}_{\dot{B}} \\ &= \varepsilon_{\dot{A}\dot{C}} (M^{*-1T})^{\dot{C}}{}_{\dot{D}} \varepsilon^{\dot{D}\dot{B}} \bar{\psi}_{\dot{B}} \end{aligned}$$

Multiplying this equation from the left by $\varepsilon^{\dot{E}\dot{A}}$ we obtain

$$\varepsilon^{\dot{E}\dot{A}} \bar{\psi}'_{\dot{A}} = \varepsilon^{\dot{E}\dot{A}} \varepsilon_{\dot{A}\dot{C}} (M^{*-1T})^{\dot{C}}{}_{\dot{D}} \varepsilon^{\dot{D}\dot{B}} \bar{\psi}_{\dot{B}}$$

$$= (M^{*-1T})^{\dot{E}}{}_{\dot{D}}\, \varepsilon^{\dot{D}\dot{B}}\, \bar{\psi}_{\dot{B}} \qquad \text{(with (1.71))}$$

Defining dotted spinors with contravariant indices by

$$\bar{\psi}^{\dot{A}} := \varepsilon^{\dot{A}\dot{B}}\, \bar{\psi}_{\dot{B}} \tag{1.73}$$

we conclude that dotted spinors with contravariant indices transform under $M^{*\,-1\,T}$ according to

$$\bar{\psi}'^{\dot{A}} = (M^{*-1T})^{\dot{A}}{}_{\dot{B}}\, \bar{\psi}^{\dot{B}} \tag{1.74}$$

Summarizing we have: For any $M \in SL(2,C)$ the matrix M, its complex conjugate M^* , the inverse of its transpose $M^{T\,-1}$ and the inverse of its hermitian conjugate $(M^+)^{-1}$ all represent SL(2,C) . Two-component spinors with covariant and contravariant dotted or undotted spinor indices transform under SL(2,C) as follows:

$$\psi'_A = M_A{}^B\, \psi_B \tag{1.75a}$$

$$\psi'^A = (M^{-1T})^A{}_B\, \psi^B \tag{1.75b}$$

$$\bar{\psi}'_{\dot{A}} = M^*{}_{\dot{A}}{}^{\dot{B}}\, \bar{\psi}_{\dot{B}} \tag{1.75c}$$

$$\bar{\psi}'^{\dot{A}} = (M^{*-1T})^{\dot{A}}{}_{\dot{B}}\, \bar{\psi}^{\dot{B}} \tag{1.75d}$$

The raising and lowering of spinor indices has to be understood in the following way:

i) Given any spinor which transforms under SL(2,C) in the self-representation, ψ_A , we can construct a contravariant spinor ψ^A which transforms under $M^{-1\,T}$:

$$\psi^A = \varepsilon^{AB}\, \psi_B = -\, \psi_B\, \varepsilon^{BA} \tag{1.76a}$$

ii) For a Weyl spinor which transforms under $M^{-1\,T}$, ψ^A

$$\psi_A = \varepsilon_{AB}\, \psi^B \tag{1.76b}$$

iii) Given any Weyl spinor which transforms under SL(2,C) according to M^{*}, $\bar{\psi}_{\dot{A}}$

$$\bar{\psi}^{\dot{A}} = \varepsilon^{\dot{A}\dot{B}}\,\bar{\psi}_{\dot{B}} = -\,\bar{\psi}_{\dot{B}}\,\varepsilon^{\dot{B}\dot{A}} \tag{1.76c}$$

iv) For a Weyl spinor in the representation (0, 1/2) which transforms under $(M^{+\,-1})$

$$\bar{\psi}_{\dot{A}} = \varepsilon_{\dot{A}\dot{B}}\,\bar{\psi}^{\dot{B}} \tag{1.76d}$$

The raising and lowering of spinor indices is performed with the help of the matrices $\varepsilon, \varepsilon^{-1}$ which also connect the two equivalent representations M and $M^{-1\,T}$. Hence the matrix ε plays the role of a metric in the spinor space F. Explicitly the raising and lowering of the spinor indices is given by

i) $$\psi^{A} = \varepsilon^{AB}\psi_{B}$$

$$\psi^{1} = \varepsilon^{1B}\psi_{B} = \varepsilon^{12}\psi_{2} = \psi_{2}$$

since $\varepsilon^{12} =: +1$ (see (1.60)),

$$\psi^{2} = \varepsilon^{2B}\psi_{B} = \varepsilon^{21}\psi_{1} = -\psi_{1}$$

ii) $$\psi_{A} = \varepsilon_{AB}\psi^{B}$$

$$\psi_{1} = \varepsilon_{1B}\psi^{B} = \varepsilon_{12}\psi^{2} = -\psi^{2}$$

since $\varepsilon_{12} = -1$,

$$\psi_{2} = \varepsilon_{21}\psi^{1} = \psi^{1}$$

iii) $$\bar{\psi}^{\dot{A}} = \varepsilon^{\dot{A}\dot{B}}\,\bar{\psi}_{\dot{B}}$$

$$\bar{\psi}^{\dot{1}} = \varepsilon^{\dot{1}\dot{B}}\,\bar{\psi}_{\dot{B}} = \varepsilon^{\dot{1}\dot{2}}\,\bar{\psi}_{\dot{2}} = \bar{\psi}_{\dot{2}}$$

where $\varepsilon^{\dot{1}\dot{2}} =: +1$ (see (1.69))

$$\bar{\psi}^{\dot{2}} = \varepsilon^{\dot{2}\dot{1}}\,\bar{\psi}_{\dot{1}} = -\,\bar{\psi}_{\dot{1}}$$

iv) $\bar{\psi}_{\dot{A}} = \varepsilon_{\dot{A}\dot{B}}\, \bar{\psi}^{\dot{B}}$

$\bar{\psi}_{\dot{1}} = \varepsilon_{\dot{1}\dot{B}}\, \bar{\psi}^{\dot{B}} = \varepsilon_{\dot{1}\dot{2}}\, \bar{\psi}^{\dot{2}} = -\bar{\psi}^{\dot{2}}$

$\bar{\psi}_{\dot{2}} = \varepsilon_{\dot{2}\dot{B}}\, \bar{\psi}^{\dot{B}} = \varepsilon_{\dot{2}\dot{1}}\, \bar{\psi}^{\dot{1}} = \bar{\psi}^{\dot{1}}$

since $\varepsilon_{\dot{1}\dot{2}} = -1,\ \varepsilon_{\dot{2}\dot{1}} = +1.$

This is a consistent set of equations.

We now define dual spaces F^*, $\dot{F}^*$ and establish the connection between ψ^* and $\bar{\psi}$. As explained in Section 1.3.1 we have four types of two-component Weyl spinors transforming according to four different representations of SL(2,C), where M, $M^{-1\,T}$, and M^*, $M^{*\,-1\,T}$ describe equivalent representations.

i) Any $\psi \in$ F transforms under the self-representation (1.52) as

$$\psi'_A = M_A{}^B \psi_B \tag{1.53}$$

Such spinors are characterized by a lower index.

ii) From (1.57) we know that spinors transforming according to

$$\psi'^A = (M^{-1T})^A{}_B\, \psi^B \tag{1.66}$$

form an equivalent representation of SL(2,C). Such spinors carry upper undotted indices and belong to F^*.

iii) A spinor transforming according to the complex conjugate self-representation is given by $\bar{\psi}_{\dot{A}} \in \dot{F}$ and is characterized by a lower dotted index,

$$\bar{\psi}'_{\dot{A}} = (M^*)_{\dot{A}}{}^{\dot{B}}\, \bar{\psi}_{\dot{B}} \tag{1.55}$$

iv) Finally an equivalent representation of SL(2,C) is given by spinors which transform according to

$$\bar{\psi}'^{\dot{A}} = (M^{*-1T})^{\dot{A}}{}_{\dot{B}}\, \bar{\psi}^{\dot{B}} \tag{1.74}$$

Such spinors carry upper dotted indices and belong to $\dot{F}^*$. A mathematical framework for these different types of spinors is given by the following considerations.

Considering F as the vector space of two-component spinors ψ_A, we may construct the dual space F^* in the following way. According to linear algebra, the elements of F^* are linear maps ϕ from F to C :

$$\phi : F \to C$$

such that for all $\psi \in F$

$$\phi(\psi) := \phi^A \psi_A \in C \tag{1.76e}$$

Hence according to our index convention we may interpret two-component spinors with upper (undotted) indices as elements of the dual space F^*, i.e.

$$\psi_A \in F\,, \quad \psi^A \in F^*$$

The asterisk on F^* should not be confused with complex conjugation; it is simply the mathematical symbol for the dual space. In addition we know from (1.65) how we can correlate an element of F to the corresponding element of F^*. The ε-matrix may be considered as a map

$$(\varepsilon^{AB}) : F \longrightarrow F^*$$

$$\psi_A \longrightarrow \psi^A = \varepsilon^{AB}\psi_B$$

The inverse map is, of course, given by the inverse matrix, i.e. the ε -matrix with lower indices

$$(\varepsilon_{AB}) : F^* \longrightarrow F$$

$$\psi^A \longrightarrow \psi_A = \varepsilon_{AB}\psi^B$$

Interpreting the composition on the right hand side of (1.76e) as matrix multiplication, spinors with upper undotted indices represent rows and those with lower indices represent columns.

In the same way one can consider the space $\dot{F}^*$ as the vector space of two-component dotted spinors with upper indices $\bar{\Psi}^{\dot{A}} \in \dot{F}^*$ and we have

$$(\bar{\Psi}_{\dot{A}}) : \dot{F}^* \longrightarrow C$$

$$\bar{\Psi}(\bar{\Phi}) = \bar{\Psi}_{\dot{A}} \bar{\Phi}^{\dot{A}} \in C \tag{1.76f}$$

Hence

$$\bar{\Psi}_{\dot{A}} \in \dot{F} (\cong \dot{F}^{**}) , \quad \bar{\Phi}^{\dot{A}} \in \dot{F}^*$$

Thus for dotted spinors we have to assign to $\bar{\Psi}_{\dot{A}}$ a row and to spinors with upper dotted indices columns. We then have

$$\varepsilon_{\dot{A}\dot{B}} : \dot{F}^* \longrightarrow \dot{F}$$

$$\bar{\Psi}^{\dot{B}} \longrightarrow \bar{\Psi}_{\dot{A}} = \varepsilon_{\dot{A}\dot{B}} \bar{\Psi}^{\dot{B}}$$

and

$$\varepsilon^{\dot{A}\dot{B}} : \dot{F} \longrightarrow \dot{F}^*$$

$$\bar{\Psi}_{\dot{B}} \longrightarrow \bar{\Psi}^{\dot{A}} = \varepsilon^{\dot{A}\dot{B}} \bar{\Psi}_{\dot{B}}$$

Furthermore the correct description of the transition from F to $\dot{F}^*$ is given by complex conjugation and multiplication by the matrix $\bar{\sigma}^0$ (see (1.176) or Section 1.3.3 where the σ -matrices are introduced)

$$(\bar{\sigma}^0)^{\dot{A}B} : F \longrightarrow \dot{F}^*$$

So we set

$$\bar{\sigma}^{0\,\dot{A}A} (\Psi_A)^* = \bar{\Psi}^{\dot{A}} \tag{1.76g}$$

The inverse map is easily found to be

$$\sigma^0{}_{A\dot{B}} (\bar{\Psi}^{\dot{B}})^* = \Psi_A \tag{1.76h}$$

In agreement with (1.88) (to be given later) we also have

$$\Psi^A = \bar{\Psi}_{\dot{B}}^* \bar{\sigma}^{0\,\dot{B}A}$$

and

$$\bar{\psi}_{\dot{A}} = \psi^{B*} \sigma^{0}{}_{B\dot{A}}$$

These relations show the connection between ψ_A and $\bar{\psi}^{\dot{A}}$. Thus the four complex numbers ψ_A, $\bar{\psi}^{\dot{A}}$, A = 1 , 2; $\dot{A}$ = $\dot{1}$, $\dot{2}$, do,in fact,define only four real independent numbers which we can take to be real ψ_A, $\bar{\psi}^{\dot{A}}$, A = 1,2; $\dot{A}$ = 1,2 .

We now consider dual maps. From linear algebra we recall the following results. Let V and W be vector spaces over a field K. For every linear map

$$\phi : V \longrightarrow W$$

we may construct the socalled "dual map"

$$\phi^* : W^* \longrightarrow V^*$$

by the following prescription: Let $\psi \in W^*$ so that ψ is a map

$$\psi : W \rightarrow K (=R, C)$$

Then we define the dual map by

$$\phi^*(\psi) = \psi \circ \phi$$

This is to be understood in the following way. Applying the left hand side onto a vector $v \in V$ we have

$$v \in V \xrightarrow{\phi^*(\psi)} C$$

where $\phi^*(\psi)$ is by construction an element of V^* and this element maps $v \in V$ onto the real or complex numbers. The right hand side of the equation then gives

$$v \in V \xrightarrow{\phi : V \rightarrow W} \phi(v) \in W \xrightarrow{\psi \in W^*} (\psi \circ \phi)(v) \in C$$

Hence the dual map ϕ^* is defined in such a way that the left and right hand sides of the equation give the same element in K when applied to an element of V. Furthermore, a linear map ϕ then implies a linear dual map ϕ^*.

It is important to note that if ϕ is a map from V to W, then the dual map ϕ^* maps W^* onto V^*, i.e. dual maps have the "opposite" direction. Describing ϕ and ϕ^* as matrices, we see that if ϕ corresponds to a matrix A then the dual map ϕ^* is given by the transposed matrix A^T. For a proof of this statement we refer to books on linear algebra.

With the help of the mathematical concept of a dual map we may explain the transformation properties of the various types of two-component Weyl spinors; within this formalism we find a natural explanation of (1.53), (1.66), (1.55) and (1.74).

We start by considering (1.53). Formally

$$M : F \longrightarrow F$$

$$\psi_B \longrightarrow \psi'_A = M_A{}^B \psi_B$$

i.e. M is a map from the vector space F to the vector space F. According to the above formalism the dual map is given by

$$\tilde{M}^{-1} : F^* \longrightarrow F^*$$

$$\tilde{M}^{-1} = M^{-1T} \qquad \text{as stated earlier}$$

(the dual map is the transposed matrix). Here we choose the inverse of the dual map instead of the dual map itself as shown in Fig.1. The figure shows that M^{-1T} can be re-expressed in terms of a sequence of maps. We start with $\psi^A \in F^*$. Applying the matrix ε_{AB} we obtain the corresponding element $\psi_A = \varepsilon_{AB}\psi^B$ of F. Then with $\psi'_A = M_A{}^B\psi_B$ we obtain the transformed element $\psi'_A = M_A{}^B\psi_B = M_A{}^B\varepsilon_{BC}\psi^C$ of F. Again applying the metric ε^{AB} we finally obtain the transformed element of F^*. The result of this composition of maps has to be the same as applying M^{-1T} to $\psi^A \in F^*$. Hence we conclude

$$(M^{-1T})^A{}_D = \varepsilon^{AB} M_B{}^C \varepsilon_{CD}$$

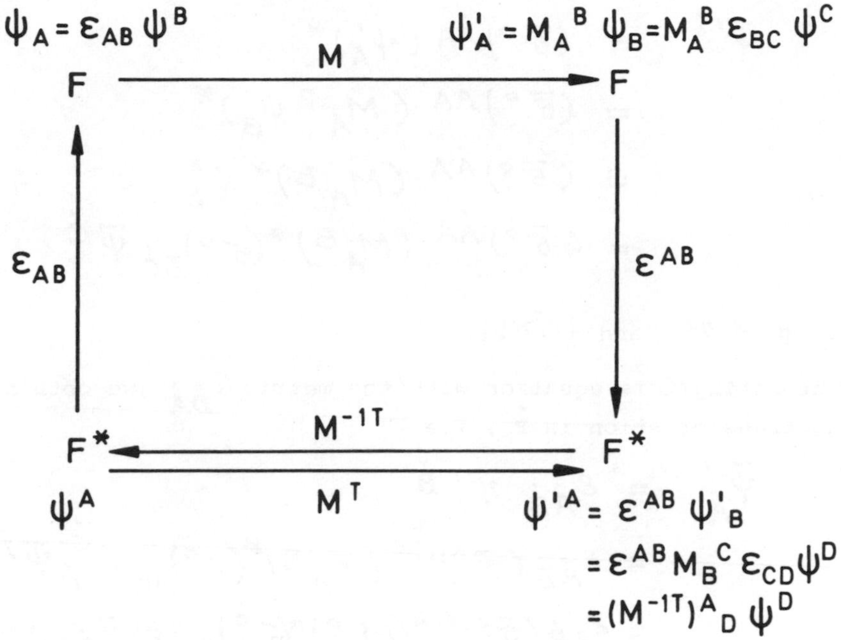

Fig. 1 : Directions of dual maps

The map $M : F \rightarrow F$ induces a corresponding map in the space $\dot{F}$ which can be constructed from M using the general prescription of complex conjugation, multiplication by the matrix $\bar{\sigma}^0$ and contraction with the metric ε.

Starting with

$$\psi'_A = M_A{}^B \psi_B \quad , \quad \psi_B \in F$$

we obtain the corresponding transformation in $\dot{F}^*$:

$$\bar{\psi}'^{\dot{A}} = (\bar{\sigma}^0)^{\dot{A}A} (\psi'_A)^*$$
$$= (\bar{\sigma}^0)^{\dot{A}A} (M_A{}^B \psi_B)^*$$
$$= (\bar{\sigma}^0)^{\dot{A}A} (M_A{}^B)^* \psi_B^*$$
$$= (\bar{\sigma}^0)^{\dot{A}A} (M_A{}^B)^* (\sigma^0)_{B\dot{C}} \bar{\psi}^{\dot{C}}$$

using (1.76g) and (1.76h).

Contracting this equation with the metric $\varepsilon_{\dot{D}\dot{A}}$,we obtain the transformation in $\dot{F}$, i.e.

$$\bar{\psi}'_{\dot{A}} = \varepsilon_{\dot{A}\dot{B}} \bar{\psi}'^{\dot{B}}$$
$$= \varepsilon_{\dot{A}\dot{B}} (\bar{\sigma}^0)^{\dot{B}A} (M_A{}^B)^* (\sigma^0)_{B\dot{C}} \delta^{\dot{C}}_{\dot{D}} \bar{\psi}^{\dot{D}}$$
$$= \varepsilon_{\dot{A}\dot{B}} (\bar{\sigma}^0)^{\dot{B}A} (M_A{}^B)^* (\sigma^0)_{B\dot{C}} \varepsilon^{\dot{C}\dot{E}} \varepsilon_{\dot{E}\dot{D}} \bar{\psi}^{\dot{D}}$$
$$= \varepsilon_{\dot{A}\dot{B}} (\bar{\sigma}^0)^{\dot{B}A} (M_A{}^B)^* (\sigma^0)_{B\dot{C}} \varepsilon^{\dot{C}\dot{E}} \bar{\psi}_{\dot{E}}$$

Defining

$$(M^*)_{\dot{A}}{}^{\dot{B}} := \varepsilon_{\dot{A}\dot{C}} (\bar{\sigma}^0)^{\dot{C}A} (M_A{}^B)^* (\sigma^0)_{B\dot{D}} \varepsilon^{\dot{D}\dot{B}}$$

then

$$\bar{\psi}'_{\dot{A}} = (M^*)_{\dot{A}}{}^{\dot{B}} \bar{\psi}_{\dot{B}} \quad , \quad M^* : \dot{F} \longrightarrow \dot{F}$$

in agreement with (1.55). The relation between the various Weyl spinors and their respective representation spaces is shown in Fig. 2 .

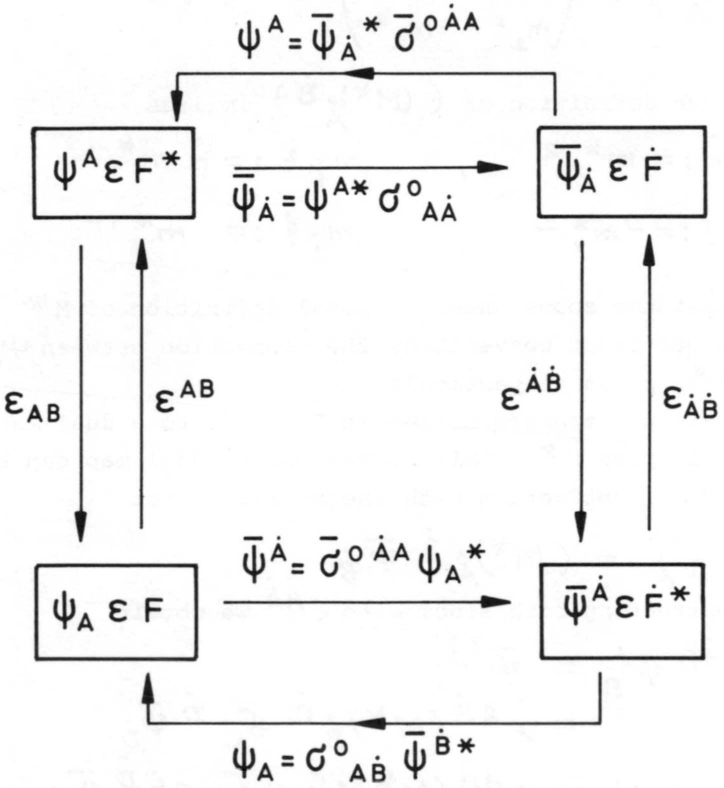

Fig. 2 : Weyl spinors and their respective representation spaces

Remark: For the matrix

$$(M_A{}^B) = \begin{pmatrix} m_1{}^1 & m_1{}^2 \\ m_2{}^1 & m_2{}^2 \end{pmatrix} \in SL(2,C)$$

the above definition of $((M^*)_{\dot{A}}{}^{\dot{B}})$ implies

$$m_{\dot{1}}{}^{\dot{1}} := m^*{}_2{}^2 \quad , \quad m_{\dot{1}}{}^{\dot{2}} := -\, m^*{}_2{}^1$$

$$m_{\dot{2}}{}^{\dot{1}} := -m^*{}_1{}^2 \quad , \quad m_{\dot{2}}{}^{\dot{2}} := m^*{}_1{}^1$$

Note that the above unconventional definition of M^* is due to our index convention; the connection between ψ_A and $\bar{\psi}^{\dot{A}}$ plays a particular role.

Again the transformation in $\dot{F}$ leads to a dual map in the dual space $\dot{F}^*$. This corresponding dual map can be derived by contraction with the metric. Given

$$\bar{\psi}'_{\dot{A}} = (M^*)_{\dot{A}}{}^{\dot{B}}\, \bar{\psi}_{\dot{B}} \tag{1.55}$$

and contracting both sides with $\varepsilon^{\dot{A}\dot{B}}$ we obtain

$$\begin{aligned}
\varepsilon^{\dot{A}\dot{B}}\, \bar{\psi}'_{\dot{B}} &= \bar{\psi}'^{\dot{A}} \\
&= \varepsilon^{\dot{A}\dot{B}} (M^*)_{\dot{B}}{}^{\dot{C}}\, \delta_{\dot{C}}{}^{\dot{D}}\, \bar{\psi}_{\dot{D}} \\
&= \varepsilon^{\dot{A}\dot{B}} (M^*)_{\dot{B}}{}^{\dot{C}}\, \varepsilon_{\dot{C}\dot{E}}\, \varepsilon^{\dot{E}\dot{D}}\, \bar{\psi}_{\dot{D}} \\
&= \varepsilon^{\dot{A}\dot{B}} (M^*)_{\dot{B}}{}^{\dot{C}}\, \varepsilon_{\dot{C}\dot{E}}\, \bar{\psi}^{\dot{E}} \\
&=: (M^{*-1T})^{\dot{A}}{}_{\dot{E}}\, \bar{\psi}^{\dot{E}}
\end{aligned}$$

where

$$(M^{*-1T})^{\dot{A}}{}_{\dot{B}} := \varepsilon^{\dot{A}\dot{C}} (M^*)_{\dot{C}}{}^{\dot{D}}\, \varepsilon_{\dot{D}\dot{B}}$$

$$M^{*-1T} : \dot{F}^* \longrightarrow \dot{F}^*$$

This is in agreement with (1.74). Again for $M \in SL(2,C)$ with

$$(M_A{}^B) = \begin{pmatrix} m_1{}^1 & m_1{}^2 \\ m_2{}^1 & m_2{}^2 \end{pmatrix}$$

the elements of the matrix M^{*-1T} are given by

$$m^{\dot 1}{}_{\dot 1} := m^*{}_1{}^1 \quad , \quad m^{\dot 1}{}_{\dot 2} := m^*{}_1{}^2$$
$$m^{\dot 2}{}_{\dot 1} := m^*{}_2{}^1 \quad , \quad m^{\dot 2}{}_{\dot 2} := m^*{}_2{}^2$$

where

$$\left((M^{*-1T})^{\dot A}{}_{\dot B} \right) = \begin{pmatrix} m^{\dot 1}{}_{\dot 1} & m^{\dot 1}{}_{\dot 2} \\ m^{\dot 2}{}_{\dot 1} & m^{\dot 2}{}_{\dot 2} \end{pmatrix}$$

In deriving the matrix M^* we see that we cannot describe complex conjugation as a <u>linear</u> map from F to $\dot F$. This can, of course, also be shown explicitly by considering a similarity transformation as in the previous equivalence proofs, and showing that the appropriate matrix does not exist. (We shall see later that such a linear transformation exists in the four-dimensional Dirac formalism). Therefore F and $\dot F$ define two inequivalent representation spaces for SL(2,C). On the other hand elements of the corresponding dual spaces F^* and $\dot F^*$ can be obtained with the metrics ε_{AB} and $\varepsilon_{\dot A \dot B}$ respectively. These are matrix representations of linear maps and therefore F and F^* are equivalent representation spaces as well as $\dot F$ and $\dot F^*$.

1.3.2 Calculations with Spinors

We now consider some explicit calculations with spinors and derive several useful formulae.

i) Proposition: The quadratic form

$$(\psi\chi) := \psi^A \chi_A \tag{1.77a}$$

is invariant under transformations of SL(2,C).

Proof: We have to show that

$$(\psi'\chi') = (\psi\chi)$$

Consider (using (1.75a) and (1.75b))

$$\begin{aligned}(\psi'\chi') &= \psi'^A \chi'_A \\ &= (M^{-1T})^A{}_B \psi^B M_A{}^C \chi_C \\ &= \psi^B (M^{-1})_B{}^A M_A{}^C \chi_C \\ &= \psi^B \delta_B{}^C \chi_C \\ &= \psi^B \chi_B \\ &= (\psi\chi)\end{aligned}$$

Similarly

$$\psi'_A \chi'^A = \psi_A \chi^A$$

ii) Proposition: The quadratic form

$$(\bar\psi\bar\chi) := \bar\psi_{\dot A} \bar\chi^{\dot A} \tag{1.77b}$$

is invariant under SL(2,C).

Proof: We have (using (1.75c) and (1.75d))

$$\begin{aligned}(\bar\psi'\bar\chi') &= \bar\psi'_{\dot A} \bar\chi'^{\dot A} \\ &= (M^*)_{\dot A}{}^{\dot B} \bar\psi_{\dot B} (M^{*-1T})^{\dot A}{}_{\dot C} \bar\chi^{\dot C} \\ &= \bar\psi_{\dot B} (M^{*T})^{\dot B}{}_{\dot A} (M^{*T-1})^{\dot A}{}_{\dot C} \bar\chi^{\dot C} \\ &= \bar\psi_{\dot B} \delta^{\dot B}{}_{\dot C} \bar\chi^{\dot C} \\ &= \bar\psi_{\dot B} \bar\chi^{\dot B}\end{aligned}$$

$$= (\bar{\psi}\bar{\chi}) \quad \text{(using (1.77b))}$$

<u>Remark</u>: For calculations with Weyl spinors it is a useful convention to sum undotted indices from the upper left to the lower right ("north-west to south-east") and dotted indices from the lower left to the upper right ("south-west to north-east") as indicated in (1.77a) and (1.77b). Thus we write

$$(\psi\chi) := \psi^A \chi_A \quad \text{and} \quad (\bar{\psi}\bar{\chi}) := \bar{\psi}_{\dot{A}} \bar{\chi}^{\dot{A}} \tag{1.78}$$

However, expressions such as (1.76b), i.e. $\psi_A = \varepsilon_{AB}\psi^B$ show that in general summations in the other directions are also possible.

iii) <u>Postulate</u>: We postulate that the components of spinors are Grassmann variables (also called Grassmann numbers or Grassmann parameters) which are not to be confused with anticommuting operators such as the spinor charges Q_A to be defined later; i.e. we demand

$$\{\psi_A, \psi^B\} = \{\psi_A, \psi_B\} = \{\psi^A, \psi^B\} = 0$$

and (1.79)

$$\{\bar{\chi}_{\dot{A}}, \bar{\chi}^{\dot{B}}\} = \{\bar{\chi}_{\dot{A}}, \bar{\chi}_{\dot{B}}\} = \{\bar{\chi}^{\dot{A}}, \bar{\chi}^{\dot{B}}\} = 0$$

If we require the ψ_A's to anticommute, an expression like $(\psi\psi)$ does not vanish. Thus

$$\begin{aligned}
(\psi\psi) &= \psi^A \psi_A \quad &\text{(with (1.77))}\\
&= \varepsilon^{AB}\psi_B\psi_A \quad &\text{(with (1.76a))}\\
&= \varepsilon^{12}\psi_2\psi_1 + \varepsilon^{21}\psi_1\psi_2 &\\
&= \psi_2\psi_1 - \psi_1\psi_2 &
\end{aligned}$$

since $\varepsilon^{12} = -\varepsilon^{12} = +1$

If the components ψ_A were required to commute, i.e. if $[\psi_1, \psi_2] = 0$, then $(\psi\psi) = 0$. Instead we assume that the ψ_A 's anticommute with each other and with other Grassmann variables such as fermion fields and spinor charges. This requirement is, of course, nothing but an application of the spin statistics theorem: half-integer spin quantities obey Fermi-Dirac statistics whereas integer spin quantities obey Bose-Einstein statistics.

iv) Proposition: For anticommuting ψ 's the quadratic form (1.77) is symmetric on exchange of spinors, i.e.

$$(\psi\chi) = (\chi\psi) \tag{1.80}$$

Proof: We have

$$\begin{aligned}
(\psi\chi) &= \psi^A \chi_A && \text{(using (1.77))}\\
&= -\chi_A \psi^A && \text{(using (1,79a))}\\
&= -\varepsilon_{AB}\chi^B \varepsilon^{AC}\psi_C && \text{(using (1.76a,b))}\\
&= -\chi^B (\varepsilon^T)_{BA}\varepsilon^{AC}\psi_C &&\\
&= \chi^B \delta_B{}^C \psi_C && \text{(using (1.61))}\\
&= \chi^B \psi_B && \text{(using (1.77))}\\
&= (\chi\psi) &&
\end{aligned}$$

v) Proposition: The quadratic form for dotted spinors, i.e. (1.78), is symmetric on exchange, i.e.

$$(\bar{\psi}\bar{\chi}) = (\bar{\chi}\bar{\psi}) \tag{1.81}$$

if $\bar{\psi}$ and $\bar{\chi}$ are Grassmann variables.

Proof: We have

$$\begin{aligned}
(\bar{\psi}\bar{\chi}) &= \bar{\psi}_{\dot{A}}\bar{\chi}^{\dot{A}} && \text{(using (1.78))}\\
&= -\bar{\chi}^{\dot{A}}\bar{\psi}_{\dot{A}} && \text{(using (1.79b))}\\
&= -\varepsilon^{\dot{A}\dot{B}}\bar{\chi}_{\dot{B}}\varepsilon_{\dot{A}\dot{C}}\bar{\psi}^{\dot{C}} && \text{(using (1.76c,d))}
\end{aligned}$$

$$= -\bar{\chi}_{\dot{B}} (\varepsilon^T)^{\dot{B}\dot{A}} \varepsilon_{\dot{A}\dot{C}} \bar{\psi}^{\dot{C}}$$
$$= \bar{\chi}_{\dot{B}} \varepsilon^{\dot{B}\dot{A}} \varepsilon_{\dot{A}\dot{C}} \bar{\psi}^{\dot{C}}$$
$$= \bar{\chi}_{\dot{B}} \delta^{\dot{B}}{}_{\dot{C}} \bar{\psi}^{\dot{C}} \qquad \text{(using (1.71))}$$
$$= \bar{\chi}_{\dot{B}} \bar{\psi}^{\dot{B}}$$
$$= (\bar{\chi}\bar{\psi}) \qquad \text{(using (1.78))}$$

vi) <u>Proposition</u>: If θ is a Grassmann variable

$$\theta^2 := (\theta\theta) = -2\theta^1\theta^2$$
$$\bar{\theta}^2 := (\bar{\theta}\bar{\theta}) = 2\bar{\theta}_{\dot{1}}\bar{\theta}_{\dot{2}} \qquad (1.82)$$

and hence $(\theta\theta)\theta^A = 0$ etc.

<u>Proof</u>: We have

$$\theta^2 = (\theta\theta) = \theta^A\theta_A \qquad \text{(using (1.77))}$$
$$= \theta^A\varepsilon_{AB}\theta^B \qquad \text{(using (1.76b))}$$
$$= \theta^1\varepsilon_{12}\theta^2 + \theta^2\varepsilon_{21}\theta^1$$
$$= -\theta^1\theta^2 + \theta^2\theta^1$$
$$= -2\theta^1\theta^2 \qquad \text{(using (1.79))}$$

(Observe:" θ squared" on the left, component " θ two" on the right in our notation!)

Similarly

$$\bar{\theta}_{\dot{A}}\bar{\theta}^{\dot{A}} = \bar{\theta}_{\dot{A}}\varepsilon^{\dot{A}\dot{B}}\bar{\theta}_{\dot{B}} = 2\bar{\theta}_{\dot{1}}\bar{\theta}_{\dot{2}}$$

using first (1.76c) and then (1.69) .

vii) Proposition: For Grassmann variables θ and $\bar{\theta}$ we have

$$\theta^A \theta^B = -\frac{1}{2}\varepsilon^{AB}(\theta\theta) \tag{1.83a}$$

$$\theta_A \theta_B = \frac{1}{2}\varepsilon_{AB}(\theta\theta) \tag{1.83b}$$

$$\bar{\theta}^{\dot{A}}\bar{\theta}^{\dot{B}} = \frac{1}{2}\varepsilon^{\dot{A}\dot{B}}(\bar{\theta}\bar{\theta}) \tag{1.83c}$$

$$\bar{\theta}_{\dot{A}}\bar{\theta}_{\dot{B}} = -\frac{1}{2}\varepsilon_{\dot{A}\dot{B}}(\bar{\theta}\bar{\theta}) \tag{1.83d}$$

Proof: We shall see later (cf. (1.129c,d)) that

$$\varepsilon_{AB}\varepsilon^{DC} = \delta_A{}^C\delta_B{}^D - \delta_A{}^D\delta_B{}^C \tag{1.129c}$$

and

$$\varepsilon_{\dot{A}\dot{B}}\varepsilon^{\dot{D}\dot{C}} = \delta_{\dot{A}}{}^{\dot{C}}\delta_{\dot{B}}{}^{\dot{D}} - \delta_{\dot{A}}{}^{\dot{D}}\delta_{\dot{B}}{}^{\dot{C}} \tag{1.129d}$$

These formulae are needed in proving (1.83 a,b,c,d).

i) Consider the expression on the right hand side of (1.83a):

$$
\begin{aligned}
-\frac{1}{2}\varepsilon^{AB}(\theta\theta) &= -\frac{1}{2}\varepsilon^{AB}\theta^C\theta_C \\
&= -\frac{1}{2}\varepsilon^{AB}\varepsilon_{CD}\theta^C\theta^D \\
&= -\frac{1}{2}(\delta_C{}^B\delta_D{}^A - \delta_C{}^A\delta_D{}^B)\theta^C\theta^D \\
&= -\frac{1}{2}(\theta^B\theta^A - \theta^A\theta^B) \\
&= \theta^A\theta^B
\end{aligned}
$$

ii) Consider the expression on the left hand side of (1.83b) :

$$
\begin{aligned}
\theta_A\theta_B &= \varepsilon_{AC}\varepsilon_{BD}\theta^C\theta^D \\
&= -\frac{1}{2}\varepsilon_{AC}\varepsilon_{BD}\varepsilon^{CD}(\theta\theta) \\
&= \frac{1}{2}\varepsilon_{AB}(\theta\theta)
\end{aligned}
$$

Multiplying by ε^{CA} we obtain another formula

$$\theta^C\theta_B = \frac{1}{2}\,\delta^C{}_B\,(\theta\theta) \tag{1.83e}$$

In a similar manner we have

iii) $$\frac{1}{2}\varepsilon^{\dot A\dot B}(\bar\theta\bar\theta) = \frac{1}{2}\varepsilon^{\dot A\dot B}\bar\theta_{\dot C}\bar\theta^{\dot C}$$
$$= \frac{1}{2}\varepsilon^{\dot A\dot B}\varepsilon_{\dot C\dot D}\bar\theta^{\dot D}\bar\theta^{\dot C}$$
$$= \frac{1}{2}(\delta^{\dot B}_{\dot C}\delta^{\dot A}_{\dot D} - \delta^{\dot A}_{\dot C}\delta^{\dot B}_{\dot D})\bar\theta^{\dot D}\bar\theta^{\dot C}$$
$$= \frac{1}{2}(\bar\theta^{\dot A}\bar\theta^{\dot B} - \bar\theta^{\dot B}\bar\theta^{\dot A}) = \bar\theta^{\dot A}\bar\theta^{\dot B}$$

iv) $$\bar\theta_{\dot A}\bar\theta_{\dot B} = \varepsilon_{\dot A\dot C}\varepsilon_{\dot B\dot D}\bar\theta^{\dot C}\bar\theta^{\dot D}$$
$$= \frac{1}{2}\varepsilon_{\dot A\dot C}\varepsilon_{\dot B\dot D}\varepsilon^{\dot C\dot D}(\bar\theta\bar\theta)$$
$$= \frac{1}{2}\varepsilon_{\dot B\dot A}(\bar\theta\bar\theta)$$
$$= -\frac{1}{2}\varepsilon_{\dot A\dot B}(\bar\theta\bar\theta)$$

Multiplying this result by $\varepsilon^{\dot C\dot A}$ we obtain the formula

$$\bar\theta^{\dot C}\bar\theta_{\dot B} = -\frac{1}{2}\,\delta^{\dot C}{}_{\dot B}\,(\bar\theta\bar\theta) \tag{1.83f}$$

viii) Proposition: Let θ, ϕ, ψ be Grassmann variables. Then

$$(\theta\phi)(\theta\psi) = -\frac{1}{2}(\phi\psi)(\theta\theta) = -\frac{1}{2}(\theta\theta)(\phi\psi) \tag{1.84}$$

This result is a Fierz reordering formula for Weyl spinors. (For other analogous relations discussed later see (1.114), (1.118), (7.81) and (7.91)).

Proof: Manipulating the expression on the left hand side we obtain

$$(\theta\phi)(\theta\psi) = (\phi\theta)(\theta\psi) \qquad \text{(with (1.80))}$$
$$= \phi^A \theta_A \theta^B \psi_B \qquad \text{(with (1.77))}$$
$$= \phi^A \varepsilon_{AC} \theta^C \theta^B \psi_B \qquad \text{(with (1.76b))}$$
$$= \phi^A \varepsilon_{AC} (-\tfrac{1}{2}\varepsilon^{CB} \theta\theta) \psi_B \qquad \text{(with (1.83a))}$$
$$= -\tfrac{1}{2} \phi^A \delta_A{}^B (\theta\theta) \psi_B$$
$$= -\tfrac{1}{2} \phi^A \psi_A (\theta\theta)$$
$$= -\tfrac{1}{2} (\phi\psi)(\theta\theta)$$

applying twice the Grassmann property.

1.3.3 Connection between SL(2,C) and $L_+^\uparrow$

We now investigate the connection between SL(2,C) and the restricted Lorentz group $L_+^\uparrow$ more closely and study the spinor representations of SL(2,C). We shall show that $L_+^\uparrow$ is homomorphic to SL(2,C); i.e. for any $M \in$ SL(2,C) there is a Lorentz matrix $\Lambda = \Lambda(M) \in L_+^\uparrow$ such that

$$\Lambda(M_1)\Lambda(M_2) = \Lambda(M_1 M_2) \qquad (1.85)$$

In order to obtain a connection between spinor calculus and the Lorentz four-vector formalism we start by introducing a set of four matrices

$$\sigma^\mu := (1, \vec{\sigma}) = (\sigma^0, \vec{\sigma}) \qquad (1.86a)$$

where σ^0 is the 2x2-unit matrix and the σ^i are the usual Pauli matrices, i.e.

$$\sigma^1 := \begin{pmatrix} 0 & 1 \\ 1 & 0 \end{pmatrix}, \quad \sigma^2 := \begin{pmatrix} 0 & -i \\ i & 0 \end{pmatrix}, \quad \sigma^3 := \begin{pmatrix} 1 & 0 \\ 0 & -1 \end{pmatrix} \qquad (1.86b)$$

The spinor index structure of the matrices σ^μ is such that

$$\sigma^\mu := (\sigma^\mu{}_{A\dot{A}}) \tag{1.87}$$

as can be shown from the adjoint representation of SL(2,C) (see (1.99)). The indices may be raised by application of the ε -tensor as in (1.76a) and (1.76c) leading to a new set of matrices , i.e.

$$\bar{\sigma}^{\mu\dot{A}A} := \varepsilon^{AB}\varepsilon^{\dot{A}\dot{B}}\sigma^\mu_{B\dot{B}} \tag{1.88a}$$

or

$$\sigma^\mu_{A\dot{A}} = \varepsilon_{AB}\varepsilon_{\dot{A}\dot{B}}(\bar{\sigma}^\mu)^{\dot{B}B} \tag{1.88b}$$

In another formulation (1.88a) is

$$(\bar{\sigma}^\mu)^{\dot{A}A} = (\bar{\sigma}^{\mu T})^{A\dot{A}} = -\varepsilon^{AB}\sigma^\mu_{B\dot{B}}\varepsilon^{\dot{B}\dot{A}} \tag{1.89}$$

Evaluating (1.88) we find that

$$\bar{\sigma}^0 = \sigma^0, \; \bar{\sigma}^i = -\sigma^i, \; i = 1,2,3 \tag{1.90}$$

Proof: We demonstrate (1.90) explicitly for i = 1. In this case

$$(\bar{\sigma}^1)^{\dot{A}A} = \varepsilon^{AB}\varepsilon^{\dot{A}\dot{B}}\sigma^1_{B\dot{B}} \quad \text{where } \sigma^1_{1\dot{1}} = \sigma^1_{2\dot{2}} = 0, \quad \sigma^1_{1\dot{2}} = \sigma^1_{2\dot{1}} = 1$$

i.e.

$$(\bar{\sigma}^1)^{\dot{1}1} = \varepsilon^{12}\varepsilon^{\dot{1}\dot{2}}\sigma^1_{2\dot{2}} = 0, \; (\bar{\sigma}^1)^{\dot{1}2} = \varepsilon^{21}\varepsilon^{\dot{1}\dot{2}}\sigma^1_{1\dot{2}} = -1$$

$$(\bar{\sigma}^1)^{\dot{2}1} = \varepsilon^{12}\varepsilon^{\dot{2}\dot{1}}\sigma^1_{2\dot{1}} = -1, \; (\bar{\sigma}^1)^{\dot{2}2} = \varepsilon^{21}\varepsilon^{\dot{2}\dot{1}}\sigma^1_{2\dot{2}} = 0$$

so that

$$\bar{\sigma}^1 = \begin{pmatrix} 0 & -1 \\ -1 & 0 \end{pmatrix} = -\sigma^1$$

Proposition: The following formulae can be shown to hold:

a) $$\mathrm{Tr}\,[\sigma^\mu \bar{\sigma}^\nu] = 2\eta^{\mu\nu}$$

where the metric tensor is

$$(\eta^{\mu\nu}) = \text{diag}\,(+1, -1, -1, -1)$$

and (1.91)

b)
$$\sigma^\mu \bar{\sigma}^\rho + \sigma^\rho \bar{\sigma}^\mu = 2\eta^{\mu\rho} 1_{2\times 2}$$

Proof: Consider

a) $\mu = \nu = 0$:

$$Tr\,(\sigma^0 \bar{\sigma}^0) = Tr\,(1.1) = 2 = 2\eta^{00}$$

$\mu = 0, \nu = i = 1,2,3$:

$$Tr\,(\sigma^0 \bar{\sigma}^i) = Tr\,(1 \bar{\sigma}^i) = -Tr(\sigma^i) = 0$$

since the Pauli matrices are traceless,

$\mu = i, \nu = j, i,j = 1,2,3$:

$$\begin{aligned} Tr(\sigma^i \bar{\sigma}^j) &= -Tr(\sigma^i \sigma^j) \\ &= -Tr\,(\delta^{ij} 1_{2\times 2} + \varepsilon^{ijk} \sigma^k) \\ &= -\delta^{ij}\, Tr\, 1_{2\times 2} - \varepsilon^{ijk}\, Tr\, \sigma^k \\ &= 2\eta^{ij} \end{aligned}$$

b) $\mu = \rho$: $2\sigma^\mu \bar{\sigma}^\mu = 2\eta^{\mu\mu} 1_{2\times 2}$

$\mu \neq \rho$:

i) $\mu = 0, \rho \neq 0$: $\sigma^0 \bar{\sigma}^i + \sigma^i \bar{\sigma}^0 = 0$

ii) $\mu = i, \rho = j$: $\sigma^i \bar{\sigma}^j + \sigma^j \bar{\sigma}^i = 0$

Proposition: The matrices σ_μ form a complete set in the sense that any complex 2x2-matrix can be expressed as a linear combination of them. In particular the completeness relation is

$$(\sigma^\mu)_{A\dot{A}} (\bar{\sigma}_\mu)^{\dot{B}B} = 2\delta_A{}^B \delta_{\dot{A}}{}^{\dot{B}} \qquad (1.92)$$

Proof: Equation (1.92) has to be shown explicitly:

$$(\sigma^\mu)_{A\dot{A}}(\bar{\sigma}_\mu)^{\dot{B}B} = \eta_{\mu\nu}(\sigma^\mu)_{A\dot{A}}(\bar{\sigma}^\nu)^{\dot{B}B}$$
$$= (\sigma^0)_{A\dot{A}}(\bar{\sigma}^0)^{\dot{B}B} - (\sigma^1)_{A\dot{A}}(\bar{\sigma}^1)^{\dot{B}B}$$
$$- (\sigma^2)_{A\dot{A}}(\bar{\sigma}^2)^{\dot{B}B} - (\sigma^3)_{A\dot{A}}(\bar{\sigma}^3)^{\dot{B}B}$$

$A = 1 = B$, $\dot{B} = \dot{A} = \dot{1}$:

$$(\sigma^\mu)_{1\dot{1}}(\bar{\sigma}_\mu)^{\dot{1}1} = (\sigma^0)_{1\dot{1}}(\bar{\sigma}^0)^{\dot{1}1} - (\sigma^1)_{1\dot{1}}(\bar{\sigma}^1)^{\dot{1}1}$$
$$- (\sigma^2)_{1\dot{1}}(\bar{\sigma}^2)^{\dot{1}1} - (\sigma^3)_{1\dot{1}}(\bar{\sigma}^3)^{\dot{1}1}$$
$$= 2\delta_1{}^1\delta_{\dot{1}}{}^{\dot{1}}$$

$A = B = 2$, $\dot{A} = \dot{B} = \dot{2}$:

$$(\sigma^\mu)_{2\dot{2}}(\bar{\sigma}_\mu)^{\dot{2}2} = (\sigma^0)_{2\dot{2}}(\bar{\sigma}^0)^{\dot{2}2} - (\sigma^1)_{2\dot{2}}(\bar{\sigma}^1)^{\dot{2}2}$$
$$- (\sigma^2)_{2\dot{2}}(\bar{\sigma}^2)^{\dot{2}2} - (\sigma^3)_{2\dot{2}}(\bar{\sigma}^3)^{\dot{2}2}$$
$$= 2\delta_2{}^2\delta_{\dot{2}}{}^{\dot{2}}$$

$A = B = 1$, $\dot{A} = \dot{B} = \dot{2}$:

$$(\sigma^\mu)_{1\dot{2}}(\bar{\sigma}_\mu)^{\dot{2}1} = (\sigma^0)_{1\dot{2}}(\bar{\sigma}^0)^{\dot{2}1} - (\sigma^1)_{1\dot{2}}(\bar{\sigma}^1)^{\dot{2}1}$$
$$- (\sigma^2)_{1\dot{2}}(\bar{\sigma}^2)^{\dot{2}1} - (\sigma^3)_{1\dot{2}}(\bar{\sigma}^3)^{\dot{2}1}$$
$$= 2\delta_1{}^1\delta_{\dot{2}}{}^{\dot{2}}$$

$A = B = 2$, $\dot{A} = \dot{B} = \dot{1}$:

$$(\sigma^\mu)_{2\dot{1}}(\bar{\sigma}_\mu)^{\dot{1}2} = (\sigma^0)_{2\dot{1}}(\bar{\sigma}^0)^{\dot{1}2} - (\sigma^1)_{2\dot{1}}(\bar{\sigma}^1)^{\dot{1}2}$$

$$-(\sigma^2)_{2i}(\bar{\sigma}^2)^{i2}-(\sigma^3)_{2i}(\bar{\sigma}^3)^{i2}$$
$$=2\delta_2^{\ 2}\delta_i^{\ i}$$

$A = 1,\ B = 2,\ \dot{A} = \dot{B} = 1$:

$$(\sigma^\mu)_{1i}(\bar{\sigma}_\mu)^{i2}=(\sigma^0)_{1i}(\bar{\sigma}^0)^{i2}-(\sigma^1)_{1i}(\bar{\sigma}^1)^{i2}$$
$$-(\sigma^2)_{1i}(\bar{\sigma}^2)^{i2}-(\sigma^3)_{1i}(\bar{\sigma}^3)^{i2}$$
$$=0$$

In the same way one can show that all other expressions vanish identically.

We are now in a position to construct the group homomorphism between SL(2,C) and the restricted Lorentz group $L_+^\uparrow$. First, however, we construct a map from Minkowski space M_4 to the set of hermitian, complex 2x2-matrices H(2,C) (H for hermitian). Thus, denoting the map by ρ, we have

$$\rho : M_4 \longrightarrow H(2,C)$$
$$x^\mu \longrightarrow \rho(x^\mu) = x_\mu \sigma^\mu$$
$$= \begin{pmatrix} x^0 - x^3 & x^1 + ix^2 \\ x^1 - ix^2 & x^0 + x^3 \end{pmatrix}$$
$$= X \tag{1.93}$$

where

$$(x^\mu) = (x^0, \vec{x}),\quad (x_\mu) = (x^0, -\vec{x})$$

Proposition: The inverse map

$$\rho^{-1} : H(2,C) \longrightarrow M_4$$

is given by the following trace relation

$$\rho^{-1} : H(2,C) \longrightarrow M_4$$
$$X \longrightarrow \rho^{-1}(X)$$
$$= x^\mu$$
$$= \frac{1}{2} Tr[X \bar{\sigma}^\mu] \tag{1.94}$$

Proof: Using (1.93) and (1.91) we obtain

$$\frac{1}{2} Tr[X\bar{\sigma}^\mu] = \frac{1}{2} Tr[x_\nu \sigma^\nu \bar{\sigma}^\mu]$$
$$= \frac{1}{2} Tr[\sigma^\nu \bar{\sigma}^\mu] x_\nu$$
$$= \frac{1}{2} \cdot 2 \eta^{\mu\nu} x_\nu$$
$$= x^\mu$$

Proposition: The determinant of $\rho(x_\mu) \in$ H(2,C) is given by

$$\det X = \det \rho(x_\mu) = x_\mu x^\mu = x^2 \tag{1.95}$$

Proof: Using (1.93) we have

$$\det \rho(x^\mu) = \det X$$
$$= \det \begin{pmatrix} x^0 - x^3 & x^1 + ix^2 \\ x^1 - ix^2 & x^0 + x^3 \end{pmatrix}$$
$$= (x^0)^2 - (x^i)^2$$
$$= \eta_{\mu\nu} x^\mu x^\nu$$
$$= x^2$$

The hermiticity of the matrix $\rho(x^\mu)$ is seen immediately from the hermiticity of the Pauli matrices

$$X = X^+ = X^{*T} \tag{1.96}$$

We now consider the action of SL(2,C) on H(2,C), which is called the adjoint representation of the group SL(2,C). This representation is defined in the following way:

$$\begin{aligned} \text{ad}: \quad SL(2,C) &\longrightarrow \text{Aut}\ (H(2,C)) \\ M &\longrightarrow \text{ad}\ M \end{aligned} \tag{1.97}$$

with

$$M' \equiv \text{ad}\, M(X) := M X M^{+}, \quad M, M^{+} \in SL(2,C) \tag{1.98}$$

where Aut (H(2,C)) is the automorphism group of H(2,C) which is isomorphic to GL(H(2,C))[30]. We have to verify therefore, that ad M(X) $\in$ H(2,C), i.e. that ad M(X) is hermitian. Using (1.98) we have

$$\begin{aligned} [\text{ad}\, M(X)]^{+} &= (M X M^{+})^{+} = M X^{+} M^{+} \\ &= M X M^{+} \qquad (X \in H(2,C)) \\ &= \text{ad}\, M(X) \end{aligned}$$

From the adjoint representation (1.98) we can derive the index structure of the Pauli matrices (1.87). We had

$$\rho(x^{\mu}) = X = x_{\mu}\sigma^{\mu}$$

which transforms under the adjoint representation of SL(2,C) as

$$X' = M X M^{+}$$

or

$$x'_{\mu}\sigma^{\mu} = M \sigma^{\nu} x_{\nu} M^{+}$$

Using the index notation of the SL(2,C)-matrices M and M^{+} we conclude

$$x'_{\mu}(\sigma^{\mu})_{A\dot{A}} = M_{A}{}^{B} x_{\nu} (\sigma^{\nu})_{B\dot{B}} (M^{*})^{\dot{B}}{}_{\dot{A}} \tag{1.99}$$

and we see that the σ-matrices must carry a dotted and an undotted index. The matrices σ^{μ} therefore map $\dot{F}$ into F , and similarly the matrices $\bar{\sigma}^{\mu}$ map F into $\dot{F}$. We now calculate the determinant of a transformed hermitian matrix. Using (1.93) and (1.98) we have

$$
\begin{aligned}
x'_\mu x'^\mu &= \det X' \\
&= \det (MXM^+) \\
&= \det(M) \det(X) \det(M^+) \\
&= \det X \\
&= x_\mu x^\mu
\end{aligned}
$$

since $M \in SL(2,C) : \det M = 1 .$ (1.100)

Thus the quadratic form $x_\mu x^\mu = \det X$ is left invariant under transformations of the adjoint representation of SL(2,C).

We now have the following construction. Starting from a Minkowski four-vector $x^\mu \in M_4$ we construct an hermitian 2x2-matrix $X \in H(2,C)$. The determinant of this matrix is simply the Minkowski product of x_μ with itself. We then go to the adjoint representation of SL(2,C) on H(2,C) and derive the result that the determinant is left invariant under SL(2,C). If we apply the map ρ^{-1} to the transformed hermitian matrix, we obtain a four-vector x'_μ such that

$$x'_\mu x'^\mu = x_\mu x^\mu$$

We can demonstrate these steps schematically in the following way:

$$M_4 \xrightarrow{\ \rho\ } H \xrightarrow{\ \text{ad } M\ } H \xrightarrow{\ \rho^{-1}\ } M_4$$

i.e. (1.101)

$$
\begin{aligned}
x_\mu \longrightarrow \rho(x_\mu) = x_\mu \sigma^\mu &\longrightarrow \text{ad}\, M(X) \longrightarrow \rho^{-1}(X') \\
= X \qquad &\qquad = MXM^+ \qquad = x'_\mu \\
&\qquad = X'
\end{aligned}
$$

This transformation from M_4 to M_4 is, of course, simply a Lorentz transformation which transforms four-vectors into four-vectors leaving the Minkowski square, i.e. line-element, invariant. Now from (1.1) we have

$$x'^{\mu} = \Lambda^{\mu}{}_{\nu} x^{\nu} \tag{1.1}$$

On the other hand we can also write, using (1.94), (1.98) and (1.93),

$$\begin{aligned} x'^{\mu} &= \tfrac{1}{2} Tr[X' \bar{\sigma}^{\mu}] \\ &= \tfrac{1}{2} Tr[M X M^{+} \bar{\sigma}^{\mu}] \\ &= \tfrac{1}{2} Tr[M x_{\nu} \sigma^{\nu} M^{+} \bar{\sigma}^{\mu}] \\ &= \tfrac{1}{2} Tr[M \sigma^{\nu} M^{+} \bar{\sigma}^{\mu}] x_{\nu} \\ &= \tfrac{1}{2} Tr[\bar{\sigma}^{\mu} M \sigma_{\nu} M^{+}] x^{\nu} \end{aligned} \tag{1.102}$$

Comparing this expression with (1.1) above we obtain

$$\Lambda^{\mu}{}_{\nu}(M) = \tfrac{1}{2} Tr[\bar{\sigma}^{\mu} M \sigma_{\nu} M^{+}] \tag{1.103}$$

This result is the explicit form of the group homomorphism

$$SL(2,C) \longrightarrow L^{\uparrow}_{+}$$

<u>Proposition</u>: The following properties hold:

i) Equation (1.103) is a group homomorphism, i.e.

$$\Lambda^{\mu}{}_{\nu}(M_1) \Lambda^{\nu}{}_{\rho}(M_2) = \Lambda^{\mu}{}_{\rho}(M_1 M_2) \qquad \forall M_1, M_2 \in SL(2,C) \tag{1.104}$$

ii) $\Lambda^{\mu}{}_{\nu}(M) \in L^{\uparrow}_{+}$, i.e. we have to verify that

a) $$\det[\Lambda^{\mu}{}_{\nu}(M)] = 1 \tag{1.105}$$

and

b) $$\Lambda^{0}{}_{0}(M) \geq 1 \tag{1.106}$$

<u>Proof</u> : We first prove (1.104). Using (1.103) we have

$$\Lambda^{\mu}{}_{\nu}(M_1) \Lambda^{\nu}{}_{\rho}(M_2) = \tfrac{1}{4} Tr[\bar{\sigma}^{\mu} M_1 \sigma_{\nu} M_1^{+}] \, .$$

$$\cdot \, Tr[\bar{\sigma}^{\nu} M_2 \sigma_{\rho} M_2^{+}]$$

$$= \frac{1}{4} Tr[M_1^{+} \bar{\sigma}^{\mu} M_1 \sigma_{\nu}] Tr[\bar{\sigma}^{\nu} M_2 \sigma_{\rho} M_2^{+}]$$

(since the trace is cyclic)

$$= \frac{1}{4}[M_1^{+} \bar{\sigma}^{\mu} M_1 \sigma_{\nu}]^{\dot{B}}{}_{\dot{B}} [\bar{\sigma}^{\nu} M_2 \sigma_{\rho} M_2^{+}]^{\dot{A}}{}_{\dot{A}}$$

$$= \frac{1}{4}[M_1^{+} \bar{\sigma}^{\mu} M_1]^{\dot{B}B} (\sigma_{\nu})_{B\dot{B}} (\bar{\sigma}^{\nu})^{\dot{A}A} \cdot$$

$$\cdot [M_2 \sigma_{\rho} M_2^{+}]_{A\dot{A}}$$

$$= \frac{1}{2}[M_1^{+} \bar{\sigma}^{\mu} M_1]^{\dot{B}B} \delta_B{}^A \delta_{\dot{B}}{}^{\dot{A}} [M_2 \sigma_{\rho} M_2^{+}]_{A\dot{A}}$$

(using (1.92))

$$= \frac{1}{2}[M_1^{+} \bar{\sigma}^{\mu} M_1]^{\dot{A}A} [M_2 \sigma_{\rho} M_2^{+}]_{A\dot{A}}$$

$$= \frac{1}{2}[M_1^{+} \bar{\sigma}^{\mu} M_1 M_2 \sigma_{\rho} M_2^{+}]^{\dot{A}}{}_{\dot{A}}$$

$$= \frac{1}{2} Tr[M_1^{+} \bar{\sigma}^{\mu} M_1 M_2 \sigma_{\rho} M_2^{+}]$$

$$= \frac{1}{2} Tr[\bar{\sigma}^{\mu} (M_1 M_2) \sigma_{\rho} (M_1 M_2)^{+}]$$

$$= \Lambda^{\mu}{}_{\rho}(M_1 M_2)$$

again using (1.103). A direct consequence of (1.104) is the following result:

$$(\Lambda^{-1})^{\mu}{}_{\nu}(M) = \Lambda^{\mu}{}_{\nu}(M^{-1}) \qquad (1.107)$$

which we can prove as follows.Using (1.104) and (1.91):

$$\Lambda^{\mu}{}_{\nu}(M) \Lambda^{\nu}{}_{\rho}(M^{-1})$$

$$= \Lambda^{\mu}{}_{\rho}(M M^{-1}) \qquad \text{(with (1.104))}$$

$$= \Lambda^{\mu}{}_{\rho}(1_{SL(2,C)})$$

$$= \frac{1}{2} Tr[\bar{\sigma}^{\mu} \sigma_{\rho}]$$

$$= \frac{1}{2} 2 \eta^{\mu}{}_{\rho} \qquad \text{(using (1.91))}$$

$$= \delta^{\mu}{}_{\rho}$$

In order to demonstrate (1.105) it suffices to recall that

$$\Lambda^{\mu}{}_{\nu}(1_{SL(2,C)}) = \delta^{\mu}{}_{\nu}$$

and

$$det(\Lambda^{\mu}{}_{\nu}(1_{SL(2,C)})) = det(\delta^{\mu}{}_{\nu}) = 1$$

Furthermore, det is a continuous function of $\Lambda^{\mu}{}_{\nu}$,so that

$$det(\Lambda^{\mu}{}_{\nu}(M))$$

must be +1 as M runs through SL(2,C). Finally (1.106) may be shown as follows (introducing a unitary matrix U which diagonalizes M).Using (1.103) we have

$$\Lambda^{0}{}_{0}(M) = \frac{1}{2} Tr[\bar{\sigma}^{0} M \sigma_{0} M^{+}]$$

$$= \frac{1}{2} Tr[M M^{+}]$$

since

$$\bar{\sigma}^{0} = \sigma^{0} = \sigma_{0} = 1_{2\times 2}$$

Now let $M \in SL(2,C)$ and take $U \in U(2,C)$ such that

$$U M U^{+} = D$$

where D is diagonal. Then

$$(U M U^{+})^{+} = D^{+}$$

or

$$U M^{+} U^{+} = D^{+}$$

Now

$$1 = det\, M = det(U M U^{+}) = det\, D = a_1 a_2$$

where $a_1, a_2 \in C$ are the eigenvalues of M. Similarly

$$1 = det\, M^{+} = det(U M^{+} U^{+}) = det\, D^{+} = a_1^* a_2^*$$

Using the fact that a trace is invariant under similarity transformations we can write

$$\begin{aligned} Tr\, MM^+ &= Tr(\mathcal{U} MM^+ \mathcal{U}^+) \\ &= Tr(\mathcal{U} M \mathcal{U}^+ \mathcal{U} M^+ \mathcal{U}^+) \\ &= Tr(DD^+) \\ &= |a_1|^2 + |a_2|^2 \end{aligned}$$

Taking into account $a_1 a_2 = 1 = a_1^* a_2^*$ we can write

$$Tr\, MM^+ = |a_1|^2 + \frac{1}{|a_1|^2} = \frac{|a_1|^4 + 1}{|a_1|^2}$$

We now set $c := |a_1|^2 \in R_+$. Then

$$Tr\, MM^+ = \frac{c^2+1}{c}$$

With this we can show that $Tr\, MM^+ \geqslant 2$. Suppose we have the opposite, i.e.

$$Tr\, MM^+ < 2$$

Then

$$\frac{c^2+1}{c} < 2$$

which implies

$$(c-1)^2 < 0$$

But this is impossible for any $c \in R_+$. Hence we conclude

$$Tr\, MM^+ \geqslant 2$$

and hence

$$\Lambda^0{}_0(M) = \frac{1}{2} Tr(MM^+) \geqslant 1$$

Thus $\Lambda^\mu{}_\nu(M)$ as defined by (1.103) is, together with (1.105), a restricted Lorentz transformation.

Our next step is to derive the inverse to (1.103); i.e.

given any Lorentz matrix $\Lambda^{\mu}{}_{\nu}$ the question is: how can we construct the corresponding SL(2,C)-matrix M = M(Λ).
We start by considering the action of the Lorentz group on M_4:

$$x'^{\mu} = \Lambda^{\mu}{}_{\nu} x^{\nu}, \quad x^{\nu} \in M_4, \quad \Lambda \in L_+^{\uparrow} \tag{1.108}$$

However, we know that the action of SL(2,C) on hermitian 2x2-matrices is given by the adjoint representation:

$$X' = M X M^{+}, \quad M \in SL(2,C), \quad X \in H$$

or, using (1.93),

$$x'_{\mu} \sigma^{\mu} = M x_{\nu} \sigma^{\nu} M^{+}, \quad x'^{\mu} \sigma_{\mu} = M x^{\nu} \sigma_{\nu} M^{+}$$

Using (1.108)

$$\Lambda^{\mu}{}_{\nu} x^{\nu} \sigma_{\mu} = M x^{\nu} \sigma_{\nu} M^{+}$$

so that

$$\Lambda^{\mu}{}_{\nu} \sigma_{\mu} = M \sigma_{\nu} M^{+} \tag{1.109}$$

<u>Proposition</u>: For every M $\in$ SL(2,C)

$$\sigma_{\mu} M \bar{\sigma}^{\mu} = 2(Tr\, M).1_{2\times 2} \tag{1.110}$$

This can be shown by explicit verification.
Then multiplying (1.109) by $\bar{\sigma}^{\nu}$ and summing over ν, thereby using (1.110), we obtain for M, $M^{+} \in$ SL(2,C)

$$\begin{aligned} \Lambda^{\mu}{}_{\nu} \sigma_{\mu} \bar{\sigma}^{\nu} &= M \sigma_{\nu} M^{+} \bar{\sigma}^{\nu} \\ &= M [2\, Tr\, M^{+}] 1_{2\times 2} \end{aligned} \tag{1.111}$$

and so

$$M \equiv M(\Lambda) = \frac{1}{2 Tr\, M^{+}} \Lambda^{\mu}{}_{\nu} \sigma_{\mu} \bar{\sigma}^{\nu}$$

Taking the determinant of (1.111) we have

$$\det(\Lambda^{\mu}{}_{\nu} \sigma_{\mu} \bar{\sigma}^{\nu})$$

$$= \det [M(2\,\mathrm{Tr}\,M^+)]$$
$$= \det M . (2\,\mathrm{Tr}\,M^+)^2 . \det 1_{2\times 2}$$

Using det M = 1 since $M \in SL(2,C)$ we can express Tr M^+ as a function of Λ , i.e.

$$2\,\mathrm{Tr}\,M^+ = \pm \left[\det(\Lambda^\mu{}_\nu \sigma_\mu \bar{\sigma}^\nu)\right]^{\frac{1}{2}}$$

The final expression for M is therefore

$$M(\Lambda) = \pm \frac{1}{[\det(\Lambda^\mu{}_\nu \sigma_\mu \bar{\sigma}^\nu)]^{1/2}} \Lambda^\mu{}_\nu \sigma_\mu \bar{\sigma}^\nu \qquad (1.112)$$

In components (1.112) is

$$M_A{}^B(\Lambda) = \pm \frac{1}{[\det(\Lambda^\mu{}_\nu \sigma_\mu \bar{\sigma}^\nu)]^{1/2}} \Lambda^\mu{}_\nu (\sigma_\mu)_{A\dot{B}} (\bar{\sigma}^\nu)^{\dot{B}B}$$

Thus the Lorentz indices μ, ν are effectively exchanged via σ, $\bar{\sigma}$ for spinor indices A,B.

Equations (1.103) and (1.112) show the connection between the two groups SL(2,C) and $L_+^\uparrow$. An important point is that both SL(2,C)-matrices M and -M lead to the same Lorentz transformation as can be seen from formula (1.103). In equation (1.112) this fact corresponds to the $\pm$ signs. Thus the correspondence $\Lambda \longleftrightarrow \pm M$ defines a two-valued representation of the restricted Lorentz group which is called the spinor representation. This leads us to the identification

$$L_+^\uparrow \simeq SL(2,C)\ /\ Z_2 \qquad (1.113)$$

and so SL(2,C) is the universal covering group of $L_+^\uparrow$ which means that SL(2,C) is simply connected[35].

1.3.4 The Fierz-Reordering Formula

The Fierz-reordering formula exchanges the order of anticommuting spinors.

Proposition: The following relation holds

$$(\phi\psi)\bar{\chi}_{\dot{A}} = \frac{1}{2}(\phi\sigma^{\mu}\bar{\chi})(\psi\sigma_{\mu})_{\dot{A}} \tag{1.114}$$

Proof: We consider the left hand side of (1.114) and use (1.77), (1.76 b,d), (1.92) and (1.88) in this order. Then

$$\begin{aligned}
(\phi\psi)\bar{\chi}_{\dot{A}} &= \phi^{B}\psi_{B}\bar{\chi}_{\dot{A}} \\
&= \phi^{B}\varepsilon_{BC}\psi^{C}\varepsilon_{\dot{A}\dot{D}}\bar{\chi}^{\dot{D}} \\
&= -\varepsilon_{CB}\phi^{B}\psi^{C}\varepsilon_{\dot{A}\dot{D}}\bar{\chi}^{\dot{D}} \\
&= \varepsilon_{CB}\psi^{C}\phi^{B}\varepsilon_{\dot{A}\dot{D}}\bar{\chi}^{\dot{D}} \\
&= \delta_{A}{}^{C}\delta_{\dot{A}}{}^{\dot{E}}\varepsilon_{CB}\varepsilon_{\dot{E}\dot{D}}\psi^{A}\phi^{B}\bar{\chi}^{\dot{D}} \\
&= \frac{1}{2}\sigma^{\mu}_{A\dot{A}}\bar{\sigma}_{\mu}^{\dot{E}C}\varepsilon_{CB}\varepsilon_{\dot{E}\dot{D}}\psi^{A}\phi^{B}\bar{\chi}^{\dot{D}} \\
&= \frac{1}{2}\sigma^{\mu}_{A\dot{A}}\varepsilon^{CF}\varepsilon^{\dot{E}\dot{H}}\sigma_{\mu F\dot{H}}\varepsilon_{CB}\cdot \\
&\qquad\cdot\varepsilon_{\dot{E}\dot{D}}\psi^{A}\phi^{B}\bar{\chi}^{\dot{D}} \\
&= \frac{1}{2}\sigma^{\mu}_{A\dot{A}}\sigma_{\mu F\dot{H}}\delta^{F}{}_{B}\delta^{\dot{H}}{}_{\dot{D}}\psi^{A}\phi^{B}\bar{\chi}^{\dot{D}}
\end{aligned}$$

(two Grassmann exchanges)

$$\begin{aligned}
&= \frac{1}{2}\phi^{B}\sigma_{\mu B\dot{D}}\bar{\chi}^{\dot{D}}\psi^{A}\sigma^{\mu}_{A\dot{A}} \\
&= \frac{1}{2}(\phi\sigma_{\mu}\bar{\chi})(\psi\sigma^{\mu})_{\dot{A}}
\end{aligned}$$

For other Fierz-reordering formulae see (1.84), (7.89), (7.90) and (7.91).

1.3.5 Further Calculations with Spinors

We begin by deriving some more formulae which are very useful in calculations.

i) Proposition:

$$(\phi\sigma^\mu\bar{\chi}) = -(\bar{\chi}\bar{\sigma}^\mu\phi) \tag{1.115}$$

Proof: We start with the expression on the left:

$$(\phi\sigma^\mu\bar{\chi})$$

$$= \phi^A(\sigma^\mu)_{A\dot{B}}\bar{\chi}^{\dot{B}} \qquad \text{(using (1.77) and (1.78))}$$

$$= \varepsilon^{AB}\phi_B(\sigma^\mu)_{A\dot{B}}\varepsilon^{\dot{B}\dot{C}}\bar{\chi}_{\dot{C}} \qquad \text{(using (1.76 a,c))}$$

$$= -\phi_B\varepsilon^{BA}(\sigma^\mu)_{A\dot{B}}\varepsilon^{\dot{B}\dot{C}}\bar{\chi}_{\dot{C}}$$

$$= -\phi_B(-\bar{\sigma}^\mu)^{\dot{C}B}\bar{\chi}_{\dot{C}} \qquad \text{(using (1.88))}$$

$$= -\bar{\chi}_{\dot{C}}(\bar{\sigma}^\mu)^{\dot{C}B}\phi_B \qquad \text{(using (1.79))}$$

$$= -(\bar{\chi}\bar{\sigma}^\mu\phi)$$

ii) Proposition:

$$(\phi\sigma^\mu\bar{\chi})^+ = (\chi\sigma^\mu\bar{\phi}) \tag{1.116a}$$

and similarly

$$(\theta\phi)^+ = \bar{\theta}\bar{\phi} \tag{1.116b}$$

Proof: The first expression is SL(2,C) covariant. Therefore with the definition of the operation "+"

$$(\phi\sigma^\mu\bar{\chi})^+ = (\bar{\chi}^+\sigma^{\mu+}\phi^+)$$

$$= \bar{\chi}^{+A}(\sigma^\mu)^+_{A\dot{B}}\,\phi^{+\dot{B}}$$

$$= \chi^{A} (\sigma^{\mu})_{A\dot{B}} \bar{\phi}^{\dot{B}}$$
$$= (\chi \sigma^{\mu} \bar{\phi})$$

with

$$\bar{\chi}^{+A} := \chi^{A} \qquad \text{and} \qquad \phi^{+\dot{B}} := \bar{\phi}^{\dot{B}}$$

It may be observed that the Grassmann property (1.79) has not been used because $(\phi\bar{\chi})^{+}$ is defined as $\bar{\chi}^{+}\phi^{+}$. Similarly we have

$$(\theta\phi)^{+} = (\phi\theta)^{+} \qquad \text{(using (1.80))}$$
$$= \theta^{+}_{\dot{B}} \phi^{+\dot{B}}$$
$$= (\bar{\theta}\bar{\phi})$$

From (1.116) we see that $\psi\sigma^{\mu}\bar{\psi}$ is hermitian.

iii) <u>Proposition</u>: As can be expected

$$\phi\sigma^{\mu}\bar{\chi}$$

transforms as a four-vector under the Lorentz group, i.e.

$$(\phi'\sigma^{\mu}\bar{\chi}') = \Lambda^{\mu}{}_{\nu}(M)\,(\phi\sigma^{\nu}\bar{\chi}) \qquad (1.117)$$

<u>Proof</u>: We have

$$(\phi'\sigma^{\mu}\bar{\chi}') = -(\bar{\chi}'\bar{\sigma}^{\mu}\phi') \qquad \text{(with (1.115))}$$
$$= -\bar{\chi}'_{\dot{A}}(\bar{\sigma}^{\mu})^{\dot{A}B}\phi'_{B} \qquad \text{(with (1.77), (1.78))}$$
$$= -M^{*}{}_{\dot{A}}{}^{\dot{B}}\bar{\chi}_{\dot{B}}(\bar{\sigma}^{\mu})^{\dot{A}B}M_{B}{}^{C}\phi_{C} \qquad \text{(with (1.75 a,c))}$$
$$= -\bar{\chi}_{\dot{B}}(M^{+})^{\dot{B}}{}_{\dot{A}}(\bar{\sigma}^{\mu})^{\dot{A}B}M_{B}{}^{C}\phi_{C}$$
$$= -\bar{\chi}_{\dot{B}}\delta^{\dot{B}}{}_{\dot{D}}(M^{+})^{\dot{D}}{}_{\dot{A}}(\bar{\sigma}^{\mu})^{\dot{A}B}M_{B}{}^{C}\delta_{C}{}^{D}\phi_{D}$$
$$= -\frac{1}{2}\bar{\chi}_{\dot{B}}(\bar{\sigma}^{\nu})^{\dot{B}D}(\sigma_{\nu})_{C\dot{D}}(M^{+})^{\dot{D}}{}_{A}\,.$$

$$\cdot(\bar{\sigma}^\mu)^{\dot{A}B} M_B{}^C \phi_D \quad \text{(using (1.92))}$$

$$= -\frac{1}{2}\left[(M^+)^{\dot{D}}{}_{\dot{A}} (\bar{\sigma}^\mu)^{\dot{A}B} M_B{}^C (\sigma_\nu)_{C\dot{D}}\right].$$

$$\cdot\left(\bar{\chi}_{\dot{B}} (\bar{\sigma}^\nu)^{\dot{B}D} \phi_D\right)$$

$$= -\frac{1}{2} Tr\left[M^+ \bar{\sigma}^\mu M \sigma_\nu\right](\bar{\chi}\bar{\sigma}^\nu\phi)$$

$$= \Lambda^\mu{}_\nu(M)(\phi\sigma^\nu\bar{\chi})$$

using (1.103) and (1.115).

iv) <u>Proposition</u>: The following relations can be shown to hold:

$$\sigma^\mu_{A\dot{A}} \bar{\theta}^{\dot{A}} (\theta\sigma^\nu\bar{\theta}) = \frac{1}{2}\eta^{\mu\nu}\theta_A(\bar{\theta}\bar{\theta})$$

$$(\theta\sigma^\mu\bar{\theta})(\theta\sigma^\nu\bar{\theta}) = \frac{1}{2}\eta^{\mu\nu}(\theta\theta)(\bar{\theta}\bar{\theta}) \qquad (1.118)$$

$$\sigma^\mu_{A\dot{A}} \bar{\theta}^{\dot{A}} (\theta\sigma^\nu\bar{\theta}) = \frac{1}{2}\eta^{\mu\nu}\theta_A(\bar{\theta}\bar{\theta})$$

<u>Proof</u>: Using (1.77) and (1.78) we have

$$(\theta\sigma^\mu\bar{\theta})(\theta\sigma^\nu\bar{\theta})$$

$$= \theta^A(\sigma^\mu)_{A\dot{B}}\bar{\theta}^{\dot{B}}\theta^C(\sigma^\nu)_{C\dot{D}}\bar{\theta}^{\dot{D}}$$

$$= -\theta^A(\sigma^\mu)_{A\dot{B}}\theta^C\bar{\theta}^{\dot{B}}(\sigma^\nu)_{C\dot{D}}\bar{\theta}^{\dot{D}}$$

$$= -(\sigma^\mu)_{A\dot{B}}(\sigma^\nu)_{C\dot{D}}\theta^A\theta^C\bar{\theta}^{\dot{B}}\bar{\theta}^{\dot{D}}$$

$$= \frac{1}{4}(\sigma^\mu)_{A\dot{B}}(\sigma^\nu)_{C\dot{D}}\varepsilon^{AC}\varepsilon^{\dot{B}\dot{D}}(\theta\theta)(\bar{\theta}\bar{\theta})$$

(with (1.83 a,c))

$$= \frac{1}{4}(\sigma^\mu)_{A\dot{B}}(\bar{\sigma}^\nu)^{\dot{B}A}(\theta\theta)(\bar{\theta}\bar{\theta})$$

(with (1.88))

$$= \frac{1}{2}\eta^{\mu\nu}(\theta\theta)(\bar{\theta}\bar{\theta}) \quad \text{(with (1.91))}$$

v) <u>Proposition</u>: The generators of SL(2,C) in the spinor representations (1/2, 0) and (0, 1/2) are given by (see (1.112))

$$\sigma^{\mu\nu} := \frac{i}{4}(\sigma^{\mu}\bar{\sigma}^{\nu} - \sigma^{\nu}\bar{\sigma}^{\mu}) \tag{1.119a}$$

$$\bar{\sigma}^{\mu\nu} := \frac{i}{4}(\bar{\sigma}^{\mu}\sigma^{\nu} - \bar{\sigma}^{\nu}\sigma^{\mu}) \tag{1.119b}$$

$$\sigma^{\mu\nu +} = \bar{\sigma}^{\mu\nu} \tag{1.119c}$$

We will need these two-dimensional spinor representations of the Lorentz generators later for the construction of the complex two-dimensional grading of the Poincaré algebra. In components the equations (1.119) read

$$(\sigma^{\mu\nu})_A{}^B = \frac{i}{4}(\sigma^{\mu}_{A\dot{A}}\bar{\sigma}^{\nu\dot{A}B} - \sigma^{\nu}_{A\dot{A}}\bar{\sigma}^{\mu\dot{A}B})$$

and

$$(\bar{\sigma}^{\mu\nu})^{\dot{A}}{}_{\dot{B}} = \frac{i}{4}(\bar{\sigma}^{\mu\dot{A}A}\sigma^{\nu}_{A\dot{B}} - \bar{\sigma}^{\nu\dot{A}A}\sigma^{\mu}_{A\dot{B}})$$

<u>Proof of (1.119)</u>: For an infinitesimal Lorentz transformation (ω_{ij} infinitesimal)

$$\Lambda^{\mu}{}_{\nu}\sigma_{\mu}\bar{\sigma}^{\nu} = \Lambda_{\mu\nu}\sigma^{\mu}\bar{\sigma}^{\nu}$$

$$= \begin{pmatrix} 1 & \omega_{01} & \omega_{02} & \omega_{03} \\ -\omega_{01} & -1 & \omega_{12} & \omega_{13} \\ -\omega_{02} & -\omega_{12} & -1 & \omega_{23} \\ -\omega_{03} & -\omega_{13} & -\omega_{23} & -1 \end{pmatrix}_{\mu\nu} \sigma^{\mu}\bar{\sigma}^{\nu}$$

$$= \begin{pmatrix} 4-2(\omega_{03}+i\omega_{12}) & -2(\omega_{01}+i\omega_{23}) \\ & +2i(\omega_{02}+i\omega_{31}) \\ -2(\omega_{01}+i\omega_{23}) & 4+2(\omega_{03}+i\omega_{12}) \\ -2i(\omega_{02}+i\omega_{31}) & \end{pmatrix}$$

Hence

$$\det(\Lambda^\mu{}_\nu \sigma_\mu \bar{\sigma}^\nu) = 16 + O(\omega^2)$$

and (omitting terms of order ω^2)

$$N = [\det(\Lambda^\mu{}_\nu \sigma_\mu \bar{\sigma}^\nu)]^{\frac{1}{2}} = 4$$

The relationship between element $\Lambda^\rho{}_\tau$ of the Lie group SO(1,3) and element $M^{\mu\nu}$ of the algebra so(1,3) is

$$\Lambda^\rho{}_\tau(\omega) = [\exp(-\tfrac{i}{2}\omega_{\mu\nu} M^{\mu\nu})]^\rho{}_\tau$$

For infinitesimal $\omega_{\mu\nu}$, element M(Λ) $\in$ SL(2,C) is

$$\begin{aligned} M(\Lambda) &= \frac{1}{N} \Lambda^\rho{}_\tau(\omega) \sigma_\rho \bar{\sigma}^\tau \\ &= \frac{1}{N} (\eta_{\rho\tau} - \tfrac{i}{2}\omega_{\mu\nu} (M^{\mu\nu})_{\rho\tau}) \sigma^\rho \bar{\sigma}^\tau \\ &= \frac{1}{N} \sigma^\mu \bar{\sigma}_\mu - \frac{i}{2} \omega_{\mu\nu} \sigma^{\mu\nu} \end{aligned}$$

where

$$\sigma^{\mu\nu} := \frac{1}{N} (M^{\mu\nu})_{\rho\tau} \sigma^\rho \bar{\sigma}^\tau$$

Now

$$x'_\rho = \Lambda_{\rho\tau} x^\tau = (\eta_{\rho\tau} + \omega_{\rho\tau}) x^\tau$$

and since every homogeneous linear transformation group possesses the trivial self-representation in the four-space spanned by the vector x_μ , we have

$$-\frac{i}{2}\omega_{\mu\nu}(M^{\mu\nu})_{\rho\tau} = \omega_{\rho\tau}$$

This equation is satisfied by

$$(M^{\mu\nu})^{\rho}{}_{\tau} = i(\eta^{\mu\rho}\delta^{\nu}{}_{\tau} - \eta^{\nu\rho}\delta^{\mu}{}_{\tau})$$

Of course, from (1.11) we know that the generators of Lorentz transformations have this form. Hence

$$\sigma^{\mu\nu} = \frac{i}{N}(\eta^{\mu\rho}\delta^{\nu}{}_{\sigma} - \eta^{\nu\rho}\delta^{\mu}{}_{\sigma})\sigma_{\rho}\bar{\sigma}^{\sigma}$$

$$= \frac{i}{4}(\sigma^{\mu}\bar{\sigma}^{\nu} - \sigma^{\nu}\bar{\sigma}^{\mu})$$

In a similar way $\bar{\sigma}^{\mu\nu}$ is obtained by considering M^* as a function of Λ. The matrices $\sigma^{\mu\nu}$, $\bar{\sigma}^{\mu\nu}$ obey the commutation relation of the Lorentz algebra, i.e. (1.13).

Proposition: The generators of rotations $\tilde{J}_i$ and Lorentz boosts $\tilde{K}_i$, defined as in (1.17) and (1.18), i.e.

$$J_i = \frac{1}{2}\varepsilon_{ijk}M^{jk}, \qquad K_i = M_{0i}$$

can be written

$$\tilde{J}^i := \frac{1}{2}\varepsilon^i{}_{jk}\sigma^{jk} = \frac{1}{2}\sigma^i\bar{\sigma}^0 \tag{1.120a}$$

$$\tilde{K}^i := \sigma^{0i} = -\frac{i}{2}\sigma^i\bar{\sigma}^0 \tag{1.120b}$$

Proof:

i) $$\tilde{J}_i := \frac{1}{2}\varepsilon_{ijk}\sigma^{jk}$$

$$= \frac{1}{2}\varepsilon_{ijk}\frac{i}{4}(\sigma^j\bar{\sigma}^k - \sigma^k\bar{\sigma}^j)$$

$$= \frac{i}{8}\varepsilon_{ijk}(-\sigma^j\sigma^k + \sigma^k\sigma^j) \quad \text{(with (1.90))}$$

$$= -\frac{i}{8}\varepsilon_{ijk}[\sigma^j, \sigma^k]$$

$$= -\frac{i}{8}\varepsilon_{ijk} \cdot 2i\,\varepsilon^{jkl}\sigma_l$$

$$= \frac{1}{4}\varepsilon_{jki}\,\varepsilon^{jkl}\sigma_l$$

$$= \frac{1}{2}\delta_i{}^l\sigma_l$$

$$= \frac{1}{2}\sigma_i$$

where we used

$$\sum_{j,k=1}^{3}\varepsilon_{jki}\varepsilon_{jkl} = 2\cdot\delta_{il}$$

ii)

$$\tilde{K}_i := \sigma^{0i} = \frac{i}{4}(\sigma^0\bar{\sigma}^i - \sigma^i\bar{\sigma}^0)$$

$$= \frac{i}{4}(-\sigma^i - \sigma^i)$$

$$= -\frac{i}{2}\sigma^i$$

where we used again (1.90).

<u>Proposition</u>: The $\sigma^{\mu\nu}$ are selfdual, i.e.

$$\sigma^{\mu\nu} = \frac{1}{2i}\varepsilon^{\mu\nu\rho\sigma}\sigma_{\rho\sigma} \tag{1.121a}$$

whereas the $\bar{\sigma}^{\mu\nu}$ are antiselfdual, i.e.

$$\bar{\sigma}^{\mu\nu} = -\frac{1}{2i}\varepsilon^{\mu\nu\rho\sigma}\bar{\sigma}_{\rho\sigma} \tag{1.121b}$$

<u>Proof</u>: In order to prove the first of these relations consider

i) $\mu = 0$, $\nu = i$, $i = 1,2,3$:

With (1.119a)

$$\sigma^{0i} = \frac{i}{4}(\sigma^0\bar{\sigma}^i - \sigma^i\bar{\sigma}^0)$$

$$= \frac{i}{4}(-\sigma^0\sigma^i - \sigma^i\sigma^0) \quad \text{(with (1.90))}$$

$$= -\frac{i}{2}\sigma^i$$

$$= \frac{1}{2i}\sigma^i$$

On the other hand (with i,j,k = 1,2,3)

$$\frac{1}{2i}\varepsilon^{0i\rho\sigma}\sigma_{\rho\sigma} = \frac{1}{2i}\varepsilon^{0ijk}\sigma_{jk}$$

$$= \frac{1}{2i}\varepsilon^{0ijk}\frac{i}{4}(\sigma_j\bar{\sigma}_k - \sigma_k\bar{\sigma}_j)$$

(using (1.119 a))

$$= \frac{1}{8}\varepsilon^{0ijk}(-\sigma_j\sigma_k + \sigma_k\sigma_j) \quad \text{(with (1.90))}$$

$$= -\frac{1}{8}\varepsilon^{0ijk}[\sigma_j, \sigma_k]$$

$$= -\frac{1}{8}\varepsilon^{0ijk}.2i\varepsilon_{jkl}\sigma^l$$

$$= \frac{1}{4i}\varepsilon^{0ijk}\varepsilon_{jkl}\sigma^l$$

$$= \frac{1}{4i}2\delta^i{}_l\sigma^l$$

$$= \frac{1}{2i}\sigma^i$$

$$= \sigma^{0i}$$

as shown above. Now consider

μ = i, ν = j, i,j = 1,2,3 :

With (1.119 a)

$$\sigma^{ij} = \frac{i}{4}(\sigma^i\bar{\sigma}^j - \sigma^j\bar{\sigma}^i)$$

$$= -\frac{i}{4}(\sigma^i \sigma^j - \sigma^j \sigma^i) \quad \text{(with (1.90))}$$
$$= -\frac{i}{4}[\sigma^i, \sigma^j]$$
$$= -\frac{i}{4} 2i \varepsilon^{ijk} \sigma_k$$
$$= \frac{1}{2} \varepsilon^{ijk} \sigma_k$$

$$\frac{1}{2i} \varepsilon^{ij\rho\sigma} \sigma_{\rho\sigma} = \frac{1}{2i}(\varepsilon^{ijko} \sigma_{ko} + \varepsilon^{ijok} \sigma_{ok})$$
$$= \frac{1}{2i}(\varepsilon^{ijko} \sigma_{ko} + \varepsilon^{ijko} \sigma_{ko})$$
$$= \frac{1}{i} \varepsilon^{ijko} \sigma_{ko}$$
$$= \frac{1}{i} \varepsilon^{ijko} \frac{i}{4}(\sigma_k \bar{\sigma}_o - \sigma_o \bar{\sigma}_k)$$
$$= \frac{1}{2} \varepsilon^{ijko} \sigma_k$$
$$= \sigma^{ij}$$

ii) $\mu = 0$, $\nu = i = 1,2,3$:
With (1.119 b)

$$\bar{\sigma}^{oi} = \frac{i}{4}(\bar{\sigma}^o \sigma^i - \bar{\sigma}^i \sigma^o)$$
$$= \frac{i}{4}(2\sigma^i) \quad \text{(with (1.90))}$$
$$= -\frac{1}{2i} \sigma^i$$

On the other hand

$$-\frac{1}{2i} \varepsilon^{oi\rho\sigma} \bar{\sigma}_{\rho\sigma} = -\frac{1}{2i} \varepsilon^{oijk} \bar{\sigma}_{jk}$$
$$= -\frac{1}{2i} \varepsilon^{oijk} \frac{i}{4}(\bar{\sigma}_j \sigma_k - \bar{\sigma}_k \sigma_j)$$
(using (1.119 b)

$$= \frac{1}{8}\varepsilon^{0ijk}(\sigma_j\sigma_k - \sigma_k\sigma_j)$$
$$= \frac{1}{8}\varepsilon^{0ijk}[\sigma_j, \sigma_k]$$
$$= \frac{1}{8}\varepsilon^{0ijk}\cdot 2i\,\varepsilon_{jkl}\,\sigma^l$$
$$= -\frac{1}{4i}\varepsilon^{0ijk}\,\varepsilon_{jkl}\,\sigma^l$$
$$= -\frac{1}{4i}2\,\delta^i{}_l\,\sigma^l$$
$$= -\frac{1}{2i}\sigma^i$$
$$= \bar{\sigma}^{0i}$$

$\mu = i$, $\nu = j$, $i,j = 1,2,3$:

Using (1.119b)

$$\bar{\sigma}^{ij} = \frac{i}{4}(\bar{\sigma}^i\sigma^j - \bar{\sigma}^j\sigma^i)$$
$$= -\frac{i}{4}[\sigma^i, \sigma^j]$$
$$= -\frac{i}{4}\cdot 2i\,\varepsilon^{ijk}\sigma_k$$
$$= \frac{1}{2}\varepsilon^{ijk}\sigma_k$$

$$-\frac{1}{2i}\varepsilon^{ij\rho\sigma}\bar{\bar{\sigma}}_{\rho\sigma} = -\frac{1}{2i}(\varepsilon^{ijko}\bar{\bar{\sigma}}_{ko} + \varepsilon^{ijok}\bar{\bar{\sigma}}_{ok})$$
$$= -\frac{1}{2i}2\,\varepsilon^{ijko}\bar{\bar{\sigma}}_{ko}$$
$$= i\,\varepsilon^{ijko}\,\frac{i}{4}(\bar{\bar{\sigma}}_k\sigma_o - \bar{\bar{\sigma}}_o\sigma_k)$$
$$= -\frac{1}{4}\varepsilon^{ijko}(-2\sigma_k)$$
$$= \frac{1}{2}\varepsilon^{ijko}\sigma_k$$
$$= \bar{\sigma}^{ij}$$

Proposition: The matrices σ^μ and $\bar{\sigma}^\mu$ defined by (1.87) and (1.88) obey the following relations

$$\sigma^\mu \bar{\sigma}^\nu + \sigma^\nu \bar{\sigma}^\mu = 2\eta^{\mu\nu} 1_{2\times 2} \tag{1.122a}$$

$$\bar{\sigma}^\mu \sigma^\nu + \bar{\sigma}^\nu \sigma^\mu = 2\eta^{\mu\nu} 1_{2\times 2} \tag{1.122b}$$

In components these equations read

$$(\sigma^\mu)_{A\dot{B}} (\bar{\sigma}^\nu)^{\dot{B}B} + (\sigma^\nu)_{A\dot{B}} (\bar{\sigma}^\mu)^{\dot{B}B} = 2\eta^{\mu\nu} \delta_A{}^B$$

$$(\bar{\sigma}^\mu)^{\dot{A}B} (\sigma^\nu)_{B\dot{B}} + (\bar{\sigma}^\nu)^{\dot{A}B} (\sigma^\mu)_{B\dot{B}} = 2\eta^{\mu\nu} \delta^{\dot{A}}{}_{\dot{B}}$$

Proof: From the definitions (1.86a) and (1.90) we have

$$(\sigma^\mu) = (1, \vec{\sigma}) \quad \text{and} \quad (\bar{\sigma}^\mu) = (1, -\vec{\sigma})$$

We verify (1.122) by considering the following cases:

i) $\mu = \nu = 0$:

$$(\sigma^0 \bar{\sigma}^0 + \sigma^0 \bar{\sigma}^0)_A{}^B = 2(1)_A{}^B = 2\eta^{00} \delta_A{}^B$$

ii) $\mu = 0$, $\nu = k = 1,2,3$:

$$(\sigma^0 \bar{\sigma}^k + \sigma^k \bar{\sigma}^0)_A{}^B = (1(-\sigma^k) + \sigma^k(1))_A{}^B$$
$$= (-\sigma^k + \sigma^k)_A{}^B = 0$$

iii) $\mu = i$, $\nu = j$, $i,j = 1,2,3$:

$$(\sigma^i \bar{\sigma}^j + \sigma^j \bar{\sigma}^i)_A{}^B = -(\sigma^i \sigma^j + \sigma^j \sigma^i)_A{}^B$$
$$= -(\{\sigma^i, \sigma^j\})_A{}^B$$
$$= -2\delta^{ij} (1)_A{}^B$$
$$= -2\eta^{ij} \delta_A{}^B$$

In later calculations we will frequently use (1.122a,b) in the form

$$\sigma^\mu \bar{\sigma}^\nu = 2\eta^{\mu\nu} 1_{2\times 2} - \sigma^\nu \bar{\sigma}^\mu \tag{1.123}$$

or

$$\bar{\sigma}^\mu \sigma^\nu = 2\eta^{\mu\nu} 1_{2\times 2} - \bar{\sigma}^\nu \sigma^\mu \tag{1.124}$$

Proposition: The generators $\sigma^{\mu\nu}$ of SL(2,C) in the spinor representation obey the following trace relation

$$\mathrm{Tr}[\sigma^{\mu\nu}\sigma^{\rho\sigma}] = \frac{1}{2}(\eta^{\mu\rho}\eta^{\nu\sigma} - \eta^{\mu\sigma}\eta^{\nu\rho}) + \frac{i}{2}\varepsilon^{\mu\nu\rho\sigma} \tag{1.125}$$

Similarly (see (7.190) later)

$$\mathrm{Tr}[\bar{\sigma}^{\mu\nu}\bar{\sigma}^{\rho\sigma}] = \frac{1}{2}(\eta^{\mu\rho}\eta^{\nu\sigma} - \eta^{\mu\sigma}\eta^{\nu\rho}) - \frac{i}{2}\varepsilon^{\mu\nu\rho\sigma} \tag{1.125'}$$

Proof: Using (1.119a) we have

$$\begin{aligned}
&\mathrm{Tr}[\sigma^{\mu\nu}\sigma^{\rho\sigma}] \\
&= -\frac{1}{16}\mathrm{Tr}\{(\sigma^\mu\bar{\sigma}^\nu - \sigma^\nu\bar{\sigma}^\mu)(\sigma^\rho\bar{\sigma}^\sigma - \sigma^\sigma\bar{\sigma}^\rho)\} \\
&= -\frac{1}{16}\mathrm{Tr}\{\sigma^\mu\bar{\sigma}^\nu\sigma^\rho\bar{\sigma}^\sigma - \sigma^\mu\bar{\sigma}^\nu\sigma^\sigma\bar{\sigma}^\rho \\
&\qquad - \sigma^\nu\bar{\sigma}^\mu\sigma^\rho\bar{\sigma}^\sigma + \sigma^\nu\bar{\sigma}^\mu\sigma^\sigma\bar{\sigma}^\rho\} \\
&= -\frac{1}{16}\mathrm{Tr}\{\sigma^\mu\bar{\sigma}^\nu\sigma^\rho\bar{\sigma}^\sigma\} + \frac{1}{16}\mathrm{Tr}\{\sigma^\mu\bar{\sigma}^\nu\sigma^\sigma\bar{\sigma}^\rho\} \\
&\quad + \frac{1}{16}\mathrm{Tr}\{\sigma^\nu\bar{\sigma}^\mu\sigma^\rho\bar{\sigma}^\sigma\} - \frac{1}{16}\mathrm{Tr}\{\sigma^\nu\bar{\sigma}^\mu\sigma^\sigma\bar{\sigma}^\rho\}
\end{aligned}$$

Using now (to be shown below)

$$\begin{aligned}
&\frac{1}{2}\mathrm{Tr}\{\sigma^\mu\bar{\sigma}^\nu\sigma^\rho\bar{\sigma}^\sigma\} \\
&\quad = \eta^{\mu\nu}\eta^{\rho\sigma} + \eta^{\nu\rho}\eta^{\mu\sigma} - \eta^{\mu\rho}\eta^{\nu\sigma} - i\varepsilon^{\mu\nu\rho\sigma}
\end{aligned} \tag{1.126}$$

we obtain

$$\mathrm{Tr}[\sigma^{\mu\nu}\sigma^{\rho\sigma}]$$

$$= -\frac{1}{8}\{\eta^{\mu\nu}\eta^{\rho\sigma}+\eta^{\nu\rho}\eta^{\mu\sigma}-\eta^{\mu\rho}\eta^{\nu\sigma}-i\varepsilon^{\mu\nu\rho\sigma}\}$$
$$+\frac{1}{8}\{\eta^{\mu\nu}\eta^{\sigma\rho}+\eta^{\mu\rho}\eta^{\nu\sigma}-\eta^{\mu\sigma}\eta^{\nu\rho}-i\varepsilon^{\mu\nu\sigma\rho}\}$$
$$+\frac{1}{8}\{\eta^{\nu\mu}\eta^{\rho\sigma}+\eta^{\mu\rho}\eta^{\nu\sigma}-\eta^{\nu\rho}\eta^{\mu\sigma}-i\varepsilon^{\nu\mu\rho\sigma}\}$$
$$-\frac{1}{8}\{\eta^{\nu\mu}\eta^{\sigma\rho}+\eta^{\mu\sigma}\eta^{\nu\rho}-\eta^{\nu\sigma}\eta^{\mu\rho}-i\varepsilon^{\nu\mu\sigma\rho}\}$$
$$= \frac{1}{2}\eta^{\mu\rho}\eta^{\nu\sigma}-\frac{1}{2}\eta^{\mu\sigma}\eta^{\nu\rho}+\frac{i}{2}\varepsilon^{\mu\nu\rho\sigma}$$

taking into account the total antisymmetry of the tensor $\varepsilon^{\mu\nu\rho\sigma}$. Hence

$$Tr[\sigma^{\mu\nu}\sigma^{\rho\sigma}]=\frac{1}{2}(\eta^{\mu\rho}\eta^{\nu\sigma}-\eta^{\mu\sigma}\eta^{\nu\rho})+\frac{i}{2}\varepsilon^{\mu\nu\rho\sigma}$$

as had to be shown. To complete the proof we have to demonstrate the validity of (1.126).

<u>Proposition</u>: The following relation holds

$$\frac{1}{2}Tr[\sigma^\mu\bar\sigma^\nu\sigma^\rho\bar\sigma^\sigma]=\eta^{\mu\nu}\eta^{\rho\sigma}+\eta^{\nu\rho}\eta^{\mu\sigma}-\eta^{\mu\rho}\eta^{\nu\sigma}-i\varepsilon^{\mu\nu\rho\sigma}$$

<u>Proof</u>: We decompose the left hand side of (1.126) in the following way:

$$\frac{1}{2}Tr(\sigma^\mu\bar\sigma^\nu\sigma^\rho\bar\sigma^\sigma)$$
$$=\frac{1}{2}Tr\{\frac{1}{2}(\sigma^\mu\bar\sigma^\nu\sigma^\rho\bar\sigma^\sigma+\sigma^\rho\bar\sigma^\nu\sigma^\mu\bar\sigma^\sigma)$$
$$+\frac{1}{2}(\sigma^\mu\bar\sigma^\nu\sigma^\rho\bar\sigma^\sigma-\sigma^\rho\bar\sigma^\nu\sigma^\mu\bar\sigma^\sigma)\}$$

The first part of the trace is symmetric in μ and ρ, whereas the second part is antisymmetric in μ and ρ. Consider first the symmetric part

$$\frac{1}{4}Tr\{\sigma^\mu\bar\sigma^\nu\sigma^\rho\bar\sigma^\sigma+\sigma^\rho\bar\sigma^\nu\sigma^\mu\bar\sigma^\sigma\}$$
$$=\frac{1}{4}Tr[(2\eta^{\mu\nu}-\sigma^\nu\bar\sigma^\mu)\sigma^\rho\bar\sigma^\sigma+\sigma^\rho\bar\sigma^\nu\sigma^\mu\bar\sigma^\sigma]$$

(using (1.123))

$$= \frac{1}{4}\mathrm{Tr}\left[2\eta^{\mu\nu}\sigma^\rho\bar\sigma^\sigma - \sigma^\nu\bar\sigma^\mu\sigma^\rho\bar\sigma^\sigma + \sigma^\rho\bar\sigma^\nu\sigma^\mu\bar\sigma^\sigma\right]$$

$$= \frac{1}{2}\eta^{\mu\nu}\mathrm{Tr}(\sigma^\rho\bar\sigma^\sigma) + \frac{1}{4}\mathrm{Tr}\left[-\sigma^\nu(2\eta^{\mu\rho} - \bar\sigma^\rho\sigma^\mu)\bar\sigma^\sigma + \sigma^\rho\bar\sigma^\nu\sigma^\mu\bar\sigma^\sigma\right]$$

(with (1.124))

$$= \eta^{\mu\nu}\eta^{\rho\sigma} - \frac{1}{2}\eta^{\mu\rho}\mathrm{Tr}(\sigma^\nu\bar\sigma^\sigma) + \frac{1}{4}\mathrm{Tr}\left[\sigma^\nu\bar\sigma^\rho\sigma^\mu\bar\sigma^\sigma + \sigma^\rho\bar\sigma^\nu\sigma^\mu\bar\sigma^\sigma\right]$$

(with (1.91))

$$= \eta^{\mu\nu}\eta^{\rho\sigma} - \eta^{\mu\rho}\eta^{\nu\sigma} + \frac{1}{4}\mathrm{Tr}\left[(2\eta^{\nu\rho} - \sigma^\rho\bar\sigma^\nu)\sigma^\mu\bar\sigma^\sigma + \sigma^\rho\bar\sigma^\nu\sigma^\mu\bar\sigma^\sigma\right]$$

(using (1.91) and (1.123))

$$= \eta^{\mu\nu}\eta^{\rho\sigma} - \eta^{\mu\rho}\eta^{\nu\sigma} + \eta^{\nu\rho}\eta^{\mu\sigma} + \frac{1}{4}\mathrm{Tr}\left[-\sigma^\rho\bar\sigma^\nu\sigma^\mu\bar\sigma^\sigma + \sigma^\rho\bar\sigma^\nu\sigma^\mu\bar\sigma^\sigma\right]$$

$$= \eta^{\mu\nu}\eta^{\rho\sigma} + \eta^{\mu\sigma}\eta^{\nu\rho} - \eta^{\mu\rho}\eta^{\nu\sigma}$$

Hence

$$\frac{1}{2}\mathrm{Tr}\left[\sigma^\mu\bar\sigma^\nu\sigma^\rho\bar\sigma^\sigma\right] = \eta^{\mu\nu}\eta^{\rho\sigma} + \eta^{\mu\sigma}\eta^{\nu\rho} - \eta^{\mu\rho}\eta^{\nu\sigma} + \frac{1}{4}\mathrm{Tr}\left[\sigma^\mu\bar\sigma^\nu\sigma^\rho\bar\sigma^\sigma - \sigma^\rho\bar\sigma^\nu\sigma^\mu\bar\sigma^\sigma\right]$$

The following result will be needed later.

<u>Proposition</u>: The tensor

$$A^{\mu\nu\rho\sigma} := \frac{1}{4}\mathrm{Tr}\left[\sigma^\mu\bar\sigma^\nu\sigma^\rho\bar\sigma^\sigma - \sigma^\rho\bar\sigma^\nu\sigma^\mu\bar\sigma^\sigma\right]$$

is totally antisymmetric in the indices μ, ν, ρ, σ and therefore proportional to the totally antisymmetric tensor in four-space $\varepsilon^{\mu\nu\rho\sigma}$, i.e.

$$A^{\mu\nu\rho\sigma} = c\,\varepsilon^{\mu\nu\rho\sigma}, \quad c = -i$$

Proof: We demonstrate the validity of the six possibilities separately.

i) $A^{\mu\nu\rho\sigma} = -A^{\nu\mu\rho\sigma}$

Consider

$$A^{\nu\mu\rho\sigma} = \frac{1}{4}[\sigma^\nu\bar{\sigma}^\mu\sigma^\rho\bar{\sigma}^\sigma - \sigma^\rho\bar{\sigma}^\mu\sigma^\nu\bar{\sigma}^\sigma]$$

$$= \frac{1}{4}\mathrm{Tr}[(2\eta^{\mu\nu} - \sigma^\mu\bar{\sigma}^\nu)\sigma^\rho\bar{\sigma}^\sigma - \sigma^\rho(2\eta^{\mu\nu} - \bar{\sigma}^\nu\sigma^\mu)\bar{\sigma}^\sigma]$$

(using (1.123) and (1.124))

$$= \frac{1}{2}\eta^{\mu\nu}\mathrm{Tr}(\sigma^\rho\bar{\sigma}^\sigma) - \frac{1}{2}\eta^{\mu\nu}\mathrm{Tr}(\sigma^\rho\bar{\sigma}^\sigma) - \frac{1}{4}[\sigma^\mu\bar{\sigma}^\nu\sigma^\rho\bar{\sigma}^\sigma - \sigma^\rho\bar{\sigma}^\nu\sigma^\mu\bar{\sigma}^\sigma]$$

$$= -\frac{1}{4}\mathrm{Tr}[\sigma^\mu\bar{\sigma}^\nu\sigma^\rho\bar{\sigma}^\sigma - \sigma^\rho\bar{\sigma}^\nu\sigma^\mu\bar{\sigma}^\sigma]$$

$$= -A^{\mu\nu\rho\sigma}$$

ii) $A^{\mu\nu\rho\sigma} = -A^{\mu\nu\sigma\rho}$

$$A^{\mu\nu\sigma\rho} = \frac{1}{4}\mathrm{Tr}[\sigma^\mu\bar{\sigma}^\nu\sigma^\sigma\bar{\sigma}^\rho - \sigma^\sigma\bar{\sigma}^\nu\sigma^\mu\bar{\sigma}^\rho]$$

$$= \frac{1}{4}\mathrm{Tr}\{\sigma^\mu\bar{\sigma}^\nu(2\eta^{\sigma\rho} - \sigma^\rho\bar{\sigma}^\sigma) - \bar{\sigma}^\nu\sigma^\mu\bar{\sigma}^\rho\sigma^\sigma\}$$

(using (1.123) and that the trace is cyclic)

$$= \frac{1}{2}\eta^{\sigma\rho}\mathrm{Tr}(\sigma^\mu\bar{\sigma}^\nu) + \frac{1}{4}\mathrm{Tr}[-\sigma^\mu\bar{\sigma}^\nu\sigma^\rho\bar{\sigma}^\sigma - \bar{\sigma}^\nu\sigma^\mu(2\eta^{\rho\sigma} - \bar{\sigma}^\sigma\sigma^\rho)]$$

$$= \frac{1}{2}\eta^{\sigma\rho}\, Tr(\sigma^\mu \bar{\sigma}^\nu) - \frac{1}{2}\eta^{\rho\sigma}\, Tr\,(\bar{\sigma}^\nu \sigma^\mu)$$

$$- \frac{1}{4} Tr\,[\sigma^\mu \bar{\sigma}^\nu \sigma^\rho \bar{\sigma}^\sigma - \sigma^\rho \bar{\sigma}^\nu \sigma^\mu \bar{\sigma}^\sigma]$$

$$= -\frac{1}{4} Tr\,[\sigma^\mu \bar{\sigma}^\nu \sigma^\rho \bar{\sigma}^\sigma - \sigma^\rho \bar{\sigma}^\nu \sigma^\mu \bar{\sigma}^\sigma]$$

$$= -A^{\mu\nu\rho\sigma}$$

iii) $A^{\mu\nu\rho\sigma} = -A^{\rho\nu\mu\sigma}$

This holds by construction.

iv) $A^{\mu\nu\rho\sigma} = -A^{\sigma\nu\rho\mu}$

We have

$$A^{\sigma\nu\rho\mu}$$

$$= \frac{1}{4} Tr\,[\sigma^\sigma \bar{\sigma}^\nu \sigma^\rho \bar{\sigma}^\mu - \sigma^\rho \bar{\sigma}^\nu \sigma^\sigma \bar{\sigma}^\mu]$$

$$= \frac{1}{4} Tr\,[\bar{\sigma}^\mu \sigma^\sigma \bar{\sigma}^\nu \sigma^\rho - \sigma^\rho \bar{\sigma}^\nu (2\eta^{\sigma\mu} - \sigma^\mu \bar{\sigma}^\sigma)]$$

(using (1.123))

$$= \frac{1}{4} Tr\,[(2\eta^{\mu\sigma} - \bar{\sigma}^\sigma \sigma^\mu) \bar{\sigma}^\nu \sigma^\rho + \sigma^\rho \bar{\sigma}^\nu \sigma^\mu \bar{\sigma}^\sigma]$$

$$- \frac{1}{2}\eta^{\sigma\mu}\, Tr\,(\sigma^\rho \bar{\sigma}^\nu)$$

$$= \frac{1}{2}\eta^{\mu\sigma}\, Tr(\bar{\sigma}^\nu \sigma^\rho) - \frac{1}{2}\eta^{\sigma\mu}\, Tr\,(\sigma^\rho \bar{\sigma}^\nu)$$

$$- \frac{1}{4} Tr\,[\sigma^\mu \bar{\sigma}^\nu \sigma^\rho \bar{\sigma}^\sigma - \sigma^\rho \bar{\sigma}^\nu \sigma^\mu \bar{\sigma}^\sigma]$$

$$= -\frac{1}{4} Tr\,[\sigma^\mu \bar{\sigma}^\nu \sigma^\rho \bar{\sigma}^\sigma - \sigma^\rho \bar{\sigma}^\nu \sigma^\mu \bar{\sigma}^\sigma]$$

$$= -A^{\mu\nu\rho\sigma}$$

v) $A^{\mu\nu\rho\sigma} = -A^{\mu\rho\nu\sigma}$

We have

$$A^{\mu\rho\nu\sigma}$$

$$= \frac{1}{4} Tr\,[\sigma^\mu \bar{\sigma}^\rho \sigma^\nu \bar{\sigma}^\sigma - \sigma^\nu \bar{\sigma}^\rho \sigma^\mu \bar{\sigma}^\sigma]$$

$$= \frac{1}{4} Tr [\sigma^\mu (2\eta^{\rho\nu} - \bar{\sigma}^\nu \sigma^\rho) \bar{\sigma}^\sigma - (2\eta^{\nu\rho} - \sigma^\rho \bar{\sigma}^\nu) \sigma^\mu \bar{\sigma}^\sigma]$$

$$= \frac{1}{2} \eta^{\rho\nu} Tr (\sigma^\mu \bar{\sigma}^\sigma) - \frac{1}{2} \eta^{\nu\rho} Tr (\sigma^\mu \bar{\sigma}^\sigma) - \frac{1}{4} Tr [\sigma^\mu \bar{\sigma}^\nu \sigma^\rho \bar{\sigma}^\sigma - \sigma^\rho \bar{\sigma}^\nu \sigma^\mu \bar{\sigma}^\sigma]$$

$$= -A^{\mu\nu\rho\sigma}$$

vi) $A^{\mu\nu\rho\sigma} = -A^{\mu\sigma\rho\nu}$

We have

$$A^{\mu\sigma\rho\nu} = \frac{1}{4} Tr [\sigma^\mu \bar{\sigma}^\sigma \sigma^\rho \bar{\sigma}^\nu - \sigma^\rho \bar{\sigma}^\sigma \sigma^\mu \bar{\sigma}^\nu]$$

$$= -\frac{1}{4} Tr [-\sigma^\rho \bar{\sigma}^\nu \sigma^\mu \bar{\sigma}^\sigma + \sigma^\mu \bar{\sigma}^\nu \sigma^\rho \bar{\sigma}^\sigma]$$

$$= -A^{\mu\nu\rho\sigma}$$

For a quantity with n indices there are n(n-1)/2 possibilities of interchanging two indices. $A^{\mu\nu\rho\sigma}$ has four indices; hence there are six possible exchanges of any two indices. We have thus shown that $A^{\mu\nu\rho\sigma}$ is antisymmetric in interchanging any two of its indices. Hence $A^{\mu\nu\rho\sigma}$ is totally antisymmetric. $A^{\mu\nu\rho\sigma}$ must be proportional to the totally antisymmetric four-tensor $\varepsilon^{\mu\nu\rho\sigma}$ since $\varepsilon^{\mu\nu\rho\sigma}$ is the only tensor in four-space with this property. Hence

$$A^{\mu\nu\rho\sigma} = c \cdot \varepsilon^{\mu\nu\rho\sigma}$$

where c is a constant.

<u>Proposition</u>: The constant c in the above result is given by

$$c = -i$$

Proof: We use the following formula

$$\varepsilon_{\mu\nu\rho\sigma}\,\varepsilon^{\mu\nu\rho\sigma} = -4! = -24$$

Then

$$\varepsilon_{\mu\nu\rho\sigma} A^{\mu\nu\rho\sigma} = c \,.\, \varepsilon_{\mu\nu\rho\sigma}\,\varepsilon^{\mu\nu\rho\sigma} = -24c$$

and

$$\begin{aligned}
&\varepsilon_{\mu\nu\rho\sigma} A^{\mu\nu\rho\sigma} \\
&= \tfrac{1}{4}\varepsilon_{\mu\nu\rho\sigma} \, Tr[\sigma^{\mu}\bar{\sigma}^{\nu}\sigma^{\rho}\bar{\sigma}^{\sigma} - \sigma^{\rho}\bar{\sigma}^{\nu}\sigma^{\mu}\bar{\sigma}^{\sigma}] \\
&= \tfrac{1}{4} Tr[\varepsilon_{\mu\nu\rho\sigma}\sigma^{\mu}\bar{\sigma}^{\nu}\sigma^{\rho}\bar{\sigma}^{\sigma} \\
&\qquad - \varepsilon_{\mu\nu\rho\sigma}\sigma^{\rho}\bar{\sigma}^{\nu}\sigma^{\mu}\bar{\sigma}^{\sigma}] \\
&= \tfrac{1}{4} Tr[\varepsilon_{\mu\nu\rho\sigma}\sigma^{\mu}\bar{\sigma}^{\nu}\sigma^{\rho}\bar{\sigma}^{\sigma} \\
&\qquad - \varepsilon_{\rho\nu\mu\sigma}\sigma^{\mu}\bar{\sigma}^{\nu}\sigma^{\rho}\bar{\sigma}^{\sigma}] \\
&= \tfrac{1}{4} Tr[\varepsilon_{\mu\nu\rho\sigma}\sigma^{\mu}\bar{\sigma}^{\nu}\sigma^{\rho}\bar{\sigma}^{\sigma} \\
&\qquad + \varepsilon_{\mu\nu\rho\sigma}\sigma^{\mu}\bar{\sigma}^{\nu}\sigma^{\rho}\bar{\sigma}^{\sigma}] \\
&= \tfrac{1}{2}\varepsilon_{\mu\nu\rho\sigma} \, Tr[\sigma^{\mu}\bar{\sigma}^{\nu}\sigma^{\rho}\bar{\sigma}^{\sigma}]
\end{aligned}$$

It is not difficult to show by an explicit calculation that this expression is simply 24 i . Hence c = - i .

Proposition: The product

$$\varepsilon_{AC}\,(\sigma^{\mu\nu})_{B}{}^{C}$$

is symmetric in A and B. Since

$$\varepsilon_{AC}\,(\sigma^{\mu\nu})_{B}{}^{C} = (\sigma^{\mu\nu}\varepsilon^{T})_{BA}$$

we demonstrate that

$$(\sigma^{\mu\nu}\varepsilon^{T})_{BA} = (\varepsilon\,\sigma^{\mu\nu T})_{BA} \qquad (1.127)$$

Similarly

$$(\varepsilon\,\bar{\sigma}^{\mu\nu})_{\dot{A}\dot{C}} = (\varepsilon\,\bar{\sigma}^{\mu\nu})_{\dot{C}\dot{A}} \qquad (1.127')$$

Proof: We have to demonstrate that

$$(\sigma^{\mu\nu}\varepsilon^T)^T = \sigma^{\mu\nu}\varepsilon^T$$

Consider the expression on the left hand side

$$(\sigma^{\mu\nu}\varepsilon^T)^T = \varepsilon(\sigma^{\mu\nu})^T$$

$$= \frac{i}{4}\varepsilon(\sigma^\mu\bar{\sigma}^\nu - \sigma^\nu\bar{\sigma}^\mu)^T \quad \text{(using (1.119a))}$$

$$= \frac{i}{4}\varepsilon(\sigma^\mu\varepsilon\sigma^{\nu T}\varepsilon^T - \sigma^\nu\varepsilon\sigma^{\mu T}\varepsilon^T)^T$$

(using (1.89))

$$= \frac{i}{4}\varepsilon(\varepsilon\sigma^\nu\varepsilon^T\sigma^{\mu T} - \varepsilon\sigma^\mu\varepsilon^T\sigma^{\nu T})$$

$$= \frac{i}{4}\varepsilon[(-\varepsilon^T)\sigma^\nu(-\varepsilon)\sigma^{\mu T} + \varepsilon^T\sigma^\mu(-\varepsilon)\sigma^{\nu T}](-\varepsilon^T)\varepsilon^T$$

(using (1.59), i.e. $\varepsilon = -\varepsilon^T, \varepsilon\varepsilon^T = 1$)

$$= \frac{i}{4}(\sigma^\mu\varepsilon\sigma^{\nu T}\varepsilon^T - \sigma^\nu\varepsilon\sigma^{\mu T}\varepsilon^T)\varepsilon^T$$

$$= \frac{i}{4}(\sigma^\mu\bar{\sigma}^\nu - \sigma^\nu\bar{\sigma}^\mu)\varepsilon^T \quad \text{(using (1.89))}$$

$$= \sigma^{\mu\nu}\varepsilon^T$$

Proposition: The following relation holds

$$(\phi\sigma^{\mu\nu}\chi) = -(\chi\sigma^{\mu\nu}\phi) \tag{1.128}$$

In particular

$$(\phi\sigma^{\mu\nu}\phi) = 0$$

Proof: We have

$$(\phi\sigma^{\mu\nu}\chi) = \phi^A(\sigma^{\mu\nu})_A{}^B\chi_B$$

$$= (\varepsilon^{AC}\phi_C)(\sigma^{\mu\nu})_A{}^B(\varepsilon_{BD}\chi^D)$$

$$= -\chi^D(\sigma^{\mu\nu}\varepsilon)_{AD}\,\varepsilon^{AC}\phi_C \qquad \text{(with (1.79))}$$
$$= \chi^D(\sigma^{\mu\nu}\varepsilon^T)_{AD}\,\varepsilon^{AC}\phi_C \qquad \text{(with (1.60))}$$
$$= \chi^D(\sigma^{\mu\nu}\varepsilon^T)_{DA}\,\varepsilon^{AC}\phi_C \qquad \text{(with (1.127))}$$
$$= \chi^D(\sigma^{\mu\nu})_D{}^B\,\varepsilon^T_{BA}\,\varepsilon^{AC}\phi_C$$
$$= -\chi^D(\sigma^{\mu\nu})_D{}^B\,\varepsilon_{BA}\,\varepsilon^{AC}\phi_C \qquad \text{(with (1.60))}$$
$$= -\chi^D(\sigma^{\mu\nu})_D{}^B\,\delta_B{}^C\phi_C \qquad \text{(with (1.61))}$$
$$= -(\chi\sigma^{\mu\nu}\phi)$$

As a consequence an expression like $(\psi\sigma^{\mu\nu}\psi)$ vanishes identically.

1.3.6 Higher Order Weyl Spinors and their Representations

Multiplying elementary Weyl spinors by one another, i.e. $\psi_A, \psi^B, \bar{\psi}_{\dot{A}}, \bar{\psi}^{\dot{B}}$ one can form higher order spinors or spinors which carry both dotted and undotted indices. In general these spinors transform under SL(2,C) as the product of individual spinors (of course, not every higher order spinor is expressible as the product of elementary spinors). From previous considerations we know that undotted spinors ψ_A describe the (1/2, 0) representation of SL(2,C) and dotted spinors $\bar{\chi}^{\dot{A}}$ the (0, 1/2) representation. The Kronecker product of the two representations

$$\left(\frac{1}{2}, 0\right) \otimes \left(0, \frac{1}{2}\right) = \left(\frac{1}{2}, \frac{1}{2}\right)$$

gives a $(2.\frac{1}{2} + 1)^2 = 4$-dimensional representation of SL(2,C) which can be shown to be equivalent to the four-dimensional self-representation of SO(1,3) in the space spanned by the x_μ, μ = 0,1,2,3, i.e. the four components of

$$\psi_A \otimes \bar{\chi}_{\dot{A}} \equiv \psi_A \bar{\chi}_{\dot{A}} = \frac{1}{2}(\psi \sigma_\mu \bar{\chi}) \sigma^\mu_{A\dot{A}} \tag{1.129a}$$

form a four vector V_μ. We can find the components of this vector by comparing (1.129a) with (1.93), i.e.

$$(X)_{A\dot{A}} = (x_\mu \sigma^\mu)_{A\dot{A}}$$

where

$$X_{1\dot{1}} = x^0 - x^3 , \quad X_{1\dot{2}} = x^1 + i x^2$$

$$X_{2\dot{1}} = x^1 - i x^2 , \quad X_{2\dot{2}} = x^0 + x^3$$

Thus

$$V_\mu = \frac{1}{2}(\psi \sigma_\mu \bar{\chi})$$

(An explicit proof of the vector character of $\psi_A \otimes \bar{\chi}_{\dot{A}}$ can be found in the literature[38]).

Proof of (1.129a): Using the summation conventions (1.77) and (1.78) we have

$$\frac{1}{2}(\psi \sigma^\mu \bar{\chi})(\sigma_\mu)_{A\dot{A}}$$

$$= \frac{1}{2} \psi^B (\sigma^\mu)_{B\dot{B}} \bar{\chi}^{\dot{B}} (\sigma_\mu)_{A\dot{A}}$$

$$= \frac{1}{2} \varepsilon^{BC} \psi_C (\sigma^\mu)_{B\dot{B}} \varepsilon^{\dot{B}\dot{D}} \bar{\chi}_{\dot{D}} (\sigma_\mu)_{A\dot{A}}$$

(using (1.76a,c))

$$= -\frac{1}{2} \psi_C \varepsilon^{BC} \varepsilon^{\dot{D}\dot{B}} \sigma^\mu_{B\dot{B}} \bar{\chi}_{\dot{D}} (\sigma_\mu)_{A\dot{A}}$$

$$= \frac{1}{2} \psi_C (\varepsilon^{CB} \varepsilon^{\dot{D}\dot{B}} \sigma^\mu_{B\dot{B}}) \bar{\chi}_{\dot{D}} (\sigma_\mu)_{A\dot{A}}$$

(using $\varepsilon^T = -\varepsilon$)

$$= \frac{1}{2} \psi_C \bar{\chi}_{\dot{D}} (\bar{\sigma}^\mu)^{\dot{D}C} (\sigma_\mu)_{A\dot{A}} \quad \text{(with (1.88))}$$

$$= \psi_C \bar{\chi}_{\dot{D}} \, \delta^{\dot{D}}{}_{\dot{A}} \, \delta^C{}_A = \psi_A \bar{\chi}_{\dot{A}} \qquad \text{(with (1.92))}$$

Equation (1.129a) can also be demonstrated with the help of the Fierz reordering formula (1.114).

Proposition: The Kronecker product of two (1/2, 0) Weyl spinors gives

$$\left(\frac{1}{2}, 0\right) \otimes \left(\frac{1}{2}, 0\right) = (0, 0) \oplus (1, 0)$$

or

$$\psi_A \otimes \chi_B = \frac{1}{2} \varepsilon_{AB} (\psi\chi) + \frac{1}{2} (\sigma^{\mu\nu} \varepsilon^T)_{AB} (\psi \sigma_{\mu\nu} \chi) \tag{1.129b}$$

corresponding to a scalar (0,0) and a selfdual second rank tensor ($\sigma_{\mu\nu}$ is selfdual according to (1.121a)).

Proof: As usual, we decompose the tensor product $\psi_A \otimes \chi_B$ into a symmetric and an antisymmetric part, i.e. we write

$$\psi_A \otimes \chi_B = \frac{1}{2} (\psi_A \chi_B + \psi_B \chi_A) + \frac{1}{2} (\psi_A \chi_B - \psi_B \chi_A)$$

Consider first the antisymmetric part. Using (1.83b) we have

$$\frac{1}{2} \{\psi_A \chi_B - \psi_B \chi_A\} = \frac{1}{2} \{(\psi\chi) \varepsilon_{AB} - \frac{1}{2} (\psi\chi) \varepsilon_{BA}\}$$
$$= \frac{1}{2} \varepsilon_{AB} (\psi\chi)$$

Now consider the symmetric part. We show first that

$$(\sigma^{\mu\nu})_A{}^B (\sigma_{\mu\nu})_C{}^D = \varepsilon_{AC} \varepsilon^{BD} + \delta_A{}^D \delta^B{}_C$$

We have

$$(\sigma^{\mu\nu})_A{}^B (\sigma_{\mu\nu})_C{}^D$$
$$= -\frac{1}{16} (\sigma^\mu \bar{\sigma}^\nu - \sigma^\nu \bar{\sigma}^\mu)_A{}^B (\sigma_\mu \bar{\sigma}_\nu - \sigma_\nu \bar{\sigma}_\mu)_C{}^D$$

(using (1.119a))

$$= -\frac{1}{8} \{(\sigma^\mu \bar{\sigma}^\nu)_A{}^B (\sigma_\mu \bar{\sigma}_\nu)_C{}^D - (\sigma^\nu \bar{\sigma}^\mu)_A{}^B (\sigma_\mu \bar{\sigma}_\nu)_C{}^D\}$$

Now

$$(\sigma^{\mu}\bar{\sigma}^{\nu})_A{}^B(\sigma_\mu\bar{\sigma}_\nu)_C{}^D$$

$$= \sigma^{\mu}_{A\dot{A}}\bar{\sigma}^{\nu\dot{A}B}\sigma_{\mu C\dot{B}}\bar{\sigma}_\nu{}^{\dot{B}D}$$

$$= \sigma^{\mu}_{A\dot{A}}\sigma_{\mu C\dot{B}}\bar{\sigma}^{\nu\dot{A}B}\bar{\sigma}_\nu{}^{\dot{B}D}$$

$$= \sigma^{\mu}_{A\dot{A}}\varepsilon_{CE}\varepsilon_{\dot{B}\dot{C}}\bar{\sigma}_\mu{}^{\dot{C}E}\varepsilon^{\dot{A}\dot{F}}\varepsilon^{BF}\sigma^{\nu}_{F\dot{F}}\bar{\sigma}_\nu{}^{\dot{B}D}$$

(using (1.88))

$$= \sigma^{\mu}_{A\dot{A}}\bar{\sigma}_\mu{}^{\dot{C}E}\sigma^{\nu}_{F\dot{F}}\bar{\sigma}_\nu{}^{\dot{B}D}\varepsilon_{CE}\varepsilon_{\dot{B}\dot{C}}\varepsilon^{\dot{A}\dot{F}}\varepsilon^{BF}$$

$$= 4\delta_A{}^E\delta_{\dot{A}}{}^{\dot{C}}\delta_F{}^D\delta_{\dot{F}}{}^{\dot{B}}\varepsilon_{CE}\varepsilon_{\dot{B}\dot{C}}\varepsilon^{\dot{A}\dot{F}}\varepsilon^{BF}$$

(using (1.92))

$$= 4\varepsilon_{CA}\varepsilon_{\dot{F}\dot{C}}\varepsilon^{\dot{C}\dot{F}}\varepsilon^{BD}$$

$$= 4\varepsilon_{CA}\varepsilon^{BD}\delta_{\dot{F}}{}^{\dot{F}} \qquad \text{(using (1.71))}$$

$$= -8\varepsilon_{AC}\varepsilon^{BD}$$

Furthermore

$$(\sigma^{\nu}\bar{\sigma}^{\mu})_A{}^B(\sigma_\mu\bar{\sigma}_\nu)_C{}^D$$

$$= \sigma^{\nu}_{A\dot{A}}\bar{\sigma}^{\mu\dot{A}B}\sigma_{\mu C\dot{B}}\bar{\sigma}_\nu{}^{\dot{B}D}$$

$$= \sigma^{\nu}_{A\dot{A}}\bar{\sigma}_\nu{}^{\dot{B}D}\sigma_{\mu C\dot{B}}\bar{\sigma}^{\mu\dot{A}B}$$

$$= 4\delta_A{}^D\delta_{\dot{A}}{}^{\dot{B}}\delta_C{}^B\delta_{\dot{B}}{}^{\dot{A}} \qquad \text{(using (1.92))}$$

$$= 8\delta_A{}^D\delta^B{}_C$$

Hence

$$(\sigma^{\mu\nu})_A{}^B (\sigma_{\mu\nu})_C{}^D$$
$$= -\frac{1}{8} \{ (\sigma^\mu \bar{\sigma}^\nu)_A{}^B (\sigma_\mu \bar{\sigma}_\nu)_C{}^D$$
$$- (\sigma^\nu \bar{\sigma}^\mu)_A{}^B (\sigma_\mu \bar{\sigma}_\nu)_C{}^D \}$$
$$= \varepsilon_{AC} \varepsilon^{BD} + \delta_A{}^D \delta^B{}_C$$

With this result we obtain

$$\frac{1}{2} (\sigma^{\mu\nu} \varepsilon^T)_{AB} (\psi \sigma_{\mu\nu} \chi)$$
$$= \frac{1}{2} (\sigma^{\mu\nu})_A{}^C (\varepsilon^T)_{CB} \psi^D (\sigma_{\mu\nu})_D{}^F \chi_F$$
$$= \frac{1}{2} \psi^D (\sigma^{\mu\nu})_A{}^C (\sigma_{\mu\nu})_D{}^F (\varepsilon^T)_{CB} \chi_F$$
$$= \frac{1}{2} \psi^D \{ \varepsilon_{AD} \varepsilon^{CF} + \delta_A{}^F \delta^C{}_D \} \varepsilon^T_{CB} \chi_F$$
$$= \frac{1}{2} \psi^D \varepsilon_{AD} \varepsilon^{FC} \varepsilon_{CB} \chi_F$$
$$+ \frac{1}{2} \psi^D \delta_A{}^F \delta^C{}_D \varepsilon^T_{CB} \chi_F$$
$$= \frac{1}{2} \psi_A \delta^F{}_B \chi_F + \frac{1}{2} \psi^C \varepsilon^T_{CB} \chi_A$$

(using (1.76b))

$$= \frac{1}{2} (\psi_A \chi_B + \psi_B \chi_A)$$

Hence

$$\psi_A \chi_B = \frac{1}{2} (\psi_A \chi_B + \psi_B \chi_A) + \frac{1}{2} (\psi_A \chi_B - \psi_B \chi_A)$$
$$= \frac{1}{2} (\sigma^{\mu\nu} \varepsilon^T)_{AB} (\psi \sigma_{\mu\nu} \chi)$$
$$+ \frac{1}{2} \varepsilon_{AB} (\psi \chi)$$

q.e.d.

The first term of (1.129b) transforms as a scalar since ε_{AB} transforms as an invariant spinor under SL(2,C). The second term transforms as a three-vector under SL(2,C) since the representation (1,0) is equivalent to the space of selfdual antisymmetric tensors of rank two. This second term is just the symmetric part of $\psi_A \otimes \chi_B$ according to (1.127), and symmetric spinors

$$T_{AB} := \frac{1}{2}(\psi_A \otimes \chi_B + \psi_B \otimes \chi_A) = T_{BA}$$

transform according to (1,0) as can be shown by an analogous construction following (1.93). The odd and even parts of this product are recognized as the one-component scalar and three-component vector quantities which are well known from the construction of the spin part of the wave function of two spin-$\frac{1}{2}$ particles. However, we can see this also as follows. From (1.119) we infer that

$$\{\sigma^{\mu\nu} \mid \mu,\nu = 0,1,2,3\}_A{}^B \Rightarrow \{\sigma^i \bar{\sigma}^0 \mid i = 1,2,3\}_A{}^B$$

$$\{\bar{\sigma}^{\mu\nu} \mid \mu,\nu = 0,1,2,3\}^{\dot{A}}{}_{\dot{B}} \Rightarrow \{\bar{\sigma}^i \sigma^0 \mid i = 1,2,3\}^{\dot{A}}{}_{\dot{B}}$$

Then

$$(\sigma^{\mu\nu}\varepsilon^T)_{AB}(\psi\sigma_{\mu\nu}\chi) \Rightarrow \begin{pmatrix} -\psi\sigma^1\chi + i\sigma^2\chi & \psi\sigma^3\chi \\ \psi\sigma^3\chi & \psi\sigma^1\chi + i\psi\sigma^2\chi \end{pmatrix}_{AB}$$

and after a rotation the three-component vector is seen to be[39]

$$\psi\vec{\sigma}\chi$$

We also observe that

$$\det\begin{pmatrix} -x + iy & z \\ z & x + iy \end{pmatrix} = -(x^2 + y^2 + z^2)$$

Remark: In later calculations we shall need some further relations involving spinors of rank two. It is convenient to derive these relations here.

Let Φ_{AB}, A, B = 1,2 be any spinor of rank two which is antisymmetric in A and B, i.e. $\Phi_{AB} \in F \oplus F$ and

$$\Phi_{AB} = -\Phi_{BA}$$

Then this spinor must be proportional to the metric ε_{AB} since Φ_{AB} has only one independent component, i.e.

$$(\Phi_{AB}) = \begin{pmatrix} 0 & \phi_{12} \\ \phi_{21} & 0 \end{pmatrix}$$

and since $\Phi_{AB} = -\Phi_{BA}$ we conclude $\phi_{12} = -\phi_{21}$. Therefore

$$\Phi_{AB} = c\,\varepsilon_{AB}$$

Now, considering

$$\varepsilon^{BA}\Phi_{AB} = c\,\varepsilon^{BA}\varepsilon_{AB} = c\,\delta^{B}{}_{B} = 2c$$

we have

$$c = \frac{1}{2}\varepsilon^{BA}\Phi_{AB} = \frac{1}{2}\Phi^{B}{}_{B}$$

and hence

$$\Phi_{AB} = \frac{1}{2}\Phi^{C}{}_{C}\,\varepsilon_{AB}$$

Then we have for an arbitrary $\Phi_{AB} \in F \oplus F$

$$\Phi_{AB} - \Phi_{BA} = \frac{1}{2}\Phi^{D}{}_{D}\,\varepsilon_{AB} - \frac{1}{2}\Phi^{D}{}_{D}\,\varepsilon_{BA} = \Phi^{D}{}_{D}\,\varepsilon_{AB}$$

$$\Phi_{AB} - \Phi_{BA} = \delta_{A}{}^{C}\delta_{B}{}^{D}\Phi_{CD} - \delta_{B}{}^{C}\delta_{A}{}^{D}\Phi_{CD} = \{\delta_{A}{}^{C}\delta_{B}{}^{D} - \delta_{B}{}^{C}\delta_{A}{}^{D}\}\Phi_{CD} \overset{!}{=} \Phi^{D}{}_{D}\,\varepsilon_{AB} \qquad (1.129c)$$

$$= \varepsilon_{AB}\, \varepsilon^{DC}\, \Phi_{CD}$$

Hence

$$\varepsilon_{AB}\, \varepsilon^{DC} = \delta_A{}^C \delta_B{}^D - \delta_A{}^D \delta_B{}^C \tag{1.129d}$$

Similarly we obtain

$$\varepsilon_{\dot{A}\dot{B}}\, \varepsilon^{\dot{D}\dot{C}} = \delta_{\dot{A}}{}^{\dot{C}} \delta_{\dot{B}}{}^{\dot{D}} - \delta_{\dot{A}}{}^{\dot{D}} \delta_{\dot{B}}{}^{\dot{C}} \tag{1.129e}$$

Multiplying (1.129c) by $\varepsilon^{CA}\varepsilon^{DB}$ and using (1.76) we obtain

$$\begin{aligned}\Phi^{CD} - \Phi^{DC} &= \varepsilon^{CA}\varepsilon^{DB}\, \Phi^E{}_E\, \varepsilon_{AB} \\ &= \delta^C{}_B\, \delta^{DB}\, \Phi^E{}_E \\ &= \varepsilon^{DC}\, \Phi^E{}_E \\ &= -\Phi^E{}_E\, \varepsilon^{CD}\end{aligned} \tag{1.129f}$$

Corresponding expressions hold for dotted spinors

$$\begin{aligned}\Phi_{\dot{A}\dot{B}} - \Phi_{\dot{B}\dot{A}} &= \Phi_{\dot{C}}{}^{\dot{C}}\, \varepsilon_{\dot{A}\dot{B}} \\ \Phi^{\dot{A}\dot{B}} - \Phi^{\dot{B}\dot{A}} &= -\Phi_{\dot{C}}{}^{\dot{C}}\, \varepsilon^{\dot{A}\dot{B}}\end{aligned} \tag{1.129g}$$

These expressions will be needed in later calculations.

1.4 Dirac and Majorana Spinors

In equations (1.20) and (1.21) we defined two sets of operators, i.e.

$$N_i := \frac{1}{2}(J_i + iK_i)$$

$$N_i^+ := \frac{1}{2}(J_i - iK_i)$$

From these operators we constructed two Casimir operators $N_i N_i$ and $N_i^+ N_i^+$ whose eigenvalues n(n+1) and m(m+1), n, m = 0, $\frac{1}{2}$, 1, ... determine finite-dimensional, non-unitary representations of the Lorentz group labelled (n, m). Under space reflection, i.e. the parity transformation, the generators of rotations remain invariant whereas the boost generators change sign, i.e. $K_i \rightarrow -K_i$. Hence the parity operation is formally equivalent to the transformation

$$N_i \longrightarrow N_i^+ , \qquad N_i^+ \longrightarrow N_i$$

i.e. the parity transformation corresponds to complex conjugation. However, we have seen with (1.52) and (1.54) that the operation of complex conjugation transforms the (1/2,0) representation into the (0,1/2) representation. In order to write such a map as a linear transformation it is necessary to double the dimension of the spinor representation. Previously we defined

F as the two-dimensional complex representation space of SL(2,C) whose elements are undotted, left-handed Weyl spinors, and

$\dot{F}^*$ as the two-dimensional complex representation space of the complex conjugate representation of SL(2,C) whose elements are dotted or socalled right-handed Weyl spinors.

We now define the direct sum of F and $\dot{F}^*$ as

$$E := F \oplus \dot{F}^*$$

E is the four-dimensional complex representation space of Dirac spinors. Let

$$\phi \in F \quad \text{and} \quad \bar{\psi} \in \dot{F}^*$$

Then

$$\Psi := \begin{pmatrix} \phi \\ \bar{\psi} \end{pmatrix} \in E \tag{1.130}$$

is a Dirac four-spinor. We define a representation of SL(2,C) on E by the map

$$M \in SL(2,C) \longrightarrow S(M) := \underset{\in Aut(E)}{\begin{pmatrix} M & 0 \\ 0 & M^{*-1} \end{pmatrix}} \tag{1.131}$$

where Aut(E) is the automorphism group of E. This representation acts on Dirac spinors $\Psi \in$ E as follows.Using (1.130) and (1.131)

$$\Psi' = S(M)\Psi = \begin{pmatrix} M & 0 \\ 0 & M^{*-1} \end{pmatrix} \begin{pmatrix} \phi \\ \bar{\psi} \end{pmatrix}$$

$$= \begin{pmatrix} M\phi \\ M^{*-1}\bar{\psi} \end{pmatrix} \tag{1.132}$$

From this relation we can obtain the index structure of Dirac spinors as well as that of matrices acting on Dirac spinors. In view of (1.53) and (1.74) we obtain

$$\Psi_a = \begin{pmatrix} \phi_A \\ \bar{\psi}^{\dot{A}} \end{pmatrix} \tag{1.133}$$

where $A = 1,2$ and $\dot{A} = \dot{1},\dot{2}$, and $a = 1,2,3,4$, and

$$S_{ab}(M) = \begin{pmatrix} M_A{}^B & 0 \\ 0 & (M^{*-1})^{\dot{A}}{}_{\dot{B}} \end{pmatrix}$$

(1.134a)

Any 4x4-matrix acting on Dirac four-spinors must therefore have the following index structure

$$\Gamma_{ab} = \begin{pmatrix} A_A{}^B & B_{A\dot{B}} \\ C^{\dot{A}B} & D^{\dot{A}}{}_{\dot{B}} \end{pmatrix} , \qquad \begin{array}{l} A, B = 1, 2 \\ \dot{A}, \dot{B} = \dot{1}, \dot{2} \\ a, b = 1, 2, 3, 4 \end{array}$$

(1.134b)

where A, B, C, D are 2x2 complex submatrices. The off-diagonal submatrices B and C must have a dotted as well as an undotted index since by construction B is a map from $\dot{F}^{*}$ to F and C is a map from F to $\dot{F}^{*}$. Applying the matrix Γ to a Dirac spinor we see that

$$\Gamma_{ab}\Psi_b = \begin{pmatrix} A_A{}^B & B_{A\dot{B}} \\ C^{\dot{A}B} & D^{\dot{A}}{}_{\dot{B}} \end{pmatrix} \begin{pmatrix} \phi_B \\ \overline{\psi}^{\dot{B}} \end{pmatrix}$$

$$= \begin{pmatrix} A_A{}^B \phi_B + B_{A\dot{B}} \overline{\psi}^{\dot{B}} \\ C^{\dot{A}B}\phi_B + D^{\dot{A}}{}_{\dot{B}} \overline{\psi}^{\dot{B}} \end{pmatrix} \begin{array}{l} \in F \\ \in \dot{F}^{*} \end{array}$$

This index picture is consistent with the summation convention for Weyl spinors.

A representation matrix for the parity operation in E is (as can be shown but will not be proven here)

$$S_R = i \begin{pmatrix} 0 & \sigma^0 \\ \overline{\sigma}^0 & 0 \end{pmatrix}$$

(the factor i is inserted as a matter of convention). Thus

$$S_R \Psi = i\begin{pmatrix} 0 & \sigma^0 \\ \bar{\sigma}^0 & 0 \end{pmatrix}\begin{pmatrix} \phi \\ \bar{\Psi} \end{pmatrix} = i\begin{pmatrix} \sigma^0 \bar{\Psi} \\ \bar{\sigma}^0 \phi \end{pmatrix} \tag{1.135}$$

In components we have

$$(S_R)_{ab} \Psi_b = i \begin{pmatrix} 0 & \sigma^0{}_{A\dot{B}} \\ \bar{\sigma}^{0\dot{A}B} & 0 \end{pmatrix}\begin{pmatrix} \phi_B \\ \bar{\Psi}^{\dot{B}} \end{pmatrix}$$
$$= i \begin{pmatrix} \sigma^0{}_{A\dot{B}} \bar{\Psi}^{\dot{B}} \\ \bar{\sigma}^{0\dot{A}B} \phi_B \end{pmatrix} = \Psi'_a$$

where $\Psi'_a \in \dot{F}^* \oplus F$ since the σ^μ map $\dot{F}^*$ into F and the $\bar{\sigma}^\mu$ F into $\dot{F}^*$. This transformation of Ψ into Ψ' demonstrates explicitly the irreducibility of the representation space E under the parity operation S_R. We conclude therefore that if parity is of interest, one has to use the Dirac spinor formalism in a relativistic theory.

We now discuss several representations of Dirac γ^μ matrices.

1.4.1 The Weyl Basis or Chiral Representation

If we take the following realization of γ matrices

$$\gamma^\mu_W := \begin{pmatrix} 0 & \sigma^\mu \\ \bar{\sigma}^\mu & 0 \end{pmatrix} \tag{1.136}$$

we have a direct relation between two-component and four-component spinors. The representation in which γ^μ matrices have the form (1.136) is called the "Weyl basis". As in any other representation we have the Clifford algebra relation

$$\{\gamma^\mu_W, \gamma^\nu_W\} = 2\eta^{\mu\nu} 1_{4\times 4} \tag{1.137}$$

We can use this to verify our two-component index formalism. Thus

$$\{\gamma^\mu_W, \gamma^\nu_W\}_{ac} := (\gamma^\mu_W)_{ab}(\gamma^\nu_W)_{bc} + (\gamma^\nu_W)_{ab}(\gamma^\mu_W)_{bc}$$

$$= \begin{pmatrix} 0 & \sigma^\mu_{A\dot{B}} \\ \bar{\sigma}^{\mu\dot{A}B} & 0 \end{pmatrix} \begin{pmatrix} 0 & \sigma^\nu_{B\dot{C}} \\ \bar{\sigma}^{\nu\dot{B}C} & 0 \end{pmatrix}$$

$$+ \begin{pmatrix} 0 & \sigma^\nu_{A\dot{B}} \\ \bar{\sigma}^{\nu\dot{A}B} & 0 \end{pmatrix} \begin{pmatrix} 0 & \sigma^\mu_{B\dot{C}} \\ \bar{\sigma}^{\mu\dot{B}C} & 0 \end{pmatrix}$$

$$= \begin{pmatrix} \sigma^\mu_{A\dot{B}}\bar{\sigma}^{\nu\dot{B}C} + \sigma^\nu_{A\dot{B}}\bar{\sigma}^{\mu\dot{B}C} & 0 \\ 0 & \bar{\sigma}^{\mu\dot{A}B}\sigma^\nu_{B\dot{C}} + \bar{\sigma}^{\nu\dot{A}B}\sigma^\mu_{B\dot{C}} \end{pmatrix}$$

$$= \begin{pmatrix} (\sigma^\mu\bar{\sigma}^\nu + \sigma^\nu\bar{\sigma}^\mu)_A{}^C & 0 \\ 0 & (\bar{\sigma}^\mu\sigma^\nu + \bar{\sigma}^\nu\sigma^\mu)^{\dot{A}}{}_{\dot{C}} \end{pmatrix}$$

$$= \begin{pmatrix} 2\eta^{\mu\nu}\delta_A{}^C & 0 \\ 0 & 2\eta^{\mu\nu}\delta^{\dot{A}}{}_{\dot{C}} \end{pmatrix}$$

(using (1.122a,b)

$$= 2\eta^{\mu\nu} \begin{pmatrix} \delta_A{}^C & 0 \\ 0 & \delta^{\dot{A}}{}_{\dot{C}} \end{pmatrix}$$

$$= 2\eta^{\mu\nu}\delta_{ac}$$

or in matrix notation

$$\{\gamma^\mu_W, \gamma^\nu_W\} = 2\eta^{\mu\nu}\, 1_{4\times 4}$$

Remark: The 4x4 unit matrix δ_{ab} has the index structure

$$\delta_{ab} = \begin{pmatrix} \delta_A{}^B & 0 \\ 0 & \delta^{\dot{A}}{}_{\dot{B}} \end{pmatrix} \tag{1.138}$$

because (1.134) reduces to (1.138) for $M = 1_{SL(2,C)}$, where $1_{SL(2,C)}$ is the unit element of SL(2,C).
We define

$$\gamma^5_W := i\gamma^0_W \gamma^1_W \gamma^2_W \gamma^3_W \tag{1.139a}$$

Proposition: We have

$$\gamma^5_W = \begin{pmatrix} -1_{2\times 2} & 0 \\ 0 & 1_{2\times 2} \end{pmatrix} \tag{1.139b}$$

Proof: From the definition (1.139a)

$$(\gamma^5_W)_{ab} = i(\gamma^0_W \gamma^1_W \gamma^2_W \gamma^3_W)_{ab}$$

$$= i(\gamma^0_W)_{ac}(\gamma^1_W)_{cd}(\gamma^2_W)_{de}(\gamma^3_W)_{eb}$$

$$= i\begin{pmatrix} 0 & \sigma^0{}_{A\dot{C}} \\ \bar{\sigma}^{0\dot{A}C} & 0 \end{pmatrix}\begin{pmatrix} 0 & \sigma^1{}_{C\dot{D}} \\ \bar{\sigma}^{1\dot{C}D} & 0 \end{pmatrix}\begin{pmatrix} 0 & \sigma^2{}_{D\dot{E}} \\ \bar{\sigma}^{2\dot{D}E} & 0 \end{pmatrix}\begin{pmatrix} 0 & \sigma^3{}_{E\dot{B}} \\ \bar{\sigma}^{3\dot{E}B} & 0 \end{pmatrix}$$

$$= i\begin{pmatrix} \sigma^0_{A\dot{C}}\bar{\sigma}^{1\dot{C}D} & 0 \\ 0 & \bar{\sigma}^{0\dot{A}C}\sigma^1_{C\dot{D}} \end{pmatrix}\begin{pmatrix} \sigma^2_{D\dot{E}}\bar{\sigma}^{3\dot{E}B} & 0 \\ 0 & \bar{\sigma}^{2\dot{D}E}\sigma^3_{E\dot{B}} \end{pmatrix}$$

$$= i\begin{pmatrix} \sigma^0_{A\dot{C}}\bar{\sigma}^{1\dot{C}D}\sigma^2_{D\dot{E}}\bar{\sigma}^{3\dot{E}B} & 0 \\ 0 & \bar{\sigma}^{0\dot{A}C}\sigma^1_{C\dot{D}}\bar{\sigma}^{2\dot{D}E}\sigma^3_{E\dot{B}} \end{pmatrix}$$

$$= i\begin{pmatrix} (\sigma^0\bar{\sigma}^1\sigma^2\bar{\sigma}^3)_A{}^B & 0 \\ 0 & (\bar{\sigma}^0\sigma^1\bar{\sigma}^2\sigma^3)^{\dot{A}}{}_{\dot{B}} \end{pmatrix}$$

$$= i\begin{pmatrix} (\sigma^0\sigma^1\sigma^2\sigma^3)_A{}^B & 0 \\ 0 & -(\sigma^0\sigma^1\sigma^2\sigma^3)^{\dot{A}}{}_{\dot{B}} \end{pmatrix}$$

using (1.90). Evaluation of the product of σ matrices yields

$$(\gamma^5_W)_{ab} = \begin{pmatrix} -\delta_A{}^B & 0 \\ 0 & \delta^{\dot{A}}{}_{\dot{B}} \end{pmatrix} \qquad (1.139c)$$

The following properties of the matrix γ^5_W are independent of the particular representation

i) $$(\gamma^5_W)^2 = 1_{4\times 4} \qquad (1.140a)$$

ii) $$\{\gamma^5_W, \gamma^\mu_W\} = 0 \qquad (1.140b)$$

<u>Proposition</u>: In the Weyl representation we have

i) $\gamma^0_W = \gamma^{0+}_W$ i.e. γ^0_W is hermitian

ii) $\gamma^5_W = \gamma^{5+}_W$ i.e. γ^5_W is hermitian

iii) $\gamma^i_W = -\gamma^{i+}_W$ i.e. γ^i_W is antihermitian, i = 1,2,3

(1.141)

<u>Proof</u>: We first demonstrate the validity of (1.141) in the matrix notation.

i) Using (1.136) we have

$$(\gamma_W^0)^+ = \begin{pmatrix} 0 & \sigma^0 \\ \bar{\sigma}^0 & 0 \end{pmatrix}^+ = \begin{pmatrix} 0 & \bar{\sigma}^{0+} \\ \sigma^{0+} & 0 \end{pmatrix}$$

$$= \begin{pmatrix} 0 & \sigma^0 \\ \bar{\sigma}^0 & 0 \end{pmatrix} = \gamma_W^0$$

using the fact that according to (1.90) and (1.86a) σ^0 and $\bar{\sigma}^0$ are 2x2 unit matrices.

ii) Using (1.139) we have

$$(\gamma_W^5)^+ = (i\gamma_W^0\gamma_W^1\gamma_W^2\gamma_W^3)^+$$

$$= -i(\gamma_W^3)^+(\gamma_W^2)^+(\gamma_W^1)^+(\gamma_W^0)^+$$

$$= i\gamma_W^3\gamma_W^2\gamma_W^1\gamma_W^0$$

(using (1.141 i)) and (1.141 iii)))

$$= i(\gamma_W^0\gamma_W^1\gamma_W^2\gamma_W^3)$$

$$= \gamma_W^5$$

where we made frequent use of (1.137) .

iii) Using (1.136) and (1.90) we have

$$(\gamma_W^i)^+ = \begin{pmatrix} 0 & \sigma^i \\ \bar{\sigma}^i & 0 \end{pmatrix}^+ = \begin{pmatrix} 0 & \bar{\sigma}^{i+} \\ \sigma^{i+} & 0 \end{pmatrix}$$

$$= \begin{pmatrix} 0 & -\sigma^{i+} \\ \sigma^{i+} & 0 \end{pmatrix} = \begin{pmatrix} 0 & -\sigma^i \\ \sigma^i & 0 \end{pmatrix}$$

$$= -\begin{pmatrix} 0 & \sigma^i \\ \bar{\sigma}^i & 0 \end{pmatrix}$$

$$= -\gamma_W^i, \quad i = 1,2,3.$$

We now verify (1.141) in terms of our index notation. First we have to clarify what we mean by transposition and hermitian conjugation of γ matrices in this context. The hermitian conjugate of $(\gamma_W^{\mu})_{ab}$ is given by

$$(\gamma_W^{\mu +})_{ab} = \begin{pmatrix} 0 & \bar{\sigma}^{\mu +}{}_{A\dot{B}} \\ \sigma^{\mu +\dot{A}B} & 0 \end{pmatrix}$$

with

$$(\gamma_W^{\mu})_{ab} = \begin{pmatrix} 0 & \sigma^{\mu}{}_{A\dot{B}} \\ \bar{\sigma}^{\mu\dot{A}B} & 0 \end{pmatrix} \tag{1.142}$$

This can be verified as follows. From Dirac theory we know that in any representation the matrix γ^0 has the property

$$\gamma^0\gamma^\mu\gamma^0 = \gamma^{\mu +}$$

In the Weyl representation this is (see (1.136).)

$$(\gamma_W^0)_{ab}(\gamma_W^\mu)_{bc}(\gamma_W^0)_{cd}$$

$$= \begin{pmatrix} 0 & \sigma^0{}_{A\dot{B}} \\ \bar{\sigma}^{0\dot{A}B} & 0 \end{pmatrix}\begin{pmatrix} 0 & \sigma^\mu{}_{B\dot{C}} \\ \bar{\sigma}^{\mu\dot{B}C} & 0 \end{pmatrix}\begin{pmatrix} 0 & \sigma^0{}_{C\dot{D}} \\ \bar{\sigma}^{0\dot{C}D} & 0 \end{pmatrix}$$

$$= \begin{pmatrix} 0 & \sigma^0_{A\dot{B}}\,\bar{\sigma}^{\mu\dot{B}C}\,\sigma^0{}_{C\dot{D}} \\ \bar{\sigma}^{0\dot{A}B}\,\sigma^\mu_{B\dot{C}}\,\bar{\sigma}^{0\dot{C}D} & 0 \end{pmatrix}$$

$$= \begin{pmatrix} 0 & (\sigma^0\bar{\sigma}^\mu\sigma^0)_{A\dot{D}} \\ (\bar{\sigma}^0\sigma^\mu\bar{\sigma}^0)^{\dot{A}D} & 0 \end{pmatrix}$$

$$=: \begin{pmatrix} 0 & (\bar{\sigma}^\mu)_{A\dot{D}} \\ (\sigma^\mu)^{\dot{A}D} & 0 \end{pmatrix}$$

where we define

$$(\bar{\sigma}^{\mu})_{A\dot{D}} := \sigma^{0}_{A\dot{B}}\, \bar{\sigma}^{\mu\,\dot{B}C}\, \sigma^{0}_{C\dot{D}}$$

$$(\sigma^{\mu})^{\dot{A}D} := \bar{\sigma}^{0\,\dot{A}B}\, \sigma^{\mu}_{B\dot{C}}\, \bar{\sigma}^{0\,\dot{C}D}$$

This definition is consistent with our discussion following (1.134). Using now the hermiticity of the Pauli matrices we obtain

$$(\gamma^{0}_{W})_{ab}(\gamma^{\mu}_{W})_{bc}(\gamma^{0}_{W})_{cd} = \begin{pmatrix} 0 & \bar{\sigma}^{\mu+}{}_{A\dot{D}} \\ \sigma^{\mu+\,\dot{A}D} & 0 \end{pmatrix} = (\gamma^{\mu+}_{W})_{ad}$$

The consistency means that $\gamma^{\mu+}_{W}$ as γ matrices in the Weyl representation act on Dirac spinors (1.133):

$$(\gamma^{\mu+}_{W})_{ab}\, \Psi_{b} = \begin{pmatrix} 0 & \bar{\sigma}^{\mu+}{}_{A\dot{B}} \\ \sigma^{\mu+\,\dot{A}B} & 0 \end{pmatrix} \begin{pmatrix} \phi_{B} \\ \bar{\psi}^{\dot{B}} \end{pmatrix}$$

$$= \begin{pmatrix} \bar{\sigma}^{\mu+}{}_{A\dot{B}}\, \bar{\psi}^{\dot{B}} \\ \sigma^{\mu+\,\dot{A}B}\, \phi_{B} \end{pmatrix}$$

The upper right 2x2-submatrix must always have the index structure exhibited in (1.142) since by construction this submatrix maps elements from $\dot{F}^{*}$ into F, whereas the lower left submatrix maps F into $\dot{F}^{*}$.

It is now easy to demonstrate (1.141) in our index notation.

i) For $\mu = 0$ we have

$$(\gamma^{0+}_{W})_{ab} = \begin{pmatrix} 0 & \bar{\sigma}^{0+}{}_{A\dot{B}} \\ \sigma^{0+\,\dot{A}B} & 0 \end{pmatrix} = \begin{pmatrix} 0 & \bar{\sigma}^{0}{}_{A\dot{B}} \\ \sigma^{0\,\dot{A}B} & 0 \end{pmatrix}$$

$$= \begin{pmatrix} 0 & \sigma^{0}_{A\dot{B}} \\ \bar{\sigma}^{0\,\dot{A}B} & 0 \end{pmatrix} = (\gamma^{0}_{W})_{ab}$$

ii) μ = i (=1,2,3):

$$\left(\gamma_W^{i+}\right)_{ab} = \begin{pmatrix} 0 & (\bar{\sigma}^{i+})_{A\dot{B}} \\ (\sigma^{i+})^{\dot{A}B} & 0 \end{pmatrix}$$

$$\begin{pmatrix} 0 & (\bar{\sigma}^{i})_{A\dot{B}} \\ (\sigma^{i})^{\dot{A}B} & 0 \end{pmatrix} = -\begin{pmatrix} 0 & (\sigma^{i})_{A\dot{B}} \\ (\bar{\sigma}^{i})^{\dot{A}B} & 0 \end{pmatrix} = -\left(\gamma_W^{i}\right)_{ab}$$

with the help of (1.90) and (1.142). The Weyl representation is a useful representation when studying the extreme relativistic limit of Dirac theory.

The Dirac equation is

$$(i\partial\!\!\!/ - m)\Psi = (i\gamma^\mu\partial_\mu - m\,1_{4\times4})\Psi = 0 \tag{1.143a}$$

In the Weyl representation we have

$$(i\gamma_W^\mu\partial_\mu - m)\Psi_W = 0 \tag{1.143b}$$

where Ψ_W is a Dirac four-spinor in the Weyl representation as in (1.130). In the extreme relativistic limit, i.e. when m = 0, we obtain

$$\gamma_W^\mu\partial_\mu\Psi_W = 0 \tag{1.144a}$$

which can be written,using (1.136) and (1.130)

$$\begin{pmatrix} 0 & (\sigma^\mu\partial_\mu)_{A\dot{B}} \\ (\bar{\sigma}^\mu\partial_\mu)^{\dot{A}B} & 0 \end{pmatrix}\begin{pmatrix} \phi_B \\ \Psi^{\dot{B}} \end{pmatrix} = 0$$

and therefore the four-by-four matrix equation (1.143) decouples into two two-by-two matrix equations

$$i(\sigma^\mu\partial_\mu)\Psi - m\phi = 0$$

(1.144b)

$$i(\bar{\sigma}^{\mu}\partial_{\mu})\phi - m\bar{\Psi} = 0 \qquad (1.144b)$$

i.e. for m = 0 we have

$$\left.\begin{aligned}(\sigma^{\mu}\partial_{\mu})_{A\dot{B}}\,\bar{\Psi}^{\dot{B}} &= 0\\ (\bar{\sigma}^{\mu}\partial_{\mu})^{\dot{A}B}\,\phi_{B} &= 0\end{aligned}\right\} \qquad (1.145)$$

These equations can be written in a compact form as

$$\begin{aligned}(i\partial_t + \vec{\sigma}\cdot\vec{p})\bar{\Psi} &= 0\\ (i\partial_t - \vec{\sigma}\cdot\vec{p})\phi &= 0\end{aligned} \qquad (1.146)$$

Equations (1.146) are wave equations for spin 1/2 particles and are called Weyl equations. From previous considerations it is clear that these equations are not invariant under the parity transformation, whereas the Dirac equation is.

As a result of experiments one usually calls massless neutrinos left-handed and massless antineutrinos right-handed (this is a convention). Furthermore, neutrinos are spin 1/2 particles obeying Fermi-Dirac statistics.

We conclude therefore that we can describe massless neutrinos by two-component Weyl spinors, transforming under SL(2, C) according to the (1/2, 0) representation ("left-handed Weyl spinors"), and massless antineutrinos are represented by two-component Weyl spinors transforming under SL(2, C) according to the (0, 1/2) representation ("right-handed Weyl spinors").

In order to link the Weyl equations with helicity eigenstates, we observe that plane wave solutions of the first of equations (1.146) for positive energy eigenstates, proportional to exp(ipx) where $E = |\vec{p}|$ (massless case !), satisfy

$$\vec{\sigma}\cdot\hat{\vec{p}}\;\Psi = \Psi$$

where
$$\hat{\vec{p}} := \frac{\vec{p}}{E} = \frac{\vec{p}}{|\vec{p}|}$$

In order to see this consider $\Psi \sim \exp(ipx)$. Then

$$i\partial_t \Psi = -p_0 \Psi = -E\Psi = -|\vec{p}|\Psi$$

Inserting this into the first Weyl equation we obtain

$$-E\Psi + \vec{\sigma}\cdot\vec{p}\;\Psi = 0$$

which is equivalent to

$$\frac{1}{2}\vec{\sigma}\cdot\hat{\vec{p}}\;\Psi = +\frac{1}{2}\Psi \tag{1.147}$$

Likewise the plane-wave eigensolution of the second Weyl equation for positive energy eigenstates, proportional to $\exp(-ipx)$ where $p^0 = |\vec{p}| = E$, obeys

$$\frac{1}{2}\vec{\sigma}\cdot\hat{\vec{p}}\;\phi = -\frac{1}{2}\phi \tag{1.148}$$

In equation (1.44) we defined the helicity operator as the projection of total angular momentum onto the momentum direction. The eigenvalues of this operator determine the helicity of the state. As shown in Section 1.2, helicity is a Poincaré invariant quantity; its eigenvalues λ determine various representations for the massless case. The sign of the eigenvalue λ determines the polarization state of particles with spin $|\lambda|$. The invariance of the helicity operator under Lorentz transformations implies that left-handed neutrinos are left-handed in any inertial system (left-handed means $\lambda = -1/2$, right-handed means $\lambda = +1/2$). Hence we conclude that ϕ is an eigenfunction of the helicity operator with eigenvalue $-1/2$, and Ψ is an eigenfunction of this operator with eigenvalue $+1/2$.

1.4.2 The Canonical Basis or Dirac Representation

The canonical basis for Dirac matrices is defined by

$$\gamma^0_D := \begin{pmatrix} 1_{2\times 2} & 0 \\ 0 & -1_{2\times 2} \end{pmatrix}, \quad \gamma^i_D := \begin{pmatrix} 0 & \sigma^i \\ \bar{\sigma}^i & 0 \end{pmatrix}$$

$$\gamma^5_D := \begin{pmatrix} 0 & \sigma^0 \\ \bar{\sigma}^0 & 0 \end{pmatrix} \tag{1.149}$$

Proposition: The Weyl representation (1.136) and the Dirac representation (1.149) are connected by a similarity transformation, i.e.

$$\Gamma_W = X \Gamma_D X^{-1}, \quad X \in GL(4,C) \tag{1.150a}$$

where Γ_W is any γ matrix in the Weyl representation and Γ_D is any γ matrix in the Dirac representation. Furthermore

$$X = \frac{1}{2^{1/2}} \begin{pmatrix} -1_{2\times 2} & \sigma^0 \\ -\bar{\sigma}^0 & -1_{2\times 2} \end{pmatrix}, \quad X^{-1} = \frac{1}{2^{1/2}} \begin{pmatrix} -1_{2\times 2} & -\sigma^0 \\ \bar{\sigma}^0 & -1_{2\times 2} \end{pmatrix} \tag{1.150b}$$

Proof: Since the γ matrices in the Weyl representation and also those in the canonical basis obey the same Clifford algebra relation (1.137), they are related by a similarity transformation. We first verify (1.150b) as an exercise in the handling of indices. Thus

$$X_{ab}X^{-1}_{bc} = \frac{1}{2}\begin{pmatrix} -\delta_A{}^B & \sigma^0_{A\dot{B}} \\ -\bar{\sigma}^{0\dot{A}B} & -\delta^{\dot{A}}{}_{\dot{B}} \end{pmatrix}\begin{pmatrix} -\delta_B{}^C & -\sigma^0_{B\dot{C}} \\ \bar{\sigma}^{0\dot{B}C} & -\delta^{\dot{B}}{}_{\dot{C}} \end{pmatrix}$$

$$= \frac{1}{2}\begin{pmatrix} \delta_A{}^B\delta_B{}^C + \sigma^0_{A\dot{B}}\bar{\sigma}^{0\dot{B}C} & \delta_A{}^B\sigma^0_{B\dot{C}} - \sigma^0_{A\dot{B}}\,\delta^{\dot{B}}{}_{\dot{C}} \\ \bar{\sigma}^{0\dot{A}B}\delta_B{}^C - \delta^{\dot{A}}{}_{\dot{B}}\,\bar{\sigma}^{0\dot{B}C} & \bar{\sigma}^{0\dot{A}B}\sigma^0_{B\dot{C}} + \delta^{\dot{A}}{}_{\dot{B}}\,\delta^{\dot{B}}{}_{\dot{C}} \end{pmatrix}$$

$$= \frac{1}{2}\begin{pmatrix} \delta_A{}^C + \delta_A{}^C & \sigma^0_{A\dot{C}} - \sigma^0_{A\dot{C}} \\ \bar{\sigma}^{0\dot{A}C} - \bar{\sigma}^{0\dot{A}C} & \delta^{\dot{A}}{}_{\dot{C}} + \delta^{\dot{A}}{}_{\dot{C}} \end{pmatrix}$$

$$= \begin{pmatrix} \delta_A{}^C & 0 \\ 0 & \delta^{\dot{A}}{}_{\dot{C}} \end{pmatrix} = (1_{4\times4})_{ac}$$

We now calculate the Dirac representation of the γ matrices from those of the Weyl representation (1.150a). Now

$$\Gamma_W = \gamma^0_W = \begin{pmatrix} 0 & \sigma^0 \\ \bar{\sigma}^0 & 0 \end{pmatrix}$$

and so

$$(\gamma^0_D)_{ad} = (X^{-1})_{ab}\,(\gamma^0_W)_{bc}\,(X)_{cd}$$

$$= \frac{1}{2}\begin{pmatrix} -\delta_A{}^B & -\sigma^0_{A\dot{B}} \\ \bar{\sigma}^{0\dot{A}B} & -\delta^{\dot{A}}{}_{\dot{B}} \end{pmatrix}\begin{pmatrix} 0 & \sigma^0_{B\dot{C}} \\ \bar{\sigma}^{0\dot{B}C} & 0 \end{pmatrix}\begin{pmatrix} -\delta_C{}^D & \sigma^0_{C\dot{D}} \\ -\bar{\sigma}^{0\dot{C}D} & -\delta^{\dot{C}}{}_{\dot{D}} \end{pmatrix}$$

$$= \frac{1}{2}\begin{pmatrix} \delta_A{}^B \sigma^0_{B\dot{C}} \bar{\sigma}^{0\dot{C}D} + \sigma^0_{A\dot{B}} \bar{\sigma}^{0\dot{B}C} \delta_C{}^D & \delta_A{}^B \sigma^0_{B\dot{C}} \delta^{\dot{C}}{}_{\dot{D}} - \sigma^0_{A\dot{B}} \bar{\sigma}^{0\dot{B}C} \sigma^0_{C\dot{D}} \\ -\bar{\sigma}^{0\dot{A}B} \sigma^0_{B\dot{C}} \bar{\sigma}^{0\dot{C}D} + \delta^{\dot{A}}{}_{\dot{B}} \bar{\sigma}^{0\dot{B}C} \delta_C{}^D & -\bar{\sigma}^{0\dot{A}B} \sigma^0_{B\dot{C}} \delta^{\dot{C}}{}_{\dot{D}} - \delta^{\dot{A}}{}_{\dot{B}} \bar{\sigma}^{0\dot{B}C} \sigma^0_{C\dot{D}} \end{pmatrix}$$

$$= \frac{1}{2}\begin{pmatrix} \sigma^0_{A\dot{C}} \bar{\sigma}^{0\dot{C}D} + \sigma^0_{A\dot{B}} \bar{\sigma}^{0\dot{B}D} & \sigma^0_{A\dot{D}} - \sigma^0_{A\dot{B}} \bar{\sigma}^{0\dot{B}C} \sigma^0_{C\dot{D}} \\ -\bar{\sigma}^{0\dot{A}B} \sigma^0_{B\dot{C}} \bar{\sigma}^{0\dot{C}D} + \bar{\sigma}^{0\dot{A}D} & -\bar{\sigma}^{0\dot{A}B} \sigma^0_{B\dot{D}} - \bar{\sigma}^{0\dot{A}C} \sigma^0_{C\dot{D}} \end{pmatrix}$$

$$= \frac{1}{2}\begin{pmatrix} \delta_A{}^D + \delta_A{}^D & \sigma^0_{A\dot{D}} - \sigma^0_{A\dot{B}} \delta^{\dot{B}}{}_{\dot{D}} \\ -\bar{\sigma}^{0\dot{A}B} \delta_B{}^D + \bar{\sigma}^{0\dot{A}D} & -\delta^{\dot{A}}{}_{\dot{D}} - \delta^{\dot{A}}{}_{\dot{D}} \end{pmatrix}$$

$$= \begin{pmatrix} \delta_A{}^D & 0 \\ 0 & -\delta^{\dot{A}}{}_{\dot{D}} \end{pmatrix} = \begin{pmatrix} 1_{2\times 2} & 0 \\ 0 & -1_{2\times 2} \end{pmatrix}_{ad}$$

where we used the relations

$$(\bar{\sigma}^0)^{\dot{A}B} (\sigma^0)_{B\dot{C}} = \delta^{\dot{A}}{}_{\dot{C}}$$

$$(\sigma^0)_{A\dot{B}} (\bar{\sigma}^0)^{\dot{B}C} = \delta_A{}^C$$

These can be understood as follows. $\bar{\sigma}^0$ is a map, the unit map, from F to $\dot{F}^*$ and σ^0 is the unit map from $\dot{F}^*$ to F. Therefore the combined expression $(\bar{\sigma}^0 \sigma^0)^{\dot{A}}{}_{\dot{C}}$ is a map starting from $\dot{F}^*$ and going to F and back to $\dot{F}^*$. Since $\bar{\sigma}^0$ and σ^0 are unit matrices, an expression like $\bar{\sigma}^0 \sigma^0$ gives the unit map in $\dot{F}^*$; hence

$$(\bar{\sigma}^0 \sigma^0)^{\dot{A}}{}_{\dot{B}} = \delta^{\dot{A}}{}_{\dot{B}} = 1_{\dot{F}^*}$$

and a corresponding consideration applies to $\sigma^0 \bar{\sigma}^0 = 1_F$.

Next we transform γ^i_W, i = 1,2,3 , into the Dirac representation, i.e.

$$\gamma^i_D = X^{-1} \gamma^i_W X$$

In components

$$(\gamma^i_D)_{ad} = (X^{-1})_{ab} (\gamma^i_W)_{bc} (X)_{cd}$$

$$= \frac{1}{2}\begin{pmatrix} -\delta_A{}^B & \sigma^0_{A\dot{B}} \\ \bar{\sigma}^{0\dot{A}B} & -\delta^{\dot{A}}{}_{\dot{B}} \end{pmatrix}\begin{pmatrix} 0 & \sigma^i_{B\dot{C}} \\ \bar{\sigma}^{i\dot{B}C} & 0 \end{pmatrix}\begin{pmatrix} -\delta_C{}^D & \sigma^0_{C\dot{D}} \\ -\bar{\sigma}^{0\dot{C}D} & -\delta^{\dot{C}}{}_{\dot{D}} \end{pmatrix}$$

$$= \frac{1}{2}\begin{pmatrix} \delta_A{}^B \sigma^i_{B\dot{C}} \bar{\sigma}^{0\dot{C}D} + \sigma^0_{A\dot{B}} \bar{\sigma}^{i\dot{B}D} & \sigma^i_{A\dot{D}} - \sigma^0_{A\dot{B}} \bar{\sigma}^{i\dot{B}C} \sigma^0_{C\dot{D}} \\ -\bar{\sigma}^{0\dot{A}B} \sigma^i_{B\dot{C}} \bar{\sigma}^{0\dot{C}D} + \bar{\sigma}^{i\dot{A}D} & -\bar{\sigma}^{0\dot{A}B} \sigma^i_{B\dot{D}} - \bar{\sigma}^{i\dot{A}C} \sigma^0_{C\dot{D}} \end{pmatrix}$$

$$= \frac{1}{2}\begin{pmatrix} 0 & \sigma^i_{A\dot{D}} - \bar{\sigma}^i_{A\dot{D}} \\ -\sigma^{i\dot{A}D} + \bar{\sigma}^{i\dot{A}D} & 0 \end{pmatrix}$$

(with (1.90))

$$= \begin{pmatrix} 0 & \sigma^{i} \\ \bar{\sigma}^{i} & 0 \end{pmatrix}_{ad}$$

where we used (with (1.90))

$$\sigma^{i}_{A\dot{C}}\,\bar{\sigma}^{0\,\dot{C}D} = \sigma^{i}{}_{A}{}^{D}$$

$$\sigma^{0}_{A\dot{B}}\,\bar{\sigma}^{i\,\dot{B}D} = \bar{\sigma}^{i}{}_{A}{}^{D} = -\sigma^{i}{}_{A}{}^{D}$$

$$-\bar{\sigma}^{0\dot{A}B}\,\sigma^{i}_{B\dot{C}}\,\bar{\sigma}^{0\,\dot{C}D} = -\sigma^{i\dot{A}D} = \bar{\sigma}^{i\dot{A}D}$$

and similar formulae. Taking into account that σ^{0} and $\bar{\sigma}^{0}$ are unit matrices which transform dotted indices into undotted indices and vice versa.

Finally we demonstrate that

$$\gamma^{5}_{D} = X^{-1}\gamma^{5}_{W}X$$

We have

$$(\gamma^{5}_{D})_{ad} = (X^{-1})_{ab}(\gamma^{5}_{W})_{bc}(X)_{cd}$$

$$= \frac{1}{2}\begin{pmatrix} -\delta_{A}{}^{B} & -\sigma^{0}_{A\dot{B}} \\ \bar{\sigma}^{0\dot{A}B} & -\delta^{\dot{A}}{}_{\dot{B}} \end{pmatrix}\begin{pmatrix} -\delta_{B}{}^{C} & 0 \\ 0 & \delta^{\dot{B}}{}_{\dot{C}} \end{pmatrix}\begin{pmatrix} -\delta_{C}{}^{D} & \sigma^{0}_{C\dot{D}} \\ -\bar{\sigma}^{0\dot{C}D} & -\delta^{\dot{C}}{}_{\dot{D}} \end{pmatrix}$$

$$= \frac{1}{2}\begin{pmatrix} -\delta_{A}{}^{B}\delta_{B}{}^{C}\delta_{C}{}^{D} + \sigma^{0}_{A\dot{B}}\,\delta^{\dot{B}}{}_{\dot{C}}\,\bar{\sigma}^{0\dot{C}D} & \delta_{A}{}^{B}\delta_{B}{}^{C}\sigma^{0}_{C\dot{D}} + \sigma^{0}_{A\dot{B}}\,\delta^{\dot{B}}{}_{\dot{C}}\,\delta^{\dot{C}}{}_{\dot{D}} \\ \bar{\sigma}^{0\dot{A}B}\delta_{B}{}^{C}\delta_{C}{}^{D} + \delta^{\dot{A}}{}_{\dot{B}}\,\delta^{\dot{B}}{}_{\dot{C}}\,\bar{\sigma}^{0\dot{C}D} & -\bar{\sigma}^{0\dot{A}B}\delta_{B}{}^{C}\sigma^{0}_{C\dot{D}} + \delta^{\dot{A}}{}_{\dot{B}}\,\delta^{\dot{B}}{}_{\dot{C}}\,\delta^{\dot{C}}{}_{\dot{D}} \end{pmatrix}$$

$$= \frac{1}{2}\begin{pmatrix} -\delta_A{}^D + \delta_A{}^D & \sigma^0_{A\dot{D}} + \sigma^0_{A\dot{D}} \\ \bar{\sigma}^{0\dot{A}D} + \bar{\sigma}^{0\dot{A}D} & -\delta^{\dot{A}}{}_{\dot{D}} + \delta^{\dot{A}}{}_{\dot{D}} \end{pmatrix}$$

$$= \begin{pmatrix} 0 & \sigma^0 \\ \bar{\sigma}^0 & 0 \end{pmatrix}_{ad}$$

The canonical basis or Dirac representation of γ^μ matrices has the unique property of all possible representations that it diagonalizes the energy via the matrix

$$\gamma^0_D = \begin{pmatrix} 1 & 0 \\ 0 & -1 \end{pmatrix}$$

in the nonrelativistic limit.

1.4.3 The Majorana Representation

Of all possible equivalent representations of γ^μ matrices obtained by a nonsingular transformation $\gamma^\mu \to X\gamma^\mu X^{-1}$ the Majorana representation plays a particular role. It is constructed so as to make the Dirac equation real. This can be seen as follows. In the form originally proposed by Dirac the Dirac equation is[40]

$$\left(i\frac{\partial}{\partial t} - \frac{1}{i}\vec{\alpha}\cdot\vec{\nabla} - \beta m\right)\Psi = 0 \tag{1.151}$$

where α^i, i = 1,2,3,and β are hermitian 4x4 matrices. These matrices are related to the γ^μ matrices by

$$\gamma^0 = \beta\,, \quad \gamma^i = \beta\alpha^i, \quad i = 1,2,3 \tag{1.152}$$

In the Dirac representation (1.149) we have

$$\beta = \begin{pmatrix} 1_{2\times2} & 0 \\ 0 & -1_{2\times2} \end{pmatrix},\ \alpha^i = \begin{pmatrix} 0 & \sigma^i \\ -\bar{\sigma}^i & 0 \end{pmatrix} \tag{1.153}$$

Hence in order to satisfy the reality property we multiply (1.151) by -i and obtain

$$\left(\frac{\partial}{\partial t} + \vec{\hat{\alpha}}\cdot\vec{\nabla} + i\hat{\beta}m\right)\psi = 0 \tag{1.154}$$

This is a real equation if and only if $\vec{\hat{\alpha}}$ are real 4x4 matrices and if $\hat{\beta}$ is purely imaginary. Hence if we set

$$\hat{\beta} = \alpha^2, \quad \hat{\alpha}^1 = -\alpha^1, \quad \hat{\alpha}^2 = \beta, \quad \hat{\alpha}^3 = -\alpha^3$$

then (1.154) is real. Following the prescription (1.152) we obtain the γ matrices in the Majorana representation:

$$\gamma^0_M = \hat{\beta} = \begin{pmatrix} 0 & \sigma^2 \\ -\sigma^2 & 0 \end{pmatrix}, \quad \gamma^1_M = \hat{\beta}\hat{\alpha}^1 = \begin{pmatrix} i\sigma^3 & 0 \\ 0 & i\sigma^3 \end{pmatrix}$$

$$\gamma^2_M = \hat{\beta}\hat{\alpha}^2 = \begin{pmatrix} 0 & -\sigma^2 \\ -\sigma^2 & 0 \end{pmatrix}, \quad \gamma^3_M = \hat{\beta}\hat{\alpha}^3 = \begin{pmatrix} -i\sigma^1 & 0 \\ 0 & -i\sigma^1 \end{pmatrix} \tag{1.155}$$

and

$$\gamma^5_M = i\gamma^0_M\gamma^1_M\gamma^2_M\gamma^3_M = \begin{pmatrix} \sigma^2 & 0 \\ 0 & -\sigma^2 \end{pmatrix} \tag{1.156}$$

<u>Proposition</u>: The Dirac representation and the Majorana representation are connected by the following similarity transformation

$$\Gamma_M = Y\,\Gamma_D\,Y^{-1} \tag{1.157}$$

where Γ_M is any γ matrix in the Majorana representation and Γ_D is the corresponding γ matrix in the Dirac representation (1.149). Furthermore

$$Y = \frac{1}{2^{1/2}}\begin{pmatrix} 1_{2\times 2} & \sigma^2 \\ -\sigma^2 & -1_{2\times 2} \end{pmatrix} = Y^{-1} \tag{1.158}$$

Proof: We verify that

$$Y_{ab} Y^{-1}_{bc} = \delta_{ac}$$

Thus

$$Y_{ab} Y^{-1}_{bc}$$

$$= \frac{1}{2}\begin{pmatrix} \delta_A{}^B & \sigma^2_{A\dot{B}} \\ -\bar{\sigma}^{2\dot{A}B} & -\delta^{\dot{A}}{}_{\dot{B}} \end{pmatrix}\begin{pmatrix} \delta_B{}^C & \sigma^2_{B\dot{C}} \\ -\bar{\sigma}^{2\dot{B}C} & -\delta^{\dot{B}}{}_{\dot{C}} \end{pmatrix}$$

$$= \frac{1}{2}\begin{pmatrix} \delta_A{}^B \delta_B{}^C - \sigma^2_{A\dot{B}}\bar{\sigma}^{2\dot{B}C}, & \delta_A{}^B \sigma^2_{B\dot{C}} - \sigma^2_{A\dot{B}}\delta^{\dot{B}}{}_{\dot{C}} \\ -\bar{\sigma}^{2\dot{A}B}\delta_B{}^C + \delta^{\dot{A}}{}_{\dot{B}}\bar{\sigma}^{2\dot{B}C}, & -\bar{\sigma}^{2\dot{A}B}\sigma^2_{B\dot{C}} + \delta^{\dot{A}}{}_{\dot{B}}\delta^{\dot{B}}{}_{\dot{C}} \end{pmatrix}$$

$$= \frac{1}{2}\begin{pmatrix} \delta_A{}^C - (\sigma^2\bar{\sigma}^2)_A{}^C & \sigma^2_{A\dot{C}} - \sigma^2_{A\dot{C}} \\ -\bar{\sigma}^{2\dot{A}C} + \bar{\sigma}^{2\dot{A}C} & -(\bar{\sigma}^2\sigma^2)^{\dot{A}}{}_{\dot{C}} + \delta^{\dot{A}}{}_{\dot{C}} \end{pmatrix}$$

Using $\bar{\sigma}^2 = -\sigma^2$ (as matrices) and $(\sigma^2)^2 = 1$ we obtain

$$Y_{ab} Y^{-1}_{bc} = \frac{1}{2}\begin{pmatrix} \delta_A{}^C + \delta_A{}^C & 0 \\ 0 & \delta^{\dot{A}}{}_{\dot{C}} + \delta^{\dot{A}}{}_{\dot{C}} \end{pmatrix} = \delta_{ac}$$

We now consider the individual γ matrices.

$\underline{\mu = 0}$:

$$(\gamma^0_M)_{ad} = Y_{ab}(\gamma^0_D)_{bc} Y^{-1}_{cd}$$

$$= \frac{1}{2}\begin{pmatrix} \delta_A{}^B & \sigma^2_{A\dot{B}} \\ -\bar{\sigma}^{2\dot{A}B} & -\delta^{\dot{A}}{}_{\dot{B}} \end{pmatrix}\begin{pmatrix} \delta_B{}^C & 0 \\ 0 & -\delta^{\dot{B}}{}_{\dot{C}} \end{pmatrix}\begin{pmatrix} \delta_C{}^D & \sigma^2_{C\dot{D}} \\ -\bar{\sigma}^{2\dot{C}D} & -\delta^{\dot{C}}{}_{\dot{D}} \end{pmatrix}$$

$$= \frac{1}{2}\begin{pmatrix} \delta_A{}^D + \sigma^2_{A\dot{C}}\bar{\sigma}^{2\dot{C}D} & \sigma^2_{A\dot{D}} + \sigma^2_{A\dot{D}} \\ -\bar{\sigma}^{2\dot{A}D} - \bar{\sigma}^{2\dot{A}D} & -\bar{\sigma}^{2\dot{A}C}\sigma^2_{C\dot{D}} - \delta^{\dot{A}}{}_{\dot{D}} \end{pmatrix}$$

$$= \begin{pmatrix} 0 & \sigma^2_{A\dot{D}} \\ -\bar{\sigma}^{2\dot{A}D} & 0 \end{pmatrix} = \begin{pmatrix} 0 & \sigma^2 \\ -\bar{\sigma}^2 & 0 \end{pmatrix}_{ad}$$

again using (1.90), i.e. $\bar{\sigma}^2 = -\sigma^2$ and $(\sigma^2)^2 = 1_{2\times 2}$.

$\underline{\mu = 1}$:

$$(\gamma^1_M)_{ad} = Y_{ab}\,(\gamma^1_D)_{bc}\,(Y^{-1})_{cd}$$

$$= \frac{1}{2}\begin{pmatrix} \delta_A{}^B & \sigma^2_{A\dot{B}} \\ -\bar{\sigma}^{2\dot{A}B} & -\delta^{\dot{A}}{}_{\dot{B}} \end{pmatrix} \begin{pmatrix} 0 & \sigma^1_{B\dot{C}} \\ \bar{\sigma}^{1\dot{B}C} & 0 \end{pmatrix} \begin{pmatrix} \delta_C{}^D & \sigma^2_{C\dot{D}} \\ -\bar{\sigma}^{2\dot{C}D} & -\delta^{\dot{C}}{}_{\dot{D}} \end{pmatrix}$$

$$= \frac{1}{2}\begin{pmatrix} \sigma^2_{A\dot{B}}\bar{\sigma}^{1\dot{B}D} - \sigma^1_{A\dot{C}}\bar{\sigma}^{2\dot{C}D} & \sigma^2_{A\dot{B}}\bar{\sigma}^{1\dot{B}C}\sigma^2_{C\dot{D}} - \sigma^1_{A\dot{D}} \\ -\bar{\sigma}^{1\dot{A}D} + \bar{\sigma}^{2\dot{A}B}\sigma^1_{B\dot{C}}\bar{\sigma}^{2\dot{C}D} & -\bar{\sigma}^{1\dot{A}C}\sigma^2_{C\dot{D}} + \bar{\sigma}^{2\dot{A}B}\sigma^1_{B\dot{D}} \end{pmatrix}$$

$$= \frac{1}{2}\begin{pmatrix} (\sigma^2\bar{\sigma}^1)_A{}^D - (\sigma^1\bar{\sigma}^2)_A{}^D, & (\sigma^2\bar{\sigma}^1\sigma^2)_{A\dot{D}} - \sigma^1_{A\dot{D}} \\ -\bar{\sigma}^{1\dot{A}D} + (\bar{\sigma}^2\sigma^1\bar{\sigma}^2)^{\dot{A}D}, & -(\bar{\sigma}^1\sigma^2)^{\dot{A}}{}_{\dot{D}} + (\bar{\sigma}^2\sigma^1)^{\dot{A}}{}_{\dot{D}} \end{pmatrix}$$

$$= \frac{1}{2}\begin{pmatrix} (\sigma^1\sigma^2 - \sigma^2\sigma^1)_A{}^D & (\sigma^2\bar{\sigma}^1\sigma^2 - \sigma^1)_{A\dot{D}} \\ (-\bar{\sigma}^1 + \bar{\sigma}^2\sigma^1\bar{\sigma}^2)^{\dot{A}D} & (\sigma^1\sigma^2 - \sigma^2\sigma^1)^{\dot{A}}{}_{\dot{D}} \end{pmatrix}$$

$$= \frac{1}{2}\begin{pmatrix} 2i(\sigma^3)_A{}^D & -(\sigma^2\sigma^1\sigma^2 + \sigma^1)_{A\dot{D}} \\ (\sigma^1 + \sigma^2\sigma^1\sigma^2)^{\dot{A}D} & 2i(\sigma^3)^{\dot{A}}{}_{\dot{D}} \end{pmatrix}$$

$$= \frac{1}{2}\begin{pmatrix} 2i\,\sigma^{3}{}_{A}{}^{D} & (\sigma^1 - \sigma^1)_{A\dot{D}} \\ (\sigma^1 - \sigma^1)^{\dot{A}D} & 2i\,\sigma^{3}{}^{\dot{A}}{}_{\dot{D}} \end{pmatrix}$$

$$= \begin{pmatrix} i\sigma^3 & 0 \\ 0 & i\sigma^3 \end{pmatrix}_{ad}$$

$\underline{\mu = 2}$:

$$(\gamma^2_M)_{ad} = Y_{ab}(\gamma^2_D)_{bc}Y^{-1}_{cd}$$

$$= \frac{1}{2}\begin{pmatrix} \delta_A{}^B & \sigma^2_{A\dot{B}} \\ -\bar{\sigma}^{2\dot{A}B} & -\delta^{\dot{A}}{}_{\dot{B}} \end{pmatrix}\begin{pmatrix} 0 & \sigma^2_{B\dot{C}} \\ \bar{\sigma}^{2\dot{B}C} & 0 \end{pmatrix}\begin{pmatrix} \delta_C{}^D & \sigma^2_{C\dot{D}} \\ -\bar{\sigma}^{2\dot{C}D} & -\delta^{\dot{C}}{}_{\dot{D}} \end{pmatrix}$$

$$= \frac{1}{2}\begin{pmatrix} \sigma^2_{A\dot{B}}\bar{\sigma}^{2\dot{B}D} - \sigma^2_{A\dot{C}}\bar{\sigma}^{2\dot{C}D} & \sigma^2_{A\dot{B}}\bar{\sigma}^{2\dot{B}C}\sigma^2_{C\dot{D}} - \sigma^2_{A\dot{D}} \\ -\bar{\sigma}^{2\dot{A}D} + \bar{\sigma}^{2\dot{A}B}\sigma^2_{B\dot{C}}\bar{\sigma}^{2\dot{C}D} & -\bar{\sigma}^{2\dot{A}C}\sigma^2_{C\dot{D}} + \bar{\sigma}^{2\dot{A}B}\sigma^2_{B\dot{D}} \end{pmatrix}$$

$$= \frac{1}{2}\begin{pmatrix} 0 & -(\sigma^2 - \sigma^2\bar{\sigma}^2\sigma^2)_{A\dot{D}} \\ -(\bar{\sigma}^2 - \bar{\sigma}^2\sigma^2\bar{\sigma}^2)^{\dot{A}D} & 0 \end{pmatrix}$$

Again using (1.90) we obtain $-\sigma^2\bar{\sigma}^2\sigma^2 = \sigma^2(\sigma^2)^2 = \sigma^2$ since $(\sigma^2)^2 = 1_{2\times 2}$. Therefore

$$(\gamma^2_M)_{ad} = \frac{1}{2}\begin{pmatrix} 0 & -2\,\sigma^2_{A\dot{D}} \\ -2\bar{\sigma}^{2\dot{A}D} & 0 \end{pmatrix} = \begin{pmatrix} 0 & -\sigma^2 \\ -\bar{\sigma}^2 & 0 \end{pmatrix}_{ad}$$

$\mu = 3$:

$$(\gamma^3_M)_{ad} = Y_{ab}(\gamma^3_D)_{bc}Y^{-1}_{cd}$$

$$= \frac{1}{2}\begin{pmatrix} \delta_A{}^B & \sigma^2_{A\dot{B}} \\ -\bar{\sigma}^{2\dot{A}B} & -\delta^{\dot{A}}{}_{\dot{B}} \end{pmatrix}\begin{pmatrix} 0 & \sigma^3_{B\dot{C}} \\ \bar{\sigma}^{3\dot{B}C} & 0 \end{pmatrix}\begin{pmatrix} \delta_C{}^D & \sigma^2_{C\dot{D}} \\ -\bar{\sigma}^{2\dot{C}D} & -\delta^{\dot{C}}{}_{\dot{D}} \end{pmatrix}$$

$$= \frac{1}{2}\begin{pmatrix} (\sigma^2\bar{\sigma}^3)_A{}^D - (\sigma^3\bar{\sigma}^2)_A{}^D & (\sigma^2\bar{\sigma}^3\sigma^2)_{A\dot{D}} - \sigma^3_{A\dot{D}} \\ -\bar{\sigma}^{3\dot{A}D} + (\bar{\sigma}^2\sigma^3\bar{\sigma}^2)^{\dot{A}D} & -(\bar{\sigma}^3\sigma^2)^{\dot{A}}{}_{\dot{D}} + (\bar{\sigma}^2\sigma^3)^{\dot{A}}{}_{\dot{D}} \end{pmatrix}$$

$$= \frac{1}{2}\begin{pmatrix} -([\sigma^2,\sigma^3])_A{}^D & 0 \\ 0 & -([\sigma^2,\sigma^3])^{\dot{A}}{}_{\dot{D}} \end{pmatrix}$$

$$= \frac{1}{2}\begin{pmatrix} -2i\,\sigma^1{}_A{}^D & 0 \\ 0 & -2i\,\sigma^{1\dot{A}}{}_{\dot{D}} \end{pmatrix} = \begin{pmatrix} -i\sigma^1 & 0 \\ 0 & -i\sigma^1 \end{pmatrix}_{ad}$$

Now the connection between the Weyl representation and the Dirac representation is given by (1.150a),

$$\Gamma_D = X^{-1}\Gamma_W X$$

and from the Dirac representation we obtain the Majorana representation from (1.157),

$$\Gamma_M = Y\Gamma_D Y^{-1}$$

Thus

$$\Gamma_M = YX^{-1}\Gamma_W XY^{-1}$$

or using $Y = Y^{-1}$ we obtain the similarity transformation that connects the Weyl representation and the Majorana representation:

$$\Gamma_M = (XY)^{-1}\Gamma_W (XY) \tag{1.159}$$

With

$$X = \frac{1}{2^{1/2}}\begin{pmatrix} -1_{2\times 2} & \sigma^0 \\ -\bar{\sigma}^0 & -1_{2\times 2}\end{pmatrix} = (X^{-1})^T = (X^{-1})^+$$

$$Y = \frac{1}{2^{1/2}}\begin{pmatrix} 1_{2\times 2} & \sigma^2 \\ -\bar{\sigma}^2 & -1_{2\times 2}\end{pmatrix} = Y^+ = Y^{-1}$$

we obtain

$$X_{ab}Y_{bc} = \frac{1}{2}\begin{pmatrix} -\delta_A{}^B & \sigma^0_{A\dot{B}} \\ -\bar{\sigma}^{0\dot{A}B} & -\delta^{\dot{A}}{}_{\dot{B}}\end{pmatrix}\begin{pmatrix} \delta_B{}^C & \sigma^2_{B\dot{C}} \\ -\bar{\sigma}^{2\dot{B}C} & \delta^{\dot{B}}{}_{\dot{C}}\end{pmatrix}$$

$$= \frac{1}{2}\begin{pmatrix} -\delta_A{}^B\delta_B{}^C - \sigma^0_{A\dot{B}}\bar{\sigma}^{2\dot{B}C} & -\delta_A{}^B\sigma^2_{B\dot{C}} - \sigma^0_{A\dot{B}}\delta^{\dot{B}}{}_{\dot{C}} \\ -\bar{\sigma}^{0\dot{A}B}\delta_B{}^C + \delta^{\dot{A}}{}_{\dot{B}}\bar{\sigma}^{2\dot{B}C} & -\bar{\sigma}^{0\dot{A}B}\sigma^2_{B\dot{C}} + \delta^{\dot{A}}{}_{\dot{B}}\delta^{\dot{B}}{}_{\dot{C}}\end{pmatrix}$$

$$= \frac{1}{2}\begin{pmatrix} -\delta_A{}^C + \sigma^2_A{}^C & -\sigma^2_{A\dot{C}} - \sigma^0_{A\dot{C}} \\ -\bar{\sigma}^{0\dot{A}C} + \bar{\sigma}^{2\dot{A}C} & -\sigma^{2\dot{A}}{}_{\dot{C}} + \delta^{\dot{A}}{}_{\dot{C}}\end{pmatrix}$$

which is usually written in matrix form as

$$XY = \frac{1}{2}\begin{pmatrix} -1+\sigma^2 & -\sigma^2-1 \\ -1-\sigma^2 & -\sigma^2+1\end{pmatrix} \tag{1.160}$$

The inverse $(XY)^{-1}$ is

$$(XY)^{-1} = Y^{-1}X^{-1} = \frac{1}{2}\begin{pmatrix} 1 & \sigma^2 \\ -\bar{\sigma}^2 & -1 \end{pmatrix}\begin{pmatrix} -1 & -\sigma^0 \\ \bar{\sigma}^0 & -1 \end{pmatrix}$$

$$= \frac{1}{2}\begin{pmatrix} -1 + \sigma^2\bar{\sigma}^0 & -\sigma^0 - \sigma^2 \\ \bar{\sigma}^2 - \bar{\sigma}^0 & \bar{\sigma}^2\sigma^0 + 1 \end{pmatrix}$$

(1.161)

which is usually written

$$(XY)^{-1} = \frac{1}{2}\begin{pmatrix} -1+\sigma^2 & -1-\sigma^2 \\ -\sigma^2-1 & -\sigma^2+1 \end{pmatrix} = (XY)^+$$

(1.162)

or

$$(XY)^{-1} = Y^{-1}X^{-1} = Y^+X^+ = (XY)^+$$

but for a clear index structure it seems to be advantageous to use the explicit form (1.161). Of course, (1.162) is the same matrix as (1.161) since as matrices

$$\bar{\sigma}^0 = 1, \quad \bar{\sigma}^2 = -\sigma^2$$

and so on. Next we check that

$$\gamma^0_M = (XY)^{-1}\gamma^0_W(XY)$$

with

$$\gamma^0_W = \begin{pmatrix} 0 & \sigma^0 \\ \bar{\sigma}^0 & 0 \end{pmatrix}$$

We do this in the compact matrix form of (1.162):

$$(XY)^{-1}\gamma^0_W(XY)$$

$$= \frac{1}{4}\begin{pmatrix} -1+\sigma^2 & -1-\sigma^2 \\ -\sigma^2-1 & 1-\sigma^2 \end{pmatrix}\begin{pmatrix} 0 & 1 \\ 1 & 0 \end{pmatrix}\begin{pmatrix} \sigma^2-1 & -\sigma^2-1 \\ -1-\sigma^2 & 1-\sigma^2 \end{pmatrix}$$

$$= \frac{1}{4}\begin{pmatrix} (-1-\sigma^2)(\sigma^2-1)+(-1+\sigma^2)(-1-\sigma^2) & -(1+\sigma^2)(-\sigma^2-1)+(-1+\sigma^2)(1-\sigma^2) \\ (1-\sigma^2)(\sigma^2-1)+(\sigma^2+1)^2 & (1-\sigma^2)(-\sigma^2-1)+(-\sigma^2-1)(1-\sigma^2) \end{pmatrix}$$

$$= \frac{1}{4}\begin{pmatrix} 0 & 4\sigma^2 \\ 4\sigma^2 & 0 \end{pmatrix} = \begin{pmatrix} 0 & \sigma^2 \\ -\bar{\sigma}^2 & 0 \end{pmatrix} = \gamma_M^0$$

(see (1.155)). The other γ^μ matrices can be checked in a similar fashion.

1.4.4 Charge Conjugation, Dirac and Weyl Representations

The charge conjugation matrix appears in Dirac theory in the following way. The Dirac theory implies the existence of electrons and positrons, particles with the same mass but opposite charges, which obey the same equation. The Dirac equation must therefore admit a symmetry corresponding to the interchange of particles and antiparticles. We thus seek a transformation $\Psi \rightarrow \Psi^C$ which reverses the sign of the charge, so that the Dirac spinor Ψ obeys the Dirac equation

$$(i\partial\!\!\!/ - eA\!\!\!/ - m)\Psi = 0, \quad A\!\!\!/ \equiv A_\mu \gamma^\mu \tag{1.163}$$

in the presence of the electromagnetic vector potential A_μ whereas the charge conjugated spinor Ψ^C obeys[34]

$$(i\partial\!\!\!/ + eA\!\!\!/ - m)\Psi^C = 0 \tag{1.164}$$

The Dirac equation coupled minimally to the electromagnetic field is

$$[\gamma^\mu(i\partial_\mu - eA_\mu) - m]\psi = 0$$

Taking the complex conjugate we obtain

$$[\gamma^{\mu *}(-i\partial_\mu - eA_\mu) - m]\psi^* = 0$$

Transposition yields

$$\psi^+[\gamma^{\mu +}(-i\overleftarrow{\partial}_\mu - eA_\mu) - m] = 0$$

$$\psi^+\gamma^0[\gamma^0\gamma^{\mu +}\gamma^0(-i\overleftarrow{\partial}_\mu - eA_\mu) - m] = 0$$

$$\overline{\psi}[\gamma^\mu(-i\overleftarrow{\partial}_\mu - eA_\mu) - m] = 0$$

where
$$\overline{\psi} := \psi^+\gamma^0$$

is the Dirac adjoint and

$$\gamma^0\gamma^{\mu +}\gamma^0 = \gamma^\mu$$

Taking again the transpose we obtain

$$[-\gamma^{\mu T}(i\partial_\mu + eA_\mu) - m]\overline{\psi}^T = 0$$

Multiplying this equation by a 4x4-matrix C from the left and inserting $C^{-1}C$ in front of $\overline{\psi}^T$ we get

$$C[-\gamma^{T\mu}(i\partial_\mu + eA_\mu) - m]C^{-1}C\overline{\psi}^T = 0$$

$$[-C\gamma^{T\mu}C^{-1}(i\partial_\mu + eA_\mu) - m]C\overline{\psi}^T = 0$$

This equation can be identified with (1.164) provided we set

$$\psi^C := C\overline{\psi}^T \tag{1.165}$$

except for a phase factor, and we have to demand that in

any representation of γ matrices

$$C\gamma^{\mu T}C^{-1} = -\gamma^{\mu} \tag{1.166}$$

The matrix C is called the charge conjugation matrix. It suffices to construct the charge conjugation matrix in some particular representation; the unitary transformation which transforms to another representation then gives the matrix C in this new representation. We consider several representations.

The Dirac representation

In the Dirac representation (1.149) the charge conjugation matrix C may be taken as

$$C_D = i\gamma_D^2\gamma_D^0 = i\begin{pmatrix} 0 & -\sigma^2 \\ \bar{\sigma}^2 & 0 \end{pmatrix} \tag{1.167}$$

This matrix possesses the following properties

$$C_D = -C_D^{-1} = -C_D^{+} = -C_D^{T} \tag{1.168}$$

Now, we first verify

$$C_D = -C_D^{-1}$$

Consider

$$-(C_D)_{ab}(C_D)_{bc}$$

$$= \begin{pmatrix} 0 & -\sigma^2_{A\dot{B}} \\ \bar{\sigma}^{2\dot{A}B} & 0 \end{pmatrix}\begin{pmatrix} 0 & -\sigma^2_{B\dot{C}} \\ \bar{\sigma}^{2\dot{B}C} & 0 \end{pmatrix}$$

$$= \begin{pmatrix} -\sigma^2_{A\dot{B}}\bar{\sigma}^{2\dot{B}C} & 0 \\ 0 & -\bar{\sigma}^{2\dot{A}B}\sigma^2_{B\dot{C}} \end{pmatrix}$$

$$= \begin{pmatrix} -(\sigma^2\bar{\sigma}^2)_A{}^C & 0 \\ 0 & -(\bar{\sigma}^2\sigma^2)^{\dot{A}}{}_{\dot{C}} \end{pmatrix}$$

$$= \begin{pmatrix} [(\sigma^2)^2]_A{}^C & 0 \\ 0 & [(\sigma^2)^2]^{\dot{A}}{}_{\dot{C}} \end{pmatrix}$$

(using (1.90))

$$= \begin{pmatrix} \delta_A{}^C & 0 \\ 0 & \delta^{\dot{A}}{}_{\dot{C}} \end{pmatrix} = (1_{4\times 4})_{ac}$$

$$(C_D^+)_{ab} = \left[i \begin{pmatrix} 0 & -\sigma^2_{A\dot{B}} \\ \bar{\sigma}^{2\dot{A}B} & 0 \end{pmatrix} \right]^+$$

$$= -i \begin{pmatrix} 0 & (\bar{\sigma}^{2+})_{A\dot{B}} \\ -(\sigma^{2+})^{\dot{A}B} & 0 \end{pmatrix} = -i \begin{pmatrix} 0 & -\sigma^2_{A\dot{B}} \\ \bar{\sigma}^{2\dot{A}B} & 0 \end{pmatrix}$$

$$= -(C_D)_{ab}$$

using the hermiticity of the Pauli matrices and $\bar{\sigma}^2 = -\sigma^2$ from (1.90). Analogously

$$(C_D^T)_{ab}$$

$$= i \begin{pmatrix} 0 & -\sigma^2_{A\dot{B}} \\ \bar{\sigma}^{2\dot{A}B} & 0 \end{pmatrix}^T = i \begin{pmatrix} 0 & (\bar{\sigma}^2)^T_{A\dot{B}} \\ -(\sigma^2)^{T\dot{A}B} & 0 \end{pmatrix}$$

$$= i \begin{pmatrix} 0 & -(\sigma^{2T})_{A\dot{B}} \\ (\bar{\sigma}^{2T})^{\dot{A}B} & 0 \end{pmatrix} = i \begin{pmatrix} 0 & \sigma^2_{A\dot{B}} \\ -\bar{\sigma}^{2\dot{A}B} & 0 \end{pmatrix}$$

$$= -i \begin{pmatrix} 0 & -\sigma^2_{A\dot{B}} \\ \bar{\sigma}^{2\dot{A}B} & 0 \end{pmatrix} = -(C_D)_{ab}$$

where we made use of (1.90) and $\sigma^{2T} = -\sigma^2$.
We now verify (1.166) in the Dirac representation using the explicit index notation. First we transform (1.166) into the following form

$$C_D \gamma_D^{\mu T} C_D^{-1} = -\gamma_D^{\mu}$$

With (1.168) this is

$$-C_D \gamma_D^{\mu T} C_D = -\gamma_D^{\mu}$$

Transposition gives

$$C_D^T \gamma_D^{\mu} C_D^T = \gamma_D^{\mu T}$$

and using (1.168) again

$$C_D \gamma_D^{\mu} C_D = \gamma_D^{\mu T}$$

$\underline{\mu = 0}$:

$$(\gamma_D^0)_{ab} = \begin{pmatrix} \delta_A{}^B & 0 \\ 0 & -\delta^{\dot{A}}{}_{\dot{B}} \end{pmatrix}$$

Using (1.167) we have

$$(C_D)_{ab} (\gamma_D^0)_{bc} (C_D)_{cd}$$

$$= -\begin{pmatrix} 0 & -\sigma^2_{A\dot{B}} \\ \bar{\sigma}^{2\dot{A}B} & 0 \end{pmatrix} \begin{pmatrix} \delta_B{}^C & 0 \\ 0 & -\delta^{\dot{B}}{}_{\dot{C}} \end{pmatrix} \begin{pmatrix} 0 & -\sigma^2_{C\dot{D}} \\ \bar{\sigma}^{2\dot{C}D} & 0 \end{pmatrix}$$

$$= -\begin{pmatrix} \sigma^2_{A\dot{B}} \bar{\sigma}^{2\dot{B}D} & 0 \\ 0 & -\bar{\sigma}^{2\dot{A}B} \sigma^2_{B\dot{D}} \end{pmatrix}$$

$$= -\begin{pmatrix} (\sigma^2\bar{\sigma}^2)_A{}^D & 0 \\ 0 & -(\bar{\sigma}^2\sigma^2)^{\dot{A}}{}_{\dot{D}} \end{pmatrix}$$

$$= -\begin{pmatrix} -[(\sigma^2)^2]_A{}^D & 0 \\ 0 & [(\sigma^2)^2]^{\dot{A}}{}_{\dot{D}} \end{pmatrix}$$

$$= -\begin{pmatrix} -\delta_A{}^D & 0 \\ 0 & \delta^{\dot A}{}_{\dot D} \end{pmatrix}$$

$$= \begin{pmatrix} (1^T_{2\times 2})_A{}^D & 0 \\ 0 & -(1^T_{2\times 2})^{\dot A}{}_{\dot D} \end{pmatrix}$$

$$= (\gamma^{0T})_{ad}$$

<u>$\mu = 1$</u> :

$$(\gamma_D^1)_{ab} = \begin{pmatrix} 0 & \sigma^1_{A\dot B} \\ \bar\sigma^{1\dot A B} & 0 \end{pmatrix}$$

Using (1.167) we have

$$(C_D)_{ab}(\gamma_D^1)_{bc}(C_D)_{cd}$$

$$= -\begin{pmatrix} 0 & -\sigma^2_{A\dot B} \\ \bar\sigma^{2\dot A B} & 0 \end{pmatrix}\begin{pmatrix} 0 & \sigma^1_{B\dot C} \\ \bar\sigma^{1\dot B C} & 0 \end{pmatrix}\begin{pmatrix} 0 & -\sigma^2_{C\dot D} \\ \bar\sigma^{2\dot C D} & 0 \end{pmatrix}$$

$$= -\begin{pmatrix} 0 & \sigma^2_{A\dot B}\,\bar\sigma^{1\dot B C}\,\sigma^2_{C\dot D} \\ \bar\sigma^{2\dot A B}\,\sigma^1_{B\dot C}\,\bar\sigma^{2\dot C D} & 0 \end{pmatrix}$$

$$= -\begin{pmatrix} 0 & (\sigma^2\bar\sigma^1\sigma^2)_{A\dot D} \\ (\bar\sigma^2\sigma^1\bar\sigma^2)^{\dot A D} & 0 \end{pmatrix}$$

$$= -\begin{pmatrix} 0 & -(\sigma^2\sigma^1\sigma^2)_{A\dot D} \\ (\sigma^2\sigma^1\sigma^2)^{\dot A D} & 0 \end{pmatrix}$$

(using (1.90))

$$= -\begin{pmatrix} 0 & \sigma^1_{A\dot{D}} \\ -\bar{\sigma}^{1\dot{A}D} & 0 \end{pmatrix} \qquad (\sigma^2\sigma^i\sigma^2 = -\sigma^i)$$

$$= \begin{pmatrix} 0 & -(\sigma^{1T})_{A\dot{D}} \\ -(\bar{\sigma}^{1T})^{\dot{A}D} & 0 \end{pmatrix} \qquad (\sigma^1 = \sigma^{1T})$$

$$= (\gamma_D^{1\,T})_{ad}$$

since

$$(\gamma_D^1)^T_{ad} = \left[\begin{pmatrix} 0 & \sigma^1_{A\dot{D}} \\ \bar{\sigma}^{1\dot{A}D} & 0 \end{pmatrix}\right]^T$$

$$= \begin{pmatrix} 0 & (\bar{\sigma}^{1T})_{A\dot{D}} \\ (\sigma^{1T})^{\dot{A}D} & 0 \end{pmatrix} = -\begin{pmatrix} 0 & (\sigma^{1T})_{A\dot{D}} \\ (\bar{\sigma}^{1T})^{\dot{A}D} & 0 \end{pmatrix}$$

Hence

$$C_D \gamma_D^1 C_D^{-1} = -(\gamma_D^1)^T$$

$\underline{\mu = 2}$:

$$\gamma_D^2 = \begin{pmatrix} 0 & \sigma^2 \\ \bar{\sigma}^2 & 0 \end{pmatrix}$$

Using (1.167),

$$C_{ab}(\gamma_D^2)_{bc}C_{cd}$$

$$= - \begin{pmatrix} 0 & -\sigma^2_{A\dot{B}} \\ \bar{\sigma}^{2\dot{A}B} & 0 \end{pmatrix} \begin{pmatrix} 0 & \sigma^2_{B\dot{C}} \\ \bar{\sigma}^{2\dot{B}C} & 0 \end{pmatrix} \begin{pmatrix} 0 & -\sigma^2_{C\dot{D}} \\ \bar{\sigma}^{2\dot{C}D} & 0 \end{pmatrix}$$

$$= - \begin{pmatrix} 0 & \sigma^2_{A\dot{B}}\, \bar{\sigma}^{2\dot{B}C}\, \sigma^2_{C\dot{D}} \\ \bar{\sigma}^{2\dot{A}B}\, \sigma^2_{B\dot{C}}\, \bar{\sigma}^{2\dot{C}D} & 0 \end{pmatrix}$$

$$= - \begin{pmatrix} 0 & (\sigma^2 \bar{\sigma}^2 \sigma^2)_{A\dot{D}} \\ (\bar{\sigma}^2 \sigma^2 \bar{\sigma}^2)^{\dot{A}D} & 0 \end{pmatrix}$$

$$= \begin{pmatrix} 0 & \sigma^2_{A\dot{D}} \\ \bar{\sigma}^{2\dot{A}D} & 0 \end{pmatrix}$$

Now the transpose of γ^2_D is

$$(\gamma^{2T}_D)_{ad} = \begin{pmatrix} 0 & \sigma^2_{A\dot{D}} \\ \bar{\sigma}^{2\dot{A}D} & 0 \end{pmatrix}^T$$

$$= \begin{pmatrix} 0 & (\bar{\sigma}^{2T})_{A\dot{D}} \\ (\sigma^{2T})^{\dot{A}D} & 0 \end{pmatrix} = \begin{pmatrix} 0 & (\sigma^2)_{A\dot{D}} \\ (\bar{\sigma}^2)^{\dot{A}D} & 0 \end{pmatrix}$$

since with (1.90)

$$\bar{\sigma}^{2T} = -\sigma^{2T} = \sigma^2 .$$

$\underline{\mu = 3}$:

$$\gamma^3_D = \begin{pmatrix} 0 & \sigma^3 \\ \bar{\sigma}^3 & 0 \end{pmatrix}$$

Then $C_{ab}(\gamma^3_D)_{bc}C_{cd}$

$$= - \begin{pmatrix} 0 & -\sigma^2_{A\dot{B}} \\ \bar{\sigma}^{2\dot{A}B} & 0 \end{pmatrix} \begin{pmatrix} 0 & \sigma^3_{B\dot{C}} \\ \bar{\sigma}^{3\dot{B}C} & 0 \end{pmatrix} \begin{pmatrix} 0 & -\sigma^2_{C\dot{D}} \\ \bar{\sigma}^{2\dot{C}D} & 0 \end{pmatrix}$$

$$= -\begin{pmatrix} 0 & \sigma^2_{A\dot{B}}\,\bar{\sigma}^{3\dot{B}C}\,\sigma^2_{C\dot{D}} \\ \bar{\sigma}^{2\dot{A}B}\,\sigma^3_{B\dot{C}}\,\bar{\sigma}^{2\dot{C}D} & 0 \end{pmatrix}$$

$$= -\begin{pmatrix} 0 & (\sigma^2\bar{\sigma}^3\sigma^2)_{A\dot{D}} \\ (\bar{\sigma}^2\sigma^3\bar{\sigma}^2)^{\dot{A}D} & 0 \end{pmatrix}$$

$$= -\begin{pmatrix} 0 & -\bar{\sigma}^3_{A\dot{D}} \\ -\sigma^{3\dot{A}D} & 0 \end{pmatrix} = -\begin{pmatrix} 0 & \sigma^3_{A\dot{D}} \\ \bar{\sigma}^{3\dot{A}D} & 0 \end{pmatrix}$$

Now

$$(\gamma_D^{3T})_{ad} = \begin{pmatrix} 0 & \sigma^3_{A\dot{D}} \\ \bar{\sigma}^{3\dot{A}D} & 0 \end{pmatrix}^T = \begin{pmatrix} 0 & (\bar{\sigma}^{3T})_{A\dot{D}} \\ (\sigma^{3T})^{\dot{A}D} & 0 \end{pmatrix}$$

$$= \begin{pmatrix} 0 & -(\sigma^{3T})_{A\dot{D}} \\ -(\bar{\sigma}^{3T})^{\dot{A}D} & 0 \end{pmatrix} = -\begin{pmatrix} 0 & \sigma^3_{A\dot{D}} \\ \bar{\sigma}^{3\dot{A}D} & 0 \end{pmatrix}$$

using (1.90) and $\sigma^3 = \sigma^{3T}$.

Hence

$$C\gamma_D^3 C^{-1} = -\gamma^{3T}$$

as had to be shown.

The Weyl representation

We transform the charge conjugation matrix (1.167) from the Dirac representation to the Weyl representation by using (1.150); thus

$$C_W = X C_D X^{-1} = iX\gamma_D^2\gamma_D^0 X^{-1}$$

$$= iX\gamma_D^2 X^{-1} X\gamma_D^0 X^{-1} = i\gamma_W^2\gamma_W^0$$

$$= \begin{pmatrix} i\sigma^2 & 0 \\ 0 & -i\sigma^2 \end{pmatrix} \tag{1.169}$$

We prove (1.169) in the submatrix formulation, avoiding cumbersome indices. Thus

$$C_W = \frac{i}{2}\begin{pmatrix} -1 & \sigma^0 \\ -\sigma^0 & -1 \end{pmatrix}\begin{pmatrix} 0 & -\sigma^2 \\ \bar{\sigma}^2 & 0 \end{pmatrix}\begin{pmatrix} -1 & -\sigma^0 \\ \bar{\sigma}^0 & -1 \end{pmatrix}$$

$$= \frac{i}{2}\begin{pmatrix} \sigma^2\bar{\sigma}^0 - \sigma^0\bar{\sigma}^2 & -\sigma^2 - \sigma^0\bar{\sigma}^2\sigma^0 \\ \bar{\sigma}^0\sigma^2\bar{\sigma}^0 + \bar{\sigma}^2 & -\bar{\sigma}^0\sigma^2 + \bar{\sigma}^2\sigma^0 \end{pmatrix}$$

$$= \frac{i}{2}\begin{pmatrix} 2\sigma^2 & 0 \\ 0 & -2\sigma^2 \end{pmatrix} = \begin{pmatrix} i\sigma^2 & 0 \\ 0 & -i\sigma^2 \end{pmatrix}$$

Remark: The correct form of the charge conjugation matrix in the Weyl representation, possessing the correct index structure, is

$$C_W = \begin{pmatrix} i\sigma^2\bar{\sigma}^0 & 0 \\ 0 & i\bar{\sigma}^2\sigma^0 \end{pmatrix} \tag{1.170}$$

with

$$(C_W)_{ab} = \begin{pmatrix} (i\sigma^2\bar{\sigma}^0)_A{}^B & 0 \\ 0 & (i\bar{\sigma}^2\sigma^0)^{\dot{A}}{}_{\dot{B}} \end{pmatrix}$$

Of course, the matrix (1.170) is the same as (1.169), since $\bar{\sigma}^0$ and σ^0 are unit matrices, which are usually deleted in the literature.

The charge conjugation matrix C_W in the Weyl representation also satisfies

$$C_W = -C_W^{-1} = -C_W^T = -C_W^{+} \tag{1.171}$$

and

$$C_W\gamma_W^\mu C_W^{-1} = -\gamma_W^{\mu T} \tag{1.172}$$

as can be checked directly, using (1.169) and (1.136), or alternatively with the help of (1.150). Thus using first (1.169) and then (1.168) we have

$$C_W = X C_D X^{-1} = -X C_D^{-1} X^{-1}$$
$$= -(X C_D X^{-1})^{-1} = -C_W^{-1}$$

Also

$$C_W = X C_D X^{-1} \quad \text{(with (1.164))}$$
$$= -X C_D^T X^{-1} \quad \text{(with (1.168))}$$
$$= -(X^{-1T} C_D X^T)^T$$
$$= -(X C_D X^{-1})^T \quad \text{(with (1.159))}$$
$$= -C_W^T \quad \text{(with (1.169))}$$

and

$$C_W = X C_D X^{-1} = -X C_D^+ X^{-1}$$
$$= -(X^{-1+} C_D X^+)^+$$
$$= -(X C_D X^{-1})^+$$
$$= -C_W^+$$

Finally we verify equation (1.172). We know that

$$C_D \gamma_D^\mu C_D^{-1} = -\gamma_D^{\mu T}$$

Hence

$$X (C_D \gamma_D^\mu C_D^{-1}) X^{-1} = -X (\gamma_D^{\mu T}) X^{-1}$$

$$X C_D X^{-1} X \gamma_D^\mu X^{-1} X C_D^{-1} X^{-1} = -X (\gamma_D^{\mu T}) X^{-1}$$

$$X C_D X^{-1} X \gamma_D^\mu X^{-1} (X C_D X^{-1})^{-1} = -(X \gamma_D^\mu X^{-1})^T$$

and so

$$C_W \gamma^\mu_W C_W^{-1} = -\gamma^{\mu\,T}_W$$

Further properties of the C matrix

i)
$$C\gamma^5 C^{-1} = \gamma^{5\,T} \qquad (1.173)$$

ii)
$$C(\gamma^5\gamma^\mu)C^{-1} = (\gamma^5\gamma^\mu)^T \qquad (1.174)$$

We first prove equation (1.173). Thus, since (1.139) is independent of the particular representation we have

$$\begin{aligned}
C\gamma^5 C^{-1} &= i\,C\gamma^0\gamma^1\gamma^2\gamma^3 C^{-1} \\
&= i\,C\gamma^0 C^{-1} C\gamma^1 C^{-1} C\gamma^2 C^{-1} C\gamma^3 C^{-1} \\
&= i\,\gamma^{0T}\gamma^{1T}\gamma^{2T}\gamma^{3T} \\
&\qquad \text{(with (1.161))} \\
&= i\,(\gamma^3\gamma^2\gamma^1\gamma^0)^T \\
&= i\,(\gamma^0\gamma^1\gamma^2\gamma^3)^T \\
&= \gamma^{5\,T}
\end{aligned}$$

using (1.137) which is valid in any representation. Equation (1.174) can be shown in a similar way:

$$\begin{aligned}
C\gamma^5\gamma^\mu C^{-1} &= C\gamma^5 C^{-1} C\gamma^\mu C^{-1} \\
&= -(\gamma^5)^T(\gamma^\mu)^T \\
&= -(\gamma^\mu\gamma^5)^T \\
&= (\gamma^5\gamma^\mu)^T
\end{aligned}$$

with (1.141).

1.4.5 Majorana Spinors

We consider a Dirac four-spinor in the Weyl representation. According to (1.133) we have

$$(\Psi)_a = \begin{pmatrix} \phi \\ \bar{\Psi} \end{pmatrix}_a = \begin{pmatrix} \phi_A \\ \bar{\Psi}_{\dot{A}} \end{pmatrix} \in F \oplus \dot{F}^* \tag{1.175}$$

For the index calculus to make sense, rows and columns must have the index structure of

$$\left(\psi_W^T\right)_b = (\phi^B, \bar{\Psi}_{\dot{B}}) \quad \text{and} \quad (\psi_W)_b = \begin{pmatrix} \phi_A \\ \bar{\Psi}^{\dot{A}} \end{pmatrix}$$

respectively. Thus

$$\begin{pmatrix} M_B{}^C & M_{B\dot{C}} \\ M^{\dot{B}C} & M^{\dot{B}}{}_{\dot{C}} \end{pmatrix} \begin{pmatrix} \phi_C \\ \bar{\Psi}^{\dot{C}} \end{pmatrix} = \begin{pmatrix} M_B{}^C \phi_C + M_{B\dot{C}} \bar{\Psi}^{\dot{C}} \\ M^{\dot{B}C} \phi_C + M^{\dot{B}}{}_{\dot{C}} \bar{\Psi}^{\dot{C}} \end{pmatrix}$$

and

$$(\phi^B, \bar{\Psi}_{\dot{B}}) \begin{pmatrix} M_B{}^C & M_{B\dot{C}} \\ M^{\dot{B}C} & M^{\dot{B}}{}_{\dot{C}} \end{pmatrix}$$

$$= (\phi^B M_B{}^C + \bar{\Psi}_{\dot{B}} M^{\dot{B}C}, \phi^B M_{B\dot{C}} + \bar{\Psi}_{\dot{B}} M^{\dot{B}}{}_{\dot{C}})$$

We now observe that the relationship between left-handed and right-handed Weyl spinors may be written

$$\psi^B = \bar{\Psi}^*_{\dot{A}} \bar{\sigma}^{0\dot{A}B}, \quad \bar{\Phi}_{\dot{B}} = \phi^{A*} \sigma^0_{A\dot{B}} \tag{1.176}$$

The consistency of these relations with (1.56), i.e.

$$\psi^A = \bar{\Psi}^*_{\dot{A}} \tag{1.177}$$

follows from the fact that $\sigma^0 = \bar{\sigma}^0 = 1_{2\times 2}$. It may be noted that ψ_A is a two-component column vector, whereas ψ^A is a two-component row; similarly $\bar{\Phi}^{\dot{A}}$ is a column and $\bar{\Phi}_{\dot{A}}$ a row.

The consistency of (1.176) and (1.177) can also be seen by calculating the Dirac conjugate of ψ, i.e.

$$\bar{\Psi}_W = \psi_W^+ \gamma_W^0$$

Using (1.177) we have

$$\bar{\Psi}_W = \psi_W^+ \gamma_W^0 = (\phi^{A*}, \bar{\Psi}_{\dot{A}}^*) \begin{pmatrix} 0 & 1 \\ 1 & 0 \end{pmatrix}$$

$$= (\bar{\Psi}_{\dot{A}}^*, \phi^{A*}) = (\psi^A, \bar{\Phi}_{\dot{A}})$$

(note that since (1.177) does not preserve a consistent index structure the matrix representaion of γ^0 must be used). On the other hand, using (1.176) we have

$$\bar{\Psi}_W = \psi_W^+ \gamma_W^0 = (\phi^{A*}, \bar{\Psi}_{\dot{A}}^*) \begin{pmatrix} 0 & \sigma^0_{A\dot{B}} \\ \bar{\sigma}^{0\dot{A}B} & 0 \end{pmatrix}$$

$$= (\bar{\Psi}_{\dot{A}}^* \bar{\sigma}^{0\dot{A}B}, \phi^{A*} \sigma^0_{A\dot{B}})$$

$$= (\psi^B, \bar{\Phi}_{\dot{B}}) \qquad \text{(with (1.176))}$$

(1.178a)

in agreement with the previous result. We note that

$$\bar{\Psi}_W^T = \begin{pmatrix} \psi_B \\ \bar{\Phi}^{\dot{B}} \end{pmatrix} \tag{1.178b}$$

The charge conjugate of a Dirac spinor in the Weyl representation is defined as in (1.165) with the charge conjugation matrix (1.169), i.e.

$$\psi_W^c = C_W \bar{\Psi}_W^T$$

Thus using (1.170) and (1.178) we have

$$\begin{aligned}(\psi_W^c)_a &= C_{ab}(\bar{\Psi}_W^T)_b \\ &= \begin{pmatrix} (i\sigma^2\bar{\sigma}^0)_A{}^B & 0 \\ 0 & (i\bar{\sigma}^2\sigma^0)^{\dot{A}}{}_{\dot{B}} \end{pmatrix} \begin{pmatrix} \psi_B \\ \bar{\Phi}^{\dot{B}} \end{pmatrix} \\ &= \begin{pmatrix} (i\sigma^2\bar{\sigma}^0)_A{}^B \psi_B \\ (i\bar{\sigma}^2\sigma^0)^{\dot{A}}{}_{\dot{B}} \bar{\Phi}^{\dot{B}} \end{pmatrix}\end{aligned}$$

Now

$$(i\sigma^2\bar{\sigma}^0)^{AB} = \begin{pmatrix} 0 & 1 \\ -1 & 0 \end{pmatrix}^{AB} = (\varepsilon^{AB})$$

and

$$(i\bar{\sigma}^2\sigma^0)_{\dot{A}\dot{B}} = \begin{pmatrix} 0 & -1 \\ 1 & 0 \end{pmatrix}_{\dot{A}\dot{B}} = (\varepsilon_{\dot{A}\dot{B}})$$

Hence

$$\begin{aligned}(i\sigma^2\bar{\sigma}^0)_A{}^B &= \varepsilon_{AC}(i\sigma^2\bar{\sigma}^0)^{CB} \\ &= \varepsilon_{AC}\,\varepsilon^{CB} = \delta_A{}^B\end{aligned}$$

and

$$\begin{aligned}(i\bar{\sigma}^2\sigma^0)^{\dot{A}}{}_{\dot{B}} &= \varepsilon^{\dot{A}\dot{C}}(i\bar{\sigma}^2\sigma^0)_{\dot{C}\dot{B}} \\ &= \varepsilon^{\dot{A}\dot{C}}\,\varepsilon_{\dot{C}\dot{B}} = \delta^{\dot{A}}{}_{\dot{B}}\end{aligned}$$

Hence

$$\left(\psi_W^c\right)_a = \begin{pmatrix} \psi_A \\ \bar{\phi}^{\dot{A}} \end{pmatrix} \tag{1.179}$$

Thus charge conjugation flips ϕ and ψ .

A Majorana spinor is a four-component Dirac spinor which satisfies (here in the Weyl representation)

$$\Psi_W = \Psi_W^C \tag{1.180}$$

Using (1.179) and the explicit form of a Dirac spinor in the Weyl representation, i.e. (1.133), we obtain

$$\Psi_W = \begin{pmatrix} \phi_A \\ \bar{\psi}^{\dot{A}} \end{pmatrix} \overset{!}{=} \begin{pmatrix} \psi_A \\ \bar{\phi}^{\dot{A}} \end{pmatrix} = \Psi_W^C$$

Thus for a Majorana spinor in the Weyl representation we can write

$$\Psi_W^M = \begin{pmatrix} \phi_A \\ \bar{\phi}^{\dot{A}} \end{pmatrix} \tag{1.181}$$

Thus a Majorana spinor has only two independent complex components and is therefore equivalent to a two-component Weyl spinor or a real Dirac spinor.

1.4.6 Calculations with Dirac Spinors

It is useful to know the connection between the four-component Dirac formalism and the two-component Weyl formalism. The use of the Weyl representation of the four-component Dirac formalism has certain advantages.

We use the following notation for Dirac spinors

$$\Psi = \begin{pmatrix} \psi_{+A} \\ \bar{\psi}_-^{\dot{A}} \end{pmatrix} \tag{1.182}$$

where the subscripts $\pm$ distinguish between Weyl spinors which are elements of representation spaces F (+) and

$\dot{F}^*$ (-) respectively. Then according to (1.178a), the Dirac conjugate of (1.182) is

$$\bar{\Psi} = (\psi_-^A, \bar{\psi}_{+\dot{A}}) \tag{1.183}$$

<u>Proposition</u>: The following relations hold

i) $$(\bar{\Psi}\chi)_4 = (\psi_-\chi_+)_2 + (\bar{\psi}_+\bar{\chi}_-)_2 \tag{1.184a}$$

ii) $$(\bar{\Psi}\gamma^5\chi)_4 = -(\psi_-\chi_+)_2 + (\bar{\psi}_+\bar{\chi}_-)_2 \tag{1.184b}$$

iii) $$(\bar{\Psi}\gamma^\mu\chi)_4 = (\bar{\psi}_+\bar{\sigma}^\mu\chi_+)_2 + (\psi_-\sigma^\mu\bar{\chi}_-)_2 \tag{1.184c}$$

iv) $$(\bar{\Psi}\gamma^\mu\gamma^5\chi)_4 = (\psi_-\sigma^\mu\bar{\chi}_-)_2 - (\bar{\psi}_+\bar{\sigma}^\mu\chi_+)_2 \tag{1.184d}$$

v) $$(\bar{\Psi}\sigma^{\mu\nu}\chi)_4 = (\psi_-\sigma^{\mu\nu}\chi_+)_2 + (\bar{\psi}_+\bar{\sigma}^{\mu\nu}\bar{\chi}_-)_2 \tag{1.184e}$$

where

$$\sigma_4^{\mu\nu} = \frac{i}{4}[\gamma^\mu, \gamma^\nu] \tag{1.184f}$$

<u>Proof</u>:

i) $$(\bar{\Psi}\chi)_4 = (\psi_-^A, \bar{\psi}_{+\dot{A}})\begin{pmatrix}\chi_{+A}\\ \bar{\chi}_-^{\dot{A}}\end{pmatrix}$$
$$= \psi_-^A\chi_{+A} + \bar{\psi}_{+\dot{A}}\bar{\chi}_-^{\dot{A}}$$
$$= (\psi_-\chi_+)_2 + (\bar{\psi}_+\bar{\chi}_-)_2$$

on using (1.77) and (1.78).

ii) $$(\bar{\Psi}\gamma^5\chi)_4 = \bar{\Psi}_a\gamma^5_{ab}\chi_b$$
$$= (\psi_-^A, \bar{\psi}_{+\dot{A}})\begin{pmatrix}-\delta_A{}^B & 0\\ 0 & \delta^{\dot{A}}{}_{\dot{B}}\end{pmatrix}\begin{pmatrix}\chi_{+B}\\ \bar{\chi}_-^{\dot{B}}\end{pmatrix}$$

(using (1.139))

$$= (\psi_-^A, \bar{\psi}_{+\dot{A}}) \begin{pmatrix} -\chi_{+A} \\ \bar{\chi}_-^{\dot{A}} \end{pmatrix}$$

$$= -\psi_-^A \chi_{+A} + \bar{\psi}_{+\dot{A}} \bar{\chi}_-^{\dot{A}}$$

$$= (\bar{\psi}_+ \bar{\chi}_-)_2 - (\psi_- \chi_+)_2$$

again using (1.77) and (1.78).

iii) $$(\bar{\psi} \gamma^\mu \chi)_4 = \bar{\psi}_a \gamma^\mu_{ab} \chi_b$$

$$= (\psi_-^A, \bar{\psi}_{+\dot{A}}) \begin{pmatrix} 0 & \sigma^\mu_{A\dot{B}} \\ \bar{\sigma}^{\mu\dot{A}B} & 0 \end{pmatrix} \begin{pmatrix} \chi_{+B} \\ \bar{\chi}_-^{\dot{B}} \end{pmatrix}$$

(using (1.136))

$$= (\psi_-^A, \bar{\psi}_{+\dot{A}}) \begin{pmatrix} \sigma^\mu_{A\dot{B}} \bar{\chi}_-^{\dot{B}} \\ \bar{\sigma}^{\mu\dot{A}B} \chi_{+B} \end{pmatrix}$$

$$= \psi_-^A \sigma^\mu_{A\dot{B}} \bar{\chi}_-^{\dot{B}} + \bar{\psi}_{+\dot{A}} \bar{\sigma}^{\mu\dot{A}B} \chi_{+B}$$

$$= (\psi_- \sigma^\mu \bar{\chi}_-)_2 + (\bar{\psi}_+ \bar{\sigma}^\mu \chi_+)_2$$

again using (1.77) and (1.78).

iv) $$(\bar{\psi} \gamma^\mu \gamma^5 \chi)_4 = \bar{\psi}_a \gamma^\mu_{ab} \gamma^5_{bc} \chi_c$$

$$= (\psi_-^A, \bar{\psi}_{+\dot{A}}) \begin{pmatrix} 0 & \sigma^\mu_{A\dot{B}} \\ \bar{\sigma}^{\mu\dot{A}B} & 0 \end{pmatrix} \begin{pmatrix} -\delta_B{}^C & 0 \\ 0 & \delta^{\dot{B}}{}_{\dot{C}} \end{pmatrix} \begin{pmatrix} \chi_{+C} \\ \bar{\chi}_-^{\dot{C}} \end{pmatrix}$$

$$= (\psi_-^A, \bar{\psi}_{+\dot{A}}) \begin{pmatrix} 0 & \sigma^\mu_{A\dot{B}} \\ \bar{\sigma}^{\mu\dot{A}B} & 0 \end{pmatrix} \begin{pmatrix} -\chi_{+B} \\ +\bar{\chi}_-^{\dot{B}} \end{pmatrix}$$

$$= (\psi_-^A, \bar{\psi}_{+\dot{A}}) \begin{pmatrix} \sigma^\mu_{A\dot{B}} \bar{\chi}_-^{\dot{B}} \\ -\bar{\sigma}^{\mu\dot{A}B} \chi_{+B} \end{pmatrix}$$

$$= \psi_-^A \sigma^\mu_{A\dot{B}} \bar{\chi}_-^{\dot{B}} - \bar{\psi}_{+\dot{A}} \bar{\sigma}^{\mu\dot{A}B} \chi_{+B}$$

$$= (\psi_- \sigma^\mu \bar{\chi}_-)_2 - (\bar{\psi}_+ \bar{\sigma}^\mu \chi_+)_2$$

again using (1.77) and (1.78).
Before we demonstrate (1.184e) we discuss (1.184f), i.e.

$$(\sigma_4^{\mu\nu})_{ab} := \frac{i}{4}([\gamma^\mu, \gamma^\nu])_{ab}$$

$$= \frac{i}{4}(\gamma^\mu_{ac}\gamma^\nu_{cb} - \gamma^\nu_{ac}\gamma^\mu_{cb})$$

$$= \frac{i}{4}\left[\begin{pmatrix} 0 & \sigma^\mu_{A\dot{C}} \\ \bar{\sigma}^{\mu\dot{A}C} & 0 \end{pmatrix} \begin{pmatrix} 0 & \sigma^\nu_{C\dot{B}} \\ \bar{\sigma}^{\nu\dot{C}B} & 0 \end{pmatrix} - \begin{pmatrix} 0 & \sigma^\nu_{A\dot{C}} \\ \bar{\sigma}^{\nu\dot{A}C} & 0 \end{pmatrix} \begin{pmatrix} 0 & \sigma^\mu_{C\dot{B}} \\ \bar{\sigma}^{\mu\dot{C}B} & 0 \end{pmatrix} \right]$$

(using (1.136))

$$= \frac{i}{4} \begin{pmatrix} (\sigma^\mu\bar{\sigma}^\nu - \sigma^\nu\bar{\sigma}^\mu)_A{}^B & 0 \\ 0 & (\bar{\sigma}^\mu\sigma^\nu - \bar{\sigma}^\nu\sigma^\mu)^{\dot{A}}{}_{\dot{B}} \end{pmatrix}$$

$$= \begin{pmatrix} (\sigma_2^{\mu\nu})_A{}^B & 0 \\ 0 & (\bar{\sigma}_2^{\mu\nu})^{\dot{A}}{}_{\dot{B}} \end{pmatrix}$$

using (1.119 a,b). With this expression it is easy to

verify equation (1.184e):

$$(\overline{\Psi}\sigma^{\mu\nu}X)_4$$

$$= (\psi_-^A, \overline{\psi}_{+\dot{A}}) \begin{pmatrix} (\sigma_2^{\mu\nu})_A{}^B & 0 \\ 0 & (\overline{\sigma}_2^{\mu\nu})^{\dot{A}}{}_{\dot{B}} \end{pmatrix} \begin{pmatrix} \chi_{+B} \\ \overline{\chi}_-^{\dot{B}} \end{pmatrix}$$

$$= (\psi_-^A, \overline{\psi}_{+\dot{A}}) \begin{pmatrix} (\sigma_2^{\mu\nu})_A{}^B \chi_{+B} \\ (\overline{\sigma}_2^{\mu\nu})^{\dot{A}}{}_{\dot{B}} \overline{\chi}_-^{\dot{B}} \end{pmatrix}$$

$$= \psi_-^A (\sigma_2^{\mu\nu})_A{}^B \chi_{+B} + \overline{\psi}_{+\dot{A}} (\overline{\sigma}_2^{\mu\nu})^{\dot{A}}{}_{\dot{B}} \overline{\chi}_-^{\dot{B}}$$

$$= (\psi_- \sigma_2^{\mu\nu} \chi_+)_2 + (\overline{\psi}_+ \overline{\sigma}_2^{\mu\nu} \overline{\chi}_-)_2$$

using (1.77) and (1.78).

1.4.7 Calculations with Majorana Spinors

As explained in Section 1.4.5 a Majorana spinor has the characteristic property that we can replace the dotted Weyl spinor by the complex conjugate of the undotted one. In the terminology of the previous section this means that we can replace ψ_-^A by ψ_+^A and $\overline{\psi}_-^{\dot{A}}$ by $\overline{\psi}_+^{\dot{A}}$. Thus we can drop the suffixes $\pm$ in (1.182) and write

$$\psi_M = \begin{pmatrix} \psi_A \\ \overline{\psi}^{\dot{A}} \end{pmatrix} \tag{1.185}$$

Then equations (1.184a) to (1.184e) read

$$(\overline{\Psi}_M X_M)_4 = (\psi\chi)_2 + (\overline{\psi}\overline{\chi})_2 \tag{1.186a}$$

$$(\bar{\Psi}_M \gamma^5 X_M)_4 = -(\psi X)_2 + (\bar{\Psi}\bar{X})_2 \tag{1.186b}$$

$$(\bar{\Psi}_M \gamma^\mu X_M)_4 = (\bar{\Psi}\bar{\sigma}^\mu \chi)_2 + (\psi \sigma^\mu \bar{\chi})_2 \tag{1.186c}$$

$$(\bar{\Psi}_M \gamma^\mu \gamma^5 X_M)_4 = -(\bar{\Psi}\bar{\sigma}^\mu \chi)_2 + (\psi \sigma^\mu \bar{\chi})_2 \tag{1.186d}$$

$$(\bar{\Psi}_M \sigma_4^{\mu\nu} X_M)_4 = (\psi \sigma^{\mu\nu} \chi)_2 + (\bar{\Psi}\bar{\sigma}^{\mu\nu} \bar{\chi})_2 \tag{1.186e}$$

<u>Proposition</u>: The following relation holds

$$(\bar{\Psi}_M \gamma^\mu \gamma^5 \psi_M)(\bar{\Psi}_M \gamma^\nu \gamma^5 \psi_M) = \eta^{\mu\nu} (\bar{\Psi}_M \psi_M)_4^2 \tag{1.187}$$

<u>Proof</u>: Using (1.186d) we have

$$(\bar{\Psi}_M \gamma^\mu \gamma^5 \psi_M)(\bar{\Psi}_M \gamma^\nu \gamma^5 \psi_M)$$

$$= \{-(\bar{\Psi}\bar{\sigma}^\mu \psi)_2 + (\psi \sigma^\mu \bar{\Psi})_2\}\{-(\bar{\Psi}\bar{\sigma}^\nu \psi)_2 + (\psi \sigma^\nu \bar{\Psi})_2\}$$

Now using (1.115) we obtain

$$(\bar{\Psi}_M \gamma^\mu \gamma^5 \psi_M)(\bar{\Psi}_M \gamma^\nu \gamma^5 \psi_M)$$

$$= [(\psi \sigma^\mu \bar{\Psi})_2 + (\psi \sigma^\mu \bar{\Psi})_2][(\psi \sigma^\nu \bar{\Psi})_2 + (\psi \sigma^\nu \bar{\Psi})_2]$$

$$= 4(\psi \sigma^\mu \bar{\Psi})_2 (\psi \sigma^\nu \bar{\Psi})_2$$

$$= 2\eta^{\mu\nu} (\psi\psi)_2 (\bar{\Psi}\bar{\Psi})_2$$

using (1.118). For the present case (1.186a) is

$$(\bar{\Psi}_M \psi_M)_4 = (\psi\psi)_2 + (\bar{\Psi}\bar{\Psi})_2 \; .$$

Taking the square of this expression we obtain

$$[(\bar{\Psi}_M \Psi_M)_4]^2 = [(\psi\psi)_2 + (\bar{\Psi}\bar{\Psi})_2]^2$$

$$= (\psi\psi)_2(\psi\psi)_2 + (\psi\psi)_2(\bar{\Psi}\bar{\Psi})_2$$

$$+ (\bar{\Psi}\bar{\Psi})_2(\psi\psi)_2 + (\bar{\Psi}\bar{\Psi})_2(\bar{\Psi}\bar{\Psi})_2$$

$$= 2(\psi\psi)_2(\bar{\Psi}\bar{\Psi})_2$$

since, as will be shown below,

$$(\psi\psi)_2(\psi\psi)_2 = (\bar{\Psi}\bar{\Psi})_2(\bar{\Psi}\bar{\Psi})_2 = 0$$

and

$$(\bar{\Psi}\bar{\Psi})_2(\psi\psi)_2 = (\psi\psi)_2(\bar{\Psi}\bar{\Psi})_2$$

Hence it follows that

$$(\bar{\Psi}_M \gamma^\mu \gamma^5 \Psi_M)_4 (\bar{\Psi}_M \gamma^\nu \gamma^5 \Psi_M) = \eta^{\mu\nu}(\bar{\Psi}\Psi)_4^2$$

We now show explicitly that $(\psi\psi)_2^2 = 0$:

$$(\psi\psi)_2^2 = (\psi\psi)_2(\psi\psi)_2$$

$$= (\psi^A\psi_A)(\psi^B\psi_B) \quad \text{(using (1.77))}$$

$$= \varepsilon^{AC}\psi_C\psi_A\,\varepsilon^{BD}\psi_D\psi_B \quad \text{(using (1.76a))}$$

$$= \varepsilon^{AC}\varepsilon^{BD}\psi_C\psi_A\psi_D\psi_B$$

$$= \varepsilon^{12}\varepsilon^{12}\psi_2\psi_1\psi_2\psi_1 + \varepsilon^{12}\varepsilon^{21}\psi_2\psi_1\psi_1\psi_2$$

$$+ \varepsilon^{21}\varepsilon^{12}\psi_1\psi_2\psi_2\psi_1 + \varepsilon^{21}\varepsilon^{21}\psi_1\psi_2\psi_1\psi_2$$

$$= \psi_2\psi_1\psi_2\psi_1 - \psi_2\psi_1\psi_1\psi_2$$

$$- \psi_1\psi_2\psi_2\psi_1 + \psi_1\psi_2\psi_1\psi_2$$

Taking into account the fact that ψ_A are Grassmann numbers we obtain (cf. discussion following (1.79))

$$(\psi\psi)_2^2 = -4\,\psi_1\psi_1\psi_2\psi_2 = 0$$

since for Grassmann numbers ψ_1

$\{\psi_1, \psi_1\} = 0$ implies $\psi_1\psi_1 + \psi_1\psi_1 = 0$

i.e. $\psi_1\psi_1 = 0$

A similar calculation can be carried out for $(\overline{\Psi}\overline{\Psi})_2^2$.

We now verify that

$$(\overline{\Psi}\overline{\Psi})_2(\psi\psi)_2 = (\psi\psi)_2(\overline{\Psi}\overline{\Psi})_2$$

Consider

$$\begin{aligned}
(\overline{\Psi}\overline{\Psi})_2(\psi\psi)_2 &= \overline{\Psi}_{\dot{A}}\overline{\Psi}^{\dot{A}}\psi^A\psi_A \\
&= -\overline{\Psi}_{\dot{A}}\psi^A\overline{\Psi}^{\dot{A}}\psi_A = \psi^A\overline{\Psi}_{\dot{A}}\overline{\Psi}^{\dot{A}}\psi_A \\
&= \psi^A\psi_A\overline{\Psi}_{\dot{A}}\overline{\Psi}^{\dot{A}} \\
&= (\psi\psi)_2(\overline{\Psi}\overline{\Psi})_2
\end{aligned}$$

We prove one more formula.

<u>Proposition</u>: The following relation holds

$$(\overline{\Psi}_M\gamma^5\psi_M)_4^2 = -(\overline{\Psi}_M\psi_M)_4^2 \tag{1.188}$$

<u>Proof</u>: Using (1.186b) we have

$$\begin{aligned}
(\overline{\Psi}_M\gamma^5\psi_M)^2 &= \{-(\psi\psi)_2 + (\overline{\Psi}\overline{\Psi})_2\}^2 \\
&= (\psi\psi)_2^2 - 2(\psi\psi)_2(\overline{\Psi}\overline{\Psi})_2 + (\overline{\Psi}\overline{\Psi})_2^2 \\
&= -2(\psi\psi)_2(\overline{\Psi}\overline{\Psi})_2 \\
&= -\{(\psi\psi)_2^2 + 2(\psi\psi)_2(\overline{\Psi}\overline{\Psi})_2 + (\overline{\Psi}\overline{\Psi})_2^2\} \\
&= -(\overline{\Psi}_M\psi_M)_4^2 \qquad \text{(using (1.186a))}
\end{aligned}$$

C H A P T E R 2

NO-GO THEOREMS AND GRADED LIE ALGEBRAS

2.1 The Coleman-Mandula Theorem and the Haag-Lopuszanski-Sohnius Theorem

We now discuss the two theorems already referred to in the Introduction.[41]

The Theorem of Coleman and Mandula[25]

Let G be a connected symmetry group of the S-matrix, i.e. a group whose generators commute with the S-matrix, and make the following five assumptions:

i) (Lorentz invariance) G contains a subgroup which is locally isomorphic to the Poincaré group.

ii) (Particle finiteness) All particle types correspond to positive-energy representations of the Poincaré group. For any finite mass M, there is only a finite number of particles with mass less than M.

iii) (Weak elastic analyticity) Elastic scattering amplitudes are analytic functions of centre-of-mass energy squared s and invariant momentum transfer squared t in some neighbourhood of the physical region, except at normal thresholds.

iv) (Occurrence of scattering) Let $|p\rangle$ and $|p'\rangle$ be any two one-particle momentum eigenstates, and let $|p,p'\rangle$ be the two-particle state constructed from these. Then

$$T\,|p,p'\rangle \neq 0$$

where T is the T-matrix defined by

$$S = 1 - i\,(2\pi)^4 \delta^4(p_\mu - p'_\mu)\,T$$

except, perhaps, for certain isolated values of s. In simpler terms this means: two plane waves scatter at almost any energy.

v) (Technical assumption) The generators of G, considered as integral operators in momentum space, have distributions for their kernels.

Then the group G is locally isomorphic to the direct product of a compact symmetry group and the Poincaré group.

We recall briefly some basic results of scattering theory. The Hilbert space $\mathcal{H}$ is the direct sum of an infinite number of subspaces, i.e.

$$\mathcal{H} = \mathcal{H}^{(1)} \oplus \mathcal{H}^{(2)} \oplus \cdots$$

Here $\mathcal{H}^{(n)}$ is the n-particle subspace. It is a subspace of the direct product (symmetric or antisymmetric in accordance with the generalized exclusion principle) of n Hilbert spaces, each being isomorphic to $\mathcal{H}^{(1)}$. The S-matrix is a unitary operator on $\mathcal{H}$. A unitary operator U on $\mathcal{H}$ is said to be a symmetry transformation of the S-matrix if

i) U transforms one-particle states into one-particle states,

ii) U acts on many-particle states as if they were tensor products of one-particle states,

iii) U commutes with S.
Thus the theorem of Coleman and Mandula, stated here without proof, demonstrates that the most general Lie algebra of symmetries of the S-matrix contains the energy momentum operator P_μ, the Lorentz rotation generator $M_{\mu\nu}$ and a finite number of Lorentz scalar operators B_ℓ, i.e.

$$[P_\mu, B_\ell] = 0, \quad [M_{\mu\nu}, B_\ell] = 0$$

where the B_ℓ constitute a Lie algebra,

$$[B_\ell, B_m] = i\, c_{\ell m}{}^{k} B_k$$

and $c_{\ell m}{}^{k}$ are the structure constants of this Lie algebra of the compact internal symmetry group (e.g.SU(2)).

<u>The Theorem of Haag, Lopuszanski and Sohnius</u>[28]

Supersymmetries avoid the restrictions of the Coleman-Mandula theorem by relaxing one condition (in this sense the Haag-Lopuszanski-Sohnius theorem is a natural extension of the Coleman-Mandula theorem). Haag, Lopuszanski and Sohnius generalize the notion of a Lie algebra to include algebraic systems whose defining relations involve in addition to the usual commutators also anticommutators. These algebras are called superalgebras or graded Lie algebras. The generalization of the Poincaré algebra to a superalgebra is obtained in its simplest version by the following procedure. One adds to the Poincaré algebra a Majorana spinor charge with components Q_a, a = 1,..., 4, with the following properties

$$\{Q_a, \bar{Q}_b\} = 2\gamma^\mu_{ab} P_\mu, \quad [Q_a, P_\mu] = 0$$

$$[Q_a, M^{\mu\nu}] = \sigma^{\mu\nu}_{4\,ab}\, Q_b \tag{2.1}$$

(we shall see later that $\{Q_a, Q_b\} = -2(\gamma^\mu C)_{ab} P_\mu$)

where (cf. (1.184))

$$\sigma_4^{\mu\nu} = \frac{i}{4}[\gamma^\mu, \gamma^\nu]$$

and

$$\bar{Q}_a = (Q^+\gamma_0)_a$$

P_μ and $M^{\mu\nu}$ are the usual generators of displacements and homogeneous Lorentz transformations of space-time. In order to incorporate an internal symmetry in a nontrivial way it is often convenient to rewrite (2.1) in terms of two-component Weyl spinors Q_A, $\bar{Q}_{\dot{A}}$. Then

$$\{Q_A, Q_B\} = 0, \quad \{\bar{Q}_{\dot{A}}, \bar{Q}_{\dot{B}}\} = 0$$

$$\{Q_A, \bar{Q}_{\dot{B}}\} = 2\sigma^\mu_{A\dot{B}} P_\mu, \; [Q_A, M^{\mu\nu}] = i(\sigma_2^{\mu\nu})_A{}^B Q_B$$

$$[Q_A, P_\mu] = 0, \quad [\bar{Q}_{\dot{A}}, P_\mu] = 0 \tag{2.2}$$

$\sigma_2^{\mu\nu}$ being defined by (1.119a). In (2.2) the dotted and undotted indices assume values A = 1,2 and $\dot{A} = \dot{1}, \dot{2}$ respectively and refer to the (0, 1/2), (1/2, 0) representations of the spinor group SL(2,C). We now assume that we have a set of generators (i.e. spinor charges) Q_A^α (α = 1,..., N, where N is the dimension of the chosen representation of G) which transform according to some representation of a compact Lie group G, such as SU(3), which represents the internal symmetry group. Then the generators of G are the Lorentz scalars B_ℓ. The $\bar{Q}_{\dot{A}}^\alpha$ transform according to the complex conjugate representation of this group. Then the relations (2.2) generalize as follows

$$\{Q_A^\alpha, Q_B^\beta\} = \{\bar{Q}_{\dot{A}}^\alpha, \bar{Q}_{\dot{B}}^\beta\} = 0$$

$$\{Q_A^\alpha, \bar{Q}_{\dot{B}}^\beta\} = 2\delta^{\alpha\beta}\sigma^\mu_{A\dot{B}}P_\mu$$
$$[Q_A^\alpha, P_\mu] = [\bar{Q}_{\dot{A}}^\alpha, P_\mu] = 0$$
$$[Q_A^\alpha, B_\ell] = i S_\ell^{\alpha\beta} Q_A^\beta$$
$$[Q_A^\alpha, M^{\mu\nu}] = i(\sigma_2{}^{\mu\nu})_A{}^B Q_B^\alpha$$
$$[B_\ell, B_m] = i c_{\ell m}{}^k B_k \qquad (2.3)$$

where the $S_\ell^{\alpha\beta}$ are the hermitian representation matrices of the representation containing the charges Q_A^α , and as mentioned before, the B_ℓ are the generators of the internal symmetry group. The theorem of Haag, Lopuszanski and Sohnius now states that the maximal symmetry of the S-matrix is the direct product of an internal symmetry with the superalgebra given by relations (2.3). The only allowed extension is the possible appearance of socalled central charges in the anticommutator of two undotted spinors. Instead of the first relation of (2.3) one would then have

$$\{Q_A^\alpha, Q_B^\beta\} = \varepsilon_{AB} Z^{\alpha\beta}, \quad Z^{\alpha\beta} = -Z^{\beta\alpha}$$

where ε_{AB} is given by (1.59). Furthermore

$$[Z^{\alpha\beta}, B_\ell] = 0$$

which is the reason why the quantities $Z^{\alpha\beta}$ are called central charges.

2.2 Graded Lie Algebras

2.2.1 Lie Algebras

Before we consider graded Lie algebras it is worthwhile to recapitulate the definition of a Lie algebra.

A Lie algebra consists of a vector space L over a field (here R or C) with a composition rule called product, written o , defined as follows:

$$o \; : \; L \times L \longrightarrow L$$

If $v_1, v_2, v_3 \in L$, then the following properties define the Lie algebra

i) $v_1 \circ v_2 \in L$ (closure)

ii) $v_1 \circ (v_2 + v_3) = v_1 \circ v_2 + v_1 \circ v_3$ (linearity)

iii) $v_1 \circ v_2 = - v_2 \circ v_1$. (antisymmetry)

iv) $v_1 \circ (v_2 \circ v_3) + v_2 \circ (v_3 \circ v_1) + v_3 \circ (v_1 \circ v_2) = 0$ (Jacobi identity)

(2.4)

Example: The matrix space of complex 2x2 matrices which are traceless and antihermitian forms a Lie algebra, su(2,C), the Lie algebra of the Lie group SU(2,C), provided we define the product as

$$a \circ b := [a, b] = ab - ba, \quad \forall a, b \in su(2,C)$$

A basis of the vector space L is given by the three matrices

$$\tau_i := \frac{i}{2} \sigma_i \;, \quad \tau_i^+ = - \tau_i$$

where the σ_i are the three Pauli matrices. We verify (2.4) for su(2,C). Thus

i)
$$\tau_i \circ \tau_j := [\tau_i, \tau_j] = - \varepsilon_{ijk} \tau_k$$

where ε_{ijk} is totally antisymmetric in i, j, k, and $\varepsilon_{123} = +1$. Hence the vector space is closed for products o = $[\, , \,]$

ii)
$$\tau_i \circ (\tau_j + \tau_k) = [\tau_i, \tau_j + \tau_k]$$

$$= \tau_i (\tau_j + \tau_k) - (\tau_j + \tau_k) \tau_i$$
$$= \tau_i \tau_j + \tau_i \tau_k - \tau_j \tau_i - \tau_k \tau_i$$
$$= \tau_i \tau_j - \tau_j \tau_i + \tau_i \tau_k - \tau_k \tau_i$$
$$= [\tau_i, \tau_j] + [\tau_i, \tau_k]$$
$$= \tau_i \circ \tau_j + \tau_i \circ \tau_k$$

This verifies the linearity of the product $\circ = [,]$.

iii)
$$\tau_i \circ \tau_j = [\tau_i, \tau_j]$$
$$= \tau_i \tau_j - \tau_j \tau_i$$
$$= -(\tau_j \tau_i - \tau_i \tau_j)$$
$$= -[\tau_j, \tau_i]$$
$$= -\tau_j \circ \tau_i$$

iv) It is cumbersome to verify the Jacobi identity; however, it is well known that the following relation holds which expresses the same property

$$[\tau_1, [\tau_2, \tau_3]] + [\tau_2, [\tau_3, \tau_1]] + [\tau_3, [\tau_1, \tau_2]] = 0$$

2.2.2 Graded Algebras

We now define graded algebras. In the simplest case a graded algebra consists of a vector space L which is the direct sum of two subspaces L_0 and L_1 ; i.e.

$$L = L_0 \oplus L_1$$

and a product o,

$$\circ : L \times L \longrightarrow L$$

with the following properties:

i) $u_1 \circ u_2 \in L_o$, $\forall$ u_1 , $u_2 \in L_o$

ii) $u \circ v \in L_1, \forall u \in L_o$, $v \in L_1$

iii) $v_1 \circ v_2 \in L_o, \forall$ $v_1, v_2 \in L_1$

(2.5)

(this algebra is called a Z_2 graded algebra).
More generally, L is the direct sum of N + 1 , $N \geq 1$, subspaces L_k, i.e.

$$L = \bigoplus_{k=0}^{N} L_k$$

with a product o :

$$o : L \times L \longrightarrow L$$

such that if $u_k \in L_k$, then

$$u_j \circ u_k \in L_{j+k \mod (N+1)}$$

A product o with such a property is called a grading.

2.2.3 Graded Lie Algebras

A graded algebra becomes a graded Lie algebra if one modifies the product o in the following way. For simplicity we consider a Z_2 graded algebra. Let L_o and L_1 be vector spaces and

$$L := L_o \oplus L_1$$

L is the direct sum of L_o and L_1 . We define the product o

$$o : L \times L \longrightarrow L$$

with the following properties:

i) Grading

$\forall x_i \in L_i$, $i = 0, 1$

$$x_i \circ x_j \in L_{i+j \mod 2}$$

Then L becomes a graded algebra according to (2.5).

ii) Supersymmetrization

$$\forall x_i \in L_i,\ x_j \in L_j,\ i,j = 0,1$$

$$x_i \circ x_j = -(-1)^{ij} x_j \circ x_i$$

iii) Generalized Jacobi identities

$$\forall x_k \in L_k,\ x_m \in L_m,\ x_\ell \in L_\ell,\ k,\ell,m \in \{0,1\}$$

$$x_k \circ (x_\ell \circ x_m)(-1)^{km} + x_\ell \circ (x_m \circ x_k)(-1)^{\ell k} + x_m \circ (x_k \circ x_\ell)(-1)^{m\ell} = 0 \tag{2.6}$$

With this definition of the product, L as a vector space becomes a graded Lie algebra. It is important to note that L is not a Lie algebra, since, as defined in (2.6 ii), the product is in general not antisymmetric. To see this it is advantageous to write (2.6 ii) out explicitly:

a) i = 0, j = 0 , i.e. $x_o \in L_o$, $y_o \in L_o$; then

$$x_o \circ y_o = -(-1)^{0.0} y_o \circ x_o = -y_o \circ x_o$$

Hence in the subspace L_o the product o is antisymmetric.

b) i = 0, j = 1, $x_o \in L_o$, $y_1 \in L_1$;

$$x_o \circ y_1 = -(-1)^{0.1} y_1 \circ x_o = -y_1 \circ x_o$$

c) i = 1, j = 1, $x_1 \in L_1$, $y_1 \in L_1$;

$$x_1 \circ y_1 = -(-1)^{1.1} y_1 \circ x_1 = y_1 \circ x_1$$

Hence the product is symmetric in the subspace L_1.
With the above definition of a graded Lie algebra the subspace L_o spans an ordinary Lie algebra, because (2.4) is satisfied in the subspace L_o. The subspace L_1 is not even an algebra, since according to (2.6 i) L_1 is not closed under products o, i.e. if $x_1, y_1 \in L_1$

$$x_1 \circ y_1 \in L_{1+1 \bmod 2} = L_o$$

2.3 The Graded Lie Algebra of the Group SU(2,C)

As an example we discuss the Z_2 grading of su(2,C), the Lie algebra of SU(2,C). As stated above, the subspace L_o in the construction of (2.6) is an ordinary Lie algebra. It is natural therefore to take L_o as the Lie algebra of SU(2) with generators τ_1, τ_2, τ_3 and

$$[\tau_i, \tau_j] = -\varepsilon_{ijk}\tau_k$$

We therefore define the product

$$o : L \times L \longrightarrow L$$

on the subspace L_o as:

$$o : L_o \times L_o \longrightarrow L_o$$

$$(\tau_i, \tau_j) \longrightarrow \tau_i \circ \tau_j := [\tau_i, \tau_j] = -\varepsilon_{ijk}\tau_k \in L_o \qquad (2.7)$$

We denote the generators of L_1 by Q_a (a = 1,..., N; N = dim L_1). We now have to define the product o when multiplying any $\tau_i \in L_o$ by Q_a, and multiplying two Q_a's $\in L_1$. In the case of the former we have

$$o : L_o \times L_1 \longrightarrow L_1$$

Thus if we form the product of any $\tau_i \in L_o$ with a $Q_a \in L_1$, then according to (2.6 i) we must obtain an element of the subspace L_1. We define therefore

$$o : (\tau_i, Q_a) \longrightarrow \tau_i \circ Q_a = -(t_i)_{ab} Q_b \in L_1 \qquad (2.8)$$

Here the $(t_i)_{ab}$ are coefficients which are restricted by the generalized Jacobi identity (2.6 iii). In the present case we have

$$x_k \in L_k, \quad x_l \in L_l, \quad x_m \in L_m$$

and

$$\tau_i \in L_o, \ \tau_j \in L_o, \ Q_a \in L_1$$

and so the generalized Jacobi identity is

$$X_k \circ (X_\ell \circ X_m)(-1)^{km} + X_m \circ (X_k \circ X_\ell)(-1)^{m\ell} + X_\ell \circ (X_m \circ X_k)(-1)^{\ell k} = 0$$

so that

$$\tau_{i\cdot} \circ (\tau_{j\cdot} \circ Q_a) + Q_a \circ (\tau_{i\cdot} \circ \tau_{j\cdot}) + \tau_{j\cdot} \circ (Q_a \circ \tau_{i\cdot}) = 0$$

and using (2.7) and (2.8)

$$\tau_{i\cdot} \circ [-(t_{j\cdot})_{ab} Q_b] + Q_a \circ [-\varepsilon_{ijk} \tau_k] + \tau_{j\cdot} \circ [(t_{i\cdot})_{ab} Q_b] = 0$$

i.e.

$$-(t_{j\cdot})_{ab} (\tau_{i\cdot} \circ Q_b) - \varepsilon_{ijk} (Q_a \circ \tau_k) + (t_{i\cdot})_{ab} (\tau_{j\cdot} \circ Q_b) = 0$$

Again using (2.8)

$$(t_{j\cdot})_{ab} (t_{i\cdot})_{bc} Q_c - \varepsilon_{ijk} (t_k)_{ab} Q_b - (t_{i\cdot})_{ab} (t_{j\cdot})_{bc} Q_c = 0$$

so that

$$[(t_{i\cdot})_{ab} (t_{j\cdot})_{bc} - (t_{j\cdot})_{ab} (t_{i\cdot})_{bc}] Q_c = -\varepsilon_{ijk} (t_k)_{ac} Q_c$$

or (since the Q_c's are independent)

$$[t_i, t_j]_{ac} = -\varepsilon_{ijk} (t_k)_{ac} \qquad (2.9)$$

This relation states that the coefficients $(t_i)_{ab}$ of (2.8) constitute matrices t_i ($\dim L_1 \times \dim L_1$ matrices) which are representation matrices of the algebra L_o . If the dimension of L_1 is two, then $t_i = \tau_i$, if the dimension of L_1 is three, then $(t_i)_{ab} = -\varepsilon_{iab}$. Finally we have to define the product o on the subspace L_1. According to (2.6 i) we have for $Q_a, Q_b \in L_1$:

$$Q_a \circ Q_b \in L_{1+1} = L_2 \cong L_o$$

According to (2.6 ii) this product on the subspace L_1 has to be symmetric, i.e.

$$Q_a \circ Q_b = -(-1)^{1.1} Q_b \circ Q_a$$
$$= Q_b \circ Q_a$$

Hence we define

$$\circ : \quad L_1 \times L_1 \longrightarrow L_o$$
$$Q_a \circ Q_b \longrightarrow (h_i)_{ab} \tau_i \tag{2.10}$$

where $(h_i)_{ab} = (h_i)_{ba}$, and the matrices h_i, $i = 1,2,3$, are three symmetric dim L_1 x dim L_1 matrices. Again we use the generalized Jacobi identities to find restrictions for the matrices h_i (an arbitrary multiplicative constant factor can be absorbed in the Q's). Let

$$\tau_i \in L_o, \quad Q_a, Q_b \in L_1$$

then (2.6iii) reads for this case

$$\tau_i \circ (Q_a \circ Q_b)(-1)^{0.1} + Q_b \circ (\tau_i \circ Q_a)(-1)^{1.1}$$
$$+ Q_a \circ (Q_b \circ \tau_1)(-1)^{1.0} = 0$$

so that

$$\tau_i \circ ((h_j)_{ab} \tau_j) - Q_b \circ (-(t_i)_{ac} Q_c)$$
$$+ Q_a \circ [(t_i)_{bc} Q_c] = 0$$

or

$$(h_j)_{ab} (\tau_i \circ \tau_j) + (t_i)_{ac} (Q_b \circ Q_c)$$
$$+ (t_i)_{bc} (Q_a \circ Q_c) = 0$$

Then

$$-(h_j)_{ab}\varepsilon_{ijk}\tau_k + (t_i)_{ac}(h_j)_{bc}\tau_j + (t_i)_{bc}(h_j)_{ac}\tau_j = 0$$

$$\left((t_i)_{ac}(h_j)_{cb} + (t_i)_{bc}(h_j)_{ca}\right)\tau_j = \varepsilon_{ijk}(h_j)_{ab}\tau_k$$
$$= \varepsilon_{ikj}(h_k)_{ab}\tau_j$$

so that

$$[(t_i h_j)_{ab} + (t_i h_j)_{ba}]\tau_j = -\varepsilon_{ijk}(h_k)_{ab}\tau_j$$

and

$$t_i h_j + (t_i h_j)^T = -\varepsilon_{ijk} h_k \qquad (2.11)$$

If we take N = dim L_1 = 2, then $t_i = \tau_i$, i = 1,2,3, and the most general form of h_i is given by (remembering that the h_i are symmetric)

$$(h_i)_{ab} = a_i \delta_{ab} + b_i(\tau_3)_{ab} + c_i(\tau_1)_{ab}$$
$$= \begin{pmatrix} a_i + \frac{i}{2}b_i & \frac{i}{2}c_i \\ \frac{i}{2}c_i & a_i - \frac{i}{2}b_i \end{pmatrix}_{ab}$$
$$(a_i, b_i, c_i \in C)$$

τ_2 does not appear in this expansion of h_i, because, as stated earlier, $(h_i)_{ab}$ has to be symmetric in a and b, and τ_2 is antisymmetric. The nine coefficients a_i , b_i , c_i have to be computed from (2.11), i.e. the relation

$$\tau_i h_j + (\tau_i h_j)^T = -\varepsilon_{ijk} h_k, \quad i,j,k = 1,2,3$$

<u>Proposition</u>: The matrices h_i are given by

$$h_i = 2c_3(\tau_i \tau_2) \qquad (2.12)$$

(this result can be guessed: since h_i is a two-by-two

matrix, we expect τ_i on the right hand side; to make this symmetric we must have $\tau_i \cdot \tau_2$; fixing the remaining overall multiplicative factor is a matter of convention).

<u>Proof</u>: We consider the various cases separately.

<u>i = 1:</u>

$$\tau_1 = \frac{i}{2}\begin{pmatrix} 0 & 1 \\ 1 & 0 \end{pmatrix}$$

$$\tau_1 h_j = \frac{i}{2}\begin{pmatrix} 0 & 1 \\ 1 & 0 \end{pmatrix}\begin{pmatrix} a_j + \frac{i}{2} b_j & \frac{i}{2} c_j \\ \frac{i}{2} c_j & a_j - \frac{i}{2} b_j \end{pmatrix}$$

$$= \frac{i}{2}\begin{pmatrix} \frac{i}{2} c_j & a_j - \frac{i}{2} b_j \\ a_j + \frac{i}{2} b_j & \frac{i}{2} c_j \end{pmatrix}$$

and

$$\tau_1 h_j + (\tau_1 h_j)^T$$

$$= \frac{i}{2}\begin{pmatrix} \frac{i}{2} c_j & a_j - \frac{i}{2} b_j \\ a_j + \frac{i}{2} b_j & \frac{i}{2} c_j \end{pmatrix} + \frac{i}{2}\begin{pmatrix} \frac{i}{2} c_j & a_j + \frac{i}{2} b_j \\ a_j - \frac{i}{2} b_j & \frac{i}{2} c_j \end{pmatrix}$$

$$= \begin{pmatrix} -\frac{1}{2} c_j & i a_j \\ i a_j & -\frac{1}{2} c_j \end{pmatrix} = -\varepsilon_{1jk} h_k \qquad \text{(from (2.11))}$$

$$= -\varepsilon_{1jk}\begin{pmatrix} a_k + \frac{i}{2} b_k & \frac{i}{2} c_k \\ \frac{i}{2} c_k & a_k - \frac{i}{2} b_k \end{pmatrix}$$

Now consider:

i = 1, j = 2 :

$$\begin{pmatrix} -\frac{1}{2} c_2 & i a_2 \\ i a_2 & -\frac{1}{2} c_2 \end{pmatrix} = -\begin{pmatrix} a_3 + \frac{i}{2} b_3 & \frac{i}{2} c_3 \\ \frac{i}{2} c_3 & a_3 - \frac{i}{2} b_3 \end{pmatrix}$$

Hence $\quad -\frac{1}{2}c_2 = -a_3 - \frac{i}{2}b_3 ,$

$$ia_2 = -\frac{i}{2}c_3 , \quad -\frac{1}{2}c_2 = -a_3 + \frac{i}{2}b_2$$

so that

$$b_3 = 0 , \quad a_3 = \frac{1}{2}c_2 , \quad a_2 = -\frac{1}{2}c_3$$

i = 1, j = 3 :

$$\begin{pmatrix} -\frac{1}{2}c_3 & ia_3 \\ ia_3 & -\frac{1}{2}c_3 \end{pmatrix} = \begin{pmatrix} a_2 + \frac{i}{2}b_2 & \frac{i}{2}c_2 \\ \frac{i}{2}c_2 & a_2 - \frac{i}{2}b_2 \end{pmatrix}$$

Hence

$$-\frac{1}{2}c_3 = a_2 + \frac{i}{2}b_2 ,$$

$$ia_3 = \frac{i}{2}c_2 , \quad -\frac{1}{2}c_3 = a_2 - \frac{i}{2}b_2$$

so that

$$b_2 = 0 , \quad a_3 = \frac{1}{2}c_2 , \quad a_2 = -\frac{1}{2}c_3$$

<u>i = 2</u> : $\quad \tau_2 = \frac{i}{2}\begin{pmatrix} 0 & -i \\ i & 0 \end{pmatrix}$

$$\tau_2 h_j = \frac{1}{2}\begin{pmatrix} 0 & 1 \\ -1 & 0 \end{pmatrix}\begin{pmatrix} a_j + \frac{i}{2}b_j & \frac{i}{2}c_j \\ \frac{i}{2}c_j & a_j - \frac{i}{2}b_j \end{pmatrix}$$

$$= \frac{1}{2}\begin{pmatrix} \frac{i}{2}c_j & a_j - \frac{i}{2}b_j \\ -a_j - \frac{i}{2}b_j & -\frac{i}{2}c_j \end{pmatrix}$$

Then

$$\tau_2 h_j + (\tau_2 h_j)^T$$

$$= \frac{1}{2}\begin{pmatrix} ic_j & -ib_j \\ -ib_j & -ic_j \end{pmatrix} = \frac{i}{2}\begin{pmatrix} c_j & -b_j \\ -b_j & -c_j \end{pmatrix}$$

$i = 2$, $j = 1$: Equating the above to $-\varepsilon_{213}\, h_3$ we obtain

$$\frac{i}{2}\begin{pmatrix} c_1 & -b_1 \\ -b_1 & -c_1 \end{pmatrix} = \begin{pmatrix} a_3 + \frac{i}{2}b_3 & \frac{i}{2}c_3 \\ \frac{i}{2}c_3 & a_3 - \frac{i}{2}b_3 \end{pmatrix}$$

Hence

$$\frac{i}{2}c_1 = a_3 + \frac{i}{2}b_3 = a_3 \qquad \text{since} \qquad b_3 = 0$$

$$-\frac{i}{2}b_1 = \frac{i}{2}c_3$$

$$-\frac{i}{2}c_1 = a_3 - \frac{i}{2}b_3 = a_3 \qquad \text{since} \qquad b_3 = 0$$

Hence

$$a_3 = 0, \quad c_1 = 0, \quad b_1 = -c_3$$

and since

$$a_3 = \tfrac{1}{2}c_2, \quad c_2 = 0 .$$

$i = 2$, $j = 3$:

$$\frac{i}{2}\begin{pmatrix} c_3 & -b_3 \\ -b_3 & -c_3 \end{pmatrix} = -\begin{pmatrix} a_1 + \frac{i}{2}b_1 & \frac{i}{2}c_1 \\ \frac{i}{2}c_1 & a_1 - \frac{i}{2}b_1 \end{pmatrix}$$

Thus

$$\frac{i}{2}c_3 = -a_1 - \frac{i}{2}b_1, \quad -\frac{i}{2}b_3 = -\frac{i}{2}c_1,$$

$$-\frac{i}{2}c_3 = -a_1 + \frac{i}{2}b_1$$

so that

$$b_3 = 0 \qquad\qquad c_1 = 0, \quad a_1 = 0, \quad b_1 = -c_3 .$$

Hence

$$
\begin{array}{lll}
a_1 = 0 & b_1 = -c_3 & c_1 = 0 \\
a_2 = 0 \; -c_3/2 & b_2 = 0 & c_2 = 0 \\
a_3 = 0 & b_3 = 0 & c_3 \text{ undetermined}
\end{array}
$$

and therefore

$$h_1 = -\frac{i}{2} c_3 \begin{pmatrix} 1 & 0 \\ 0 & -1 \end{pmatrix} = 2 c_3 (\tau_1 \tau_2)$$

$$h_2 = -\frac{1}{2} c_3 \begin{pmatrix} 1 & 0 \\ 0 & 1 \end{pmatrix} = 2 c_3 (\tau_2 \tau_2)$$

$$h_3 = \frac{i}{2} c_3 \begin{pmatrix} 0 & 1 \\ 1 & 0 \end{pmatrix} = 2 c_3 (\tau_3 \tau_2)$$

i.e.

$$h_i = 2 c_3 (\tau_i \tau_2)$$

The constant c_3 can be absorbed in redefined generators Q_a as mentioned before.

We finally consider the Jacobi identity for three operators Q. In this case this relation is

$$Q_a \circ (Q_b \circ Q_c)(-1) + Q_b \circ (Q_c \circ Q_a)(-1) + Q_c \circ (Q_a \circ Q_b)(-1) = 0$$

and so

$$Q_a \circ ((h_i)_{bc} \tau_i) + Q_b \circ ((h_i)_{ca} \tau_i) + Q_c \circ ((h_i)_{ab} \tau_i) = 0$$

i.e.

$$(h_i)_{bc} (Q_a \circ \tau_i) + (h_i)_{ca} (Q_b \circ \tau_i) + (h_i)_{ab} (Q_c \circ \tau_i) = 0$$

Using (2.8) with $t_i = \tau_i$,

$$(h_i)_{bc} (\tau_i)_{ad} Q_d + (h_i)_{ca} (\tau_i)_{bd} Q_d + (h_i)_{ab} (\tau_i)_{cd} Q_d = 0$$

or, using (2.12) and the fact that the Q's are independent of one another

$$(\tau_i\tau_2)_{bc}(\tau_i)_{ad} + (\tau_i\tau_2)_{ca}(\tau_i)_{bd}$$
$$+ (\tau_i\tau_2)_{ab}(\tau_i)_{cd} = 0$$

This relation is valid, i.e. satisfied by the τ_i's, as can be checked by direct calculation.

In summary we have the following construction. The graded Lie algebra su(2,C) is

$$\tau_i \circ \tau_j = -\varepsilon_{ijk}\tau_k , \quad L_o \times L_o \longrightarrow L_o$$
$$\tau_i \circ Q_a = -(\tau_i)_{ab} Q_b , \quad L_o \times L_1 \longrightarrow L_{o+1} = L_1$$
$$Q_a \circ Q_b = (\tau_i\tau_2)_{ab}\tau_i , \quad L_1 \times L_1 \longrightarrow L_{1+1} = L_2 \simeq L_o$$

The structure of the coefficients of the product in $L_o x L_1$ is determined by generalized Jacobi identities. The only freedom one has is the choice of representation for these coefficients. The important point is that the coefficient matrix of τ_i o Q_a has to be a representation matrix of the algebra given by L_o. This result is an immediate consequence of the generalized Jacobi identity with one Q_a. Thus the graded Lie algebra su(2,C) has 5 elements: τ_i for i = 1,2,3, and Q_a for a = 1,2 and is written OSp(1/2) ("one slash two")[42]. The dimension of the chosen representation then determines the dimension of the subspace L_1 and therefore the number of operators Q_a, a = 1,..., dim L_1. Once the representation has been chosen, the coefficient matrices of the product Q_a o Q_b may be found from the Jacobi identity with two Q's. For consistency these matrices have to satisfy the generalized Jacobi identity with three Q's; this, however, is not always the case - for example, an extension of SU(2) with an n-dimensional L_1 ($n > 2$) is not possible, since the Jacobi identity for three Q's then requires the h_i to be zero.

2.4 Z_2 Graded Lie Algebras

We now discuss Z_2 graded Lie algebras in more detail. We recall first some general properties of Z_2 graded Lie algebras. A linear algebra L:= Span $\{X_\mu\}$ is given a Z_2 grading if L is the direct sum of two subspaces L_o and L_1 :

$$L = L_o \oplus L_1$$

where we set

$$L_o = \text{Span } \{E_i\}, \quad i = 1,\dots,\dim L_o$$
$$L_1 = \text{Span } \{Q_a\}, \quad a = 1,\dots,\dim L_1$$

on which a composition law

$$o : L \times L \longrightarrow L$$

acts as follows (see (2.5))

$$\left.\begin{array}{l} L_o \ o \ L_o \subset L_o \\ L_o \ o \ L_1 \subset L_1 \\ L_1 \ o \ L_1 \subset L_o \end{array}\right\} \qquad (2.13)$$

L_o is an ordinary Lie algebra.

Definition: We assign to any $X_\mu \in L$ a degree $g \in \{0,1\}$ by defining

$$g_\mu := g(X_\mu) = 0 \longleftrightarrow X_\mu \in L_o \qquad (2.14)$$

$$g_\mu := g(X_\mu) = 1 \longleftrightarrow X_\mu \in L_1 \qquad (2.15)$$

and we say $X_\mu \in L$ is even if $g_\mu = 0$

$X_\mu \in L$ is odd if $g_\mu = 1$.

With these definitions we see that the set of generators E_i, $i = 1,\ldots, \dim L_o$, which is a basis of L_o, consists of even elements, whereas the generators Q_a, which span the subspace L_1, are odd.

Definition: We define the product on L by

$$\circ : L \times L \longrightarrow L$$

$$\circ : (X_\mu, X_\nu) \longrightarrow X_\mu \circ X_\nu := X_\mu X_\nu - (-1)^{g(X_\mu) g(X_\nu)} X_\nu X_\mu \qquad (2.16)$$

We now consider this product separately on the subspaces L_o, L_1:

i) $\circ : L_o \times L_o \longrightarrow L_o$

Let $E_i, E_j \in L_o$ and according to (2.14)

$$g(E_i) = g(E_j) = 0$$

Then

$$E_i \circ E_j = E_i E_j - (-1)^{g_i g_j} E_j E_i \qquad \text{(from (2.16))}$$

$$= E_i E_j - E_j E_i$$

$$=: [E_i, E_j] \qquad (2.17)$$

With this construction we are consistent with the requirement that the product o be antisymmetric on the subspace L_o (see (2.6 ii)).

ii) $\circ : L_o \times L_1 \longrightarrow L_1$

Let $E_i \in L_o$ and $Q_a \in L_1$; then

$$g(E_i) = 0, \; g(Q_a) = 1$$

and

$$E_i \circ Q_a = E_i Q_a - (-1)^{g_i g_a} Q_a E_i$$

$$= E_i Q_a - Q_a E_i$$

$$=: [E_i, Q_a] \qquad (2.18)$$

Again the product is antisymmetric on $L_o \times L_1$, as demanded by (2.6 ii).

iii) $\circ : L_1 \times L_1 \longrightarrow L_0$

Let $Q_a, Q_b \in L_1$ and according to (2.15)

Then
$$g(Q_a) = g(Q_b) = 1$$

$$\begin{aligned} Q_a \circ Q_b &= Q_a Q_b - (-1)^{g_a g_b} Q_b Q_a \\ &= Q_a Q_b + Q_b Q_a \\ &=: \{Q_a, Q_b\} \end{aligned} \tag{2.19}$$

From (2.19) we see that the product o is symmetric on the subspace L_1.

We now introduce generalized structure constants. Since $X_\mu \circ X_\nu \in L$, the product $X_\mu \circ X_\nu$ must be a linear combination of the basis elements X_ω, i.e.

$$X_\mu \circ X_\nu = c_{\mu\nu}{}^{\omega} X_\omega \tag{2.20}$$

where
$$c_{\mu\nu}{}^{\omega} = -(-1)^{g_\mu g_\nu} c_{\nu\mu}{}^{\omega} \tag{2.21}$$

are the generalized structure constants of the graded Lie algebra. According to (2.21) the structure constants are antisymmetric for even-even and even-odd pairs of elements of L and symmetric for odd-odd pairs. We now prove the relation (2.21).

In order to prove (2.21) we start from (2.20), i.e.

$$X_\mu \circ X_\nu = c_{\mu\nu}{}^{\omega} X_\omega \quad \text{and} \quad X_\nu \circ X_\mu = c_{\nu\mu}{}^{\omega} X_\omega$$

On the other hand, using (2.16) we have

$$\begin{aligned} X_\mu \circ X_\nu &= X_\mu X_\nu - (-1)^{g_\mu g_\nu} X_\nu X_\mu \\ X_\nu \circ X_\mu &= X_\nu X_\mu - (-1)^{g_\nu g_\mu} X_\mu X_\nu \\ &= X_\nu X_\mu - (-1)^{g_\mu g_\nu} X_\mu X_\nu \end{aligned}$$

and

$$(-1)^{g_\mu g_\nu} X_\nu \circ X_\mu = (-1)^{g_\mu g_\nu} X_\nu X_\mu - [(-1)^{g_\mu g_\nu}]^2 X_\mu X_\nu$$
$$= (-1)^{g_\mu g_\nu} X_\nu X_\mu - X_\mu X_\nu$$
$$= -\{X_\mu X_\nu - (-1)^{g_\mu g_\nu} X_\nu X_\mu\}$$
$$= -(X_\mu \circ X_\nu) \quad \text{(with (2.16))}$$

Hence

$$X_\mu \circ X_\nu + (-1)^{g_\mu g_\nu} X_\nu \circ X_\mu = 0$$

and so with (2.20)

$$c_{\mu\nu}{}^{\omega} X_\omega + (-1)^{g_\mu g_\nu} c_{\nu\mu}{}^{\omega} X_\omega = 0, \ \forall X_\omega \in L$$

i.e.

$$c_{\mu\nu}{}^{\omega} = -(-1)^{g_\mu g_\nu} c_{\nu\mu}{}^{\omega}$$

since the X_ω are independent.

We now consider the generalized Jacobi identities. For $X_\mu, X_\nu, X_\rho \in$ L these are defined by

$$[X_\mu \circ (X_\nu \circ X_\rho)](-1)^{g_\mu g_\rho} + [X_\nu \circ (X_\rho \circ X_\mu)](-1)^{g_\nu g_\mu}$$
$$+ [X_\rho \circ (X_\mu \circ X_\nu)](-1)^{g_\rho g_\nu} = 0 \qquad (2.22)$$

We demonstrate that (2.16) obeys the generalized Jacobi identity (2.22):

$$X_\mu \circ (X_\nu \circ X_\rho)(-1)^{g_\mu g_\rho} + X_\rho \circ (X_\mu \circ X_\nu)(-1)^{g_\rho g_\nu}$$
$$+ X_\nu \circ (X_\rho \circ X_\mu)(-1)^{g_\nu g_\mu}$$
$$= [X_\mu (X_\nu \circ X_\rho) - (-1)^{g_\mu (g_\nu + g_\rho)} (X_\nu \circ X_\rho) X_\mu] \cdot$$
$$\cdot (-1)^{g_\mu g_\rho}$$

$$+[X_\rho(X_\mu \circ X_\nu) - (-1)^{g_\rho(g_\mu+g_\nu)}(X_\mu \circ X_\nu)X_\rho](-1)^{g_\rho g_\nu}$$

$$+[X_\nu(X_\rho \circ X_\mu) - (-1)^{g_\nu(g_\rho+g_\mu)}(X_\rho \circ X_\mu)X_\nu](-1)^{g_\nu g_\mu}$$

$$=[X_\mu(X_\nu X_\rho - (-1)^{g_\nu g_\rho}X_\rho X_\nu) - (-1)^{g_\mu(g_\nu+g_\rho)} \cdot$$

$$\cdot (X_\nu X_\rho - (-1)^{g_\nu g_\rho}X_\rho X_\nu)X_\mu](-1)^{g_\mu g_\rho}$$

$$+[X_\rho(X_\mu X_\nu - (-1)^{g_\mu g_\nu}X_\nu X_\mu) - (-1)^{g_\rho(g_\mu+g_\nu)} \cdot$$

$$\cdot (X_\mu X_\nu - (-1)^{g_\mu g_\nu}X_\nu X_\mu)X_\rho](-1)^{g_\rho g_\nu}$$

$$+[X_\nu(X_\rho X_\mu - (-1)^{g_\rho g_\mu}X_\mu X_\rho) - (-1)^{g_\nu(g_\rho+g_\mu)} \cdot$$

$$\cdot (X_\rho X_\mu - (-1)^{g_\rho g_\mu}X_\mu X_\rho)X_\nu](-1)^{g_\nu g_\mu}$$

$$=[X_\mu X_\nu X_\rho - (-1)^{g_\nu g_\rho}X_\mu X_\rho X_\nu - (-1)^{g_\mu(g_\nu+g_\rho)} \cdot$$

$$\cdot X_\nu X_\rho X_\mu + (-1)^{g_\mu g_\nu + g_\mu g_\rho + g_\nu g_\rho}X_\rho X_\nu X_\mu] \cdot$$

$$\cdot (-1)^{g_\mu g_\rho}$$

$$+[X_\rho X_\mu X_\nu - (-1)^{g_\mu g_\nu}X_\rho X_\nu X_\mu$$

$$- (-1)^{g_\rho g_\mu + g_\rho g_\nu}X_\mu X_\nu X_\rho$$

$$+ (-1)^{g_\rho g_\mu + g_\rho g_\nu + g_\mu g_\nu}X_\nu X_\mu X_\rho](-1)^{g_\rho g_\nu}$$

$$+[X_\nu X_\rho X_\mu - (-1)^{g_\rho g_\mu}X_\nu X_\mu X_\rho$$

$$- (-1)^{g_\nu g_\rho + g_\nu g_\mu}X_\rho X_\mu X_\nu$$

$$+ (-1)^{g_\nu g_\rho + g_\nu g_\mu + g_\rho g_\mu}X_\mu X_\rho X_\nu](-1)^{g_\nu g_\mu}$$

$$
\begin{aligned}
&= X_\mu X_\nu X_\rho (-1)^{g_\mu g_\rho} - X_\mu X_\rho X_\nu (-1)^{g_\rho(g_\mu+g_\nu)} \\
&\quad - X_\nu X_\rho X_\mu (-1)^{g_\mu g_\nu} + X_\rho X_\nu X_\mu (-1)^{g_\nu(g_\mu+g_\rho)} \\
&\quad + X_\rho X_\mu X_\nu (-1)^{g_\rho g_\nu} - X_\rho X_\nu X_\mu (-1)^{g_\nu(g_\mu+g_\rho)} \\
&\quad - X_\mu X_\nu X_\rho (-1)^{g_\rho g_\mu} + X_\nu X_\mu X_\rho (-1)^{g_\mu(g_\rho+g_\nu)} \\
&\quad + X_\nu X_\rho X_\mu (-1)^{g_\nu g_\mu} - X_\nu X_\mu X_\rho (-1)^{g_\mu(g_\nu+g_\rho)} \\
&\quad - X_\rho X_\mu X_\nu (-1)^{g_\nu g_\rho} + X_\mu X_\rho X_\nu (-1)^{g_\rho(g_\mu+g_\nu)} \\
&= X_\mu X_\nu X_\rho \left((-1)^{g_\mu g_\rho} - (-1)^{g_\rho g_\mu}\right) \\
&\quad + X_\mu X_\rho X_\nu \left[(-1)^{g_\rho(g_\mu+g_\nu)} - (-1)^{g_\rho(g_\nu+g_\mu)}\right] \\
&\quad + X_\rho X_\nu X_\mu \left[(-1)^{g_\nu(g_\mu+g_\rho)} - (-1)^{g_\nu(g_\mu+g_\rho)}\right] \\
&\quad + X_\rho X_\mu X_\nu \left[(-1)^{g_\rho g_\nu} - (-1)^{g_\nu g_\rho}\right] \\
&\quad + X_\nu X_\mu X_\rho \left[(-1)^{g_\mu(g_\rho+g_\nu)} - (-1)^{g_\mu(g_\rho+g_\nu)}\right] \\
&\quad + X_\nu X_\rho X_\mu \left[(-1)^{g_\nu g_\mu} - (-1)^{g_\mu g_\nu}\right] \\
&= 0
\end{aligned}
$$

Hence the composition law (2.16) is a product which obeys all conditions of the product of a graded Lie algebra as defined by (2.6). We now write out the generalized Jacobi identity for the four different possibilities.

i) $X_\mu, X_\nu, X_\rho \in L_0$: i.e. we take

$$X_\mu = E_i, \quad X_\nu = E_j, \quad X_\rho = E_k$$

and

$$(-1)^{g_\mu g_\rho} = (-1)^{g_\rho g_\nu} = (-1)^{g_\mu g_\nu} = +1$$

Then (2.22) reads in the case of three generators of L_o

$$X_\mu \circ (X_\nu \circ X_\rho)(-1)^{g_\mu g_\rho} + X_\rho \circ (X_\mu \circ X_\nu)(-1)^{g_\rho g_\nu} + X_\nu \circ (X_\rho \circ X_\mu)(-1)^{g_\mu g_\nu}$$

$$= E_i \circ (E_j \circ E_k) + E_k \circ (E_i \circ E_j) + E_j \circ (E_k \circ E_i)$$

$$= E_i (E_j \circ E_k) - (E_j \circ E_k) E_i + E_k (E_i \circ E_j) - (E_i \circ E_j) E_k + E_j (E_k \circ E_i) - (E_k \circ E_i) E_j$$

$$= E_i (E_j E_k - E_k E_j) - (E_j E_k - E_k E_j) E_i + E_k (E_i E_j - E_j E_i) - (E_i E_j - E_j E_i) E_k + E_j (E_k E_i - E_i E_k) - (E_k E_i - E_i E_k) E_j$$

$$= E_i [E_j, E_k] - [E_j, E_k] E_i + E_k [E_i, E_j] - [E_i, E_j] E_k + E_j [E_k, E_i] - [E_k, E_i] E_j$$

$$= [E_i, [E_j, E_k]] + [E_j, [E_k, E_i]] + [E_k, [E_i, E_j]]$$

$$= 0$$

Hence the generalized Jacobi identity for three E's reduces to the Jacobi identity for the Lie algebra L_o , i.e.

$$[E_i, [E_j, E_k]] + [E_j, [E_k, E_i]] + [E_k, [E_i, E_j]] = 0 \qquad (2.23)$$

ii) $\underline{X_\mu, X_\nu \in L_o , X_\rho \in L_1}$: i.e. we take

$$X_\mu = E_i \in L_o , \quad X_\nu = E_j \in L_o , \quad X_\rho = Q_a \in L_1$$

In this case equation (2.22) becomes

$$
\begin{aligned}
& X_\mu \circ (X_\nu \circ X_\rho)(-1)^{g_\mu g_\rho} + X_\rho \circ (X_\mu \circ X_\nu)(-1)^{g_\rho g_\nu} \\
&\qquad + X_\nu \circ (X_\rho \circ X_\mu)(-1)^{g_\mu g_\nu} \\
&= E_{i\cdot} \circ (E_{j\cdot} \circ Q_a)(-1)^{0\cdot 1} + Q_a \circ (E_{i\cdot} \circ E_{j\cdot})(-1)^{1\cdot 0} \\
&\qquad + E_{j\cdot} \circ (Q_a \circ E_{i\cdot})(-1)^{0\cdot 0} \\
&= E_{i\cdot}(E_{j\cdot} \circ Q_a) - (-1)^{g_{i\cdot}(g_{j\cdot}+g_a)}(E_{j\cdot} \circ Q_a)E_{i\cdot} \\
&\quad + Q_a(E_{i\cdot} \circ E_{j\cdot}) - (-1)^{g_a(g_{i\cdot}+g_{j\cdot})}(E_{i\cdot} \circ E_{j\cdot})Q_a \\
&\quad + E_{j\cdot}(Q_a \circ E_{i\cdot}) - (-1)^{g_{j\cdot}(g_a+g_{i\cdot})}(Q_a \circ E_{i\cdot})E_{j\cdot} \\
&= E_{i\cdot}\left(E_{j\cdot}Q_a - (-1)^{g_j g_a} Q_a E_{j\cdot}\right) \\
&\quad - \left(E_{j\cdot}Q_a - (-1)^{g_j g_a} Q_a E_{j\cdot}\right)E_{i\cdot} \\
&\quad + Q_a\left(E_{i\cdot}E_{j\cdot} - (-1)^{g_i g_j} E_{j\cdot}E_{i\cdot}\right) \\
&\quad - \left(E_{i\cdot}E_{j\cdot} - (-1)^{g_i g_j} E_{j\cdot}E_{i\cdot}\right)Q_a \\
&\quad + E_{j\cdot}\left(Q_a E_{i\cdot} - (-1)^{g_a g_i} E_{i\cdot}Q_a\right) \\
&\quad - \left(Q_a E_{i\cdot} - (-1)^{g_a g_i} E_{i\cdot}Q_a\right)E_{j\cdot} \\
&= E_{i\cdot}[E_{j\cdot}, Q_a] - [E_{j\cdot}, Q_a]E_{i\cdot} + Q_a[E_{i\cdot}, E_{j\cdot}] \\
&\quad - [E_{i\cdot}, E_{j\cdot}]Q_a + E_{j\cdot}[Q_a, E_{i\cdot}] - [Q_a, E_{i\cdot}]E_{j\cdot} \\
&= [E_{i\cdot}, [E_{j\cdot}, Q_a]] + [E_{j\cdot}, [Q_a, E_{i\cdot}]] \\
&\qquad + [Q_a, [E_{i\cdot}, E_{j\cdot}]] \\
&= 0
\end{aligned}
$$

Hence the generalized Jacobi identity for two E's and one Q is

$$[E_i,[E_j,Q_a]]+[E_j,[Q_a,E_i]]$$
$$+[Q_a,[E_i,E_j]]=0 \qquad (2.24)$$

iii) $\underline{X_\mu \in L_0, X_\nu, X_\rho \in L_1}$: here we set

$$X_\mu = E_i \in L_0, \quad X_\nu = Q_a \in L_1, \quad X_\rho = Q_b \in L_1$$

Then equation (2.22) becomes

$$X_\mu \circ (X_\nu \circ X_\rho)(-1)^{g_\mu g_\rho} + X_\rho \circ (X_\mu \circ X_\nu)(-1)^{g_\rho g_\nu}$$
$$+ X_\nu \circ (X_\rho \circ X_\mu)(-1)^{g_\nu g_\mu}$$
$$= E_i \circ (Q_a \circ Q_b)(-1)^{g_i g_b} + Q_b \circ (E_i \circ Q_a)(-1)^{g_b g_a}$$
$$+ Q_a \circ (Q_b \circ E_i)(-1)^{g_a g_i}$$
$$= E_i (Q_a \circ Q_b) - (-1)^{g_i(g_a+g_b)}(Q_a \circ Q_b)E_i$$
$$- Q_b(E_i \circ Q_a) + (-1)^{g_b(g_i+g_a)}(E_i \circ Q_a)Q_b$$
$$+ Q_a(Q_b \circ E_i) - (-1)^{g_a(g_b+g_i)}(Q_b \circ E_i)Q_a$$
$$= E_i(Q_a Q_b - (-1)^{g_a g_b} Q_b Q_a)$$
$$- (Q_a Q_b - (-1)^{g_a g_b} Q_b Q_a)E_i$$
$$- Q_b(E_i Q_a - (-1)^{g_i g_a} Q_a E_i)$$
$$- (E_i Q_a - (-1)^{g_i g_a} Q_a E_i)Q_b$$
$$+ Q_a(Q_b E_i - (-1)^{g_i g_b} E_i Q_b)$$
$$+ (Q_b E_i - (-1)^{g_i g_b} E_i Q_b)Q_a$$

$$= E_i \{Q_a, Q_b\} - \{Q_a, Q_b\} E_i$$
$$- Q_b [E_i, Q_a] - [E_i, Q_a] Q_b$$
$$+ Q_a [Q_b, E_i] + [Q_b, E_i] Q_a$$
$$= [E_i, \{Q_a, Q_b\}] - \{Q_b, [E_i, Q_a]\}$$
$$+ \{Q_a, [Q_b, E_i]\}$$
$$= 0$$

Hence the Jacobi relation for one E and two Q's is

$$[E_i, \{Q_a, Q_b\}] + \{Q_a, [Q_b, E_i]\}$$
$$- \{Q_b, [E_i, Q_a]\} = 0 \qquad (2.25)$$

iv) $\underline{X_\mu, X_\nu, X_\rho \in L_1}$: here we set

$$X_\mu = Q_a \in L_1, \; X_\nu = Q_b \in L_1, \; X_\rho = Q_c \in L_1$$

Then equation (2.22) becomes for this case of three Q's

$$X_\mu \circ (X_\nu \circ X_\rho)(-1)^{g_\mu g_\rho} + X_\rho \circ (X_\mu \circ X_\nu)(-1)^{g_\rho g_\nu}$$
$$+ X_\nu \circ (X_\rho \circ X_\mu)(-1)^{g_\nu g_\mu}$$
$$= Q_a \circ (Q_b \circ Q_c)(-1)^{g_a g_c} + Q_c \circ (Q_a \circ Q_b)(-1)^{g_c g_b}$$
$$+ Q_b \circ (Q_c \circ Q_a)(-1)^{g_b g_a}$$
$$= - Q_a (Q_b \circ Q_c) + (-1)^{g_a (g_b + g_c)} (Q_b \circ Q_c) Q_a$$
$$- Q_c (Q_a \circ Q_b) + (-1)^{g_c (g_a + g_b)} (Q_a \circ Q_b) Q_c$$
$$- Q_b (Q_c \circ Q_a) + (-1)^{g_b (g_c + g_a)} (Q_c \circ Q_a) Q_b$$
$$= - Q_a (Q_b Q_c - (-1)^{g_b g_c} Q_c Q_b)$$

$$
\begin{aligned}
&+(Q_b Q_c - (-1)^{g_b g_c} Q_c Q_b) Q_a \\
&- Q_c (Q_a Q_b - (-1)^{g_a g_b} Q_b Q_a) \\
&+ (Q_a Q_b - (-1)^{g_a g_b} Q_b Q_a) Q_c \\
&- Q_b (Q_c Q_a - (-1)^{g_c g_a} Q_a Q_c) \\
&+ (Q_c Q_a - (-1)^{g_c g_a} Q_a Q_c) Q_b \\
&= - Q_a \{Q_b, Q_c\} + \{Q_b, Q_c\} Q_a - Q_c \{Q_a, Q_b\} \\
&\quad - \{Q_a, Q_b\} Q_c - Q_b \{Q_c, Q_a\} + \{Q_c, Q_a\} Q_b \\
&= - [Q_a, \{Q_b, Q_c\}] - [Q_c, \{Q_a, Q_b\}] \\
&\qquad\qquad - [Q_b, \{Q_c, Q_a\}] \\
&= 0
\end{aligned}
$$

Hence for three Q's the Jacobi identity is

$$
\begin{aligned}
&[Q_a, \{Q_b, Q_c\}] + [Q_b, \{Q_c, Q_a\}] \\
&\quad + [Q_c, \{Q_a, Q_b\}] = 0 \qquad (2.26)
\end{aligned}
$$

<u>Remark</u>: It is important to observe that the commutators and anticommutators in (2.23) to (2.26) are consistent with the composition law (2.6 i) for graded Lie algebras. Consider for example the generalized Jacobi identity (2.25), i.e.

$$[E_i, \{Q_a, Q_b\}] + \{Q_a, [Q_b, E_i]\} - \{Q_b, [E_i, Q_a]\} = 0$$

According to (2.13) $\{Q_a , Q_b\} \in L_0$; but on L_0 the product has to be antisymmetric, so an expression like $[E_i, \{Q_a , Q_b\}]$ is consistent with (2.6). Similarly $[E_i , Q_a] \in L_1$ and $\{Q_b, [E_i , Q_a]\}$ has the correct bracket structure.

In conclusion we introduce the socalled generalized Killing form. The generalized Killing form on L is defined by[35,43]

$$b_{\mu\nu} = \sum_{\rho,\sigma} (-1)^{g_\sigma} c_{\mu\rho}{}^{\sigma} c_{\nu\sigma}{}^{\rho} \tag{2.27}$$

where

$$g_\sigma := g(X_\sigma), \quad [X_\mu, X_\nu] = c_{\mu\nu}{}^{\sigma} X_\sigma,$$

$$X_\sigma \in L_0 \Rightarrow g_\sigma = 0, \quad X_\sigma \in L_1 \Rightarrow g_\sigma = 1.$$

Here $b_{\mu\nu} = B(X_\mu, X_\nu)$ is a symmetric bilinear form when acting on the subspace L_0 and thus reduces to the usual Killing form for the Lie algebra L_0. We can then apply Cartan's criterion[35]:the Lie algebra L_0 is semisimple if and only if

$$B(X_\mu, X_\nu)\,|_{L_0 \times L_0}$$

is nondegenerate, i.e. $\det(b_{\mu\nu}) \neq 0$.
Furthermore L_0 is compact if and only if the generalized Killing form restricted to L_0 is negative definite. Hence a compact Lie algebra is always semisimple. In addition, the generalized Killing form is antisymmetric when acting on the subspace L_1, i.e.

$$b_{ab} = B(Q_a, Q_b) = -B(Q_b, Q_a)$$

and

$$B(X_\mu, X_\nu)\Big|_{L_0 \times L_1} = 0$$

2.5 Graded Matrices

An endomorphism $M : L \longrightarrow L$ acting on L[35] can be represented by a graded matrix which has the following matrix structure. Let $\dim L_o = n$, $\dim L_1 = m$. Then

$$M = \begin{pmatrix} A & B \\ C & D \end{pmatrix} \in \text{end } (L) \tag{2.28}$$

is an $(n+m) \times (n+m)$ matrix where

A is an nxn square matrix

B is an nxm submatrix

C is an mxn submatrix

and D is an mxm square matrix.

Since L is the direct sum of L_o and L_1, i.e. $L = L_o \oplus L_1$, a vector $v \in L$ has the structure

$$v = \begin{pmatrix} v_0 \\ v_1 \end{pmatrix}, \; v_0 \in L_0, \; v_1 \in L_1 \tag{2.29}$$

with $g(v_0) = 0$, $g(v_1) = 1$ according to (2.14) and (2.15) where v_0 is an element of the bosonic sector of L, i.e. L_o, and v_1 is an element of the fermionic sector of L, i.e. L_1. Introducing the following index notation

$$\begin{aligned} A &= (A_{ij}), \quad i,j = 1,\ldots,n \\ B &= (B_{ia}), \quad i = 1,\ldots,n; \; a = 1,\ldots,m \\ C &= (C_{ai}), \quad a = 1,\ldots,m; \; i = 1,\ldots,n \\ D &= (D_{ab}), \quad a,b = 1,\ldots,m \\ v_0 &= (v_{0i}), \quad i = 1,\ldots,n \\ v_1 &= (v_{1a}), \quad a = 1,\ldots,m \end{aligned} \tag{2.30}$$

we obtain (with w,z = 1,..., n+m)

$$M_{wz}v_z = \begin{pmatrix} A_{ij} & B_{ia} \\ C_{bj} & D_{ba} \end{pmatrix} \begin{pmatrix} v_{0j} \\ v_{1a} \end{pmatrix}$$

$$= \begin{pmatrix} A_{ij}v_{0j} + B_{ia}v_{1a} \\ C_{bj}v_{0j} + D_{ba}v_{1a} \end{pmatrix} \equiv \begin{pmatrix} v'_{0i} \\ v'_{1b} \end{pmatrix}$$

Thus

$$v_0' = Av_0 + Bv_1 \in L_0$$
$$v_1' = Cv_0 + Dv_1 \in L_1$$

Since $v_0' \in L_0$ we have $g(v_0')=0$. But $g(v_0) = 0$ and $g(v_1) = 1$. Now since $Av_0 \in L_0$ and $Bv_1 \in L_0$, we must have $g(Av_0) = 0$ and $g(Bv_1) = 0$ mod 2.

Hence

$$g(Av_0) = g(A) + g(v_0)$$

implies

$$g(A) = 0$$

and

$$g(Bv_1) = g(B) + g(v_1)$$

implies

$$g(B) = 1$$

Similarly since $v_1' \in L_1$, we have $g(v_1')=1$. But $g(v_0)=0, g(v_1)=1$, so that

$$g(C) = 1 \qquad \text{and} \qquad g(D) = 0.$$

This means that A and D must be even submatrices with degree g = 0, and B and C have to be odd with degree g = 1. It follows, the elements of B and C are anticommuting variables and therefore behave as Grassmann numbers, i.e.

$$B_{ia}C_{bj} = -C_{bj}B_{ia} \tag{2.31}$$

From this relation we obtain the transposition rule

(2.32)

by setting in (2.31) a = b = 1,...,m so that

$$(BC)_{ij} = - C_{aj} B_{ia} = - (C^T B^T)_{ji}$$

Definition: The supertrace is defined by

$$\text{S Tr } M = \text{Tr } A - \text{Tr } D \tag{2.33}$$

where A and D are submatrices as in (2.28). The supertrace is defined in such a way that

$$\text{S Tr } (M_1 M_2) = \text{S Tr } (M_2 M_1) \tag{2.34}$$

Proof: We consider

$$\text{S Tr}(M_1 M_2) = \text{S Tr} \left\{ \begin{pmatrix} A_1 & B_1 \\ C_1 & D_1 \end{pmatrix} \begin{pmatrix} A_2 & B_2 \\ C_2 & D_2 \end{pmatrix} \right\}$$

$$= \text{S Tr} \begin{pmatrix} A_1 A_2 + B_1 C_2 & A_1 B_2 + B_1 D_2 \\ C_1 A_2 + D_1 C_2 & C_1 B_2 + D_1 D_2 \end{pmatrix}$$

$$= \text{Tr}(A_1 A_2 + B_1 C_2) - \text{Tr}(C_1 B_2 + D_1 D_2)$$

(using (2.33))

$$= \text{Tr}(A_1 A_2) + \text{Tr}(B_1 C_2) - \text{Tr}(C_1 B_2) - \text{Tr}(D_1 D_2)$$

$$= \text{Tr}(A_2 A_1) - \text{Tr}(C_2 B_1) + \text{Tr}(B_2 C_1) - \text{Tr}(D_2 D_1)$$

$$= \text{Tr}(A_2 A_1 + B_2 C_1) - \text{Tr}(C_2 B_1 + D_2 D_1)$$

$$= \text{S Tr} \begin{pmatrix} A_2 A_1 + B_2 C_1 & A_2 B_1 + B_2 D_1 \\ C_2 A_1 + D_2 C_1 & C_2 B_1 + D_2 D_1 \end{pmatrix}$$

$$= \text{S Tr} \left\{ \begin{pmatrix} A_2 & B_2 \\ C_2 & D_2 \end{pmatrix} \begin{pmatrix} A_1 & B_1 \\ C_1 & D_1 \end{pmatrix} \right\}$$

$$= \text{S Tr}(M_2 M_1)$$

We observe that the supertrace is defined in such a way that it is cyclic for graded matrices paralleling the analogous property of the trace for ordinary matrices.

Definition: The determinant of the supermatrix M is the superdeterminant S det M defined by

$$\mathrm{S\,det}\, M := \exp \{ \mathrm{S\,Tr}\, \ln M \} \tag{2.35}$$

If

$$M = \exp X \tag{2.36}$$

then we have, of course,

$$\mathrm{S\,det}\, M = \exp \{ \mathrm{S\,Tr}\; X \} \tag{2.37}$$

The superdeterminant is defined in analogy to the well known result

$$\det M = \exp \mathrm{tr} \ln M$$

for ordinary square matrices M. The latter is easily verified by setting M = 1 + L. Then

$$\begin{aligned} &\mathrm{tr} \ln (1 + L) \\ = \;&\mathrm{tr} \;(L - L^2/2 + L^3/3 - \dots) \end{aligned}$$

If U is a matrix such that

$$\begin{aligned} \mathrm{tr}\, L = \mathrm{tr}\,(U L U^{-1}) &= \mathrm{tr}\,(\Lambda) \\ &= \sum_i \lambda_i \end{aligned}$$

where Λ is diagonal, then

$$\begin{aligned} &\mathrm{tr} \ln (1+L) \\ &= \sum_i \lambda_i - \frac{1}{2} \sum_i \lambda_i^2 + \frac{1}{3} \sum_i \lambda_i^3 - \dots \\ &= \sum_i \ln (1+\lambda_i) = \ln \prod_i (1+\lambda_i) \\ &= \ln \det (1+\Lambda) = \ln \det (1+L) \end{aligned}$$

so that

$$\mathrm{tr} \ln M = \ln \det M$$

It is a natural consequence of (2.35) to define unimodular graded matrices by

$$S \det M = 1$$

and

$$S\,\mathrm{Tr}\, X = 0$$

Proposition: Let M_1 and M_2 be two graded matrices. Then

$$S \det (M_1 M_2) = S \det M_1 \cdot S \det M_2 \quad (2.38)$$

Proof: Consider

$$S\det(M_1M_2) = \exp\{ S\mathrm{Tr} \ln M_1M_2\}$$

using (2.35). We now define

$$A := \ln M_1 \quad , \quad B := \ln M_2$$

and use the Baker-Campbell-Hausdorff formula[44]

$$e^A e^B = e^{A+B+\frac{1}{2}[A,B]+\frac{1}{12}[A,[A,B]] - \frac{1}{12}[B,[B,A]]+\cdots}$$

$$= M_1 M_2$$

Taking the logarithm of both sides we obtain

$$\ln(M_1M_2) = \ln(e^A e^B) = A+B+\tfrac{1}{2}[A,B]+\cdots$$

Then

$$\ln(M_1M_2) = A+B+\tfrac{1}{2}[A,B]+\cdots$$

Next we take the supertrace of both sides so that

$$\begin{aligned} S\mathrm{Tr} \ln(M_1M_2) &= S\mathrm{Tr}\, A + S\mathrm{Tr}\, B + \tfrac{1}{2} S\mathrm{Tr}([A,B]) + \cdots \\ &= S\mathrm{Tr} \ln M_1 + S\mathrm{Tr} \ln M_2 \\ &\quad + \tfrac{1}{2} S\mathrm{Tr}(AB-BA) + \cdots \\ &= S\mathrm{Tr} \ln M_1 + S\mathrm{Tr} \ln M_2 \end{aligned}$$

All commutator terms vanish because the supertrace obeys (2.34). Hence

$$\begin{aligned} S\det(M_1 M_2) &= \exp\{S\mathrm{Tr}\ln(M_1 M_2)\} \\ &= \exp\{S\mathrm{Tr}\ln M_1 + S\mathrm{Tr}\ln M_2\} \\ &= \exp\{S\mathrm{Tr}\ln M_1\}\cdot\exp\{S\mathrm{Tr}\ln M_2\} \\ &= S\det(M_1)\cdot S\det(M_2) \end{aligned}$$

in view of (2.30).

<u>Proposition</u>: The superdeterminant (2.35) can be expressed in terms of ordinary determinants by means of the following formulae. Let

$$M = \begin{pmatrix} A & B \\ C & D \end{pmatrix}$$

be a graded matrix; then

$$S\det M = \frac{\det(A - BD^{-1}C)}{\det D} \tag{2.39}$$

$$= \frac{\det A}{\det(D - CA^{-1}B)} \tag{2.40}$$

<u>Proof</u>: In order to verify (2.39) we decompose the matrix M in the following form

$$\begin{aligned} M &= \begin{pmatrix} E & F \\ 0_{m\times n} & 1_{m\times m} \end{pmatrix}\begin{pmatrix} 1_{n\times n} & 0_{n\times m} \\ G & H \end{pmatrix} \\ &= M_1 M_2 \\ &= \begin{pmatrix} E + FG & FH \\ G & H \end{pmatrix} \end{aligned}$$

Comparing this with the standard form above we obtain

$$E + FG = A \,, \quad FH = B \,, \quad C = G \,, \quad D = H$$

or

$$E = A - B D^{-1} C, \quad F = B D^{-1}$$
$$G = C \,, \quad H = D$$

Then

$$\begin{aligned} S\det M &= S\det (M_1 M_2) \\ &= S\det M_1 \,.\, S\det M_2 \\ &\qquad \text{(with (2.38))} \\ &= \exp\{S\mathrm{Tr} \ln M_1\} \exp\{S\mathrm{Tr} \ln M_2\} \end{aligned}$$

Now for the particular form of M_1 under consideration

$$S\,\mathrm{Tr} \ln M_1 = \mathrm{Tr} \ln E$$

as can be shown with the help of the power series expansion of the logarithm. Thus setting $M_1 = 1 + L$ so that

$$L = \begin{pmatrix} E-1 & F \\ 0 & 0 \end{pmatrix}, \quad L^2 = \begin{pmatrix} (E-1)^2 & (E-1)F \\ 0 & 0 \end{pmatrix},$$

$$L^3 = \begin{pmatrix} (E-1)^3 & (E-1)^2 F \\ 0 & 0 \end{pmatrix}, \ldots$$

we have

$$\begin{aligned} S\mathrm{Tr} \ln M_1 &= S\mathrm{Tr} \ln (1+L) \\ &= S\mathrm{Tr} \left[L - \tfrac{1}{2} L^2 + \tfrac{1}{3} L^3 - \cdots\right] \\ &= S\mathrm{Tr} \left[\begin{pmatrix} E-1 & F \\ 0 & 0 \end{pmatrix} \right. \\ &\qquad \left. - \tfrac{1}{2} \begin{pmatrix} (E-1)^2 & (E-1)F \\ 0 & 0 \end{pmatrix} + \cdots \right] \end{aligned}$$

$$= Tr(E-1) - \frac{1}{2} Tr(E-1)^2 + \frac{1}{3} Tr(E-1)^3 - \ldots$$
$$= Tr \ln E$$

Inserting this in our expression for S det M we obtain

$$S \det M = \exp\{Tr \ln E\} \cdot \exp\{-Tr \ln H\}$$
$$= \det E \,.\, (\det H)^{-1}$$
$$= \frac{\det(A - BD^{-1}C)}{\det D}$$

Equation (2.40) can be verified by using the decomposition

$$M = \begin{pmatrix} 1_{n\times n} & 0 \\ P & Q \end{pmatrix} \begin{pmatrix} R & S \\ 0 & 1_{m\times m} \end{pmatrix}$$
$$= \begin{pmatrix} R & S \\ PR & PS+Q \end{pmatrix} \overset{!}{=} \begin{pmatrix} A & B \\ C & D \end{pmatrix}$$

Hence

$$P = CA^{-1}, \quad Q = D - CA^{-1}B$$
$$R = A, \quad B = S$$

Then as above

$$S \det M = \det R \,.\, (\det Q)^{-1}$$
$$= \frac{\det A}{\det(D - CA^{-1}B)}$$

which had to be shown.

Definition: Supertransposition is defined by

$$M^{ST} = \begin{pmatrix} A^T & -C^T \\ B^T & D^T \end{pmatrix} \tag{2.41}$$

in order to mimic the ordinary law of transposition, i.e.

$$(M_1 M_2)^{ST} = M_2^{\ ST} M_1^{\ ST} \tag{2.42}$$

<u>Proof</u>: We have

$$M_2^{ST} M_1^{ST} = \begin{pmatrix} A_2 & B_2 \\ C_2 & D_2 \end{pmatrix}^{ST} \begin{pmatrix} A_1 & B_1 \\ C_1 & D_1 \end{pmatrix}^{ST}$$

$$= \begin{pmatrix} A_2^T & -C_2^T \\ B_2^T & D_2^T \end{pmatrix} \begin{pmatrix} A_1^T & -C_1^T \\ B_1^T & D_1^T \end{pmatrix}$$

(using (2.41))

$$= \begin{pmatrix} A_2^T A_1^T - C_2^T B_1^T & -A_2^T C_1^T - C_2^T D_1^T \\ B_2^T A_1^T + D_2^T B_1^T & -B_2^T C_1^T + D_2^T D_1^T \end{pmatrix}$$

$$= \begin{pmatrix} (A_1 A_2)^T + (B_1 C_2)^T & -(C_1 A_2)^T - (D_1 C_2)^T \\ (A_1 B_2)^T + (B_1 D_2)^T & (C_1 B_2)^T + (D_1 D_2)^T \end{pmatrix}$$

(using (2.32))

$$= \begin{pmatrix} (A_1 A_2 + B_1 C_2)^T & -(C_1 A_2 + D_1 C_2)^T \\ (A_1 B_2 + B_1 D_2)^T & (C_1 B_2 + D_1 D_2)^T \end{pmatrix}$$

$$= \begin{pmatrix} A_1 A_2 + B_1 C_2 & A_1 B_2 + B_1 D_2 \\ C_1 A_2 + D_1 C_2 & C_1 B_2 + D_1 D_2 \end{pmatrix}^{ST}$$

(using (2.41))

$$= \left\{ \begin{pmatrix} A_1 & B_1 \\ C_1 & D_1 \end{pmatrix} \begin{pmatrix} A_2 & B_2 \\ C_2 & D_2 \end{pmatrix} \right\}^{ST} = (M_1 M_2)^{ST}$$

Finally we prove the following result.

Proposition:

$$S\det(M^{ST}) = S\det(M) \tag{2.43}$$

Proof: We have (using (2.39), (2.31) and again (2.39))

$$S\det(M^{ST}) = S\det\begin{pmatrix} A^T & -C^T \\ B^T & D^T \end{pmatrix}$$

$$= \frac{\det(A^T + C^T (D^T)^{-1} B^T)}{\det(D^T)}$$

$$= \frac{\det(A^T - (BD^{-1}C)^T)}{\det D}$$

$$= \frac{\det[(A - BD^{-1}C)^T]}{\det D}$$

$$= \frac{\det(A - BD^{-1}C)}{\det D}$$

$$= S\det(M).$$

CHAPTER 3

THE SUPERSYMMETRIC EXTENSION OF THE POINCARÉ ALGEBRA

3.1 The Supersymmetric Extension of the Poincaré Algebra in the Four-Component Dirac Formulation

With equation (1.30) we obtained the commutation relations of the ten-dimensional Lie algebra of the Poincaré group, i.e.

$$[P_\mu, P_\nu] = 0$$
$$[M_{\mu\nu}, P_\lambda] = i(\eta_{\nu\lambda}P_\mu - \eta_{\mu\lambda}P_\nu)$$
$$[M_{\mu\nu}, M_{\rho\sigma}] = -i(\eta_{\mu\rho}M_{\nu\sigma} - \eta_{\mu\sigma}M_{\nu\rho} - \eta_{\nu\rho}M_{\mu\sigma} + \eta_{\nu\sigma}M_{\mu\rho}) \quad (3.1)$$

The generators P_μ and $M_{\mu\nu}$ span a ten-dimensional vector space with the composition law given by (3.1). In order to construct an extension of the Poincaré algebra to a graded Lie algebra it is natural to take the Poincaré algebra as the subspace L_0 of the Z_2 graded Lie algebra L which we want to construct. From previous considerations in Chapter 2 we know that the coefficients of the product on $L_o \times L_1$ have to form a matrix representation of L_o with dimension equal to that of L_1. In the following we consider in detail the case when L_1 is four-dimensional. Thus we have four basis elements denoted by Q_a, a = 1,2,3,4, which span the subspace L_1. We therefore search for a grading of the Poincaré algebra with the following properties:

i) L_o : Poincaré algebra (3.1)

ii) $L_1 = \text{span}\,\{Q_a\}$, a = 1,2,3,4

iii) The product o : $L_o \times L_1 \longrightarrow L_1$ is defined by

a) $P_\mu \circ Q_a = 0$

b) $M_{\mu\nu} \circ Q_a = -(\sigma^4_{\mu\nu})_{ab} Q_b$

(3.2)

(we shall find that $\sigma^4_{\mu\nu} = \frac{i}{4}[\gamma_\mu, \gamma_\nu]$). In case iii) a) implies that the Q's transform trivially under translations, whereas iii) b) implies that the Q's transform as spinors under homogeneous Lorentz transformations. We note in particular:

i) With the choice (3.2) we satisfy the requirement that the coefficients of the product on $L_o \times L_1$ form a matrix representation of L_o. In this case the coefficients are 0 and $\sigma^4_{\mu\nu}$ which together have to form a 4x4 matrix representation of the Poincaré algebra.

ii) In agreement with the general properties of the product of a graded Lie algebra (cf. (2.6)), we set (cf. also (2.16))

$$\begin{aligned} P_\mu \circ Q_a &= P_\mu Q_a - (-1)^{g(P_\mu)g(Q_a)} Q_a P_\mu \\ &= P_\mu Q_a - Q_a P_\mu \\ &= [P_\mu, Q_a] = 0 \end{aligned} \tag{3.3}$$

with

$$g(P_\mu) = 0 \quad \text{since} \quad P_\mu \in L_0$$
$$g(Q_a) = 1 \quad \text{since} \quad Q_a \in L_1$$

and furthermore

$$\begin{aligned} M_{\mu\nu} \circ Q_a &= M_{\mu\nu} Q_a - (-1)^{g(M_{\mu\nu}) g(Q_a)} Q_a M_{\mu\nu} \\ &= M_{\mu\nu} Q_a - Q_a M_{\mu\nu} \\ &= [M_{\mu\nu}, Q_a] \\ &= -(\sigma^4_{\mu\nu})_{ab} Q_b \end{aligned} \qquad (3.4)$$

where

$$g(M_{\mu\nu}) = 0 \quad \text{since} \quad M_{\mu\nu} \in L_0$$
$$g(Q_a) = 1 \quad \text{since} \quad Q_a \in L_1$$

iii) The above grading of the Poincaré algebra is by no means unique. There is an infinite number of other possible gradings, for which the Q's transform under other representations of the Poincaré group.

iv) The graded Poincaré algebra is an example of a graded Lie algebra with more structure than that given in the definition (3.2). If we take $L = L_0 \oplus L_1 \oplus L_2$ with generators

$$M_{\mu\nu} \in L_0, \quad Q_a \in L_1, \quad P_\mu \in L_2$$

it is easy to show that

$$X_k \circ X_j \in L_{k+j \bmod 3}, \quad X_k \in L_k, \; X_j \in L_j$$

Before we consider the generalized Jacobi identities in detail we prove the following result.

<u>Proposition</u>: Let $\sigma^4_{\mu\nu}$ be defined by

$$\sigma^4_{\mu\nu} := \frac{i}{4} [\gamma_\mu, \gamma_\nu] \, ;$$

then

$$[\sigma_{\mu\nu},\sigma_{\rho\sigma}] = -i(\eta_{\mu\rho}\sigma_{\nu\sigma} - \eta_{\mu\sigma}\sigma_{\nu\rho} - \eta_{\nu\rho}\sigma_{\mu\sigma} + \eta_{\nu\sigma}\sigma_{\mu\rho}) \tag{3.5}$$

i.e. the $\sigma_{\mu\nu}$'s satisfy the same commutation relation as $M_{\mu\nu}$ (cf. (3.1)), which means that the matrices $\sigma_{\mu\nu}$ form a four-dimensional representation of the Lorentz algebra.

Proof: Consider

$$\begin{aligned}
[\sigma_{\mu\nu},\sigma_{\rho\sigma}] &= -\frac{1}{16}[[\gamma_\mu,\gamma_\nu],[\gamma_\rho,\gamma_\sigma]] \\
&= -\frac{1}{16}[\gamma_\mu\gamma_\nu - \gamma_\nu\gamma_\mu, \gamma_\rho\gamma_\sigma - \gamma_\sigma\gamma_\rho] \\
&= -\frac{1}{16}\{[\gamma_\mu\gamma_\nu - \gamma_\nu\gamma_\mu, \gamma_\rho\gamma_\sigma] - [\gamma_\mu\gamma_\nu - \gamma_\nu\gamma_\mu, \gamma_\sigma\gamma_\rho]\} \\
&= -\frac{1}{16}\{[\gamma_\mu\gamma_\nu, \gamma_\rho\gamma_\sigma] - [\gamma_\nu\gamma_\mu, \gamma_\rho\gamma_\sigma] \\
&\qquad - [\gamma_\mu\gamma_\nu, \gamma_\sigma\gamma_\rho] + [\gamma_\nu\gamma_\mu, \gamma_\sigma\gamma_\rho]\}
\end{aligned}$$

Using the relation

$$[A,BC] = A[B,C] + [A,B]C$$

this expression is

$$\begin{aligned}
= -\frac{1}{16}\{ &\gamma_\mu[\gamma_\nu,\gamma_\rho]\gamma_\sigma + \gamma_\mu\gamma_\rho[\gamma_\nu,\gamma_\sigma] \\
&+ \gamma_\rho[\gamma_\mu,\gamma_\sigma]\gamma_\nu + [\gamma_\mu,\gamma_\rho]\gamma_\sigma\gamma_\nu - \gamma_\nu\gamma_\rho[\gamma_\mu,\gamma_\sigma] \\
&- \gamma_\nu[\gamma_\mu,\gamma_\rho]\gamma_\sigma - \gamma_\rho[\gamma_\nu,\gamma_\sigma]\gamma_\mu - [\gamma_\nu,\gamma_\rho]\gamma_\sigma\gamma_\mu \\
&- \gamma_\mu\gamma_\sigma[\gamma_\nu,\gamma_\rho] - \gamma_\mu[\gamma_\nu,\gamma_\sigma]\gamma_\rho - \gamma_\sigma[\gamma_\mu,\gamma_\rho]\gamma_\nu \\
&- [\gamma_\mu,\gamma_\sigma]\gamma_\rho\gamma_\nu + \gamma_\nu\gamma_\sigma[\gamma_\mu,\gamma_\rho] + \gamma_\nu[\gamma_\mu,\gamma_\sigma]\gamma_\rho \\
&+ \gamma_\sigma[\gamma_\nu,\gamma_\rho]\gamma_\mu + [\gamma_\nu,\gamma_\sigma]\gamma_\rho\gamma_\mu \}
\end{aligned}$$

$$= \frac{i}{4}\{\gamma_\mu \gamma_\rho \sigma_{\nu\sigma} + \gamma_\mu \sigma_{\nu\rho} \gamma_\sigma + \gamma_\rho \sigma_{\mu\sigma} \gamma_\nu + \sigma_{\mu\rho} \gamma_\sigma \gamma_\nu$$
$$- \gamma_\nu \gamma_\rho \sigma_{\mu\sigma} - \gamma_\nu \sigma_{\mu\rho} \gamma_\sigma - \gamma_\rho \sigma_{\nu\sigma} \gamma_\mu - \sigma_{\nu\rho} \gamma_\sigma \gamma_\mu$$
$$- \gamma_\mu \gamma_\sigma \sigma_{\nu\rho} - \gamma_\mu \sigma_{\nu\sigma} \gamma_\rho - \gamma_\sigma \sigma_{\mu\rho} \gamma_\nu - \sigma_{\mu\sigma} \gamma_\rho \gamma_\nu$$
$$+ \gamma_\nu \gamma_\sigma \sigma_{\mu\rho} + \gamma_\nu \sigma_{\mu\sigma} \gamma_\rho + \gamma_\sigma \sigma_{\nu\rho} \gamma_\mu + \sigma_{\nu\sigma} \gamma_\rho \gamma_\mu\}$$

$$= \frac{i}{4}\{\gamma_\mu \gamma_\rho \sigma_{\nu\sigma} + \sigma_{\nu\sigma} \gamma_\rho \gamma_\mu - \gamma_\rho \sigma_{\nu\sigma} \gamma_\mu - \gamma_\mu \sigma_{\nu\sigma} \gamma_\rho$$
$$+ \sigma_{\mu\rho} \gamma_\sigma \gamma_\nu + \gamma_\nu \gamma_\sigma \sigma_{\mu\rho} - \gamma_\nu \sigma_{\mu\rho} \gamma_\sigma - \gamma_\sigma \sigma_{\mu\rho} \gamma_\nu$$
$$- \sigma_{\mu\sigma} \gamma_\rho \gamma_\nu - \gamma_\nu \gamma_\rho \sigma_{\mu\sigma} + \gamma_\nu \sigma_{\mu\sigma} \gamma_\rho + \gamma_\rho \sigma_{\mu\sigma} \gamma_\nu$$
$$- \gamma_\mu \gamma_\sigma \sigma_{\nu\rho} - \sigma_{\nu\rho} \gamma_\sigma \gamma_\mu + \gamma_\mu \sigma_{\nu\rho} \gamma_\sigma + \gamma_\sigma \sigma_{\nu\rho} \gamma_\mu\}$$

Using the following result which we prove below,

$$[\sigma_{\mu\nu}, \gamma_\rho] = i(\eta_{\nu\rho} \gamma_\mu - \eta_{\mu\rho} \gamma_\nu), \qquad (3.6)$$

it is possible to rearrange various terms so that

$$[\sigma_{\mu\nu}, \sigma_{\rho\sigma}]$$
$$= \frac{i}{4}\{\gamma_\mu \gamma_\rho \sigma_{\nu\sigma} + i\eta_{\sigma\rho} \gamma_\nu \gamma_\mu - i\eta_{\nu\rho} \gamma_\sigma \gamma_\mu + \gamma_\rho \sigma_{\nu\sigma} \gamma_\mu$$
$$- \gamma_\rho \sigma_{\nu\sigma} \gamma_\mu - i\eta_{\sigma\rho} \gamma_\mu \gamma_\nu + i\eta_{\nu\rho} \gamma_\mu \gamma_\sigma - \gamma_\mu \gamma_\rho \sigma_{\nu\sigma}$$
$$+ i\eta_{\rho\sigma} \gamma_\mu \gamma_\nu - i\eta_{\mu\sigma} \gamma_\rho \gamma_\nu + \gamma_\sigma \sigma_{\mu\rho} \gamma_\nu + \gamma_\nu \gamma_\sigma \sigma_{\mu\rho}$$
$$- i\eta_{\rho\sigma} \gamma_\nu \gamma_\mu + i\eta_{\mu\sigma} \gamma_\nu \gamma_\sigma - \gamma_\nu \gamma_\sigma \sigma_{\mu\rho} - \gamma_\sigma \sigma_{\mu\rho} \gamma_\nu$$
$$- \sigma_{\mu\sigma} \gamma_\rho \gamma_\nu + i\eta_{\sigma\rho} \gamma_\nu \gamma_\mu - i\eta_{\mu\rho} \gamma_\nu \gamma_\sigma - \gamma_\nu \sigma_{\mu\sigma} \gamma_\rho$$
$$+ \gamma_\nu \sigma_{\mu\sigma} \gamma_\rho - i\eta_{\sigma\rho} \gamma_\mu \gamma_\nu + i\eta_{\mu\rho} \gamma_\sigma \gamma_\nu + \sigma_{\mu\sigma} \gamma_\rho \gamma_\nu$$

$$-\gamma_\mu\gamma_\sigma\sigma_{\nu\rho}-i\eta_{\rho\sigma}\gamma_\nu\gamma_\mu+i\eta_{\nu\sigma}\gamma_\rho\gamma_\mu-\gamma_\sigma\sigma_{\nu\rho}\gamma_\mu$$
$$+i\eta_{\rho\sigma}\gamma_\mu\gamma_\nu-i\eta_{\nu\sigma}\gamma_\mu\gamma_\rho+\gamma_\mu\gamma_\sigma\sigma_{\nu\rho}+\gamma_\sigma\sigma_{\nu\rho}\gamma_\mu\}$$
$$=\frac{i}{4}\{i\eta_{\sigma\rho}[\gamma_\nu\gamma_\mu-\gamma_\mu\gamma_\nu]+i\eta_{\nu\rho}[\gamma_\mu\gamma_\sigma-\gamma_\sigma\gamma_\mu]$$
$$+i\eta_{\sigma\rho}[\gamma_\mu\gamma_\nu-\gamma_\nu\gamma_\mu]+i\eta_{\mu\rho}[\gamma_\sigma\gamma_\nu-\gamma_\nu\gamma_\sigma]$$
$$+i\eta_{\mu\sigma}[\gamma_\nu\gamma_\rho-\gamma_\rho\gamma_\nu]+i\eta_{\nu\sigma}[\gamma_\rho\gamma_\mu-\gamma_\mu\gamma_\rho]$$
$$+i\eta_{\sigma\rho}[\gamma_\nu\gamma_\mu-\gamma_\mu\gamma_\nu]+i\eta_{\sigma\rho}[\gamma_\mu\gamma_\nu-\gamma_\nu\gamma_\mu]\}$$
$$=\frac{i}{4}\{i\eta_{\nu\rho}[\gamma_\mu,\gamma_\sigma]+i\eta_{\mu\rho}[\gamma_\sigma,\gamma_\nu]$$
$$+i\eta_{\mu\sigma}[\gamma_\nu,\gamma_\rho]+i\eta_{\nu\sigma}[\gamma_\rho,\gamma_\mu]\}$$
$$=i\eta_{\nu\rho}\sigma_{\mu\sigma}+i\eta_{\mu\rho}\sigma_{\sigma\nu}+i\eta_{\mu\sigma}\sigma_{\nu\rho}+i\eta_{\nu\sigma}\sigma_{\rho\mu}$$
$$=-i(\eta_{\mu\rho}\sigma_{\nu\sigma}-\eta_{\mu\sigma}\sigma_{\nu\rho}-\eta_{\nu\rho}\sigma_{\mu\sigma}+\eta_{\nu\sigma}\sigma_{\mu\rho})$$

We now prove the result (3.6). Consider

$$[\sigma_{\mu\nu},\gamma_\rho]=\frac{i}{4}[[\gamma_\mu,\gamma_\nu],\gamma_\rho]$$
$$=\frac{i}{4}[\gamma_\mu\gamma_\nu-\gamma_\nu\gamma_\mu,\gamma_\rho]$$
$$=\frac{i}{4}(\gamma_\mu\gamma_\nu\gamma_\rho-\gamma_\nu\gamma_\mu\gamma_\rho-\gamma_\rho\gamma_\mu\gamma_\nu+\gamma_\rho\gamma_\nu\gamma_\mu)$$

With the help of the Clifford algebra relation

$$\{\gamma_\mu,\gamma_\nu\}=2\eta_{\mu\nu}1_{4\times4}$$

we can rearrange various terms so that

$$[\sigma_{\mu\nu},\gamma_\rho]=\frac{i}{4}\{\gamma_\mu\gamma_\nu\gamma_\rho-2\eta_{\mu\nu}\gamma_\rho+\gamma_\mu\gamma_\nu\gamma_\rho$$

$$-2\eta_{\mu\rho}\gamma_\nu + \gamma_\mu\gamma_\rho\gamma_\nu + 2\eta_{\mu\nu}\gamma_\rho - \gamma_\rho\gamma_\mu\gamma_\nu \}$$
$$= \frac{i}{4}\{2\gamma_\mu\gamma_\nu\gamma_\rho - 2\eta_{\mu\rho}\gamma_\nu + 2\eta_{\rho\nu}\gamma_\mu - \gamma_\mu\gamma_\nu\gamma_\rho - 2\eta_{\mu\rho}\gamma_\nu + \gamma_\mu\gamma_\rho\gamma_\nu\}$$
$$= \frac{i}{4}\{\gamma_\mu\gamma_\nu\gamma_\rho - 4\eta_{\mu\rho}\gamma_\nu + 2\eta_{\rho\nu}\gamma_\mu + 2\eta_{\rho\nu}\gamma_\mu - \gamma_\mu\gamma_\nu\gamma_\rho\}$$
$$= i(\eta_{\rho\nu}\gamma_\mu - \eta_{\mu\rho}\gamma_\nu)$$

We now consider the generalized Jacobi identities. According to (2.22) these are defined by

$$X_\mu \circ (X_\nu \circ X_\rho)(-1)^{g_\mu g_\rho} + X_\rho \circ (X_\mu \circ X_\nu)(-1)^{g_\rho g_\nu} + X_\nu \circ (X_\rho \circ X_\mu)(-1)^{g_\mu g_\nu} = 0$$

for $X_\mu, X_\nu, X_\rho \in L$. We investigate these relations for the case when L_o is the Poincaré algebra (3.1). For $X_\mu, X_\nu, X_\rho \in L_o$ equation (2.22) reduces to the Jacobi identity of the Poincaré algebra which is automatically satisfied since L_o is a Lie algebra.

For $X_\mu, X_\nu \in L_o$, i.e. $X_\mu, X_\nu \in \{M_{\rho\sigma}, P_\lambda\}$ and $X_\rho \in L_1$, i.e. $X_\rho = Q_a$, we have

$$g(X_\mu) = g_\mu = 0, \; g(X_\nu) = g_\nu = 0,$$
$$g(Q_a) = g_a = 1$$

and (2.22) becomes (using (3.3) and (3.4))

$$X_\mu \circ (X_\nu \circ Q_a)(-1)^{g_\mu g_a} + Q_a \circ (X_\mu \circ X_\nu)(-1)^{g_a g_\nu} + X_\nu \circ (Q_a \circ X_\mu)(-1)^{g_\nu g_\mu}$$
$$= [X_\mu, [X_\nu, Q_a]] + [Q_a, [X_\mu, X_\nu]]$$

$$+ [X_\nu, [Q_a, X_\mu]]$$

$$= 0 \qquad (3.7)$$

For reasons of consistency we have to check that our choice of grading of the Poincaré algebra, i.e. (3.2), satisfies the Jacobi identities. We therefore consider these for the various cases.

i) $X_\mu = P_\mu,\ X_\nu = P_\nu,\ X_\rho = Q_a$: in this case it is easily seen that the expression on the left hand side of (3.7) vanishes so that the following Jacobi identity is satisfied

$$[P_\mu, [P_\nu, Q_a]] + [Q_a, [P_\mu, P_\nu]] + [P_\nu, [Q_a, P_\mu]] = 0 \qquad (3.8)$$

In verifying this relation one uses, of course, the relations (3.1) and (3.3). The verifications of the other Jacobi identities are handled in a similar way.

ii) $X_\mu = M_{\mu\nu},\ X_\nu = P_\rho,\ X_\rho = Q_a$: in this case we have

$$[M_{\mu\nu}, [P_\rho, Q_a]] + [Q_a, [M_{\mu\nu}, P_\rho]] + [P_\rho, [Q_a, M_{\mu\nu}]]$$

$$= [M_{\mu\nu}, 0] + [Q_a, i\eta_{\nu\rho}P_\mu - i\eta_{\mu\rho}P_\nu]$$

$$+ [P_\rho, (\sigma^4_{\mu\nu})_{ab}\, Q_b]$$

(using (3.3), (3.1), (3.4))

$$= i\eta_{\nu\rho}[Q_a, P_\mu] - i\eta_{\mu\rho}[Q_a, P_\nu] + (\sigma^4_{\mu\nu})_{ab}[P_\rho, Q_b]$$

$$= 0 \qquad \text{(with (3.3))}$$

Hence

$$[M_{\mu\nu}, [P_\rho, Q_a]] + [Q_a, [M_{\mu\nu}, P_\rho]]$$

$$+ [P_\rho, [Q_a, M_{\mu\nu}]] = 0 \qquad (3.9)$$

iii) $\underline{X_\mu = M_{\mu\nu},\ X_\nu = M_{\rho\sigma},\ X_\rho = Q_a}$: in this case

$$
\begin{aligned}
&[M_{\mu\nu},[M_{\rho\sigma},Q_a]]+[Q_a,[M_{\mu\nu},M_{\rho\sigma}]] \\
&\qquad + [M_{\rho\sigma},[Q_a,M_{\mu\nu}]] \\
&= [M_{\mu\nu},(-\sigma^4_{\rho\sigma})_{ab}Q_b]+[M_{\rho\sigma},(\sigma^4_{\mu\nu})_{ab}Q_b] \\
&\quad +[Q_a,\{-i(\eta_{\mu\rho}M_{\nu\sigma}-\eta_{\mu\sigma}M_{\nu\rho}-\eta_{\nu\rho}M_{\mu\sigma}+\eta_{\nu\sigma}M_{\mu\rho})\}]
\end{aligned}
$$

(using (3.4) and (3.1))

$$
\begin{aligned}
&= -(\sigma^4_{\rho\sigma})_{ab}[M_{\mu\nu},Q_b]-i\eta_{\mu\rho}[Q_a,M_{\nu\sigma}] \\
&\quad +i\eta_{\mu\sigma}[Q_a,M_{\nu\rho}]+i\eta_{\nu\rho}[Q_a,M_{\mu\sigma}] \\
&\quad -i\eta_{\nu\sigma}[Q_a,M_{\mu\rho}]+(\sigma^4_{\mu\nu})_{ab}[M_{\rho\sigma},Q_b] \\
&= (\sigma^4_{\rho\sigma})_{ab}(\sigma^4_{\mu\nu})_{bc}Q_c-i\eta_{\mu\rho}(\sigma^4_{\nu\sigma})_{ac}Q_c \\
&\quad +i\eta_{\mu\sigma}(\sigma^4_{\nu\rho})_{ac}Q_c+i\eta_{\nu\rho}(\sigma^4_{\mu\sigma})_{ac}Q_c \\
&\quad -i\eta_{\nu\sigma}(\sigma^4_{\mu\rho})_{ac}Q_c-(\sigma^4_{\mu\nu})_{ab}(\sigma^4_{\rho\sigma})_{bc}Q_c
\end{aligned}
$$

(using (3.4))

$$
\begin{aligned}
&= -([\sigma^4_{\mu\nu},\sigma^4_{\rho\sigma}])_{ac}Q_c \\
&\quad -i[\eta_{\mu\rho}(\sigma^4_{\nu\sigma})_{ac}-\eta_{\mu\sigma}(\sigma^4_{\nu\rho})_{ac} \\
&\qquad -\eta_{\nu\rho}(\sigma^4_{\mu\sigma})_{ac}+\eta_{\nu\sigma}(\sigma^4_{\mu\rho})_{ac}]Q_c \\
&= -([\sigma^4_{\mu\nu},\sigma^4_{\rho\sigma}])_{ac}Q_c+([\sigma^4_{\mu\nu},\sigma^4_{\rho\sigma}])_{ac}Q_c
\end{aligned}
$$

(using (3.5))

$$= 0$$

Hence

$$[M_{\mu\nu},[M_{\rho\sigma},Q_a]]+[Q_a,[M_{\mu\nu},M_{\rho\sigma}]] + [M_{\rho\sigma},[Q_a,M_{\mu\nu}]] = 0 \tag{3.10}$$

From the derivation of (3.10) we see that the essential point is the fact that the matrices $\sigma^4_{\mu\nu}$ form a representation of the Lorentz algebra, i.e. the validity of (3.5), where the $\sigma^4_{\mu\nu}$ are the coefficients of the product $M_{\mu\nu} \circ Q_a$ (see (3.2)). One can also argue the other way round as was done in Chapter 2 for SU(2) : Given the generalized Jacobi relation (3.10) (and also, of course, (3.8) and (3.9)) one can ask, what are the properties which the coefficients of the products $P_\mu \circ Q_a$ and $M_{\mu\nu} \circ Q_a$ must obey ? The conclusion is, as was demonstrated in Chapter 2, that these coefficients must be representation matrices of the Lie algebra L_o.

We have not yet defined the product on L_1. With (2.13) and (2.16) we have

$$\begin{aligned} \circ : L_1 \times L_1 &\longrightarrow L_o \\ Q_a, Q_b &\longrightarrow Q_a \circ Q_b = Q_a Q_b - (-1)^{\text{grad } b} Q_b Q_a \\ &\qquad = Q_a Q_b + Q_b Q_a \\ &\qquad =: \{Q_a, Q_b\} \end{aligned} \tag{3.11}$$

According to the general theory of graded Lie algebras this product has to be symmetric as indicated in (3.11), and it must close into L_o. Hence the most general form of (3.11) is

$$\{Q_a, Q_b\} = (h^\mu)_{ab} P_\mu + (k^{\mu\nu})_{ab} M_{\mu\nu} \tag{3.12}$$

where (h^μ) and $(k^{\mu\nu})$ are symmetric 4x4 matrices since the product, i.e. the left hand side, is symmetric, and $k^{\mu\nu}$ is antisymmetric in μ and ν .

In order to obtain more information about the matrices h^{μ} and $k^{\mu\nu}$ we choose as a basis for the 4x4 matrices the following set of γ matrices:

$$\{1_{4\times 4}, \gamma^{\mu}, \sigma^{4}_{\mu\nu}, \gamma^{5}\gamma^{\mu}, \gamma^{5}\} \tag{3.13}$$

where γ^{5} is defined by (1.139). From Section 1.4 we know that there exists a matrix C such that

$$C\gamma^{\mu T}C^{-1} = -\gamma^{\mu}$$

or equivalently

$$\gamma^{\mu}C = -(\gamma^{\mu}C^{T})^{T} \tag{3.14}$$

Furthermore C is antisymmetric (cf. (1.168)) and therefore (3.14) tells us that $(\gamma^{\mu}C) = (\gamma^{\mu}C)^{T}$, i.e. $\gamma^{\mu}C$ is symmetric. We now establish the following result.

<u>Proposition</u>: Provided

$$\gamma^{\mu}C = (\gamma^{\mu}C)^{T} \tag{3.15}$$

then

$$\sigma_{4}^{\mu\nu}C = (\sigma_{4}^{\mu\nu}C)^{T} \tag{3.16}$$

i.e. $\sigma^{\mu\nu}C$ is symmetric.

<u>Proof</u>: We have

$$(\sigma^{\mu\nu}C)^{T} = \frac{i}{4}([\gamma^{\mu},\gamma^{\nu}]C)^{T}$$

$$= \frac{i}{4}(\gamma^{\mu}\gamma^{\nu}C - \gamma^{\nu}\gamma^{\mu}C)^{T}$$

$$= \frac{i}{4}(C^{T}\gamma^{\nu T}\gamma^{\mu T} - C^{T}\gamma^{\mu T}\gamma^{\nu T})$$

$$= \frac{i}{4}((\gamma^{\nu}C)^{T}\gamma^{\mu T} - (\gamma^{\mu}C)^{T}\gamma^{\nu T})$$

$$= \frac{i}{4}(\gamma^{\nu}C\gamma^{\mu T} - \gamma^{\mu}C\gamma^{\nu T})$$

(using (3.15))

$$= \frac{i}{4}(-\gamma^{\nu}\gamma^{\mu}C + \gamma^{\mu}\gamma^{\nu}C)$$

$$= \frac{i}{4} [\gamma^\mu, \gamma^\nu] C$$
$$= \sigma^{\mu\nu} C$$

Furthermore

$$(\gamma^5 C)^T = - \gamma^5 C \tag{3.16a}$$

$$(\gamma^5 \gamma^\mu C)^T = - \gamma^5 \gamma^\mu C \tag{3.16b}$$

Thus from the set of basis elements for the 4x4 matrices (3.13) we can construct two symmetric matrices $\gamma^\mu C$ and $\sigma^{\mu\nu} C$. In addition, the symmetric 4x4 matrix $\sigma^{\mu\nu} C$ is antisymmetric in μ and ν . This is exactly what we need for h^μ and $k^{\mu\nu}$ which appear in (3.12). We conclude therefore that

$$h^\mu = a \gamma^\mu C, \quad k^{\mu\nu} = b \sigma^{\mu\nu} C \tag{3.17}$$

where a and b are constants.

We obtain further restrictions on the right hand side of (3.12) by studying the generalized Jacobi identity for two generators Q_a and Q_b . In the case of $X_\mu \in L_0, X_\nu, X_\rho \in L_1$, the generalized Jacobi identity (2.22) is, using the following values of the respective degrees

$$g(X_\mu) = 0 , \quad X_\mu \in \{P_\mu, M_{\rho\sigma}\}$$
$$g(X_\nu) = g(X_\rho) = 1, X_\nu = Q_a, X_\rho = Q_b,$$

the relation

$$\begin{aligned} & X_\mu \circ (Q_a \circ Q_b)(-1)^{g_\mu g_b} + Q_b \circ (X_\mu \circ Q_a)(-1)^{g_a g_b} \\ & \qquad + Q_a \circ (Q_b \circ X_\mu)(-1)^{g_a g_\mu} \\ & = [X_\mu, \{Q_a, Q_b\}] - \{Q_b, [X_\mu, Q_a]\} \\ & \qquad + \{Q_a, [Q_b, X_\mu]\} \\ & \overset{!}{=} 0 \end{aligned} \tag{3.18}$$

We now have two possibilities.

i) $\underline{X_\mu = P_\mu}$: in this case (3.18) is

$$0 = [P_\mu, \{Q_a, Q_b\}] - \{Q_b, [P_\mu, Q_a]\} + \{Q_a, [Q_b, P_\mu]\}$$

$$= [P_\mu, a(\gamma^\nu C)_{ab} P_\nu + b(\sigma^{\rho\sigma} C)_{ab} M_{\rho\sigma}]$$

(using (3.3), (3.12), (3.17))

$$= a(\gamma^\nu C)_{ab} [P_\mu, P_\nu] + b(\sigma^{\rho\sigma} C)_{ab} [P_\mu, M_{\rho\sigma}]$$

$$= b(\sigma^{\rho\sigma} C)_{ab} \{ i\eta_{\rho\mu} P_\sigma - i\eta_{\sigma\mu} P_\rho \}$$

$$= ib\{(\sigma^{\rho\sigma}\eta_{\rho\mu} P_\sigma - \sigma^{\rho\sigma}\eta_{\sigma\mu} P_\rho) C\}_{ab}$$

$$= ib\{(\sigma_\mu{}^\sigma P_\sigma - \sigma^\rho{}_\mu P_\rho) C\}_{ab}$$

$$= ib\{(\sigma_{\mu\nu} - \sigma_{\nu\mu}) C\}_{ab} P^\nu$$

$$= ib\{2\sigma_{\mu\nu} C\}_{ab}$$

where we used the antisymmetry of $\sigma^{\mu\nu}$ in μ and ν. Hence in order that the Jacobi identity (3.18) be satisfied the coefficient b must be zero. This result has an important consequence: The anticommutator of two Q's is a linear combination of the momentum operators P_μ. Hence the requirement that the Jacobi identity for P_μ and two operators Q be satisfied leads to the relation

$$\{Q_a, Q_b\} = a(\gamma^\mu C)_{ab} P_\mu \qquad (3.19)$$

We now consider the other case.

ii) $\underline{X_\mu = M_{\mu\nu}}$: inserting $M_{\mu\nu}$ into (3.18) we obtain

$$0 = [M_{\mu\nu}, \{Q_a, Q_b\}] - \{Q_b, [M_{\mu\nu}, Q_a]\}$$
$$+ \{Q_a, [Q_b, M_{\mu\nu}]\}$$
$$= [M_{\mu\nu}, a(\gamma^\rho C)_{ab} P_\rho] - \{Q_b, (-\sigma^4_{\mu\nu})_{ac} Q_c\}$$
$$+ \{Q_a, (\sigma^4_{\mu\nu})_{bc} Q_c\}$$
$$= a(\gamma^\rho C)_{ab} [M_{\mu\nu}, P_\rho] + (\sigma^4_{\mu\nu})_{ac} \{Q_b, Q_c\}$$
$$+ (\sigma^4_{\mu\nu})_{bc} \{Q_a, Q_c\}$$
$$= a(\gamma^\rho C)_{ab}\, i(\eta_{\nu\rho} P_\mu - \eta_{\mu\rho} P_\nu)$$
$$+ (\sigma^4_{\mu\nu})_{ac}\, a(\gamma^\rho C)_{bc} P_\rho$$
$$+ (\sigma^4_{\mu\nu})_{bc}\, a(\gamma^\rho C)_{ac} P_\rho$$

or, since a $\neq$ 0 ,

$$-i(\gamma^\rho C)_{ab} (\eta_{\nu\rho} P_\mu - \eta_{\mu\rho} P_\nu)$$
$$= [(\sigma^4_{\mu\nu} \gamma_\rho C)_{ab} + (\sigma^4_{\mu\nu} \gamma_\rho C)_{ba}] P^\rho$$

using (3.15). Now, since

$$\tfrac{1}{2}\{\sigma^4_{\mu\nu}, \gamma_\rho\} C + \tfrac{1}{2}[\sigma^4_{\mu\nu}, \gamma_\rho] C$$
$$= \tfrac{1}{2} \sigma^4_{\mu\nu} \gamma_\rho C + \tfrac{1}{2} \gamma_\rho \sigma^4_{\mu\nu} C + \tfrac{1}{2} \sigma^4_{\mu\nu} \gamma_\rho C$$
$$- \tfrac{1}{2} \gamma_\rho \sigma^4_{\mu\nu} C$$
$$= \sigma^4_{\mu\nu} \gamma_\rho C$$

we can write

$$\sigma^4_{\mu\nu} \gamma_\rho C = \tfrac{1}{2}\{\sigma^4_{\mu\nu}, \gamma_\rho\} C + \tfrac{1}{2}[\sigma^4_{\mu\nu}, \gamma_\rho] C \qquad (3.20)$$

We can rewrite the anticommutator and the commutator appearing in this formula in the form

$$\{\sigma^4_{\mu\nu}, \gamma_\rho\} = -i\,\varepsilon_{\mu\nu\rho\sigma}\,\gamma^5\gamma^\sigma \tag{3.21}$$

$$[\sigma^4_{\mu\nu}, \gamma_\rho] = i(\eta_{\nu\rho}\gamma_\mu - \eta_{\mu\rho}\gamma_\nu) \tag{3.22}$$

(see (3.6)). Then

$$-i(\gamma^\rho C)_{ab}(\eta_{\nu\rho}P_\mu - \eta_{\mu\rho}P_\nu)$$
$$= [(\sigma^4_{\mu\nu}\gamma_\rho C)_{ab} + (\sigma^4_{\mu\nu}\gamma_\rho C)_{ba}]P^\rho$$
$$= [\tfrac{1}{2}(\{\sigma^4_{\mu\nu}, \gamma_\rho\}C)_{ab} + \tfrac{1}{2}([\sigma^4_{\mu\nu}, \gamma_\rho]C)_{ab}$$
$$+ \tfrac{1}{2}(\{\sigma^4_{\mu\nu}, \gamma_\rho\}C)_{ba} + \tfrac{1}{2}([\sigma^4_{\mu\nu}, \gamma_\rho]C)_{ba}]P^\rho$$

(with (3.20))

$$= [-\tfrac{i}{2}\varepsilon_{\mu\nu\rho\sigma}(\gamma^5\gamma^\sigma C)_{ab} + \tfrac{i}{2}\eta_{\nu\rho}(\gamma_\mu C)_{ab}$$
$$- \tfrac{i}{2}\eta_{\mu\rho}(\gamma_\nu C)_{ab} - \tfrac{i}{2}\varepsilon_{\mu\nu\rho\sigma}(\gamma^5\gamma^\sigma C)_{ba}$$
$$+ \tfrac{i}{2}\eta_{\nu\rho}(\gamma_\mu C)_{ba} - \tfrac{i}{2}\eta_{\mu\rho}(\gamma_\nu C)_{ba}]P^\rho$$

(with (3.21) and (3.22))

$$= [i\eta_{\nu\rho}(\gamma_\mu C)_{ab} - i\eta_{\mu\rho}(\gamma_\nu C)_{ab}]P^\rho$$

(with (3.15), (3.16b))

$$= i(\gamma_\mu C)_{ab}P_\nu - i(\gamma_\nu C)_{ab}P_\mu$$

The left hand side of this relation is

$$-i(\gamma^\rho C)_{ab}(\eta_{\nu\rho}P_\mu - \eta_{\mu\rho}P_\nu)$$
$$= i(\gamma_\mu C)_{ab}P_\nu - i(\gamma_\nu C)_{ab}P_\mu$$

We see that the right hand side cancels exactly the left hand side. Hence the generalized Jacobi relation for

$M_{\mu\nu}$, Q_a and Q_b is satisfied.

Next we demonstrate that the Jacobi relation for three Q's is satisfied. For three Q's the relation (2.22) becomes

$$Q_a \circ (Q_b \circ Q_c)(-1)^{g_a g_c} + Q_c \circ (Q_a \circ Q_b)(-1)^{g_c g_b}$$
$$+ Q_b \circ (Q_c \circ Q_a)(-1)^{g_b g_a} = 0$$

Using (3.11) and (3.19) the left hand side can be written

$$[Q_a, \{Q_b, Q_c\}] + [Q_c, \{Q_a, Q_b\}] + [Q_b, \{Q_c, Q_a\}]$$
$$= [Q_a, a(\gamma^\mu C)_{bc} P_\mu] + [Q_c, a(\gamma^\mu C)_{ab} P_\mu]$$
$$+ [Q_b, a(\gamma^\mu C)_{ca} P_\mu]$$
$$= a(\gamma^\mu C)_{bc} [Q_a, P_\mu] + a(\gamma^\mu C)_{ab} [Q_c, P_\mu]$$
$$+ a(\gamma^\mu C)_{ca} [Q_b, P_\mu]$$
$$= 0$$

using (3.3). Hence the Jacobi identity for three Q's is satisfied.

We have therefore shown that the chosen grading (3.2) is a correct grading which is consistent with all possible Jacobi identities, and we have seen that the product on $L_1 x L_1$ is given by the relation

$$\{Q_a, Q_b\} = a(\gamma^\mu C)_{ab} P_\mu$$

The coefficient a which appears in this relation is arbitrary and can be absorbed in the Q's. We can impose one further restriction on the Q's which does not change the structure of the algebra. From the chosen grading and from previous considerations concerning the subspace L_1 of a Z_2 graded Lie algebra, in this case spanned by the Q's, it is clear that we have four complex operators Q_a and hence eight independent components. Thus besides (3.19) we would also have to specify the products

$$Q_a \circ \bar{Q}_b = \{Q_a, \bar{Q}_b\} \quad \text{and} \quad \bar{Q}_a \circ \bar{Q}_b = \{\bar{Q}_a, \bar{Q}_b\}$$

Here, however, we restrict our considerations to the particular case where Q is a Majorana spinor. In this case we know from Chapter 1.5 that Q and $\overline{Q}$ are no longer independent.

<u>Proposition</u>: If we impose the Majorana condition on the generators Q, i.e. the condition (1.180), we obtain from (3.19) the relation

$$\{Q_a, \bar{Q}_b\} = -a(\gamma^\mu)_{ab} P_\mu \tag{3.23}$$

<u>Corollary</u>: For P^0 to be positive definite the constant a has to be negative (conventionally chosen to be -2).

<u>Proof</u>: The Majorana condition (1.180) is

$$Q_a = Q_a^c = (C\bar{Q}^T)_a = C_{ab}(\bar{Q}^T)_b$$

where $\quad \bar{Q} = Q^\dagger \gamma_0$

Multiplying the last equation from the left by C_{ca} we obtain

$$\begin{aligned} C_{ca}Q_a &= C_{ca}C_{ab}(\bar{Q}^T)_b \\ &= (C^2)_{cb}(\bar{Q}^T)_b \\ &= -\delta_{cb}(\bar{Q}^T)_b \qquad \text{(using (1.168))} \\ &= -(\bar{Q}^T)_c \end{aligned}$$

from which we conclude

$$\begin{aligned} (CQ)_a &= -\bar{Q}_a^T \\ (CQ)_a^T &= -\bar{Q}_a \end{aligned}$$

$$(Q^T C^T)_a = -\bar{Q}_a$$

Again using (1.168)

$$(Q^T C)_a = \bar{Q}_a \tag{3.24a}$$

For later reference purposes we derive two further relations. From the Majorana condition for the Q's we obtain

$$\frac{\partial}{\partial \bar{Q}^T_b} Q_a = C_{ab} \tag{3.24b}$$

Now from (3.24a)

$$\bar{Q}^T_b = C^T_{bc} Q_c$$

We can verify that these relations agree with each other. Thus

$$\begin{aligned}\frac{\partial}{\partial \bar{Q}^T_b} Q_a &= \frac{\partial}{\partial \bar{Q}^T_b} \left((C^T)^{-1} C^T Q\right)_a \\ &= \frac{\partial}{\partial \bar{Q}^T_b} (C^T)^{-1}_{ac} \bar{Q}^T_c \\ &= (C^T)^{-1}_{ab} \\ &= C_{ab} \quad (\text{since } C^T_W = C^{-1}_W)\end{aligned}$$

in agreement with (3.24b). Similarly we obtain

$$\begin{aligned}\frac{\partial}{\partial Q_b} \bar{Q}_a &= \frac{\partial}{\partial Q_b} Q^T_c C_{ca} \\ &= C_{ba}\end{aligned} \tag{3.24c}$$

These relations will be needed at a later stage. Now returning to our problem and starting from (3.19) we obtain

$$Q_a Q_b + Q_b Q_a = a(\gamma^\mu)_{ac} C_{cb} P_\mu$$

Multiplying by C

$$Q_a Q_b C_{bd} + Q_b C_{bd} Q_a = a(\gamma^\mu)_{ac} C_{cb} C_{bd} P_\mu$$

so that

$$Q_a (Q^T C)_d + (Q^T C)_d Q_a = -a(\gamma^\mu)_{ad} P_\mu$$

Using (3.24a) this becomes

$$Q_a \bar{Q}_b + \bar{Q}_b Q_a = - a (\gamma^\mu)_{ab} P_\mu$$

and so

$$\{ Q_a, \bar{Q}_b \} = - a (\gamma^\mu)_{ab} P_\mu$$

Similarly one can show that

$$\{ \bar{Q}_a, \bar{Q}_b \} = - a (C^{-1} \gamma^\mu)_{ab} P_\mu$$

In order to prove the corollary we go to the rest frame in which $\vec{P} = 0$, so that

$$\{ Q_a, \bar{Q}_b \} = - a \gamma^0_{ab} P_0$$

or

$$\{ Q_a, \bar{Q}_b \} \gamma^0_{ba} = - 4a P_0$$

since $(\gamma^0)^2 = 1$. Thus

$$\begin{aligned} P_0 &= - \frac{1}{4a} (Q_a \gamma^{0T}_{ab} \bar{Q}_b + \bar{Q}_b \gamma^0_{ba} Q_a) \\ &= - \frac{1}{4a} (Q_a \gamma^{0T}_{ab} Q^+_c \gamma^0_{cb} + Q^+_c \gamma^0_{cb} \gamma^0_{ba} Q_a) \\ &= - \frac{1}{4a} (Q_b Q^+_b + Q^+_b Q_b) \end{aligned}$$

This is positive definite provided a is negative. In the following we choose $a = - 2$. Further, if we go to the Majorana representation, the Majorana spinors are real. Then

$$H \equiv P_0 = \frac{1}{8} 2 \sum_b Q_b^2 = \frac{1}{4} \sum_b Q_b^2 \geqslant 0$$

Hence for any state $|\psi\rangle$ of the Hilbert space over which H is defined

$$\langle \psi | H | \psi \rangle \geqslant 0$$

and just like

$$P|\Omega\rangle = 0, \quad P = -i\frac{\partial}{\partial x}$$

for a state which is invariant under translations in x, so a state $|\Omega\rangle$ is supersymmetric if

$$Q_a|\Omega\rangle = 0$$

Hence supersymmetry is unbroken if and only if

$$\langle\Omega|H|\Omega\rangle = 0 = \min\langle\psi|H|\psi\rangle$$

It follows that the supersymmetric state $|\Omega\rangle$ is the ground state. Further discussions of these aspects are given in Chapter 9.

3.2 The Supersymmetric Extension of the Poincaré Algebra in the Two-Component Weyl Formulation

Starting with (3.23) it is possible to express the anticommuting part of the supersymmetrically extended Poincaré algebra in terms of two-component Weyl spinors, taking into account the close connection between four-component Majorana spinors and two-component Weyl spinors. Writing (3.23) in the Weyl representation, using (1.136) and (1.181), we see that (3.23) becomes a set of four anticommutators (with the constant a in (3.19), (3.23) chosen as -2), i.e.

$$Q_a\bar{Q}_b + \bar{Q}_b Q_a = 2\gamma^\mu_{ab} P_\mu$$

so that

$$\begin{pmatrix} Q_A \\ \bar{Q}^{\dot{A}} \end{pmatrix}(Q^B, \bar{Q}_{\dot{B}}) + (Q^B, \bar{Q}_{\dot{B}})\begin{pmatrix} Q_A \\ \bar{Q}^{\dot{A}} \end{pmatrix}$$

$$= 2\begin{pmatrix} 0 & (\sigma^\mu)_{A\dot{B}} \\ (\bar{\sigma}^\mu)^{\dot{A}B} & 0 \end{pmatrix} P_\mu$$

i.e.

$$\begin{pmatrix} Q_A Q^B + Q^B Q_A & Q_A \bar{Q}_{\dot{B}} + \bar{Q}_{\dot{B}} Q_A \\ \bar{Q}^{\dot{A}} Q^B + Q^B \bar{Q}^{\dot{A}} & \bar{Q}^{\dot{A}} \bar{Q}_{\dot{B}} + \bar{Q}_{\dot{B}} \bar{Q}^{\dot{A}} \end{pmatrix}$$

$$= 2\begin{pmatrix} 0 & (\sigma^\mu)_{A\dot{B}} \\ (\bar{\sigma}^\mu)^{\dot{A}B} & 0 \end{pmatrix} P_\mu$$

so that

$$\{Q_A, Q^B\} = 0, \quad \{Q_A, \bar{Q}_{\dot{B}}\} = 2\sigma^\mu_{A\dot{B}} P_\mu$$
$$\{\bar{Q}^{\dot{A}}, Q^B\} = 2\bar{\sigma}^{\mu\dot{A}B} P_\mu, \quad \{\bar{Q}^{\dot{A}}, \bar{Q}_{\dot{B}}\} = 0 \tag{3.25}$$

Furthermore, using

$$\sigma^4_{\mu\nu} = \begin{pmatrix} (\sigma^2_{\mu\nu})_A{}^B & 0 \\ 0 & (\bar{\sigma}^2_{\mu\nu})^{\dot{A}}{}_{\dot{B}} \end{pmatrix}$$

the commutator (3.4) becomes

$$\left.\begin{aligned} [M_{\mu\nu}, Q_A] &= -(\sigma^2_{\mu\nu})_A{}^B Q_B \\ [M_{\mu\nu}, \bar{Q}^{\dot{A}}] &= -(\bar{\sigma}^2_{\mu\nu})^{\dot{A}}{}_{\dot{B}} \bar{Q}^{\dot{B}} \end{aligned}\right\} \tag{3.26}$$

CHAPTER 4

REPRESENTATIONS OF THE SUPER-POINCARÉ ALGEBRA

4.1 Casimir Operators

In order to classify representations of the Super-Poincaré algebra we use a method similar to that described in Chapter 1. We recall first the structure of the Super-Poincaré algebra as presented in the previous chapter. The chosen algebra has 14 generators, i.e.

4 generators of translations P_μ,

6 generators of Lorentz transformations $M_{\mu\nu}$,

4 spinor charges Q_a (Majorana spinors)

The Super-Poincaré algebra is given by

$$[P_\mu, P_\nu] = 0$$

$$[M_{\mu\nu}, P_\rho] = -i(\eta_{\mu\rho}P_\nu - \eta_{\nu\rho}P_\mu)$$

$$[M_{\mu\nu}, M_{\rho\sigma}] = -i(\eta_{\mu\rho}M_{\nu\sigma} + \eta_{\nu\sigma}M_{\mu\rho} - \eta_{\mu\sigma}M_{\nu\rho} - \eta_{\nu\rho}M_{\mu\sigma})$$

$$[P_\mu, Q_a] = 0$$
$$[M_{\mu\nu}, Q_a] = -(\sigma^4_{\mu\nu})_{ab} Q_b$$
$$\{Q_a, \bar{Q}_b\} = 2(\gamma^\mu)_{ab} P_\mu$$
$$\{Q_a, Q_b\} = -2(\gamma^\mu C)_{ab} P_\mu$$
$$\{\bar{Q}_a, \bar{Q}_b\} = 2(C^{-1}\gamma^\mu)_{ab} P_\mu \tag{4.1}$$

where
$$\sigma^4_{\mu\nu} = \frac{i}{4}[\gamma_\mu, \gamma_\nu]$$
and indices a and b run from 1 to 4. The algebra (4.1) satisfies all relevant generalized Jacobi identities as was demonstrated in Chapter 3.

In view of our discussion at the end of this chapter we recall the following steps in adding an internal symmetry. As explained in Chapter 2, the theorem of Haag, Lopuszanski and Sohnius allows the presence of an internal symmetry group with generators B_ℓ[28]. According to the theorem of Coleman and Mandula these generators B_ℓ commute with P_μ and $M_{\mu\nu}$.[25] Furthermore we saw that it is necessary in this case to introduce a set of N spinor charges

Q_a^α, $\alpha = 1, \ldots, N$ and $a = 1,2,3,4$,

such that the commutator of Q_a^α and B_ℓ closes into the subspace L_1, i.e.

$$[Q_a^\alpha, B_\ell] = i\, S_\ell^{\alpha\beta} Q_a^\beta \tag{4.2}$$

where the matrices S_ℓ have to be a representation of the algebra of the internal symmetry group, i.e. of

$$[B_\ell, B_m] = i\, c_{\ell m}{}^k B_k$$

The coefficients S_ℓ appearing in (4.2) are NxN-dimensional matrices where N depends on the representation one has chosen. The product of two Q's is a simple extension of (3.23), i.e.

$$\{Q_a^\alpha, \bar{Q}_b^\beta\} = 2\,\delta^{\alpha\beta}(\gamma^\mu)_{ab}P_\mu \qquad (4.3)$$

The algebra with N = 1 is called the supersymmetry algebra, and algebras with $N > 1$ are called extended supersymmetry algebras.

As mentioned before, in order to classify representations of the Super-Poincaré algebra (4.1) we use the method of Casimir operators as in Chapter 1. The basic assumption is that the generators P_μ, $M_{\mu\nu}$ and Q_a can be realized as linear operators acting on an infinite dimensional Hilbert space of state vectors. In order to classify representations we construct Casimir operators whose eigenvalues determine the representation.

From (4.1) it is easy to see that $P^2 = P_\mu P^\mu$ is again an invariant operator. It is important to note that this fact, i.e. that P^2 is a Casimir operator of the Super-Poincaré algebra (4.1), depends crucially on the grading chosen. For example, if we take L_1 five-dimensional (i.e. five generators Q) and take as five-dimensional representation of L_o the matrix representation

$$g(a,\Lambda) = \begin{pmatrix} \Lambda & a \\ 0 & 1 \end{pmatrix}, \quad \Lambda \in L_+^\uparrow, \quad a \in T_4$$

then the commutator of P and Q will be nonzero. In this case P^2 will not be an invariant operator.

<u>Proposition</u>: The square of the Pauli-Ljubanski vector

$$W^\mu = \frac{1}{2}\varepsilon^{\mu\nu\rho\sigma}M_{\nu\rho}P_\sigma \qquad (4.4)$$

is not a Casimir operator of the Super-Poincaré algebra (4.1).

<u>Proof</u>: We saw in Chapter 1 that $W^2 = W_\mu W^\mu$ commutes with all generators of L_o. Therefore the only commutator of interest is

$$[W^2, Q_a]$$

First we derive the commutation relation of W_μ with Q_a, i.e.

$$[W_\mu, Q_a] = \frac{1}{2}[\varepsilon_{\mu\nu\rho\sigma} M^{\nu\rho} P^\sigma, Q_a]$$

$$= \frac{1}{2}\varepsilon_{\mu\nu\rho\sigma}\{M^{\nu\rho}[P^\sigma, Q_a] + [M^{\nu\rho}, Q_a]P^\sigma\}$$

$$= \frac{1}{2}\varepsilon_{\mu\nu\rho\sigma}[M^{\nu\rho}, Q_a]P^\sigma \qquad \text{(with (4.1))}$$

$$= -\frac{1}{2}\varepsilon_{\mu\nu\rho\sigma}(\sigma^{\nu\rho})_{ab} Q_b P^\sigma \qquad \text{(with (4.1))}$$

$$= \frac{1}{2}\varepsilon_{\sigma\mu\nu\rho}(\sigma^{\nu\rho})_{ab} Q_b P^\sigma$$

$$= -\frac{1}{2}\varepsilon_{\mu\sigma\nu\rho}(\sigma^{\nu\rho})_{ab} Q_b P^\sigma$$

Using

$$\varepsilon_{\mu\nu\rho\sigma}\, \sigma_4^{\rho\sigma} = -2i\gamma^5 \sigma^4_{\mu\nu} \tag{4.5}$$

(this relation can be verified by using $\sigma_2^{\mu\nu}, \bar{\sigma}_2^{\mu\nu}$ of (1.121 a,b) and the expression for $\sigma_4^{\mu\nu}$ in the proof of (1.184e)) we obtain

$$[W_\mu, Q_a] = i(\gamma^5 \sigma_{\mu\sigma} Q)_a P^\sigma$$

$$= i(\sigma_{\mu\sigma}\gamma^5 Q)_a P^\sigma$$

since

$$[\gamma^5, \sigma_{\mu\nu}] = 0$$

Now

$$\gamma_\sigma \gamma_\mu = \frac{1}{2}\{\gamma_\sigma, \gamma_\mu\} + \frac{1}{2}[\gamma_\sigma, \gamma_\mu]$$
$$= \eta_{\sigma\mu} - 2i\,\sigma_{\sigma\mu}$$

so that

$$\sigma_{\mu\sigma} = -\frac{i}{2}(\gamma_\sigma \gamma_\mu - \eta_{\mu\sigma})$$

taking into account the antisymmetry of $\sigma_{\mu\nu}$. Hence

$$[W_\mu, Q_a] = i(\sigma_{\mu\sigma}\gamma^5 Q)_a P^\sigma$$
$$= \frac{1}{2}[(\gamma_\sigma \gamma_\mu - \eta_{\mu\sigma})\gamma^5 Q]_a P^\sigma$$
$$= \frac{1}{2}(\gamma_\sigma \gamma_\mu \gamma^5 Q)_a P^\sigma - \frac{1}{2}\eta_{\mu\sigma}(\gamma^5 Q)_a P^\sigma$$
$$= \frac{1}{2}(\not{P}\gamma_\mu \gamma^5 Q)_a - \frac{1}{2}P_\mu(\gamma^5 Q)_a \tag{4.6}$$

since

$$[P_\mu, Q_a] = 0$$

and where we define

$$\not{P} := P_\mu \gamma^\mu \tag{4.7}$$

We can now calculate the commutator

$$[W^2, Q_a] = [W_\mu W^\mu, Q_a]$$
$$= W^\mu [W_\mu, Q_a] + [W_\mu, Q_a]W^\mu$$
$$= W^\mu\{-\frac{1}{2}P_\mu(\gamma^5 Q)_a + \frac{1}{2}(\not{P}\gamma_\mu\gamma^5 Q)_a\}$$
$$+ \{-\frac{1}{2}P_\mu(\gamma^5 Q)_a + \frac{1}{2}(\not{P}\gamma_\mu\gamma^5 Q)_a\}W^\mu$$

(using (4.6))

$$= -\frac{1}{2}W^\mu P_\mu(\gamma^5 Q)_a + \frac{1}{2}W^\mu(\not{P}\gamma_\mu\gamma^5 Q)_a$$
$$-\frac{1}{2}P_\mu(\gamma^5 Q)_a W^\mu + \frac{1}{2}(\not{P}\gamma_\mu\gamma^5 Q)_a W^\mu$$

The following factor of the first term can be shown to vanish, i.e.

$$-\frac{1}{2}W^{\mu}P_{\mu} = -\frac{1}{4}\varepsilon^{\mu\nu\rho\sigma}M_{\nu\rho}P_{\sigma}P_{\mu}$$

$$= -\frac{1}{8}[\varepsilon^{\mu\nu\rho\sigma}M_{\nu\rho}P_{\sigma}P_{\mu} + \varepsilon^{\sigma\nu\rho\mu}M_{\nu\rho}P_{\mu}P_{\sigma}]$$

$$= -\frac{1}{8}\varepsilon^{\mu\nu\rho\sigma}M_{\nu\rho}(P_{\sigma}P_{\mu} - P_{\mu}P_{\sigma})$$

$$= 0$$

The third term also vanishes since

$$P_{\mu}W^{\mu} = 0$$

Hence

$$[W^2, Q_a]$$

$$= \frac{1}{2}W^{\mu}(\not{P}\gamma_{\mu}\gamma_5 Q)_a + \frac{1}{2}(\not{P}\gamma_{\mu}\gamma_5 Q)_a W^{\mu}$$

$$= \frac{1}{2}W^{\mu}(\not{P}\gamma_{\mu}\gamma_5 Q)_a + \frac{1}{4}\varepsilon^{\mu\nu\rho\sigma}(\not{P}\gamma_{\mu}\gamma^5 Q)_a M_{\nu\rho}P_{\sigma}$$

$$= \frac{1}{2}W^{\mu}(\not{P}\gamma_{\mu}\gamma_5 Q)_a + \frac{1}{4}\varepsilon^{\mu\nu\rho\sigma}(\not{P}\gamma_{\mu}\gamma^5)_{ab}Q_b M_{\nu\rho}P_{\sigma}$$

$$= \frac{1}{2}W^{\mu}(\not{P}\gamma_{\mu}\gamma_5 Q)_a + \frac{1}{4}\varepsilon^{\mu\nu\rho\sigma}(\not{P}\gamma_{\mu}\gamma^5)_{ab} \cdot$$

$$\cdot (M_{\nu\rho}Q_b + (\sigma_{\nu\rho})_{bc}Q_c)P_{\sigma}$$

(with (4.1))

$$= \frac{1}{2}W^{\mu}(\not{P}\gamma_{\mu}\gamma_5 Q)_a$$

$$+ \frac{1}{4}\varepsilon^{\mu\nu\rho\sigma}(P_{\lambda}\gamma^{\lambda}\gamma_{\mu}\gamma^5)_{ab}M_{\nu\rho}Q_b P_{\sigma}$$

$$+ \frac{1}{4}\varepsilon^{\mu\nu\rho\sigma}(P_{\lambda}\gamma^{\lambda}\gamma_{\mu}\gamma^5)_{ab}(\sigma_{\nu\rho})_{bc}Q_c P_{\sigma}$$

$$= \frac{1}{2}W^{\mu}(\not{P}\gamma_{\mu}\gamma_5 Q)_a$$

$$+ \frac{1}{4}\varepsilon^{\mu\nu\rho\sigma}P_{\lambda}M_{\nu\rho}P_{\sigma}(\gamma^{\lambda}\gamma_{\mu}\gamma^5 Q)_a$$

$$+ \frac{1}{4} \varepsilon^{\mu\nu\rho\sigma} P_\lambda (\gamma^\lambda \gamma_\mu \gamma^5 \sigma_{\nu\rho} Q)_a P_\sigma$$

$$= \frac{1}{2} W^\mu (\not{P} \gamma_\mu \gamma_5 Q)_a$$

$$+ \frac{1}{4} \varepsilon^{\mu\nu\rho\sigma} (M_{\nu\rho} P_\lambda + i\eta_{\nu\lambda} P_\rho - i\eta_{\rho\lambda} P_\nu) P_\sigma (\gamma^\lambda \gamma_\mu \gamma^5 Q)_a$$

$$+ \frac{1}{4} \varepsilon^{\mu\nu\rho\sigma} p^\lambda (\gamma_\lambda \gamma_\mu \gamma^5 \sigma^4_{\nu\rho} Q)_a P_\sigma$$

$$= \frac{1}{2} W^\mu (\not{P} \gamma_\mu \gamma_5 Q)_a$$

$$+ \frac{1}{4} \varepsilon^{\mu\nu\rho\sigma} M_{\nu\rho} P_\sigma P_\lambda (\gamma^\lambda \gamma_\mu \gamma^5 Q)_a$$

$$+ \frac{i}{4} \varepsilon^{\mu\nu\rho\sigma} \eta_{\nu\lambda} P_\rho P_\sigma (\gamma^\lambda \gamma_\mu \gamma^5 Q)_a$$

$$- \frac{i}{4} \varepsilon^{\mu\nu\rho\sigma} \eta_{\rho\lambda} P_\nu P_\sigma (\gamma^\lambda \gamma_\mu \gamma^5 Q)_a$$

$$+ \frac{1}{4} \varepsilon^{\mu\nu\rho\sigma} p^\lambda (\gamma_\lambda \gamma_\mu \gamma^5 \sigma^4_{\nu\rho} Q)_a P_\sigma$$

$$= \frac{1}{2} W^\mu (\not{P} \gamma_\mu \gamma_5 Q)_a + \frac{1}{2} W^\mu (\not{P} \gamma_\mu \gamma^5 Q)_a$$

$$+ \frac{1}{4} \varepsilon^{\mu\nu\rho\sigma} p^\lambda (\gamma_\lambda \gamma_\mu \gamma^5 \sigma_{\nu\rho} Q)_a P_\sigma$$

(since $\varepsilon^{\mu\nu\rho\sigma} P_\rho P_\sigma = 0$)

$$= W^\mu (\not{P} \gamma_\mu \gamma_5 Q)_a + \frac{1}{4} p^\lambda (\gamma_\lambda \gamma_\mu \gamma^5 \varepsilon^{\mu\nu\rho\sigma} \cdot$$

$$\cdot \sigma_{\nu\rho} Q)_a P_\sigma$$

$$= W^\mu (\not{P} \gamma_\mu \gamma_5 Q)_a - \frac{i}{2} p^\lambda (\gamma_\lambda \gamma_\mu (\gamma_5)^2 \sigma^{\mu\sigma} Q)_a P_\sigma$$

$$= W^\mu (\not{P} \gamma_\mu \gamma_5 Q)_a - \frac{i}{2} (\not{P} \gamma_\mu \sigma^{\mu\sigma} Q)_a P_\sigma$$

Now

$$\gamma_\mu \sigma^{\mu\sigma} = \frac{i}{4} \gamma_\mu [\gamma^\mu, \gamma^\sigma]$$

$$= \frac{i}{4} \gamma_\mu (\gamma^\mu \gamma^\sigma - \gamma^\sigma \gamma^\mu)$$

$$= \frac{i}{4}\gamma_\mu(\gamma^\mu\gamma^\sigma - 2\eta^{\mu\sigma} + \gamma^\mu\gamma^\sigma)$$
$$= \frac{i}{2}(\gamma_\mu\gamma^\mu\gamma^\sigma - \gamma^\sigma)$$
$$= \frac{i}{2}(4\gamma^\sigma - \gamma^\sigma)$$
$$= \frac{3i}{2}\gamma^\sigma$$

Hence we have

$$[W^2, Q_a]$$
$$= W^\mu(\not{P}\gamma_\mu\gamma_5 Q)_a + \frac{3}{4}(\not{P}\gamma^\sigma Q)_a P_\sigma$$
$$= W^\mu(\not{P}\gamma_\mu\gamma_5 Q)_a + \frac{3}{4}(\not{P}\not{P}Q)_a$$

or

$$[W^2, Q_a] = W^\mu(\not{P}\gamma_\mu\gamma_5 Q)_a + \frac{3}{4}P^2 Q_a \qquad (4.8)$$

Thus the commutator of W^2 and Q_a does not vanish. Our next step is the explicit construction of an invariant operator. We first define a pseudovector

$$X_\mu := \frac{1}{2}\bar{Q}\gamma_\mu\gamma_5 Q \qquad (4.9)$$

Proposition: We have

$$[X_\mu, Q_a] = -2(\not{P}\gamma_\mu\gamma_5 Q)_a \qquad (4.10)$$

Proof: Consider (using (4.9))

$$[X_\mu, Q_a] = \frac{1}{2}[\bar{Q}\gamma_\mu\gamma_5 Q, Q_a]$$
$$= \frac{1}{2}(\bar{Q}\gamma_\mu\gamma_5)_b Q_b Q_a - \frac{1}{2}Q_a(\bar{Q}\gamma_\mu\gamma_5 Q)$$
$$= -\frac{1}{2}(\bar{Q}\gamma_\mu\gamma_5)_b Q_a Q_b - (\bar{Q}\gamma_\mu\gamma_5)_b(\gamma_\lambda C)_{ba}P^\lambda$$
$$-\frac{1}{2}Q_a(\bar{Q}\gamma_\mu\gamma_5 Q) \qquad \text{(using (3.19) with } a = -2)$$

$$= -\tfrac{1}{2}\bar{Q}_b Q_a(\gamma_\mu\gamma_5 Q)_b - (\bar{Q}\gamma_\mu\gamma^5\gamma_\lambda C)_a P^\lambda$$
$$-\tfrac{1}{2} Q_a\bar{Q}_b(\gamma_\mu\gamma^5 Q)_b$$
$$= -\not{P}_{ab}(\gamma_\mu\gamma_5 Q)_b - (\bar{Q}\gamma_\mu\gamma^5\gamma_\lambda C)_a P^\lambda$$

(using (3.23) with $a = -2$)

$$= -(\not{P}\gamma_\mu\gamma_5 Q)_a - \bar{Q}_c(\gamma_\mu\gamma^5\gamma_\lambda C)_{ca}P^\lambda$$
$$= -(\not{P}\gamma_\mu\gamma_5 Q)_a - Q^T_b C_{bc}(\gamma_\mu\gamma_5\gamma_\lambda C)_{ca}P^\lambda$$

(using (3 .24a))

$$= -(\not{P}\gamma_\mu\gamma_5 Q)_a - (Q^T C\gamma_\mu\gamma_5\gamma_\lambda C)_a P^\lambda$$
$$= -(\not{P}\gamma_\mu\gamma_5 Q)_a - (Q^T\gamma_\mu^T\gamma_5^T\gamma_\lambda^T C^2)_a P^\lambda$$

(using $C\gamma_\mu = -\gamma_\mu^T C$,
$C\gamma_5 = \gamma_5^T C$)

$$= -(\not{P}\gamma_\mu\gamma_5 Q)_a + (Q^T\gamma_\mu^T\gamma_5^T\gamma_\lambda^T)_a P^\lambda$$

(since $C^2 = -1_{4\times4}$)

$$= -(\not{P}\gamma_\mu\gamma_5 Q)_a + (P_\lambda\gamma^\lambda\gamma_5\gamma_\mu Q)^T_a$$
$$= -(\not{P}\gamma_\mu\gamma_5 Q)_a - (\not{P}\gamma_\mu\gamma_5 Q)_a$$
$$= -2(\not{P}\gamma_\mu\gamma_5 Q)_a$$

as claimed by (4.10).

We now define a new vector B_μ by

$$B_\mu := W_\mu + \tfrac{1}{4}X_\mu \tag{4.11}$$

where W_μ is the Pauli-Ljubanski vector and X_μ is given by (4.9). Then

$$[B_\mu, Q_a] = [W_\mu + \tfrac{1}{4}X_\mu, Q_a]$$

$$= [W_\mu, Q_a] + \frac{1}{4}[X_\mu, Q_a]$$
$$= \frac{1}{2}(\not{P}\gamma_\mu\gamma_5 Q)_a - \frac{1}{2}P_\mu(\gamma^5 Q)_a$$
$$- \frac{1}{2}(\not{P}\gamma_\mu\gamma_5 Q)_a$$

so that

$$[B_\mu, Q_a] = -\frac{1}{2}P_\mu(\gamma^5 Q)_a \tag{4.12}$$

We now define the tensor

$$C_{\mu\nu} := B_\mu P_\nu - B_\nu P_\mu \tag{4.13}$$

Then the following result can be shown to hold.

Proposition: We have

$$[C_{\mu\nu}, Q_a] = 0 \tag{4.14}$$

Proof: We consider the commutator on the left hand side, i.e.

$$[C_{\mu\nu}, Q_a] = [B_\mu P_\nu - B_\nu P_\mu, Q_a]$$
$$= [B_\mu P_\nu, Q_a] - [B_\nu P_\mu, Q_a]$$
$$= B_\mu[P_\nu, Q_a] + [B_\mu, Q_a]P_\nu$$
$$- B_\nu[P_\mu, Q_a] - [B_\nu, Q_a]P_\mu$$
$$= [B_\mu, Q_a]P_\nu - [B_\nu, Q_a]P_\mu \quad \text{(using (4.1))}$$
$$= -\frac{1}{2}P_\mu(\gamma^5 Q)_a P_\nu + \frac{1}{2}P_\nu(\gamma^5 Q)_a P_\mu \quad \text{(using (4.12))}$$
$$= -\frac{1}{2}(\gamma^5 Q)_a P_\mu P_\nu + \frac{1}{2}(\gamma^5 Q)_a P_\nu P_\mu \quad \text{(using (4.1))}$$
$$= -\frac{1}{2}(\gamma^5 Q)_a [P_\mu, P_\nu]$$
$$= 0 \quad \text{(with (4.1))}$$

From $C_{\mu\nu}$ we can construct the Casimir operator

$$C^2 := C_{\mu\nu}C^{\mu\nu} \tag{4.15}$$

Proposition: The operator C^2 commutes with all generators of the Super-Poincaré group, i.e.

i) $$[C^2, Q_a] = 0 \tag{4.16}$$

ii) $$[C^2, P_\mu] = 0 \tag{4.17}$$

iii) $$[C^2, M_{\mu\nu}] = 0 \tag{4.18}$$

Equations (4.16) to (4.18) establish that C^2, defined by (4.15), is a Casimir operator of the Super-Poincaré algebra.

Proof:

i) Consider

$$[C^2, Q_a] = [C_{\mu\nu} C^{\mu\nu}, Q_a]$$
$$= C_{\mu\nu}[C^{\mu\nu}, Q_a] + [C_{\mu\nu}, Q_a] C^{\mu\nu}$$
$$= 0 \quad \text{(using (4.14))}$$

ii) Consider

$$[C^2, P_\rho] = C_{\mu\nu}[C^{\mu\nu}, P_\rho] + [C_{\mu\nu}, P_\rho] C^{\mu\nu}$$
$$= C_{\mu\nu}[B^\mu P^\nu - B^\nu P^\mu, P_\rho] + [B_\mu P_\nu - B_\nu P_\mu, P_\rho] C^{\mu\nu}$$
$$= C_{\mu\nu}\{B^\mu[P^\nu, P_\rho] + [B^\mu, P_\rho]P^\nu - B^\nu[P^\mu, P_\rho] - [B^\nu, P_\rho]P^\mu\}$$
$$+ \{B_\mu[P_\nu, P_\rho] + [B_\mu, P_\rho]P_\nu - B_\nu[P_\mu, P_\rho] - [B_\nu, P_\rho]P_\mu\} C^{\mu\nu}$$
$$= C_{\mu\nu}\{[B^\mu, P_\rho]P^\nu - [B^\nu, P_\rho]P^\mu\}$$
$$+ \{[B_\mu, P_\rho]P_\nu - [B_\nu, P_\rho]P_\mu\} C^{\mu\nu}$$
$$= C_{\mu\nu}\{[W^\mu + \tfrac{1}{4}X^\mu, P_\rho]P^\nu - [W^\nu + \tfrac{1}{4}X^\nu, P_\rho]P^\mu\}$$
$$+ \{[W_\mu + \tfrac{1}{4}X_\mu, P_\rho]P_\nu - [W_\nu + \tfrac{1}{4}X_\nu, P_\rho]P_\mu\} C^{\mu\nu}$$

$$= C_{\mu\nu}\{[W^\mu, P_\rho]P^\nu + \frac{1}{4}[X^\mu, P_\rho]P^\nu$$
$$- [W^\nu, P_\rho]P^\mu - \frac{1}{4}[X^\nu, P_\rho]P^\mu\}$$
$$+ \{[W_\mu, P_\rho]P_\nu + \frac{1}{4}[X_\mu, P_\rho]P_\nu$$
$$- [W_\nu, P_\rho]P_\mu - \frac{1}{4}[X_\nu, P_\rho]P_\mu\}C^{\mu\nu}$$
$$= C_{\mu\nu}\frac{1}{4}\{[X^\mu, P_\rho]P^\nu - [X^\nu, P_\rho]P^\mu\}$$
$$+ \frac{1}{4}\{[X_\mu, P_\rho]P_\nu - [X_\nu, P_\rho]P^\mu\}C^{\mu\nu}$$

(using (1.33a))

$$= 0$$

since

$$[X_\mu, P_\rho]$$
$$= \frac{1}{2}[\bar{Q}\gamma_\mu\gamma_5 Q, P_\rho] \quad \text{(with (4.9))}$$
$$= \frac{1}{2}(\bar{Q}\gamma_\mu\gamma_5)_a [Q_a, P_\rho] + \frac{1}{2}[\bar{Q}_a, P_\rho](\gamma_\mu\gamma_5 Q)_a$$
$$= 0 \quad \text{(with (4.1))}$$

iii) The operator $C^2 = C_{\mu\nu}\, C^{\mu\nu}$ commutes with $M_{\rho\sigma}$ because by construction it is a scalar under Lorentz transformations. This completes the proof. Hence we have shown that there are two Casimir operators of the Super-Poincaré algebra: P^2 and C^2.

In the following we use the eigenvalues of the Casimir operators to label the irreducible representations - and thus particle multiplets - of the Super-Poincaré algebra. The calculations will be performed in the rest frame. States boosted to a momentum P_μ are then defined by the particle's wave equation.

4.2 Classification of Irreducible Representations

4.2.1 N = 1 Supersymmetry

The irreducible representations are characterized by the eigenvalues of P^2 and C^2. Let $P^2 = m^2 > 0$, and for simplicity we choose the rest frame,

$$P_\mu = (m, \vec{0})$$

Then in the rest frame C^2 is

$$\begin{aligned} C^2 &= C_{\mu\nu} C^{\mu\nu} = (B_\mu P_\nu - B_\nu P_\mu)(B^\mu p^\nu - B^\nu p^\mu) \\ &= 2 B_\mu P_\nu B^\mu p^\nu - 2 B_\mu P_\nu B^\nu p^\mu \\ &= 2 B_\mu B^\mu m^2 - 2 B_0^2 m^2 \\ &= 2 m^2 B_k B^k \end{aligned} \tag{4.19}$$

where k = 1,2,3. Now from (4.11) with (1.40)

$$\begin{aligned} B_k &= W_k + \frac{1}{4} X_k \\ &= m S_k + \frac{1}{8} \bar{Q} \gamma_k \gamma_5 Q \\ &=: m J_k \end{aligned} \tag{4.20}$$

The operator J_k is defined by this relation. Subsequently we will use the Weyl formalism. It is therefore desirable to know the definition of J_k in terms of Weyl operators. For this purpose we use (1.136) and (1.139c) in rewriting (4.20). Then (using (1.136) and (1.139c))

$$\begin{aligned} &\bar{Q} \gamma_k \gamma_5 Q \\ &= (Q^A, \bar{Q}_{\dot{A}}) \begin{pmatrix} 0 & \sigma_{kA\dot{B}} \\ \bar{\sigma}_k^{\dot{A}B} & 0 \end{pmatrix} \begin{pmatrix} -\delta_B{}^C & 0 \\ 0 & \delta^{\dot{B}}{}_{\dot{C}} \end{pmatrix} \begin{pmatrix} Q_C \\ \bar{Q}^{\dot{C}} \end{pmatrix} \\ &= (\bar{Q}_{\dot{A}} \bar{\sigma}_k^{\dot{A}B}, Q^A \sigma_{kA\dot{B}}) \begin{pmatrix} -\delta_B{}^C Q_C \\ \delta^{\dot{B}}{}_{\dot{C}} \bar{Q}^{\dot{C}} \end{pmatrix} \end{aligned}$$

$$= -\overline{Q}_{\dot{A}}\overline{\sigma}_k^{\dot{A}B}Q_B + Q^A\sigma_{kA\dot{B}}\overline{Q}^{\dot{B}}$$
$$= -(\overline{Q}\overline{\sigma}_k Q) + (Q\sigma_k\overline{Q})$$
$$= 2(Q\sigma_k\overline{Q}) = -2(\overline{Q}\overline{\sigma}_k Q)$$

using (1.115). Hence (4.20) can be rewritten

$$mJ_k = mS_k - \frac{1}{4}(\overline{Q}\overline{\sigma}_k Q) \tag{4.20'}$$

We can now rewrite C^2 in terms of J_k. Then (4.19) becomes

$$C^2 = 2m^4 J_k J^k \tag{4.21}$$

and this implies that J_k is an angular momentum, i.e.

$$[J_k, J_l] = i\,\varepsilon_{klm}J_m \tag{4.22}$$

One can check that this relation holds for J_k defined by (4.20). Furthermore

$$[J_k, Q_a] = \frac{1}{m}[B_k, Q_a] \quad \text{(with (4.20))}$$
$$= -\frac{1}{2m}P_k(\gamma^5 Q)_a \quad \text{(with (4.12))}$$
$$= 0 \tag{4.23a}$$

since in the rest frame $P_k = 0$. Correspondingly

$$[J_k, Q_A] = 0, \quad [J_k, \overline{Q}^{\dot{A}}] = 0 \tag{4.23b}$$

by decomposing Q_a into its Weyl components.
Now according to (4.21) $J^2 = J_k J^k$ is an invariant operator with eigenvalues of the form $j(j+1)$ as in the case of ordinary angular momentum; here j is either integral or half-integral (i.e. there exist appropriate representations of J_k which satisfy for integral or half-integral spin the relation (4.22)).

The irreducible representations of the Super-Poincaré algebra are specified by the values of m^2 and $j(j+1)$, the eigenvalues of the Casimir operators. The states of the representation can then be given the labels m^2, j and j_3, the latter being the eigenvalue of J_3 which assumes the values $-j, -j+1, \ldots, j$. However, these states are not, in general, eigenstates of $\vec{S}^2$ and S_3 of ordinary spin $\vec{S}$. In order to find the spin content it is convenient to work in terms of the two-component Weyl formalism with commutation relations (3.25), (3.26). We have

$$[M_{\mu\nu}, Q_A] = -(\sigma^2_{\mu\nu})_A{}^B Q_B$$

so that

$$[M_{ij}, Q_A] = -(\sigma^2_{ij})_A{}^B Q_B$$

In the rest frame the spin operator is given by

$$S_k = \frac{1}{2}\varepsilon_{kij} M_{ij}$$

Hence the commutator of S_k and Q_a is

$$\begin{aligned}[S_k, Q_A] &= -\frac{1}{2}\varepsilon_{kij}(\sigma^2_{ij})_A{}^B Q_B \\ &= -\frac{1}{2}(\sigma_k\bar{\sigma}^0)_A{}^B Q_B\end{aligned} \tag{4.24}$$

where we used (1.120a), and the unit matrix $\bar{\sigma}^0$ has been inserted to provide the correct index structure for σ_k in agreement with (1.87). Similarly

$$[S_k, \bar{Q}^{\dot{A}}] = -\frac{1}{2}(\bar{\sigma}^k\sigma^0)^{\dot{A}}{}_{\dot{B}}\bar{Q}^{\dot{B}} \tag{4.25}$$

From (3.25) we obtain in the rest frame

$$\begin{aligned}\{Q_A, \bar{Q}_{\dot{B}}\} &= 2(\sigma^0)_{A\dot{B}} P_0 \\ &= 2m\,\sigma^0_{A\dot{B}}\end{aligned}$$

Hence

$$\{Q_1, \bar{Q}_{\dot{1}}\} = \{Q_2, \bar{Q}_{\dot{2}}\} = 2m$$

$$\{Q_1, \bar{Q}_{\dot{2}}\} = \{Q_2, \bar{Q}_{\dot{1}}\} = 0$$

$$\{Q_A, Q_B\} = \{\bar{Q}_{\dot{A}}, \bar{Q}_{\dot{B}}\} = 0 \tag{4.26}$$

These relations define a Clifford algebra for operators which we can now call creation ($\bar{Q}$) and annihilation (Q) operators.

We now consider irreducible spinor representations of the Super-Poincaré algebra characterized by m and j, and consider the states with fixed value of j_3. Among these states is a state $|\Omega\rangle$ which is annihilated by Q_A (A = 1,2), i.e.

$$Q_A|\Omega\rangle = 0 \tag{4.27}$$

This state called the Clifford vacuum is the vacuum state of spinor representations and can be either bosonic or fermionic. This state must not be confused with the quantum mechanical ground state or vacuum defined as the state of lowest energy. Instead it is an eigenstate of the Casimir operator $P_\mu P^\mu$, i.e.

$$P_\mu P^\mu|\Omega\rangle = m^2|\Omega\rangle \tag{4.28}$$

Thus the Clifford vacuum is that particular state in the irreducible representation of the Super-Poincaré algebra specified by the numbers m, j and j_3 which is such that (4.27) holds. Since by construction P^2 and C^2 are the only Casimir operators in the present case, the Clifford vacuum $|\Omega\rangle$ is nondegenerate (ignoring internal symmetries).

We now consider another state in the chosen representation, written $|\beta\rangle$, with fixed value of j_3. Then (see below)

$$|\Omega\rangle = Q_1 Q_2 |\beta\rangle$$

and
$$Q_1|\Omega> = Q_1 Q_1 Q_2 |\beta>$$
$$= 0$$
since Q_1 anticommutes with itself, and similarly
$$Q_2|\Omega> = Q_2 Q_1 Q_2 |\beta> = 0$$
Furthermore $|\Omega>$ will have the same eigenvalue of J_3 as $|\beta>$ since
$$J_3|\Omega> = J_3 Q_1 Q_2 |\beta>$$
$$= Q_1 J_3 Q_2 |\beta> \quad \text{(with (4.23a))}$$
$$= Q_1 Q_2 J_3 |\beta> \quad \text{(with (4.23a))}$$
$$= j_3 Q_1 Q_2 |\beta> = j_3 |\Omega>$$

This shows that we must have a Clifford vacuum in any representation of the Super-Poincaré algebra. Now from (4.20')
$$J_k = S_k - \frac{1}{4m}(\bar{Q}\bar{\sigma}_k Q)$$
which implies that the Clifford vacuum is an eigenstate of the operators $\vec{S}^2$ and S_3 of ordinary spin angular momentum with eigenvalues $s(s+1)$ and s_3. Hence we write
$$|\Omega> := |m, s, s_3> \tag{4.29}$$
We now consider the action of $\bar{Q}^{\dot{A}}$ on the Clifford vacuum, i.e. we consider states
$$\bar{Q}^{\dot{1}}|m, s, s_3>, \quad \bar{Q}^{\dot{2}}|m, s, s_3> \tag{4.30}$$
Proposition: The states (4.30) satisfy the relation
$$J_3 \bar{Q}^{\dot{A}}|m, s, s_3> = j_3 \bar{Q}^{\dot{A}}|m, s, s_3> \tag{4.31}$$
Proof: Using (4.23b) and (4.29) we have
$$J_3 \bar{Q}^{\dot{A}}|m, s, s_3> = \bar{Q}^{\dot{A}} J_3 |m, s, s_3>$$
$$= \bar{Q}^{\dot{A}} J_3 |\Omega>$$
$$= \bar{Q}^{\dot{A}} j_3 |\Omega>$$

$$= j_3 \bar{Q}^{\dot{A}}|\Omega\rangle$$
$$= j_3 \bar{Q}^{\dot{A}}|m, s\, s_3\rangle$$

Equation (4.31) shows that the states (4.30) are states in the representation of the Super-Poincaré algebra with eigenvalues j, m of the Casimir operators and eigenvalue j_3 of the operator J_3. In the following we use the identities

$$(\bar{\sigma}_3 \sigma^0)^{\dot{1}}{}_{\dot{1}} = -1 \ , \ (\bar{\sigma}_3 \sigma^0)^{\dot{2}}{}_{\dot{2}} = 1$$

Now, using (4.25) we have

$$S_3\, \bar{Q}^{\dot{1}}|\Omega\rangle$$
$$= \bar{Q}^{\dot{1}} S_3|\Omega\rangle - \tfrac{1}{2}(\bar{\sigma}_3 \sigma^0)^{\dot{1}}{}_{\dot{B}}\, \bar{Q}^{\dot{B}}|\Omega\rangle$$
$$= \bar{Q}^{\dot{1}} J_3|\Omega\rangle - \tfrac{1}{2}(\bar{\sigma}_3 \sigma^0)^{\dot{1}}{}_{\dot{B}}\, \bar{Q}^{\dot{B}}|\Omega\rangle$$

(using (4.20'))

$$= \bar{Q}^{\dot{1}} J_3|\Omega\rangle - \tfrac{1}{2}(\bar{\sigma}_3 \sigma^0)^{\dot{1}}{}_{\dot{1}}\, \bar{Q}^{\dot{1}}|\Omega\rangle$$
$$= \bar{Q}^{\dot{1}} j_3|\Omega\rangle + \tfrac{1}{2}\bar{Q}^{\dot{1}}|\Omega\rangle$$
$$= (j_3 + \tfrac{1}{2})\bar{Q}^{\dot{1}}|\Omega\rangle \qquad (4.32)$$

and similarly

$$S_3\, \bar{Q}^{\dot{2}}|\Omega\rangle$$
$$= \bar{Q}^{\dot{2}} S_3|\Omega\rangle - \tfrac{1}{2}(\bar{\sigma}_3 \sigma^0)^{\dot{2}}{}_{\dot{B}}\, \bar{Q}^{\dot{B}}|\Omega\rangle$$
$$= \bar{Q}^{\dot{2}} J_3|\Omega\rangle - \tfrac{1}{2}(\bar{\sigma}_3 \sigma^0)^{\dot{2}}{}_{\dot{2}}\, \bar{Q}^{\dot{2}}|\Omega\rangle$$
$$= \bar{Q}^{\dot{2}} j_3|\Omega\rangle - \tfrac{1}{2}\bar{Q}^{\dot{2}}|\Omega\rangle$$
$$= (j_3 - \tfrac{1}{2})\bar{Q}^{\dot{2}}|\Omega\rangle \qquad (4.33)$$

Equations (4.32) and (4.33) imply that $\bar{Q}^{\dot{1}}$ raises the value of j_3 by 1/2 and $\bar{Q}^{\dot{2}}$ lowers it by the same amount. Applying $\bar{Q}^{\dot{1}}\bar{Q}^{\dot{2}}$ to the Clifford vacuum we obtain yet another state:

$$
\begin{aligned}
& S_3\bar{Q}^{\dot{1}}\bar{Q}^{\dot{2}}|\Omega\rangle \\
&= \bar{Q}^{\dot{1}}S_3\bar{Q}^{\dot{2}}|\Omega\rangle - \tfrac{1}{2}(\bar{\sigma}_3\sigma^0)^{\dot{1}}{}_{\dot{B}}\bar{Q}^{\dot{B}}\bar{Q}^{\dot{2}}|\Omega\rangle \\
&\qquad \text{(using (4.25))} \\
&= \bar{Q}^{\dot{1}}S_3\bar{Q}^{\dot{2}}|\Omega\rangle + \tfrac{1}{2}\bar{Q}^{\dot{1}}\bar{Q}^{\dot{2}}|\Omega\rangle \\
&= \bar{Q}^{\dot{1}}\bar{Q}^{\dot{2}}S_3|\Omega\rangle + \tfrac{1}{2}\bar{Q}^{\dot{1}}\bar{Q}^{\dot{2}}|\Omega\rangle \\
&\qquad - \bar{Q}^{\dot{1}}\tfrac{1}{2}(\bar{\sigma}_3\sigma^0)^{\dot{2}}{}_{\dot{B}}\bar{Q}^{\dot{B}}|\Omega\rangle \\
&= \bar{Q}^{\dot{1}}\bar{Q}^{\dot{2}}J_3|\Omega\rangle + \tfrac{1}{2}\bar{Q}^{\dot{1}}\bar{Q}^{\dot{2}}|\Omega\rangle - \tfrac{1}{2}\bar{Q}^{\dot{1}}\bar{Q}^{\dot{2}}|\Omega\rangle \\
&= j_3\bar{Q}^{\dot{1}}\bar{Q}^{\dot{2}}|\Omega\rangle
\end{aligned}
\tag{4.34}
$$

This result shows that $\bar{Q}_{\dot{1}}\bar{Q}_{\dot{2}}|\Omega\rangle$ is a state whose eigenvalue is j_3. Again applying $\bar{Q}^{\dot{A}}$ gives zero and so ends the construction of eigenstates of S_3.

We now summarize our findings. For each pair of values of m and j of the Casimir operators we obtain an irreducible representation of the Super-Poincaré algebra (4.1). The states corresponding to specific values of m and j contain 2j+1 subspaces according to the possible values of j_3. For fixed j_3 each subspace contains four eigenstates of S_3 with eigenvalues $s_3 = j_3,\ j_3 + 1/2,\ j_3 - 1/2$, and again j_3, corresponding to the operators

$$\mathbb{1},\ \bar{Q}^{\dot{1}},\ \bar{Q}^{\dot{2}} \text{ and } \bar{Q}^{\dot{1}}\bar{Q}^{\dot{2}}$$

respectively. These states span an irreducible 4(2j+1)-dimensional representation of the little algebra given by (4.26). Schematically we have the following picture:

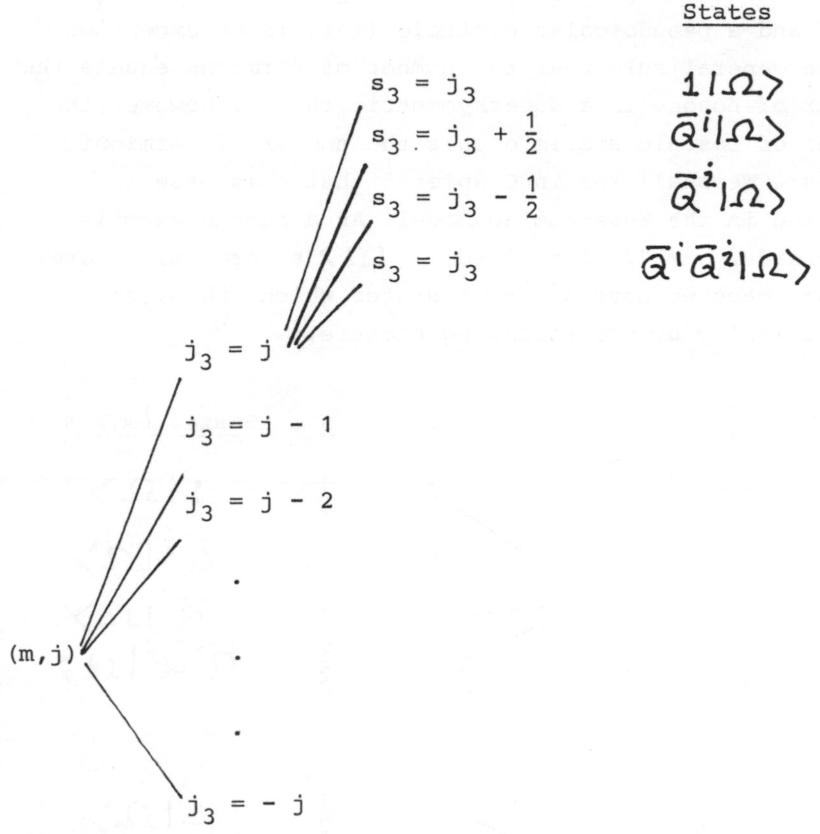

From (3.25) we deduce that any product of two Q's has negative parity. Thus if $|\Omega\rangle$ is a scalar state, $\bar{Q}\bar{Q}|\Omega\rangle$ is pseudoscalar and vice versa.

We now consider two examples. The lowest-dimensional representation is the representation with j = 0 ($|\Omega\rangle$ a bosonic state). For this representation there are only three spins: s = 0, s = $\frac{1}{2}$ and again s = 0 correspon-ding to a scalar particle, a spin-$\frac{1}{2}$ particle (with $s_3 = \pm \frac{1}{2}$) and a pseudoscalar particle (this is an exception to the general rule that the number of fermions equals the number of bosons in a supersymmetric theory; however, the number of bosonic states equals the number of fermionic states). We shall see in Chapter 5 that this case is realized in the Wess-Zumino model. As a second example we consider j = 1/2 for fixed m ($|\Omega\rangle$ a fermionic state). In this case we have a set of states which is given schematically by the following picture:

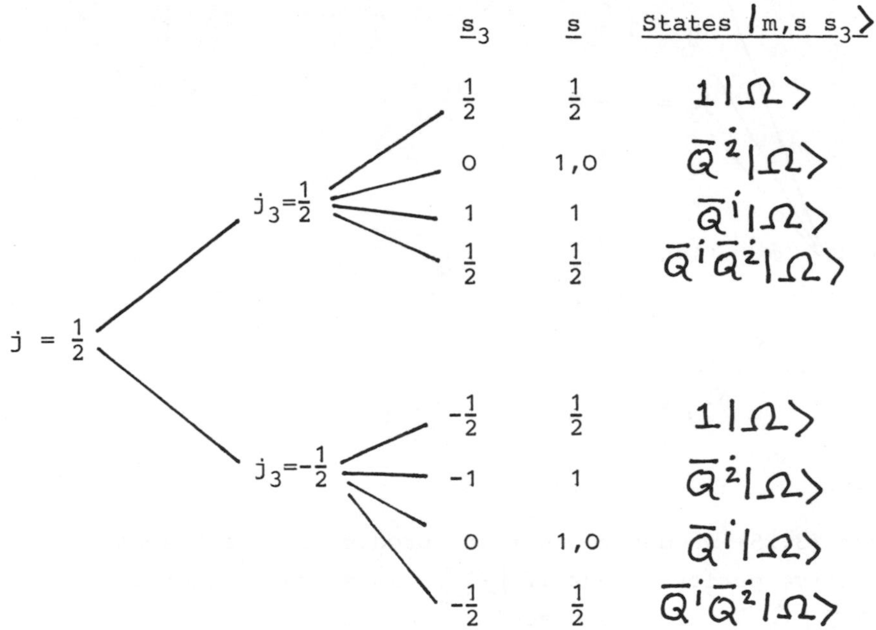

The particles described by these states are: a pseudoscalar particle ($\bar{Q}|\Omega\rangle$), a vector particle ($\bar{Q}^{\dot{1}}|\Omega\rangle$, $\bar{Q}^{\dot{2}}|\Omega\rangle$) and two spin-$\frac{1}{2}$ particles (one is described by $1|\Omega\rangle$, the other by $\bar{Q}^{\dot{1}}\bar{Q}^{\dot{2}}|\Omega\rangle$). (A pseudoscalar quantity changes sign under the parity transformation like a vector). Of course, the pseudoscalar state and the associated vector state arise from the Kronecker product of Weyl spinors, i.e.

$$(0,\tfrac{1}{2}) \otimes (0,\tfrac{1}{2}) = (0,0) + (0,1)$$

in much the same way as in (1.129b). Thus a fermionic state $|\Omega\rangle$ carries effectively an index such as $\dot{A}$ and then transforms like $\bar{Q}^{\dot{A}}$. Thus $\bar{Q}^{\dot{2}}|\Omega\rangle \sim \bar{Q}^{\dot{2}}\bar{Q}^{\dot{1}}|\,\rangle$ where $|\,\rangle$ transforms as a bosonic state. Clearly there is only one such pseudoscalar state.

The above scheme reveals an important feature of supersymmetric theories: An irreducible representation of the Super-Poincaré algebra has the number of fermionic states equal to the number of bosonic states and thus the number of fermions equal to the number of bosons (with the exception referred to above). Before we proceed to representations of the $N > 1$ Super-Poincaré algebra, we prove the following result.

<u>Proposition</u>: Every representation of the supersymmetry algebra (4.1) contains as many bosonic states as fermionic states.

<u>Proof</u>: We first introduce a fermion number operator N_F which is such that $(-1)^{N_F}$ has eigenvalue +1 when applied to bosonic states (i.e. states containing an even number of fermions n_F). Thus

$$(-1)^{N_F}|\,\rangle = (-1)^{n_F}|\,\rangle$$

when the state $|\,\rangle$ contains n_F fermions. Since Q_A and $\bar{Q}^{\dot{A}}$ annihilate or create fermions we have

$$Q_A(-1)^{N_F}|\rangle = (-1)^{N_F-1}Q_A|\rangle$$

i.e.

$$(-1)^{N_F}Q_A = -Q_A(-1)^{N_F}$$

For any finite-dimensional representation of the Super-Poincaré algebra which is such that the trace is well defined we have

$$\begin{aligned}
&Tr\left[(-1)^{N_F}\{Q_A,\bar{Q}_{\dot{B}}\}\right] \\
&= Tr\left[(-1)^{N_F}(Q_A\bar{Q}_{\dot{B}}+\bar{Q}_{\dot{B}}Q_A)\right] \\
&= Tr\left[(-1)^{N_F}Q_A\bar{Q}_{\dot{B}}\right]+Tr\left[Q_A(-1)^{N_F}\bar{Q}_{\dot{B}}\right] \\
&= 0
\end{aligned}$$

as a result of the above relation. On the other hand, using (3.25) we obtain

$$\begin{aligned}
0 &= Tr\left[(-1)^{N_F}\{Q_A,\bar{Q}_{\dot{B}}\}\right] \\
&= Tr\left[(-1)^{N_F}2(\sigma^\mu)_{A\dot{B}}P_\mu\right] \\
&= 2(\sigma^\mu)_{A\dot{B}}P_\mu\, Tr\left[(-1)^{N_F}\right]
\end{aligned}$$

For fixed nonzero momentum P_μ this reduces to

$$Tr\left[(-1)^{N_F}\right] = 0$$

i.e.

$$\begin{aligned}
&\sum_B \langle B|(-1)^{N_F}|B\rangle + \sum_F \langle F|(-1)^{N_F}|F\rangle \\
&\qquad = N_B - N_F \\
&\qquad = 0
\end{aligned}$$

This demonstrates that supersymmetric representations contain equal numbers of bosonic and fermionic states.

<u>Remark</u>: The factor $(-1)^{N_F}$ is in fact the operator used by Witten[45] to discuss qualitatively features of supersymmetry breaking. The trace of this operator is a particular example of the much more general concept of the index of an operator. Considerations of this type lead to a proof of the Atiyah-Singer index theorem in the context of supersymmetric quantum mechanics. For details we refer to D. Friedan and P. Windey[46], L. Alvarez-Gaumé[47] and E. Witten[45].

4.2.2 N > 1 Supersymmetry

We remarked at the beginning of this chapter that if we want to include an internal symmetry in the Super-Poincaré algebra in a nontrivial way, we have to extend the algebra (4.1) by introducing a set of N spinor charges Q_A^α $\bar{Q}_{\dot{A}}^\alpha$ ($\alpha = 1,\ldots, N$) where N is the dimension of a representation of the internal symmetry group. For the Super-Poincaré algebra we have in the rest frame according to (3.25)

$$\{Q_A^\alpha, Q_B^\beta\} = \{\bar{Q}_{\dot{A}}^\alpha, \bar{Q}_{\dot{B}}^\beta\} = 0 \tag{4.35}$$

$$\{Q_A^\alpha, \bar{Q}_{\dot{B}}^\beta\} = 2m(\sigma^0)_{A\dot{B}}\,\delta^{\alpha\beta} \tag{4.36}$$

Here the indices α and β run from 1 to N. Rescaling the generators Q by writing

$$a_A^\alpha := \frac{1}{(2m)^{1/2}} Q_A^\alpha , (a_A^\alpha)^+ := \frac{1}{(2m)^{1/2}} \bar{Q}_{\dot{A}}^\alpha \tag{4.37}$$

we obtain

$$\{a_A^\alpha, a_B^\beta\} = \{(a_A^\alpha)^+, (a_B^\beta)^+\} = 0 \tag{4.38}$$

$$\{a_A^\alpha, (a_B^\beta)^+\} = \delta_{AB}\,\delta^{\alpha\beta} \tag{4.39}$$

This is the algebra of 2N fermionic annihilation and creation operators a^{α}_{A} and $(a^{\alpha}_{A})^{+}$ respectively. The representations of the algebra (4.38), (4.39) can be constructed with the help of a Clifford vacuum $|\Omega\rangle$ defined by

$$a^{\alpha}_{A}|\Omega\rangle = 0, \alpha = 1,\ldots, N; A = 1,2, \qquad (4.40)$$

and as before this Clifford vacuum state is an eigenstate of the Casimir operator P^2 with eigenvalue m^2, i.e.

$$P^2|\Omega\rangle = m^2|\Omega\rangle \qquad (4.41)$$

Furthermore, the Clifford vacuum $|\Omega\rangle$ transforms according to an irreducible representation of the little group SU(2) of the momentum operator in the rest frame. In other words, if the vacuum $|\Omega\rangle$ has spin s, it belongs to a (2s+1)-dimensional representation of SU(2,C).

The states of the representation of the algebra (4.38), (4.39) are constructed by applying creation operators $(a^{\alpha}_{A})^{+}$ to $|\Omega\rangle$. Then an n-particle state is given by

$$\Phi^{(n)\alpha_1\ldots\alpha_n}_{A_1\cdots A_n} = \frac{1}{(n!)^{1/2}}(a^{\alpha_1}_{A_1})^{+}\cdots(a^{\alpha_k}_{A_n})^{+}|\Omega\rangle \qquad (4.42)$$

Now from (4.38) in this case

$$\{(a^{\alpha_i}_{A_i})^{+}, (a^{\alpha_j}_{A_j})^{+}\} = 0 \qquad (4.43)$$

so that

$$\Phi^{(n)\alpha_1\ldots\alpha_i\ldots\alpha_j\ldots\alpha_n}_{A_1\ldots A_i\ldots A_j\ldots A_n} = -\Phi^{(n)\alpha_1\ldots\alpha_j\ldots\alpha_i\ldots\alpha_n}_{A_1\ldots A_j\ldots A_i\ldots A_n}$$

Thus the state $\Phi^{(n)}$ is antisymmetric under exchange of two pairs of indices (α_i, A_i) and (α_j, A_j). Here each index pair takes 2N different values since A_i = 1,2 and α_i' = 1,..., N. Thus n must be less than or equal to 2N.

As a first example we consider the case N = 1, the case treated explicitly before. According to the above the only states which can be constructed are

$$\Phi^{(0)} = |\Omega\rangle$$
$$\Phi^{(1)} \sim \bar{Q}_{\dot{A}}|\Omega\rangle$$
$$\Phi^{(2)} \sim \bar{Q}_{\dot{A}}\bar{Q}_{\dot{B}}|\Omega\rangle$$

These are the four states which span the subspace for a given value of j_3 as in Section 4.2.1 .

In view of the anticommutator (4.43) we can construct $\binom{2N}{n}$ different states for any given value of n. In the example of N = 1 this means, for n = 0 we have $\binom{2}{0} = 1$ state, i.e. $\Phi^{(0)} = |\Omega\rangle$; for n = 1 we have $\binom{2}{1} = 2$ states, namely $\bar{Q}_{\dot{1}}|\Omega\rangle$ and $\bar{Q}_{\dot{2}}|\Omega\rangle$; finally for n = 2 there is only one possible state $\Phi^{(2)}$.

Summing over all values of n gives the dimension d of the representation, i.e.

$$d = \sum_{n=0}^{2N} \binom{2N}{n} = (1+1)^{2N} = 2^{2N} \qquad (4.44)$$

The case N = 1 has the dimension $d = 2^2 = 4$, corresponding to the four states which span the subspace for a given value of j_3.

If the Clifford vacuum $|\Omega\rangle$ is not degenerate, (4.42) is called the fundamental irreducible massive multiplet. This multiplet (with n assuming values from o to 2N) has dimension 2^{2N} with 2^{2N-1} bosonic and the same number of fermionic states. In the case N = 1 the fundamental multiplet is given by

$$\Phi^{(0)} = |\Omega\rangle$$
$$\Phi^{(1)}_{A} = (a_A)^{+}|\Omega\rangle$$
$$\Phi^{(2)}_{AB} = \frac{1}{2^{1/2}}(a_A)^{+}(a_B)^{+}|\Omega\rangle$$
$$= -\frac{1}{2^{3/2}}\varepsilon_{AB}(a_C)^{+}(a^C)^{+}|\Omega\rangle$$

(with (1.83d))and thus yields as many bosonic as fermionic states as was demonstrated in Section 4.2.1.

In the following we consider in detail the case N = 2. This means, we take spinor charges Q^{α}_{A}, $\overline{Q}^{\dot{\alpha}}_{A}$ in the fundamental representation of the internal symmetry group SU(2,C), which describes, for instance, isospin. According to (4.44) the dimension of the representation of the algebra (4.38), (4.39) is $d = 2^{2.2} = 16$. Furthermore, the fundamental multiplet is given by a 16-plet of states. For n = 0, there is $\binom{2N}{0} = \binom{4}{0} = 1$ state given by

$$\Phi^{(0)} = |\Omega\rangle$$

For n = 1, there are $\binom{2N}{1} = \binom{4}{1} = 4$ states:

$$\Phi^{(1)\alpha}_{A} = (a^{\alpha}_{A})^{+}|\Omega\rangle$$

or explicitly

$$\Phi^{(1)1}_{1} = (a^{1}_{1})^{+}|\Omega\rangle$$

$$\Phi^{(1)2}_{1} = (a^{2}_{1})^{+}|\Omega\rangle$$

$$\Phi^{(1)1}_{2} = (a^{1}_{2})^{+}|\Omega\rangle$$

$$\Phi^{(1)2}_{2} = (a^{2}_{2})^{+}|\Omega\rangle$$

For n = 2, we have $\binom{2N}{2} = \binom{4}{2} = 6$ possible states. These are given by

$$\Phi^{(2)11}_{11} = \frac{1}{2^{1/2}}(a^{1}_{1})^{+}(a^{1}_{1})^{+}|\Omega\rangle = 0$$

$$\Phi^{(2)21}_{11} = \frac{1}{2^{1/2}}(a^{2}_{1})^{+}(a^{1}_{1})^{+}|\Omega\rangle = -\Phi^{(2)12}_{11}$$

$$\Phi^{(2)22}_{11} = \frac{1}{2^{1/2}}(a^{2}_{1})^{+}(a^{2}_{1})^{+}|\Omega\rangle = 0$$

$$\Phi^{(2)11}_{21} = \frac{1}{2^{1/2}}(a^{1}_{2})^{+}(a^{1}_{1})^{+}|\Omega\rangle = -\Phi^{(2)11}_{12}$$

$$\Phi^{(2)11}_{22} = \frac{1}{2^{1/2}}(a^1_2)^+(a^1_2)^+|\Omega> = 0$$

$$\Phi^{(2)12}_{22} = \frac{1}{2^{1/2}}(a^1_2)^+(a^2_2)^+|\Omega> = -\Phi^{(2)21}_{22}$$

$$\Phi^{(2)22}_{22} = \frac{1}{2^{1/2}}(a^2_2)^+(a^2_2)^+|\Omega> = 0$$

$$\Phi^{(2)21}_{21} = \frac{1}{2^{1/2}}(a^2_2)^+(a^1_1)^+|\Omega> = -\Phi^{(2)12}_{12}$$

$$\Phi^{(2)21}_{12} = \frac{1}{2^{1/2}}(a^2_1)^+(a^1_2)^+|\Omega> = -\Phi^{(2)12}_{21}$$

$$\Phi^{(2)22}_{12} = \frac{1}{2^{1/2}}(a^2_1)^+(a^2_2)^+|\Omega> = -\Phi^{(2)21}_{22}$$

The vanishing of various states is due to the anticommutator (4.43) which is for $\Phi^{(2)22}_{22}$:

$$\{(a^2_2)^+, (a^2_2)^+\} = 2(a^2_2)^+(a^2_2)^+ = 0$$

and therefore

$$(a^2_2)^+(a^2_2)^+|\Omega> = 0$$

For n = 3, we have $\binom{2N}{3} = \binom{4}{3} = 4$ possible states. The nonvanishing states are

$$\Phi^{(3)211}_{121} = \frac{1}{(3!)^{1/2}}(a^2_1)^+(a^1_2)^+(a^1_1)^+|\Omega>$$

$$\Phi^{(3)211}_{221} = \frac{1}{(3!)^{1/2}}(a^2_2)^+(a^1_2)^+(a^1_1)^+|\Omega>$$

$$\Phi^{(3)221}_{121} = \frac{1}{(3!)^{1/2}}(a^2_1)^+(a^2_2)^+(a^1_1)^+|\Omega>$$

$$\Phi^{(3)122}_{221} = \frac{1}{(3!)^{1/2}}(a^1_2)^+(a^2_2)^+(a^2_1)^+|\Omega>$$

Finally, for n = 4, we have one nonvanishing state given by

$$\Phi^{(4)\,2211}_{\;\;\;\;\;\;2121} = \frac{1}{(4!)^{1/2}} (a^2_2)^+ (a^2_1)^+ (a^1_2)^+ (a^1_1)^+ |\Omega\rangle$$

C H A P T E R 5

THE WESS-ZUMINO MODEL

5.1 The Lagrangian and the Equations of Motion[26]

We now construct a field theoretical realization of the Super-Poincaré algebra (4.1). In this application we choose the representation corresponding to nonvanishing mass ($m^2 \neq 0$) and $j = 0$. According to the considerations of Section 4.2.1 this model describes two spin-0 fields A, B (one scalar, the other pseudoscalar) and one spin-$\frac{1}{2}$ Majorana field ψ. According to the theory of representations of the Super-Poincaré algebra, all three fields belong to the same mass multiplet. The case to be discussed here is the socalled "on-shell" case, in which the algebra of generators is treated on the basis of the equations of motion of the fields. The "off-shell" case does not require the validity of the equations of motion; we shall see later that this necessitates the introduction of further fields, called auxiliary fields, as additional degrees of freedom.

Our starting point is the following Lagrangian density

$$\mathcal{L} = \frac{1}{2}(\partial_\mu A)(\partial^\mu A) - \frac{1}{2}m^2A^2 + \frac{1}{2}(\partial_\mu B)(\partial^\mu B) - \frac{1}{2}m^2B^2 + \frac{1}{2}\bar{\Psi}(i\not{\partial} - m)\psi - mgA(A^2+B^2) - g(\bar{\Psi}\psi A + i\bar{\Psi}\gamma^5\psi B) - \frac{1}{2}g^2(A^2+B^2)^2 \tag{5.1}$$

with

$A = A^+$, $B = B^+$ and $\psi = \psi^C$ (Majorana)

We see that in contrast to nonsupersymmetric field theories all fields have the same mass m and couple with the same strength g. As explained in Chapter 4 this is due to the fact that states of a particular representation of the Super-Poincare algebra are characterized by the eigenvalue m^2 of P^2 and different values of spin s. For reasons of simplicity we consider here only the free case, i.e.

$$\mathcal{L}\big|_{g=0} = \mathcal{L}_{\text{free}}$$

The equations of motion are obtained by applying to (5.1) the Euler-Lagrange equation

$$\frac{\partial \mathcal{L}}{\partial \phi_i} - \partial_\mu \frac{\partial \mathcal{L}}{\partial \partial_\mu \phi_i} = 0 \tag{5.2}$$

where

$$\phi_i \in \{A, B, \psi, \bar{\Psi}\}$$

i) $\underline{\phi = A:}$ The equation of motion for the scalar field is

$$\frac{\partial \mathcal{L}}{\partial A} - \partial_\mu \frac{\partial \mathcal{L}}{\partial \partial_\mu A} = 0$$

where

$$\frac{\partial \mathcal{L}}{\partial A} = -m^2A - g\bar{\Psi}\psi - mg(3A^2+B^2) - 2g^2A(A^2+B^2)$$

$$\frac{\partial \mathcal{L}}{\partial \partial_\mu A} = \partial^\mu A$$

We obtain therefore

$$(\Box+m^2)A=-g[\bar{\Psi}\psi+m(3A^2+B^2) + 2gA(A^2+B^2)] \qquad (5.3)$$

ii) $\underline{\phi = B:}$ The equation of motion for the pseudoscalar field B is

$$\frac{\partial \mathcal{L}}{\partial B} - \partial_\mu \frac{\partial \mathcal{L}}{\partial \partial_\mu B} = 0$$

where

$$\frac{\partial \mathcal{L}}{\partial B} = -m^2 B - ig\bar{\Psi}\gamma^5\psi - 2mg\,AB - 2g^2 B(A^2+B^2)$$

$$\frac{\partial \mathcal{L}}{\partial \partial_\mu B} = \partial^\mu B$$

Hence

$$(\Box+m^2)B=-g[i\bar{\Psi}\gamma^5\psi+2mAB + 2gB(A^2+B^2)] \qquad (5.4)$$

iii) $\underline{\phi = \bar{\Psi}:}$ Here

$$\frac{\partial \mathcal{L}}{\partial \bar{\Psi}} - \partial_\mu \frac{\partial \mathcal{L}}{\partial \partial_\mu \bar{\Psi}} = 0$$

with

$$\frac{\partial \mathcal{L}}{\partial \bar{\Psi}} = \frac{1}{2}(i\not{\partial}-m)\psi - g\psi A - ig\gamma_5\psi B$$

$$\frac{\partial \mathcal{L}}{\partial \partial_\mu \bar{\Psi}} = 0$$

Hence

$$(i\not{\partial}-m)\psi = 2g(A+i\gamma_5 B)\psi \qquad (5.5)$$

iv) $\underline{\phi = \psi:}$ Here

$$\frac{\partial \mathcal{L}}{\partial \psi} - \partial_\mu \frac{\partial \mathcal{L}}{\partial \partial_\mu \psi} = 0$$

with

$$\partial \mathcal{L}/\partial\psi = -\frac{1}{2}m\bar{\Psi} - g\bar{\Psi}A - ig\bar{\Psi}\gamma_5 B$$

$$\partial \mathcal{L}/\partial \partial_\mu \psi = \frac{i}{2} \bar{\psi} \gamma^\mu$$

and the equation of motion for ψ is seen to be

$$i \partial_\mu \bar{\psi} \gamma^\mu + m \bar{\psi} = -2g (\bar{\psi} A + i \bar{\psi} \gamma_5 B) \quad (5.6)$$

5.2 Symmetries

We now investigate continuous transformations of the fields A, B, ψ , which are symmetry transformations of the theory defined by (5.1). Such transformations are symmetry transformations of the theory if the action

$$S = \int \mathcal{L} d^4 x \qquad (5.7)$$

is left invariant. Under supersymmetry transformations (as defined below the variation of $\mathcal{L}$ is a total derivative, i.e.

$$\mathcal{L}' - \mathcal{L} =: \delta \mathcal{L} = \partial_\mu V^\mu \neq 0 \qquad (5.8)$$

In the case of (e.g.) Lorentz transformations $\mathcal{L} = \mathcal{L}'$ and $\delta \mathcal{L} = 0$. However, for S to be invariant it suffices that $\delta \mathcal{L} = \partial_\mu V^\mu$, where $\int d^4 x \, \partial_\mu V^\mu$ can be converted into a surface integral with the help auf Gauss's theorem, and for a surface sufficiently far away from where the fields are nonzero this integral will vanish. This is, in fact, the situation in the case of supersymmetry transformations. We shall see later explicitly that socalled "superfields" (to be defined later) integrated in the Grassmann sense with respect to socalled "supercoordinates" (which together with Minkowski coordinates comprise a "superspace") yield precisely a Minkowski total derivative, and thus permit the construction of manifestly supersymmetric action integrals. The variation of the Lagrangian density under an arbitrary infinitesimal variation of the fields

$$A \longrightarrow A' = A + \delta A, \quad B \longrightarrow B' = B + \delta B,$$
$$\psi \longrightarrow \psi' = \psi + \delta\psi$$

is (with $\phi_i = A, B, \psi$)

$$\begin{aligned}
\delta\mathcal{L} &= \mathcal{L}(\phi_i', \partial_\mu \phi_i') - \mathcal{L}(\phi_i, \partial_\mu \phi_i) \\
&= \frac{\partial\mathcal{L}}{\partial\phi_i}\delta\phi_i + \frac{\partial\mathcal{L}}{\partial\partial_\mu\phi_i}\delta\partial_\mu\phi_i \\
&= \left\{\frac{\partial\mathcal{L}}{\partial\phi_i} - \partial_\mu\left(\frac{\partial\mathcal{L}}{\partial\partial_\mu\phi_i}\right)\right\}\delta\phi_i \\
&\quad + \partial_\mu\left(\frac{\partial\mathcal{L}}{\partial\partial_\mu\phi_i}\delta\phi_i\right) \\
&= \partial_\mu\left(\frac{\partial\mathcal{L}}{\partial\partial_\mu\phi_i}\delta\phi_i\right) \qquad \text{(with (5.2))} \\
&\equiv \partial_\mu V^\mu \qquad \text{(with (5.8))}
\end{aligned}$$

From this equality we see that the equations of motion define a conserved current density given by

$$j^\mu = V^\mu - \frac{\partial\mathcal{L}}{\partial\partial_\mu\phi_i}\delta\phi_i \tag{5.9}$$

From now on we consider only the free part of the Lagrangian density (5.1), i.e.

$$\mathcal{L}_{\text{free}} = \frac{1}{2}(\partial_\mu A)^2 - \frac{1}{2}m^2A^2 + \frac{1}{2}(\partial_\mu B)^2 - \frac{1}{2}m^2B^2 + \frac{1}{2}\bar{\Psi}(i\not\partial - m)\psi \tag{5.10}$$

and we consider the following variations of the fields which transform fermions and bosons into each other (these transformations will be derived later)

$$\delta A = \bar{\varepsilon}\psi(x) \tag{5.11a}$$

$$\delta B = -i\bar{\varepsilon}\gamma^5\psi(x) \tag{5.11b}$$

$$\delta\psi = -(i\not\partial + m)(A - i\gamma^5 B)\varepsilon \tag{5.11c}$$

$$\delta\bar{\psi} = \bar{\varepsilon}(A - i\gamma^5 B)(i\overleftarrow{\not\partial} - m) \tag{5.11d}$$

where ε is an x-independent Grassmann variable. Equations (5.11a) to (5.11d) constitute the infinitesimal supersymmetry transformation of the set of fields A, B and ψ . For later reference and comparison purposes we rewrite the variations (5.11) in the Weyl formulation. Thus, using our previous considerations we have

$$\delta A = \bar{\varepsilon}\psi(x)$$

$$= (\varepsilon^A, \bar{\varepsilon}_{\dot{A}}) \begin{pmatrix} \psi_A \\ \bar{\psi}^{\dot{A}} \end{pmatrix} = \varepsilon^A \psi_A + \bar{\varepsilon}_{\dot{A}} \bar{\psi}^{\dot{A}}$$

$$= (\varepsilon\psi)_2 + (\bar{\varepsilon}\bar{\psi})_2 \quad \text{(see also (1.186a))}$$

$$\delta B = -i\bar{\varepsilon}\gamma^5_W \psi(x) \quad \text{(with (1.139))}$$

$$= i(\varepsilon^A, \bar{\varepsilon}_{\dot{A}}) \begin{pmatrix} 1_{2\times 2} & 0 \\ 0 & -1_{2\times 2} \end{pmatrix} \begin{pmatrix} \psi_A \\ \bar{\psi}^{\dot{A}} \end{pmatrix}$$

$$= i(\varepsilon^A \psi_A - \bar{\varepsilon}_{\dot{A}} \bar{\psi}^{\dot{A}})$$

$$= i[(\varepsilon\psi)_2 - (\bar{\varepsilon}\bar{\psi})_2] \quad \text{(see also (1.186b))}$$

$$\delta\psi = -(i\gamma^\mu_W \partial_\mu + m)(A - i\gamma^5 B)\varepsilon$$

(with (1.139))

$$= -\left[i\begin{pmatrix} 0 & \sigma^{\mu}{}_{A\dot{A}} \\ \bar{\sigma}^{\mu\dot{A}A} & 0 \end{pmatrix}\partial_\mu + m 1_{4\times 4}\right] \cdot$$

$$\cdot\left[A\,1_{4\times4}+i\begin{pmatrix}1_{2\times2} & 0\\ 0 & -1_{2\times2}\end{pmatrix}B\right]\begin{pmatrix}\varepsilon_A\\ \bar\varepsilon^{\dot A}\end{pmatrix}$$

$$=-\begin{pmatrix} m & i\sigma^\mu\partial_\mu\\ i\bar\sigma^\mu\partial_\mu & m\end{pmatrix}\begin{pmatrix}A+iB & 0\\ 0 & A-iB\end{pmatrix}\begin{pmatrix}\varepsilon_A\\ \bar\varepsilon^{\dot A}\end{pmatrix}$$

$$=-\begin{pmatrix} m(A+iB)\varepsilon_A + i\sigma^\mu_{A\dot A}\,\partial_\mu(A-iB)\bar\varepsilon^{\dot A}\\ i\bar\sigma^{\mu\,\dot A A}\partial_\mu(A+iB)\varepsilon_A + m(A-iB)\bar\varepsilon^{\dot A}\end{pmatrix}$$

$$=-\begin{pmatrix} m(A+iB)\varepsilon_A + i(\sigma^\mu\bar\varepsilon)_A\,\partial_\mu(A-iB)\\ i(\bar\sigma^\mu\varepsilon)^{\dot A}\partial_\mu(A+iB)+m(A-iB)\bar\varepsilon^{\dot A}\end{pmatrix}$$

Hence setting

$$f(x)=\frac{1}{2}(A+iB)$$

we obtain

$$\delta f=\bar\varepsilon\bar\psi\ ,\qquad \delta f^*=\varepsilon\psi$$

and

$$\left.\begin{aligned}\delta\psi_A &= -2m f(x)\varepsilon_A + 2i(\sigma^\mu\bar\varepsilon)_A\,\partial_\mu f^*(x)\\ \delta\bar\psi^{\dot A} &= -2m f^*(x)\bar\varepsilon^{\dot A} - 2i(\bar\sigma^\mu\varepsilon)^{\dot A}\partial_\mu f(x)\end{aligned}\right\}\qquad (5.11')$$

Since A is a scalar field and B a pseudoscalar field, the complex combination f(x) transforms under the parity transformation like complex conjugation. Thus f(x) does not have a well defined parity - as expected, since the Weyl formulation does not preserve a definite parity. Thus the Wess-Zumino theory can be looked at as the field theory of a single complex scalar field and a spinor Majorana field in the Weyl formulation - and in fact, we shall recover it later in precisely this form (see the discussion following (8.36)).

As stated earlier ψ is a Majorana spinor field, so that $\overline{\psi}$ and ψ are not independent, as was demonstrated in Section 1.4.5 . Hence the transformations (5.11c) and (5.11d) are not independent; i.e. (5.11d) follows from (5.11c). We demonstrate this as an exercise. Since ψ is a Majorana spinor we have according to (1.180)

$$\psi = \psi^c = C\overline{\psi}^T$$

where C is the charge conjugation matrix and $\overline{\psi} = \psi^{\dagger}\gamma_0$. Then

$$\overline{\psi} = -\psi^T C^{-1} \qquad (5.12)$$

using the antisymmetry of C, and

$$\begin{aligned}
\delta\overline{\psi} &= -(\delta\psi)^T C^{-1} \qquad \text{(from (5.12))}\\
&= [(i\partial\!\!\!/ + m)(A - i\gamma^5 B)\varepsilon]^T C^{-1} \qquad \text{(from (5.11c))}\\
&= \varepsilon^T (A - i\gamma^5 B)^T (i\overleftarrow{\partial}_\mu \gamma^\mu + m)^T C^{-1}\\
&= \varepsilon^T C^{-1} C (A - i\gamma^{5T} B)(i\overleftarrow{\partial}_\mu \gamma^{\mu T} + m) C^{-1}\\
&= -\overline{\varepsilon} C (A - i\gamma^{5T} B)(i\overleftarrow{\partial}_\mu \gamma^{\mu T} + m) C^{-1}\\
&= -\overline{\varepsilon}(A - iC\gamma^{5T} B)(i\overleftarrow{\partial}_\mu \gamma^{\mu T} C^{-1} + m)\\
&\qquad \text{(using (1.173))}\\
&= \overline{\varepsilon}(A - i\gamma^5 B)(i\overleftarrow{\partial}_\mu \gamma^\mu - m)\\
&\qquad \text{(using (1.166))}
\end{aligned}$$

Next we show that the variation $\delta\mathcal{L}$ caused by the transformation (5.11a) to (5.11d) is a total derivative. Our starting point is the free Lagrangian density (5.10). Then

$$\begin{aligned}
\delta\mathcal{L} = &(\partial_\mu A)\,\delta(\partial^\mu A) - m^2 A\,\delta A\\
&+ (\partial_\mu B)\,\delta(\partial^\mu B) - m^2 B\,\delta B\\
&+ \tfrac{1}{2}\,\delta\overline{\psi}(i\partial\!\!\!/ - m)\psi + \tfrac{1}{2}\,\overline{\psi}(i\partial\!\!\!/ - m)\,\delta\psi
\end{aligned}$$

$$= \bar{\varepsilon}(\partial_\mu A)(\partial^\mu \psi(x)) - m^2 A \bar{\varepsilon}\psi(x)$$
$$-i\bar{\varepsilon}\gamma^5(\partial_\mu B)\partial^\mu\psi(x) + i\bar{\varepsilon}\gamma^5 m^2 B\psi(x)$$
$$+\tfrac{1}{2}\bar{\varepsilon}(A - i\gamma^5 B)(i\overleftarrow{\not\partial} - m)(i\not\partial - m)\psi$$
$$-\tfrac{1}{2}\bar{\Psi}(i\overleftarrow{\not\partial} - m)(i\not\partial + m)(A - i\gamma^5 B)\varepsilon$$

(using (5.11))

$$= \bar{\varepsilon}(\partial_\mu A)(\partial^\mu \psi(x)) - m^2 A \bar{\varepsilon}\psi(x)$$
$$-i\bar{\varepsilon}\gamma^5(\partial_\mu B)(\partial^\mu\psi(x)) + i\bar{\varepsilon}\gamma^5 m^2 B\psi(x)$$
$$+\tfrac{1}{2}\bar{\Psi}(\Box + m^2)(A - i\gamma^5 B)\varepsilon$$

(using (5.5) for g = 0)

where

$$\not\partial\not\partial = \partial_\mu\gamma^\mu\partial_\nu\gamma^\nu = \tfrac{1}{2}\partial_\mu\partial_\nu(\gamma^\mu\gamma^\nu + \gamma^\nu\gamma^\mu)$$
$$= \partial_\mu\partial_\nu\eta^{\mu\nu} = \partial_\mu\partial^\mu = \Box$$

Using the equations of motion for A and B we obtain

$$\delta\mathcal{L} = \bar{\varepsilon}(\partial_\mu A)(\partial^\mu\psi) - m^2 A\bar{\varepsilon}\psi(x)$$
$$-i\bar{\varepsilon}\gamma^5(\partial_\mu B)(\partial^\mu\psi) + i\bar{\varepsilon}\gamma^5 m^2 B\psi(x)$$
$$= \bar{\varepsilon}(\partial_\mu A)(\partial^\mu\psi) + \Box A\,\bar{\varepsilon}\psi(x)$$
$$-i\bar{\varepsilon}\gamma^5(\partial_\mu B)(\partial^\mu\psi) - i\bar{\varepsilon}\gamma^5\Box B\,\psi(x)$$
$$= \partial_\mu\{\bar{\varepsilon}(\partial^\mu A)\psi - i\bar{\varepsilon}\gamma^5(\partial^\mu B)\psi\}$$
$$= \partial_\mu\{\bar{\varepsilon}[\partial^\mu(A - i\gamma^5 B)]\psi\}$$
$$=: \partial_\mu V^\mu$$

Hence

$$V^\mu = \bar{\varepsilon}\,[\partial^\mu(A - i\gamma^5 B)]\psi \tag{5.13}$$

For the other contributions to (5.9) we consider

$$\sum_i \frac{\partial\mathcal{L}}{\partial\partial_\mu\Phi_i}\delta\Phi_i = \frac{\partial\mathcal{L}}{\partial\partial_\mu A}\delta A + \frac{\partial\mathcal{L}}{\partial\partial_\mu B}\delta B + \frac{\partial\mathcal{L}}{\partial\partial_\mu\psi}\delta\psi$$

$$= (\partial_\mu A)\,\delta A + (\partial_\mu B)\,\delta B + \frac{i}{2}\bar{\Psi}\gamma_\mu \delta\psi$$

$$= (\partial_\mu A)\,\bar{\varepsilon}\psi - i(\partial_\mu B)\,\bar{\varepsilon}\gamma^5\psi$$

$$- \frac{i}{2}\bar{\Psi}\gamma_\mu (i\not{\partial} + m)(A - i\gamma^5 B)\varepsilon$$

(with (5.11))

$$= \bar{\varepsilon}[\partial_\mu(A - i\gamma^5 B)]\psi - \frac{i}{2}\bar{\Psi}\gamma_\mu(i\not{\partial} + m)(A - i\gamma^5 B)\varepsilon$$

$$= \bar{\varepsilon}[\partial_\mu(A - i\gamma^5 B)]\psi + \frac{i}{2}\varepsilon^T(A - i\gamma^5 B)^T(i\overleftarrow{\not{\partial}} + m)^T\gamma_\mu^T\bar{\Psi}^T$$

(with $\{\bar{\Psi}, \varepsilon\} = 0$)

$$= \bar{\varepsilon}[\partial_\mu(A - i\gamma^5 B)]\psi$$

$$+ \frac{i}{2}\varepsilon^T C^{-1} C (A - i\gamma^5 B)^T (i\overleftarrow{\not{\partial}} + m)^T \gamma_\mu^T \bar{\Psi}^T$$

$$= \bar{\varepsilon}[\partial_\mu(A - i\gamma^5 B)]\psi$$

$$- \frac{i}{2}\bar{\varepsilon}(AC - iC\gamma^{5T}B)(i\overleftarrow{\partial}_\rho \gamma^{\rho T} + m)\gamma_\mu^T\bar{\Psi}^T$$

$$= \bar{\varepsilon}[\partial_\mu(A - i\gamma^5 B)]\psi$$

$$+ \frac{i}{2}\bar{\varepsilon}(A - i\gamma^5 B)(i\overleftarrow{\partial}_\rho C\gamma^{\rho T} + mC)C^{-1}\gamma_\mu C\bar{\Psi}^T$$

$$= \bar{\varepsilon}[\partial_\mu(A - i\gamma^5 B)]\psi$$

$$+ \frac{i}{2}\bar{\varepsilon}(A - i\gamma^5 B)(i\overleftarrow{\partial}_\rho C\gamma^{\rho T}C^{-1} + m)\gamma_\mu C\bar{\Psi}^T$$

$$= \bar{\varepsilon}[\partial_\mu(A - i\gamma^5 B)]\psi - \frac{i}{2}\bar{\varepsilon}(A - i\gamma^5 B)(i\overleftarrow{\partial}_\rho\gamma^\rho - m)\gamma_\mu\psi \tag{5.14}$$

This is an algebraic relation derived without explicit use of the equations of motion. Inserting this into (5.9) we obtain (using (5.13))

$$j^\mu = V^\mu - \sum_i \frac{\partial \mathcal{L}}{\partial \partial_\mu \phi_i}\delta\phi_i$$

$$= \frac{i}{2}\bar{\varepsilon}(A - i\gamma^5 B)(i\overleftarrow{\not{\partial}} - m)\gamma^\mu\psi$$

Hence

$$j^\mu = \frac{i}{2}\bar{\varepsilon}(A - i\gamma^5 B)(i\overleftarrow{\not\partial} - m)\gamma^\mu\psi \tag{5.15}$$

The vector j^μ is called the supercurrent. It is useful to define a slightly modified current, the conserved Majorana spinor current k^μ, by the relation

$$j^\mu = \frac{1}{\lambda}\bar{\varepsilon}k^\mu \tag{5.16}$$

where λ is a multiplicative real constant which has to be determined such that the spinor charge $\int k^0_a \, d^3x$ satisfies the same anticommutation relation as Q_a. Thus in the Wess-Zumino model we have the spinor current density

$$k^\mu = \frac{i}{2}\lambda(A - i\gamma^5 B)(i\overleftarrow{\not\partial} - m)\gamma^\mu\psi \tag{5.17}$$

We can verify explicitly that

$$\partial_\mu k^\mu = 0 \tag{5.18}$$

This must, of course, be valid since by construction (5.16) is a conserved current. We first rewrite (5.17),

$$\begin{aligned}
k^\mu &= \frac{i}{2}\lambda(A - i\gamma^5 B)(i\overleftarrow{\not\partial} - m)\gamma^\mu\psi \\
&= \frac{i}{2}\lambda(A - i\gamma^5 B)\, i\overleftarrow{\not\partial}\gamma^\mu\psi \\
&\qquad - \frac{i}{2}\lambda m(A - i\gamma^5 B)\gamma^\mu\psi \\
&= -\frac{i}{2}\lambda m\gamma^\mu(A + i\gamma^5 B)\psi \\
&\qquad - \frac{1}{2}\lambda[\not\partial(A + i\gamma^5 B)]\gamma^\mu\psi
\end{aligned}$$

(using (1.141))

Then

$$\begin{aligned}
\partial_\mu k^\mu = &-\frac{i}{2}\lambda m\not\partial[(A + i\gamma^5 B)\psi] \\
&- \frac{1}{2}\lambda[\not\partial\not\partial(A - i\gamma^5 B)]\psi \\
&- \frac{1}{2}\lambda[\not\partial(A + i\gamma^5 B)]\not\partial\psi
\end{aligned}$$

$$
\begin{aligned}
&= -\frac{i}{2} m\lambda(\not\partial A - i\gamma^5 \not\partial B)\psi - \frac{i}{2} m\lambda(A - i\gamma^5 B)\not\partial\psi \\
&\quad - \frac{1}{2}\lambda[\not\partial\not\partial A - i\gamma^5 \not\partial\not\partial B]\psi - \frac{1}{2}\lambda[\not\partial(A + i\gamma^5 B)]\not\partial\psi \\
&= -\frac{1}{2}\lambda(\not\partial A - i\gamma^5 \not\partial B)(\not\partial\psi + im\psi) \\
&\quad - \frac{1}{2}\lambda[\partial_\mu\partial^\mu A - i\gamma^5 \partial_\mu\partial^\mu B]\psi \\
&\quad - \frac{i}{2} m\lambda(A - i\gamma^5 B)\not\partial\psi \\
&= -\frac{1}{2}\lambda(\not\partial A - i\gamma^5 \not\partial B)(-i)(i\not\partial - m)\psi \\
&\quad - \frac{1}{2}\lambda[\Box A - i\gamma^5 \Box B]\psi \\
&\quad - \frac{1}{2} m^2\lambda(A - i\gamma^5 B)\psi \\
&= -\frac{1}{2}\lambda\{(\Box + m^2)A - i\gamma^5(\Box + m^2)B\}\psi \\
&= 0
\end{aligned}
$$

as expected.

The spinor charges Q_a are defined by

$$Q_a := \int d^3x\, k^0_a \tag{5.19}$$

where k^0_a is given by

$$k^0_a = \frac{i}{2}\lambda[\{(A - i\gamma^5 B)(i\overleftarrow{\not\partial} - m)\}\gamma^0\psi]_a \tag{5.20}$$

with a = 1,2,3,4. The important point is the following. If the Wess-Zumino model is a supersymmetric field theory, the spinor charges (5.19) with the charge density (5.20) must satisfy the commutation and anticommutation relations of the Super-Poincaré algebra (4.1). In particular we have to show (choosing λ suitably) that the spinor charges (5.19) satisfy the relation

$$\{Q_a, \overline{Q}_b\} = 2P_\mu(\gamma^\mu)_{ab}$$

If we can show this, we have shown that the Wess-Zumino model is a field theoretical realization of the supersymmetry algebra (4.1). Thus we have to search for a representation of the spinor charges (5.19) as linear operators acting on Fock space.

5.3 Plane Wave Expansions

The equation of motion of a Dirac field ψ is

$$(i\partial\!\!\!/ - m)\psi(x) = 0 \tag{5.21}$$

Thus a plane wave solution

$$\psi(x) = \exp\{-ipx\}\, u(\vec{p}) \tag{5.22}$$

satisfies

$$(p\!\!\!/ - m)\, u(\vec{p}) = 0 \tag{5.23}$$

This equation has four independent solutions,

two solutions for

$$p^0 := \omega_p = +(\vec{p}^{\,2} + m^2)^{1/2}$$

and

two solutions for

$$p^0 := -\omega_p = -(\vec{p}^{\,2} + m^2)^{1/2}$$

In the rest frame $(p_\mu) = (p_0, \vec{0}) = (m, \vec{0})$ and (5.23) becomes

$$(\gamma^0 - 1)\, u(\vec{0}) = 0 \tag{5.24}$$

so that $u(\vec{0})$ is an eigenstate of γ^0. Taking the γ matrices in the Dirac representation (1.149), i.e.

$$\gamma^0 = \begin{pmatrix} 1_{2\times 2} & 0 \\ 0 & 1_{2\times 2} \end{pmatrix}, \gamma^i = \begin{pmatrix} 0 & \sigma^i \\ \bar{\sigma}^i & 0 \end{pmatrix}, \gamma^5 = \begin{pmatrix} 0 & \sigma^0 \\ \bar{\sigma}^0 & 0 \end{pmatrix}$$

we can define two eigenvectors of γ^0 in the form

$$u(\vec{0},1) = \begin{pmatrix} 1 \\ 0 \\ 0 \\ 0 \end{pmatrix}, \quad u(\vec{0},2) = \begin{pmatrix} 0 \\ 1 \\ 0 \\ 0 \end{pmatrix} \tag{5.25}$$

In an analogous way we construct eigenvectors for

$$p_0 < 0, \ \vec{p} = 0, \ p_0 = -m,$$

so that (5.23) becomes

$$(\gamma^0 + 1)\, v(\vec{0}) = 0, \quad \gamma^0 v(\vec{0}) = -v(\vec{0})$$

and we define

$$v(\vec{0},1) = \begin{pmatrix} 0 \\ 0 \\ 0 \\ 1 \end{pmatrix}, \quad v(\vec{0},2) = \begin{pmatrix} 0 \\ 0 \\ -1 \\ 0 \end{pmatrix} \tag{5.26}$$

We could now boost these solutions (5.25), (5.26) from rest to a velocity $v = |\vec{p}|/p_0$ by a pure Lorentz transformation. However, it is easier to observe that

$$(\not{p} - m)(\not{p} + m) = p^2 - m^2 = 0$$

and therefore (with s = 1,2)

$$u(\vec{p},s) = N(p)(\not{p} + m)\, u(\vec{0},s) \tag{5.27}$$

and

$$\begin{aligned} &(\not{p} - m)\, u(\vec{p},s) \\ &= N(p)(\not{p} - m)(\not{p} + m)\, u(\vec{0},s) = 0 \end{aligned}$$

such that (5.23) is satisfied. N(p) is a normalization factor. Furthermore

$$v(\vec{p},s) = (-\not{p} + m)\, N(p)\, v(\vec{0},s) \tag{5.28}$$

such that

$$\begin{aligned} &(\not{p} + m)\, v(\vec{p},s) \\ &= (\not{p} + m)(-\not{p} + m)\, N(p)\, v(\vec{0},s) \\ &= (-p^2 + m^2)\, N(p)\, v(\vec{0},s) = 0 \end{aligned}$$

For the conjugate spinors we find

$$\bar{u}(\vec{p},s) = \bar{u}(\vec{0},s)N(p)(\not{p}+m)$$
$$\bar{v}(\vec{p},s) = \bar{v}(\vec{0},s)N(p)(-\not{p}+m) \tag{5.29}$$

The normalization factor N(p) has to be determined from the requirement

$$\left.\begin{aligned} \bar{u}(\vec{p},s)u(\vec{p},r) &= \delta_{rs} \\ \bar{v}(\vec{p},s)v(\vec{p},r) &= -\delta_{rs} \end{aligned}\right\} \tag{5.30a}$$

It is easily seen that also

$$\left.\begin{aligned} \bar{u}(\vec{p},s)v(\vec{p},r) &= 0 \\ \bar{v}(\vec{p},s)u(\vec{p},r) &= 0 \end{aligned}\right\} \tag{5.30b}$$

<u>Proposition</u>: The factor N(p) is given by

$$N(p) = [2m(m+\omega_p)]^{-1/2} \tag{5.31}$$

<u>Proof</u>: We have

$$\bar{u}(\vec{p},1)u(\vec{p},1) = 1$$

Using (5.29) and (5.27) this can be written

$$\bar{u}(\vec{0},1)N(p)(\not{p}+m)N(p)(\not{p}+m)u(\vec{0},1) = 1$$

i.e.

$$\bar{u}(\vec{0},1)(\not{p}+m)(\not{p}+m)u(\vec{0},1) = \frac{1}{N^2(p)}$$

and so

$$\bar{u}(\vec{0},1)(p^2+2m\not{p}+m^2)u(\vec{0},1) = \frac{1}{N^2(p)}$$

Using (5.25) we can write this

$$\frac{1}{N^2(p)} = u^+(\vec{0},1)\gamma^0(2m^2+2mp_\mu\gamma^\mu)u(\vec{0},1)$$

$$= (1,0,0,0)\begin{pmatrix} 1_{2\times2} & 0 \\ 0 & -1_{2\times2} \end{pmatrix}\cdot$$

$$\cdot (2m^2 1_{4\times 4} + 2m p_\mu \gamma^\mu) \begin{pmatrix} 1 \\ 0 \\ 0 \\ 0 \end{pmatrix}$$

$$= \quad (1,0,0,0) \Big[2m^2 1_{4\times 4}$$

$$+ 2m \begin{pmatrix} p_0 & 0 & p_3 & p_1 - ip_2 \\ 0 & p_0 & p_1 + ip_2 & -p_3 \\ -p_3 & -p_1 + ip_2 & -p_0 & 0 \\ -p_1 - ip_2 & p_3 & 0 & -p_0 \end{pmatrix} \Big] \begin{pmatrix} 1 \\ 0 \\ 0 \\ 0 \end{pmatrix}$$

$$= (1,0,0,0) \left[2m^2 \begin{pmatrix} 1 \\ 0 \\ 0 \\ 0 \end{pmatrix} + 2m \begin{pmatrix} p_0 \\ 0 \\ -p_3 \\ -p_1 - ip_2 \end{pmatrix} \right]$$

$$= 2m^2 + 2mp_0$$

so that
$$\frac{1}{N^2(p)} = 2m(m + \omega_p)$$

which proves (5.31). Hence the spinors u, v have the form

$$u(\vec{p}, s) = \frac{(m + \not{p})\, u(\vec{0}, s)}{[2m(m + \omega_p)]^{1/2}} \tag{5.32}$$

$$v(\vec{p}, s) = \frac{(m - \not{p})\, v(\vec{0}, s)}{[2m(m + \omega_p)]^{1/2}} \tag{5.33}$$

Second quantization is achieved by imposing equal-time anticommutation relations on the spinor fields $\psi(x)$, i.e.

$$\{\psi_a(\vec{x}, t), \psi_b^\dagger(\vec{y}, t)\} = \delta_{ab}\, \delta(\vec{x} - \vec{y})$$
$$\{\psi_a(\vec{x}, t), \psi_b(\vec{x}', t)\} = 0 = \{\psi_a^\dagger(\vec{x}, t), \psi_b^\dagger(\vec{x}', t)\} \tag{5.34}$$

The field $\psi(x)$ is now a linear operator acting on Fock space. Fourier decomposition of $\psi(x)$ leads to

$$\psi(x)=\frac{1}{(2\pi)^{3/2}}\sum_s\int d^3p\left(\frac{m}{\omega_p}\right)^{1/2}\cdot\{b(\vec{p},s)u(\vec{p},s)e^{-ipx}+d^+(\vec{p},s)v(\vec{p},s)e^{ipx}\}\tag{5.35}$$

(observe: in the exponents of the exponentials px $=p_\mu x^\mu$) where $\sum_s$ is the sum over spin states, and

$b^+(\vec{p},s)$ and $d(\vec{p},s)$

are momentum dependent creation and annihilation operators. Equation (5.35) is the plane wave expansion of the usual Dirac spinor field. We obtain some restrictions on $b(\vec{p},s)$ and $d^+(\vec{p},s)$ if we impose the Majorana property on the spinor field $\psi(x)$. The Majorana condition is

$$\psi(x)=\psi^c(x)=C[\overline{\psi}(x)]^T$$

Taking the hermitian conjugate of (5.35) we obtain

$$\psi^+(x)=\frac{1}{(2\pi)^{3/2}}\sum_s\int d^3p\left(\frac{m}{\omega_p}\right)^{1/2}\{b^+(\vec{p},s)u^+(\vec{p},s)e^{ipx}+d(\vec{p},s)v^+(\vec{p},s)e^{-ipx}\}$$

Then

$$\overline{\psi}^T(x)=(\psi^+(x)\gamma^0)^T=\frac{1}{(2\pi)^{3/2}}\sum_s\int d^3p\left(\frac{m}{\omega_p}\right)^{1/2}\cdot\{b^+(\vec{p},s)\overline{u}^T(\vec{p},s)e^{ipx}+d(\vec{p},s)\overline{v}^T(\vec{p},s)e^{-ipx}\}$$

Then the charge-conjugated spinor has the expansion

$$\psi^c(x)=C\overline{\psi}^T(x)=\frac{1}{(2\pi)^{3/2}}\sum_s\int d^3p\left(\frac{m}{\omega_p}\right)^{1/2}\cdot\{b^+(\vec{p},s)C\overline{u}^T(\vec{p},s)e^{ipx}+d(\vec{p},s)C\overline{v}^T(\vec{p},s)e^{-ipx}\}\tag{5.36}$$

<u>Proposition</u>: The following relations hold

i)
$$u^c(\vec{0},s) \equiv C\bar{u}^T(\vec{0},s) = v(\vec{0},s) \tag{5.37}$$

ii)
$$v^c(\vec{0},s) \equiv C\bar{v}^T(\vec{0},s) = u(\vec{0},s) \tag{5.38}$$

<u>Proof</u>:

i) We have

$$u(\vec{p},s) = \begin{pmatrix} u_1(\vec{p}) \\ u_2(\vec{p}) \\ u_3(\vec{p}) \\ u_4(\vec{p}) \end{pmatrix}$$

where

$$u_2(\vec{0}) = u_3(\vec{0}) = u_4(\vec{0}) = 0$$

and

$$u_1(\vec{0}) = 1 \qquad \text{for } s = 1 \text{ (see (5.25))}$$

Then $\bar{u}^T = (u^+\gamma^0)^T$

$$= \begin{pmatrix} u_1^* \\ u_2^* \\ -u_3^* \\ -u_4^* \end{pmatrix} \qquad \text{where } \gamma^0 = \begin{pmatrix} 1_{2\times 2} & 0 \\ 0 & -1_{2\times 2} \end{pmatrix}$$

Now

$$C = i\gamma^2\gamma^0 = -i\begin{pmatrix} 0 & \sigma^2 \\ \sigma^2 & 0 \end{pmatrix}$$

(using (1.167)) so that

$$u^c(\vec{p},s) = C\,\bar{u}^T(\vec{p},s)$$

$$= -\begin{pmatrix} 0 & 0 & 0 & 1 \\ 0 & 0 & -1 & 0 \\ 0 & 1 & 0 & 0 \\ -1 & 0 & 0 & 0 \end{pmatrix}\begin{pmatrix} u_1^* \\ u_2^* \\ -u_3^* \\ -u_4^* \end{pmatrix} = -\begin{pmatrix} -u_4^*(\vec{p}) \\ u_3^*(\vec{p}) \\ u_2^*(\vec{p}) \\ -u_1^*(\vec{p}) \end{pmatrix}$$

Then (using (5.26))

$$u^c(\vec{0},1) = \begin{pmatrix} 0 \\ 0 \\ 0 \\ 1 \end{pmatrix} = v(\vec{0},1),\ u^c(\vec{0},2) = \begin{pmatrix} 0 \\ 0 \\ -1 \\ 0 \end{pmatrix} = v(\vec{0},2)$$

ii) Equation (5.38) can be shown in a similar way.
We need some more results.

<u>Proposition</u>:From (5.37) and (5.38) we can obtain

i)
$$u^c(\vec{p},s) = v(\vec{p},s) \tag{5.39}$$

and

ii)
$$v^c(\vec{p},s) = u(\vec{p},s) \tag{5.40}$$

<u>Proof</u>:

i) $u^c(\vec{p},s) = C[\bar{u}(\vec{p},s)]^T$

$= N(p)C[\bar{u}(\vec{0},s)(\not{p}+m)]^T$ (with (5.29))

$= N(p)C[(\not{p}+m)^T\bar{u}^T(\vec{0},s)]$

$= N(p)(p_\mu C\gamma^{\mu T} + mC)\bar{u}^T(\vec{0},s)$

$= N(p)(-p_\mu\gamma^\mu + m)C\bar{u}^T(\vec{0},s)$

(using (1.166))

$= N(p)(-\not{p}+m)v(\vec{0},s)$ (with (5.37))

$= v(\vec{p},s)$ (with (5.33))

ii) $v^c(\vec{p},s) = C[\bar{v}(\vec{p},s)]^T$

$= N(p)C[\bar{v}(\vec{0},s)(-\not{p}+m)]^T$

(with (5.29))

$$= N(p)\, C[(-\not{p}+m)^T (\bar{v}(\vec{o},s))^T]$$
$$= N(p)\,(-p_\mu C\gamma^{\mu T} + mC)\,\bar{v}^T(\vec{o},s)$$
$$= N(p)\,(p_\mu \gamma^\mu + m)\, C\bar{v}^T(\vec{o},s)$$

(using (1.166))

$$= N(p)(\not{p}+m)\, u(\vec{o},s) \qquad \text{(with (5.38))}$$
$$= u(\vec{p},s) \qquad \text{(with (5.32))}$$

Inserting (5.39) and (5.40) into (5.36) we obtain

$$\psi^c(x) = \frac{1}{(2\pi)^{3/2}} \sum_s \int d^3p \left(\frac{m}{\omega_p}\right)^{1/2} \cdot \{ b^+(\vec{p},s)\, v(\vec{p},s)\, e^{ipx} + d(\vec{p},s)\, u(\vec{p},s)\, e^{-ipx} \}$$

Imposing the Majorana condition, this has to be equal to (cf. (5.35))

$$\psi(x) = \frac{1}{(2\pi)^{3/2}} \sum_s \int d^3p \left(\frac{m}{\omega_p}\right)^{1/2} \cdot \{ d^+(\vec{p},s)\, v(\vec{p},s)\, e^{ipx} + b(\vec{p},s)\, u(\vec{p},s)\, e^{-ipx} \}$$

Thus for Majorana spinors we have the condition

$$b(\vec{p},s) = d(\vec{p},s) \quad \text{and} \quad b^+(\vec{p},s) = d^+(\vec{p},s) \qquad (5.41)$$

Hence a Majorana spinor has the following plane wave expansion

$$\psi^M(x) = \frac{1}{(2\pi)^{3/2}} \sum_s \int d^3p \left(\frac{m}{\omega_p}\right)^{1/2} \{ d(\vec{p},s)\, u(\vec{p},s)\, e^{-ipx} + d^+(\vec{p},s)\, v(\vec{p},s)\, e^{ipx} \} \qquad (5.42)$$

with

$$\psi^M(x) = \psi^{MC}(x)$$

Thus a Majorana spinor involves creation and annihilation operators $d^+(\vec{p},s)$ and $d(\vec{p},s)$ instead of $d^+(\vec{p},s)$ and $b(\vec{p},s)$ as in the expansion of a Dirac spinor.

In order to obtain the anticommutation relations for $d(\vec{p},s)$ and $d^+(\vec{p},s)$ it is advantageous to solve (5.42) for $d(\vec{p},s)$. Then

$$d(\vec{p},s) = \frac{1}{(2\pi)^{3/2}}\left(\frac{m}{\omega_p}\right)^{1/2} \int d^3x\, e^{ipx} u^+(\vec{p},s)\psi(x) \tag{5.43}$$

Using equations (5.30) it can be shown that

$$\begin{aligned} \{d(\vec{p},r), d^+(\vec{p}\,',s)\} &= \delta_{rs}\,\delta(\vec{p}-\vec{p}\,') \\ \{d(\vec{p},r), d(\vec{p}\,',s)\} &= 0 = \{d^+(\vec{p},r), d^+(\vec{p}\,',s)\} \end{aligned} \tag{5.44}$$

This completes our consideration of the plane wave expansion of a Majorana field. For the scalar fields A and B appearing in the Wess-Zumino model, we use the following plane wave expansions

$$A(x) = \frac{1}{(2\pi)^{3/2}} \int d^3p\, (2\omega_p)^{-1/2} [a(\vec{p})e^{-ipx} + a^+(\vec{p})e^{ipx}] \tag{5.45}$$

and

$$B(x) = \frac{1}{(2\pi)^{3/2}} \int d^3p\, (2\omega_p)^{-1/2} [b(\vec{p})e^{-ipx} + b^+(\vec{p})e^{ipx}] \tag{5.46}$$

Our next step is the calculation of Q_a.

<u>Proposition</u>: Inserting (5.42), (5.45) and (5.46) into (5.20) and using the definition

$$Q_a = \int d^3x\, k_a^0$$

we obtain

$$\begin{aligned} Q_a = \frac{i\lambda}{2^{1/2}} m^{1/2} \sum_s \int d^3p\, \{ & C(\vec{p})d^+(\vec{p},s)v(\vec{p},s) \\ & - D(\vec{p})\,d(\vec{p},s)u(\vec{p},s)\}_a \end{aligned} \tag{5.47}$$

where

$$\left.\begin{aligned} C(\vec{p}) &= a(\vec{p})1_{4\times4} - i\gamma^5 b(\vec{p}) \\ D(\vec{p}) &= a^+(\vec{p})1_{4\times4} - i\gamma^5 b^+(\vec{p}) \end{aligned}\right\} \quad (5.48)$$

<u>Proof</u>: Using (5.19), (5.20), (5.42), (5.45) and (5.46) we have

$$Q_a = \int d^3x\, k_a^0$$

$$= \frac{i}{2}\lambda \int d^3x \left[\{(A - i\gamma^5 B)(i\overleftarrow{\not\partial} - m)\}\gamma^0\psi(x)\right]_a$$

$$= \frac{i}{2}\lambda \int d^3x \Big[\Big\{ \frac{1}{(2\pi)^{3/2}} \int d^3p \frac{1}{(2\omega_p)^{1/2}} \cdot$$

$$\cdot (a(\vec{p})e^{-ipx} + a^+(\vec{p})e^{ipx})$$

$$- i\gamma^5 \frac{1}{(2\pi)^{3/2}} \int d^3p \frac{1}{(2\omega_p)^{1/2}} (b(\vec{p})e^{-ipx} + b^+(\vec{p})e^{ipx})\Big\} \cdot$$

$$\cdot (i\overleftarrow{\not\partial} - m)\gamma^0 \frac{1}{(2\pi)^{3/2}} \sum_s \int d^3k \left(\frac{m}{\omega_k}\right)^{1/2} \cdot$$

$$\cdot \{d(\vec{k},s)u(\vec{k},s)e^{-ikx} + d^+(\vec{k},s)v(\vec{k},s)e^{ikx}\}\Big]_a$$

$$= \frac{i\lambda}{2(2\pi)^3} \sum_s \int d^3x \int d^3p \int d^3k \left(\frac{m}{2\omega_p\omega_k}\right)^{1/2} \cdot$$

$$\cdot \{[C(\vec{p})e^{-ipx} + D(\vec{p})e^{ipx}](i\overleftarrow{\not\partial} - m)\gamma^0 \cdot$$

$$\cdot [d(\vec{k},s)u(\vec{k},s)e^{-ikx} + d^+(\vec{k},s)v(\vec{k},s)e^{ikx}]\}_a$$

$$= \frac{i\lambda}{2(2\pi)^3} \sum_s \int d^3x \int d^3p \int d^3k \left(\frac{m}{2\omega_p\omega_k}\right)^{1/2} \cdot$$

$$\cdot \{[C(\vec{p})e^{-ipx}(\not p - m) + D(\vec{p})e^{ipx}(-\not p - m)]\gamma^0 \cdot$$

$$\cdot\left[d(\vec{k},s)u(\vec{k},s)e^{-ikx}+d^{+}(\vec{k},s)v(\vec{k},s)e^{ikx}\right]\Big\}_a$$

$$=\frac{i\lambda}{2(2\pi)^3}\sum_s\int d^3x\int d^3p\int d^3k\left(\frac{m}{2\omega_p\omega_k}\right)^{1/2}\cdot$$

$$\cdot\Big\{C(\vec{p})(\not p-m)\gamma^0 d(\vec{k},s)u(\vec{k},s)e^{-i(p+k)x}$$

$$+C(\vec{p})(\not p-m)\gamma^0 d^{+}(\vec{k},s)v(\vec{k},s)e^{-i(p-k)x}$$

$$-D(\vec{p})(\not p+m)\gamma^0 d(\vec{k},s)u(\vec{k},s)e^{i(p-k)x}$$

$$-D(\vec{p})(\not p+m)\gamma^0 d^{+}(\vec{k},s)v(\vec{k},s)e^{i(p+k)x}\Big\}_a$$

Now

$$(2\pi)^{-3}\int d^3x\, e^{i(p+k)x}$$

$$=(2\pi)^{-3}\int d^3x\, e^{i(\omega_p+\omega_k)t-i(\vec{p}+\vec{k})\cdot\vec{x}}$$

$$=\frac{e^{i(\omega_p+\omega_k)t}}{(2\pi)^3}\int d^3x\, e^{-i(\vec{p}+\vec{k})\cdot\vec{x}}$$

$$=e^{i(\omega_p+\omega_k)t}\,\delta(\vec{p}+\vec{k})$$

$$=e^{2i\omega_p t}\,\delta(\vec{p}+\vec{k})$$

Furthermore

$$(2\pi)^{-3}\int d^3x\, e^{i(p-k)x}=\delta(\vec{p}-\vec{k})$$

We obtain therefore

$$Q_a=\frac{i\lambda}{2}\sum_s\int d^3p\int d^3k\left(\frac{m}{2\omega_p\omega_k}\right)^{1/2}\cdot$$

$$\cdot\Big\{C(\vec{p})(\not p-m)\gamma^0 d(\vec{k},s)u(\vec{k},s)e^{-2i\omega_p t}\cdot$$

$$\cdot\,\delta(\vec{p}+\vec{k})$$

$$+C(\vec{p})(\not p-m)\gamma^0 d^{+}(\vec{k},s)v(\vec{k},s)\,\delta(\vec{p}-\vec{k})$$

$$- D(\vec{p})(\not p + m)\gamma^0 d(\vec{k},s)u(\vec{k},s)\,\delta(\vec{p}-\vec{k})$$
$$- D(\vec{p})(\not p + m)\gamma^0 d^+(\vec{k},s)v(\vec{k},s) \cdot$$
$$\cdot\, e^{2i\omega_p t}\,\delta(\vec{p}+\vec{k})\Big\}_a$$

$$= \frac{i\lambda}{2}\sum_s \int d^3p \left(\frac{m}{2}\right)^{1/2} \frac{1}{\omega_p} \cdot$$
$$\cdot \Big\{ C(\vec{p})(\not p - m)\gamma^0 d(-\vec{p},s)u(-\vec{p},s)e^{-2i\omega_p t}$$
$$+ C(\vec{p})(\not p - m)\gamma^0 d^+(\vec{p},s)v(\vec{p},s)$$
$$- D(\vec{p})(\not p + m)\gamma^0 d(\vec{p},s)u(\vec{p},s)$$
$$- D(\vec{p})(\not p + m)\gamma^0 d^+(-\vec{p},s)v(-\vec{p},s)e^{2i\omega_p t}\Big\}_a$$

(5.49)

Now according to (5.23) we have

$$(\not p - m)\,u(\vec{p},s) = 0$$
$$(p_0\gamma_0 - \vec{p}\cdot\vec{\gamma} - m)\,u(\vec{p},s) = 0$$
$$\gamma_0(p_0\gamma_0 - \vec{p}\cdot\vec{\gamma} - m)\,u(\vec{p},s) = 0$$
$$(p_0\gamma_0^2 - \vec{p}\cdot\gamma_0\vec{\gamma} - m\gamma_0)\,u(\vec{p},s) = 0$$

Using the Clifford algebra relation

$$\{\gamma^\mu, \gamma^\nu\} = 2\eta^{\mu\nu}$$

we obtain

$$\gamma^0\gamma^i = -\gamma^i\gamma^0$$

so that

$$(p_0\gamma_0^2 + \vec{p}\cdot\vec{\gamma}\gamma^0 - m\gamma^0)\,u(\vec{p},s) = 0$$
$$(p_0\gamma_0 + \vec{p}\cdot\vec{\gamma} - m)\,\gamma^0 u(\vec{p},s) = 0$$

Replacing $\vec{p}$ by $-\vec{p}$ we have

$$(\not{p}-m)\gamma^0 u(-\vec{p},s)=0 \tag{5.50}$$

Furthermore

$$(\not{p}+m)v(\vec{p},s)=0$$

$$(p_0\gamma_0-\vec{p}\cdot\vec{\gamma}+m)v(\vec{p},s)=0$$

$$\gamma_0(p_0\gamma_0-\vec{p}\cdot\vec{\gamma}+m)v(\vec{p},s)=0$$

$$(p_0\gamma_0^2-\vec{p}\cdot\gamma_0\vec{\gamma}+m\gamma_0)v(\vec{p},s)=0$$

$$(p_0\gamma_0+\vec{p}\cdot\vec{\gamma}+m)\gamma_0 v(\vec{p},s)=0$$

$$(\not{p}+m)\gamma_0 v(-\vec{p},s)=0 \tag{5.51}$$

Inserting (5.50) and (5.51) into (5.49) we see that the t-dependent terms vanish, so that

$$Q_a=\frac{i\lambda}{2}\sum_s\int d^3p\left(\frac{m}{2}\right)^{1/2}\frac{1}{\omega_p}\cdot$$

$$\cdot\left\{C(\vec{p})(\not{p}-m)\gamma^0 d^+(\vec{p},s)v(\vec{p},s)-D(\vec{p})(\not{p}+m)\gamma^0 d(\vec{p},s)u(\vec{p},s)\right\}_a \tag{5.52}$$

Now

$$(\not{p}-m)\gamma^0 v(\vec{p},s)$$

$$=(p_0\gamma_0-\vec{p}\cdot\vec{\gamma}-m)\gamma^0 v(\vec{p},s)$$

$$=(p_0(\gamma_0)^2-\vec{p}\cdot\vec{\gamma}\gamma_0-m\gamma^0)v(\vec{p},s) \tag{5.53}$$

and from (5.51)

$$(p_0\gamma_0^2+\vec{p}\cdot\vec{\gamma}\gamma_0+m\gamma_0)v(\vec{p},s)=0$$

i.e.

$$p_0\gamma_0^2 v(\vec{p},s)=(-\vec{p}\cdot\vec{\gamma}\gamma_0-m\gamma_0)v(\vec{p},s)$$

so that (5.53) becomes

$$(\not{p} - m)\gamma^0 v(\vec{p}, s) = 2p_0(\gamma_0)^2 v(\vec{p}, s)$$
$$= 2p_0 v(\vec{p}, s) \qquad (5.54)$$

We also have the following result.

Proposition: We can show that

$$(\not{p} + m)\gamma^0 u(\vec{p}, s) = 2p_0 u(\vec{p}, s) \qquad (5.55)$$

Proof: We have

$$(\not{p} + m)\gamma^0 u(\vec{p}, s) = (p_0\gamma_0 - \vec{p}\cdot\vec{\gamma} + m)\gamma^0 u(\vec{p}, s)$$
$$= [p_0(\gamma_0)^2 - \vec{p}\cdot\vec{\gamma}\gamma_0 + m\gamma_0]\, u(\vec{p}, s)$$

From (5.50) we obtain

$$p_0(\gamma_0)^2 u(\vec{p}, s) = (-\vec{p}\cdot\vec{\gamma}\gamma_0 + m\gamma_0)\, u(\vec{p}, s)$$

so that

$$(\not{p} + m)\gamma^0 u(\vec{p}, s) = 2p_0(\gamma_0)^2 u(\vec{p}, s)$$
$$= 2p_0 u(\vec{p}, s)$$

Inserting (5.54) and (5.55) into (5.52) we obtain

$$Q_a = \frac{i\lambda}{2}\left(\frac{m}{2}\right)^{1/2} \sum_s \int d^3p \, \frac{1}{\omega_p} \cdot$$
$$\cdot \{C(\vec{p})\, d^+(\vec{p}, s)\, 2p_0 v(\vec{p}, s)$$
$$- D(\vec{p})\, d(\vec{p}, s)\, 2p_0 u(\vec{p}, s)\}_a$$
$$= i\lambda\left(\frac{m}{2}\right)^{1/2} \sum_s \int d^3p \, \{C(\vec{p})\, d^+(\vec{p}, s)\, v(\vec{p}, s)$$
$$- D(\vec{p})\, d(\vec{p}, s)\, u(\vec{p}, s)\}_a$$

This completes the proof of (5.47).

5.4 Projection Operators

We define the operators

$$\Lambda_+ := \frac{\not{p} + m}{2m} \tag{5.56}$$

and

$$\Lambda_- := \frac{-\not{p} + m}{2m} \tag{5.57}$$

Proposition: $\Lambda_\pm$ are projection operators, i.e.

$$\left.\begin{aligned} &\text{i)} \quad (\Lambda_\pm)^2 = \Lambda_\pm \\ &\text{ii)} \quad \Lambda_\pm \Lambda_\mp = 0 \\ &\text{iii)} \quad \Lambda_+ + \Lambda_- = 1 \end{aligned}\right\} \tag{5.58}$$

Proof:

i)
$$\begin{aligned} (\Lambda_\pm)^2 &= \frac{1}{4m^2}(\not{p} \pm m)(\not{p} \pm m) \\ &= \frac{1}{4m^2}(\not{p}\not{p} \pm 2m\not{p} + m^2) \\ &= \frac{1}{4m^2}(p^2 \pm 2m\not{p} + m^2) \\ &= \frac{1}{4m^2}(2m^2 \pm 2m\not{p}) \\ &= \Lambda_\pm \end{aligned}$$

ii)
$$\begin{aligned} \Lambda_\pm \Lambda_\mp &= \frac{1}{4m^2}(\pm\not{p} + m)(\mp\not{p} + m) \\ &= \frac{1}{4m^2}(-\not{p}\not{p} + m^2) \\ &= \frac{1}{4m^2}(-p^2 + m^2) \\ &= \frac{1}{4m^2}(-m^2 + m^2) \\ &= 0 \end{aligned}$$

iii) $\Lambda_+ + \Lambda_- = \frac{\not p + m}{2m} + \frac{-\not p + m}{2m} = 1$

Proposition: The projection operators Λ_+ and Λ_- project out the positive and negative energy-states respectively, i.e.

$$\left.\begin{aligned} \Lambda_+ u(\vec{p},s) &= u(\vec{p},s) \\ \Lambda_- u(\vec{p},s) &= 0 \\ \Lambda_+ v(\vec{p},s) &= 0 \\ \Lambda_- v(\vec{p},s) &= v(\vec{p},s) \end{aligned}\right\} \quad (5.59)$$

Proof: The second and third of equations (5.59) follow immediately from (5.23). Considering the first equation we have

$$\Lambda_+ u(\vec{p},s) = \frac{1}{2m}(\not p + m)u(\vec{p},s)$$
$$= \frac{1}{2m}(p_0\gamma_0 - \vec{p}\cdot\vec{\gamma} + m)\,u(\vec{p},s)$$

Now $\quad (\not p - m)u(\vec{p},s) = 0$

so that

$$p_0\gamma_0\, u(\vec{p},s) = (\vec{p}\cdot\vec{\gamma} + m)\,u(\vec{p},s)$$

Then

$$\Lambda_+ u(\vec{p},s) = \frac{1}{2m}\,2m\,u(\vec{p},s) = u(\vec{p},s)$$

The last of equations (5.59) can be shown in a similar fashion.

Proposition: The following completeness relations hold

$$\Lambda_{+\,ab}(\vec{p}) = \sum_s u_a(\vec{p},s) \otimes \bar{u}_b(\vec{p},s) \quad (5.60)$$

$$\Lambda_{-\,ab}(\vec{p}) = -\sum_s v_a(\vec{p},s) \otimes \bar{v}_b(\vec{p},s) \quad (5.61)$$

Proof:

$$\sum_s u(\vec{p},s) \otimes \bar{u}(\vec{p},s)$$

$$= (\not{p}+m) \frac{\sum_s u(\vec{0},s) \otimes \bar{u}(\vec{0},s)}{2m(m+\omega_p)} (\not{p}+m)$$

(using (5.29) and (5.32))

$$= \frac{(\not{p}+m)}{2m(m+\omega_p)} \left[u(\vec{0},1) \otimes \bar{u}(\vec{0},1) + u(\vec{0},2) \otimes \bar{u}(\vec{0},2) \right] (\not{p}+m)$$

$$= \frac{1}{2m(m+\omega_p)} (\not{p}+m) \frac{1}{2}(1+\gamma^0)(\not{p}+m)$$

$$= \frac{1}{2m(m+\omega_p)} \frac{1}{2} \{ (\not{p}+m)(\not{p}+m) + (\not{p}+m)\gamma_0(\not{p}+m) \}$$

(using (5.25) and (5.26))

$$= \frac{1}{2m(m+\omega_p)} \frac{1}{2} \{ \not{p}^2 + 2m\not{p} + m^2 + 2\omega_p(\not{p}+m) \}$$

(using (5.62) to be shown below)

$$= \frac{1}{2m(m+\omega_p)} \frac{1}{2} \{ (2m+2\omega_p)(\not{p}+m) \}$$

(since $p^2 - m^2 = 0$)

$$= \frac{\not{p}+m}{2m}$$

$$= \Lambda_+(p)$$

where we used

$$(\not{p}+m)\gamma_0(\not{p}+m) = 2\omega_p(\not{p}+m) \tag{5.62}$$

which may be shown as follows:

$$(\not{p}+m)\gamma_0(\not{p}+m)$$

$$= \not{p}\gamma_0\not{p} + m^2\gamma_0 + m(\not{p}\gamma_0 + \gamma_0\not{p})$$

$$= \not p \gamma_0 (p_0 \gamma_0 - \vec{p}\cdot\vec{\gamma}) + m^2 \gamma_0$$
$$+ m[p_\mu \gamma^\mu \gamma_0 + \gamma_0 (p_0 \gamma_0 - \vec{p}\cdot\vec{\gamma})]$$
$$= \not p (p_0 \gamma_0^2 + \vec{p}\cdot\vec{\gamma}\gamma_0) + p^2 \gamma_0$$
$$+ m[p_0 \gamma^0 - \vec{p}\cdot\vec{\gamma} + p_0 \gamma_0 + \vec{p}\cdot\vec{\gamma}]\gamma_0$$

(since $p^2 - m^2 = 0$)

$$= (p_0 \gamma^0 - \vec{p}\cdot\vec{\gamma})(p_0 \gamma_0 + \vec{p}\cdot\vec{\gamma})\gamma_0 + p^2 \gamma_0 + 2mp_0 \gamma_0^2$$
$$= \{(p_0 \gamma^0)^2 + p_0 \gamma^0 \vec{p}\cdot\vec{\gamma} - \vec{p}\cdot\vec{\gamma} p_0 \gamma_0 - \vec{p}\cdot\vec{\gamma}\,\vec{p}\cdot\vec{\gamma}\}\gamma_0$$
$$+ (p_0 \gamma_0 - \vec{p}\cdot\vec{\gamma})^2 \gamma_0 + 2mp_0$$
$$= [p_0^2 \gamma_0^2 + p_0 \gamma^0 \vec{p}\cdot\vec{\gamma} - \vec{p}\cdot\vec{\gamma} p_0 \gamma_0 - \vec{p}\cdot\vec{\gamma}\,\vec{p}\cdot\vec{\gamma}]\gamma_0$$
$$+ [p_0^2 \gamma_0^2 - p_0 \gamma_0 \vec{p}\cdot\vec{\gamma} - \vec{p}\cdot\vec{\gamma} p_0 \gamma_0 + \vec{p}\cdot\vec{\gamma}\,\vec{p}\cdot\vec{\gamma}]\gamma_0$$
$$+ 2mp_0$$
$$= 2p_0^2 \gamma^0 - 2\vec{p}\cdot\vec{\gamma} p_0 (\gamma_0)^2 + 2mp_0$$
$$= 2p_0 (p_0 \gamma^0 - \vec{p}\cdot\vec{\gamma}) + 2mp_0$$
$$= 2p_0 \not p + 2mp_0$$
$$= 2\omega_p (\not p + m)$$

where

$$p_0 = \omega_p = (\vec{p}^{\,2} + m^2)^{1/2}$$

In a similar way one can show that

$$(\not p - m)\gamma^0 (\not p - m) = 2\omega_p (\not p - m) \qquad (5.63)$$

We have

$$(\not p - m)\gamma^0 (\not p - m) = \not p \gamma^0 \not p + m^2 \gamma_0 - m(\not p \gamma_0 + \gamma_0 \not p)$$
$$= 2p_0 \not p - 2mp_0$$

Here the last expression results in the same way as

above so that

$$(\not p - m)\gamma^0(\not p - m) = 2p_0(\not p - m)$$

which had to be shown.

We now come to the proof of (5.61). We have, using (5.33) and (5.29),

$$\sum_s v(\vec{p},s) \otimes \bar{v}(\vec{p},s)$$

$$= (m - \not p)\frac{\sum_s v(\vec{0},s) \otimes \bar{v}(\vec{0},s)}{2m(m+\omega_p)}(m - \not p)$$

$$= \frac{1}{2m(m+\omega_p)}\{(m-\not p)[v(\vec{0},1) \otimes \bar{v}(\vec{0},1) + v(\vec{0},2) \otimes \bar{v}(0,2)](m-\not p)\}$$

Using (5.26) we obtain

$$v(\vec{0},1) \otimes \bar{v}(\vec{0},1) + v(\vec{0},2) \otimes \bar{v}(\vec{0},2)$$

$$= \begin{pmatrix} 0 \\ 0 \\ 0 \\ 1 \end{pmatrix} \otimes (0\ 0\ 0\ -1) + \begin{pmatrix} 0 \\ 0 \\ -1 \\ 0 \end{pmatrix} \otimes (0\ 0\ 1\ 0)$$

$$= \begin{pmatrix} 0 & 0 & 0 & 0 \\ 0 & 0 & 0 & 0 \\ 0 & 0 & -1 & 0 \\ 0 & 0 & 0 & -1 \end{pmatrix} = -\frac{1}{2}(1_{4\times 4} - \gamma^0)$$

Hence

$$\sum_s v(\vec{p},s) \otimes \bar{v}(\vec{p},s)$$

$$= -\frac{1}{2m(m+\omega_p)}\left\{(\not p - m)\frac{1-\gamma^0}{2}(\not p - m)\right\}$$

$$= -\frac{1}{4m(m+\omega_p)}\{(\not p - m)(\not p - m) - (\not p - m)\gamma^0(\not p - m)\}$$

$$= -\frac{1}{4m(m+\omega_p)}\{2m^2 - 2m\not{p} - 2\omega_p(\not{p} - m)\}$$

(using (5.63))

$$= -\frac{1}{2m(m+\omega_p)}\{m(m-\not{p}) + \omega_p(m-\not{p})\}$$

$$= -\frac{1}{2m(m+\omega_p)}(m+\omega_p)(m-\not{p})$$

$$= -\frac{m-\not{p}}{2m}$$

$$= -\Lambda_-$$ (with (6.57))

Some other formulae which will be needed later follow immediately from (5.61), e.g.

$$\sum_a \bar{u}_a(\vec{p},s)\Lambda_{-ab}(\vec{p})$$

$$= -\sum_a \bar{u}_a(\vec{p},s)\sum_t v_a(\vec{p},t)\bar{v}_b(\vec{p},t)$$

(with (5.61))

$$= 0$$ (with (5.30b))

Thus

$$\left.\begin{array}{l} \bar{u}_a(\vec{p},s)\Lambda_{-ab}(\vec{p}) = 0 \\ \text{and similarly} \\ \bar{v}_a(\vec{p},s)\Lambda_{+ab}(\vec{p}) = 0 \end{array}\right\} \qquad (5.63')$$

5.5 Anticommutation Relations

We mentioned earlier that we want to show that the Wess-Zumino model is a supersymmetric field theory. Thus we have to show that the spinor charges (5.47) obey the anticommutation relation

$$\{Q_a, \bar{Q}_b\} = 2\not{P}_{ab}$$

We first have to derive the Dirac adjoint of the Majorana spinor charge, i.e. $\bar{Q}_a$. Now according to (5.47)

$$Q_a = i\lambda\left(\frac{m}{2}\right)^{1/2}\sum_s\int d^3p\,[C(\vec{p})\,d^+(\vec{p},s)\,v(\vec{p},s) - D(\vec{p})\,d(\vec{p},s)\,u(\vec{p},s)]_a$$

Then

$$Q_a^+ = -i\lambda\left(\frac{m}{2}\right)^{1/2}\sum_s\int d^3p\,[v^+(\vec{p},s)\,d(\vec{p},s)\,C^+(\vec{p}) - u^+(\vec{p},s)\,d^+(\vec{p},s)\,D^+(\vec{p})]_a$$

and the Dirac adjoint is

$$\begin{aligned}\bar{Q}_a &= (Q^+\gamma_0)_a \\ &= -i\lambda\left(\frac{m}{2}\right)^{1/2}\sum_s\int d^3p\,[v^+(\vec{p},s)\,d(\vec{p},s)\,C^+(\vec{p})\gamma_0 - u^+(\vec{p},s)\,d^+(\vec{p},s)\,D^+(\vec{p})\gamma_0]_a\end{aligned} \tag{5.64}$$

Now, using (5.48) we have

$$\begin{aligned}C^+(\vec{p})\gamma_0 &= (a^+(\vec{p})1 + i\gamma_5^+ b^+(\vec{p}))\gamma_0 \\ &= a^+(\vec{p})\gamma_0 - i\gamma_0\gamma_5^+ b^+(\vec{p}) \\ &= \gamma_0(a^+(\vec{p}) - i\gamma_5\, b^+(\vec{p})) \\ &= \gamma_0 D(\vec{p}) \qquad \text{(with (5.48))}\end{aligned}$$

So

$$C^{+}(\vec{p})\gamma_0 = \gamma_0 D(\vec{p}) \tag{5.65}$$

Furthermore we have

$$D^{+}(\vec{p})\gamma_0 = \gamma_0 C(\vec{p}) \tag{5.66}$$

Inserting (5.65) and (5.66) into (5.64) we obtain

$$\begin{aligned}\bar{Q}_a &= -i\lambda\left(\frac{m}{2}\right)^{1/2}\sum_s\int d^3p\,\{v^{+}(\vec{p},s)\gamma_0\, d(\vec{p},s)\, D(\vec{p}) \\ &\qquad - u^{+}(\vec{p},s)\gamma_0\, d^{+}(\vec{p},s)\, C(\vec{p})\} \\ &= -i\lambda\left(\frac{m}{2}\right)^{1/2}\sum_s\int d^3p\,\{\bar{v}(\vec{p},s)\, d(\vec{p},s)\, D(\vec{p}) \\ &\qquad - \bar{u}(\vec{p},s)\, d^{+}(\vec{p},s)\, C(\vec{p})\}\end{aligned} \tag{5.67}$$

Before we calculate the anticommutator of Q and $\bar{Q}$, we demonstrate that the following relations hold:

$$[C_{ac}(\vec{p}), D_{bd}(\vec{p}\,')] = (\delta_{ac}\delta_{bd} - \gamma^5_{ac}\gamma^5_{bd})\,\delta(\vec{p}-\vec{p}\,') \tag{5.68}$$

and

$$\left.\begin{aligned}[][C_{ac}(\vec{p}), C_{bd}(\vec{p}\,')] &= 0 \\ [C^{+}_{ac}(\vec{p}), C^{+}_{bd}(\vec{p}\,')] &= 0 \\ [D_{ac}(\vec{p}), D_{bd}(\vec{p}\,')] &= 0 \\ [D^{+}_{ac}(\vec{p}), D^{+}_{bd}(\vec{p}\,')] &= 0\end{aligned}\right\} \tag{5.69}$$

In order to prove (5.68) we use (5.48), i.e.

$$\begin{aligned}&[C_{ac}(\vec{p}), D_{bd}(\vec{p}\,')] \\ &= [a(\vec{p})\delta_{ac} - i\gamma^5_{ac}b(\vec{p}), a^{+}(\vec{p}\,')\delta_{bd} - i\gamma^5_{bd}b^{+}(\vec{p}\,')] \\ &= [a(\vec{p}), a^{+}(\vec{p}\,')]\delta_{ac}\delta_{bd} - i\gamma^5_{ac}\gamma^5_{bd}[b(\vec{p}), a^{+}(\vec{p}\,')]\end{aligned}$$

$$-i\delta_{ac}\gamma^5_{bd}[a(\vec{p}),b^+(\vec{p}\,')]-\gamma^5_{ac}\gamma^5_{bd}[b(\vec{p}),b^+(\vec{p}\,')]$$

$$=\delta_{ac}\delta_{bd}\,\delta(\vec{p}-\vec{p}\,')-\gamma^5_{ac}\gamma^5_{bd}\,\delta(\vec{p}-\vec{p}\,')$$

$$=(\delta_{ac}\delta_{bd}-\gamma^5_{ac}\gamma^5_{bd})\,\delta(\vec{p}-\vec{p}\,')$$

using the commutation relations for bosonic creation and annihilation operators

$$[a(\vec{p}),a^+(\vec{p}\,')]=\delta(\vec{p}-\vec{p}\,')$$

$$[b(\vec{p}),b^+(\vec{p}\,')]=\delta(\vec{p}-\vec{p}\,')$$

All the commutators of (5.69) are easily seen to vanish.

We can now calculate the anticommutator of Q and $\overline{Q}$. Using (5.47) and (5.67) we have

$$\{Q_a,\bar{Q}_b\}$$

$$=\Big\{i\lambda\left(\frac{m}{2}\right)^{1/2}\sum_s\int d^3p\Big(C(\vec{p})d^+(\vec{p},s)v(\vec{p},s)-D(\vec{p})d(\vec{p},s)u(\vec{p},s)\Big)_a,$$

$$-i\lambda\left(\frac{m}{2}\right)^{1/2}\sum_r\int d^3k\Big(\bar{v}(\vec{k},r)d(\vec{k},r)D(\vec{k})-\bar{u}(\vec{k},r)d^+(\vec{k},r)C(\vec{k})\Big)_b\Big\}$$

$$=\frac{1}{2}m\lambda^2\sum_{s,r}\int d^3p\int d^3k\,\cdot$$

$$\cdot\Big\{C_{ac}(\vec{p})d^+(\vec{p},s)v_c(\vec{p},s)-D_{ac}(\vec{p})d(\vec{p},s)u_c(\vec{p},s),\ \bar{v}_d(\vec{k},r)d(\vec{k},r)D_{db}(\vec{k})-\bar{u}_d(\vec{k},r)d^+(\vec{k},r)C_{db}(\vec{k})\Big\}$$

$$=\frac{1}{2}m\lambda^2\sum_{s,r}\int d^3p\int d^3k\,\cdot$$

$$\Big([C_{ac}(\vec{p})d^+(\vec{p},s)v_c(\vec{p},s)-D_{ac}(\vec{p})d(\vec{p},s)u_c(\vec{p},s)].$$

$$\cdot\left[D_{db}(\vec{k})\,d(\vec{k},r)\,\overline{v}_d(\vec{k},r) - C_{db}(\vec{k})\,d^+(\vec{k},r)\,\overline{u}_d(\vec{k},r)\right]$$

$$+$$

$$\left[D_{db}(\vec{k})\,d(\vec{k},r)\,\overline{v}_d(\vec{k},r) - C_{db}(\vec{k})\,d^+(\vec{k},r)\,\overline{u}_d(\vec{k},r)\right]\cdot$$

$$\cdot\left[C_{ac}(\vec{p})\,d^+(\vec{p},s)\,v_c(\vec{p},s) - D_{ac}(\vec{p})\,d(\vec{p},s)\,u_c(\vec{p},s)\right]\Big)$$

$$= \tfrac{1}{2} m \lambda^2 \sum_{s,r} \int d^3p \int d^3k \,\cdot$$

$$\begin{aligned}
\cdot\Big(& C_{ac}(\vec{p})\,d^+(\vec{p},s)\,v_c(\vec{p},s)\,D_{db}(\vec{k})\,d(\vec{k},r)\,\overline{v}_d(\vec{k},r) \\
& - C_{ac}(\vec{p})\,d^+(\vec{p},s)\,v_c(\vec{p},s)\,C_{db}(\vec{k})\,d^+(\vec{k},r)\,\overline{u}_d(\vec{k},r) \\
& - D_{ac}(\vec{p})\,d(\vec{p},s)\,u_c(\vec{p},s)\,D_{db}(\vec{k})\,d(\vec{k},r)\,\overline{v}_d(\vec{k},r) \\
& + D_{ac}(\vec{p})\,d(\vec{p},s)\,u_c(\vec{p},s)\,C_{db}(\vec{k})\,d^+(\vec{k},r)\,\overline{u}_d(\vec{k},r) \\
& + D_{db}(\vec{k})\,d(\vec{k},r)\,\overline{v}_d(\vec{k},r)\,C_{ac}(\vec{p})\,d^+(\vec{p},s)\,v_c(\vec{p},s) \\
& - D_{db}(\vec{k})\,d(\vec{k},r)\,\overline{v}_d(\vec{k},r)\,D_{ac}(\vec{p})\,d(\vec{p},s)\,u_c(\vec{p},s) \\
& - C_{db}(\vec{k})\,d^+(\vec{k},r)\,\overline{u}_d(\vec{k},r)\,C_{ac}(\vec{p})\,d^+(\vec{p},s)\,v_c(\vec{p},s) \\
& + C_{db}(\vec{k})\,d^+(\vec{k},r)\,\overline{u}_d(\vec{k},r)\,D_{ac}(\vec{p})\,d(\vec{p},s)\,u_c(\vec{p},s)\Big)
\end{aligned}$$

Now consider the following terms of the above expression:

$$\begin{aligned}
& - C_{ac}(\vec{p})\,d^+(\vec{p},s)\,v_c(\vec{p},s)\,C_{db}(\vec{k})\,d^+(\vec{k},r)\,\overline{u}_d(\vec{k},r) \\
& - C_{db}(\vec{k})\,d^+(\vec{k},r)\,\overline{u}_d(\vec{k},r)\,C_{ac}(\vec{p})\,d^+(\vec{p},s)\,v_c(\vec{p},s) \\
& = d^+(\vec{p},s)\,d^+(\vec{k},r)\Big[-C_{ac}(\vec{p})\,v_c(\vec{p},s)\,C_{db}(\vec{k})\,\overline{u}_d(\vec{k},r) \\
& \qquad\qquad + C_{db}(\vec{k})\,\overline{u}_d(\vec{k},r)\,C_{ac}(\vec{p})\,v_c(\vec{p},s)\Big]
\end{aligned}$$

(using (5.44))

$$= d^+(\vec{p},s)\,d^+(\vec{k},r)\Big[-C_{ac}(\vec{p})\,C_{db}(\vec{k})$$

$$+ C_{db}(\vec{k}) C_{ac}(\vec{p}) \big] v_c(\vec{p},s) \bar{u}_d(\vec{k},r)$$

$$= - d^+(\vec{p},s) d(\vec{k},r) \left[C_{ac}(\vec{p}), C_{db}(\vec{k}) \right] v_c(\vec{p},s) \bar{u}_d(\vec{k},r)$$

$$= 0 \qquad \text{(with (5.69))}$$

A similar calculation yields

$$- D_{ac}(\vec{p}) d(\vec{p},s) u_c(\vec{p},s) D_{db}(\vec{k}) d(\vec{k},r) \bar{v}_d(\vec{k},r)$$

$$- D_{db}(\vec{k}) d(\vec{k},r) \bar{v}_d(\vec{k},r) D_{ac}(\vec{p}) d(\vec{p},s) u_c(\vec{p},s)$$

$$= 0$$

Hence

$$\{Q_a, \bar{Q}_b\} = \frac{1}{2} m \lambda^2 \sum_{r,s} \int d^3p \int d^3k \cdot$$

$$\cdot \big[\{ C_{ac}(\vec{p}) d^+(\vec{p},s), D_{db}(\vec{k}) d(\vec{k},r) \} v_c(\vec{p},s) \bar{v}_d(\vec{k},r)$$

$$+ \{ D_{ac}(\vec{p}) d(\vec{p},s), C_{db}(\vec{k}) d^+(\vec{k},r) \} u_c(\vec{p},s) \bar{u}_d(\vec{k},r) \big]$$

Now

$$\begin{aligned} \{ C d^+, D d \} &= C d^+ D d + D d C d^+ \\ &= C D d^+ d + D C d d^+ \\ &= C D d^+ d + D C d d^+ + D C d^+ d - D C d^+ d \\ &= [C, D] d^+ d + D C \{ d, d^+ \} \end{aligned}$$

and we obtain

$$\{Q_a, \bar{Q}_b\} = \frac{1}{2} m \lambda^2 \sum_{r,s} \int d^3p \int d^3k \cdot$$

$$\Big(\big[[C_{ac}(\vec{p}), D_{db}(\vec{k})] d^+(\vec{p},s) d(\vec{k},r)$$

$$+ D_{db}(\vec{k}) C_{ac}(\vec{p}) \{ d(\vec{k},r), d^+(\vec{p},s) \} \big] v_c(\vec{p},s) \bar{v}_d(\vec{k},r)$$

$$+\Big[[C_{db}(\vec{k}),D_{ac}(\vec{k})]\,d^\dagger(\vec{k},r)\,d(\vec{p},s)$$

$$+D_{ac}(\vec{p})\,C_{db}(\vec{k})\{d(\vec{p},s),d^\dagger(\vec{k},r)\}\Big]\,u_c(\vec{p},s)\,\overline{u}_d(\vec{k},r)\Big)$$

$$=\frac{1}{2}m\lambda^2\sum_{r,s}\int d^3p\int d^3k\,\cdot$$

$$\cdot\Big\{\Big([\delta_{ac}\delta_{db}-\gamma^5_{ac}\gamma^5_{db}]\,\delta(\vec{p}-\vec{k})\,d^\dagger(\vec{p},s)\,d(\vec{k},r)$$

$$+D_{db}(\vec{k})\,C_{ac}(\vec{p})\,\delta_{rs}\,\delta(\vec{p}-\vec{k})\Big)\,v_c(\vec{p},s)\,\overline{v}_d(\vec{k},r)$$

$$+\Big([\delta_{db}\delta_{ac}-\gamma^5_{db}\gamma^5_{ac}]\,\delta(\vec{p}-\vec{k})\,d^\dagger(\vec{k},r)\,d(\vec{p},s)$$

$$+D_{ac}(\vec{p})\,C_{db}(\vec{k})\,\delta_{rs}\,\delta(\vec{p}-\vec{k})\Big)\,u_c(\vec{p},s)\,\overline{u}_d(\vec{k},r)\Big\}$$

(using (5.68) and (5.44))

$$=\frac{1}{2}m\lambda^2\sum_{r,s}\int d^3p\,\Big\{\Big([\delta_{ac}\delta_{db}-\gamma^5_{ac}\gamma^5_{db}]\,d^\dagger(\vec{p},s)\,d(\vec{p},r)$$

$$+D_{db}(\vec{p})\,C_{ac}(\vec{p})\,\delta_{rs}\Big)\,v_c(\vec{p},s)\,\overline{v}_d(\vec{p},r)$$

$$+\Big([\delta_{db}\delta_{ac}-\gamma^5_{db}\gamma^5_{ac}]\,d^\dagger(\vec{p},r)\,d(\vec{p},s)$$

$$+D_{ac}(\vec{p})\,C_{db}(\vec{p})\,\delta_{rs}\Big)\,u_c(\vec{p},s)\,\overline{u}_d(\vec{p},r)\Big\}$$

$$=\frac{1}{2}m\lambda^2\sum_{r,s}\int d^3p\,\Big\{d^\dagger(\vec{p},r)\,d(\vec{p},s)\,\cdot$$

$$\cdot(\delta_{ac}\delta_{db}-\gamma^5_{ac}\gamma^5_{db})\cdot\big(v_c(\vec{p},r)\,\overline{v}_d(\vec{p},s)$$

$$+u_c(\vec{p},s)\,\overline{u}_d(\vec{p},r)\big)+\delta_{rs}\big(D_{db}(\vec{p})\,C_{ac}(\vec{p})\,\cdot$$

$$\cdot\,v_c(\vec{p},r)\,\overline{v}_d(\vec{p},s)+D_{ac}(\vec{p})\,C_{db}(\vec{p})\,u_c(\vec{p},s)\,\overline{u}_d(\vec{p},r)\big)\Big\}$$

$$=\frac{1}{2}m\lambda^2\sum_{r,s}\int d^3p\;d^\dagger(\vec{p},r)\,d(\vec{p},s)\,.$$

$$\cdot\left(\delta_{ac}\delta_{db}-\gamma^5_{ac}\gamma^5_{db}\right)\left(v_c(\vec{p},r)\bar{v}_d(\vec{p},s)\right.$$
$$\left.+u_c(\vec{p},s)\bar{u}_d(\vec{p},r)\right)$$
$$+\frac{1}{2}m\lambda^2\int d^3p\left[-D_{db}(\vec{p})C_{ac}(\vec{p})(\Lambda_-)_{cd}\right.$$
$$\left.+D_{ac}(\vec{p})C_{db}(\vec{p})(\Lambda_+)_{cd}\right]$$

(using (5.60) and (5.61))

We now consider the second integral and use (5.48), (5.56) and (5.57). Then

$$\frac{1}{2}m\lambda^2\int d^3p\left[D_{ac}(\vec{p})C_{db}(\vec{p})(\Lambda_+)_{cd}\right.$$
$$\left.-D_{db}(\vec{p})C_{ac}(\vec{p})(\Lambda_-)_{cd}\right]$$
$$=\frac{1}{2}m\lambda^2\int d^3p\left[\left(a^+(\vec{p})\delta_{ac}-i\gamma^5_{ac}b^+(\vec{p})\right)\cdot\right.$$
$$\cdot\left(a(\vec{p})\delta_{db}-i\gamma^5_{db}b(\vec{p})\right)\left(\frac{\not{p}+m}{2m}\right)_{cd}$$
$$-\left(a^+(\vec{p})\delta_{db}-i\gamma^5_{db}b^+(\vec{p})\right)\cdot$$
$$\left.\cdot\left(a(\vec{p})\delta_{ac}-i\gamma^5_{ac}b(\vec{p})\right)\left(\frac{-\not{p}+m}{2m}\right)_{cd}\right]$$
$$=\frac{1}{2}m\lambda^2\int d^3p\left\{a^+(\vec{p})a(\vec{p})\left(\frac{\not{p}+m}{2m}\right)_{ab}\right.$$
$$-i\,a^+(\vec{p})b(\vec{p})\gamma^5_{db}\left(\frac{\not{p}+m}{2m}\right)_{ad}$$
$$-ib^+(\vec{p})a(\vec{p})\gamma^5_{ac}\left(\frac{\not{p}+m}{2m}\right)_{cb}-b^+(\vec{p})b(\vec{p})\gamma^5_{ac}\gamma^5_{db}\left(\frac{\not{p}+m}{2m}\right)_{cd}$$
$$-a^+(\vec{p})a(\vec{p})\left(\frac{-\not{p}+m}{2m}\right)_{ab}+i\,a^+(\vec{p})b(\vec{p})\gamma^5_{ac}\left(\frac{-\not{p}+m}{2m}\right)_{cb}$$

$$+ i\, b^+(\vec{p})\, a(\vec{p})\, \gamma^5_{db} \left(\frac{-\not{p}+m}{2m}\right)_{ad}$$

$$+ b^+(\vec{p})\, b(\vec{p})\, \gamma^5_{db}\, \gamma^5_{ac} \left(\frac{-\not{p}+m}{2m}\right)_{cd} \Big\}$$

$$= \frac{1}{2} m \lambda^2 \int d^3p \Big\{ a^+(\vec{p})\, a(\vec{p}) \left(\frac{\not{p}}{m}\right)_{ab} - i\, a^+(\vec{p})\, b(\vec{p}) \left(\frac{\not{p}+m}{2m}\gamma^5\right)_{ab}$$

$$- i\, b^+(\vec{p})\, a(\vec{p}) \left(\gamma^5 \frac{\not{p}+m}{2m}\right)_{ab} - b^+(\vec{p})\, b(\vec{p}) \left(\gamma^5 \frac{\not{p}+m}{2m}\gamma^5\right)_{ab}$$

$$+ i\, a^+(\vec{p})\, b(\vec{p}) \left(\gamma^5 \frac{-\not{p}+m}{2m}\right)_{ab} + i\, b^+(\vec{p})\, a(\vec{p}) \left(\frac{-\not{p}+m}{2m}\gamma^5\right)_{ab}$$

$$+ b^+(\vec{p})\, b(\vec{p}) \left(\gamma^5 \frac{-\not{p}+m}{2m}\gamma^5\right)_{ab} \Big\}$$

$$= \frac{1}{2} m \lambda^2 \int d^3p \Big\{ a^+(\vec{p})\, a(\vec{p}) \left(\frac{\not{p}}{m}\right)_{ab} + b^+(\vec{p})\, b(\vec{p}) \left(\frac{\not{p}}{m}\right)_{ab} \Big\}$$

$$= \frac{1}{2} \lambda^2 \int d^3p\, \not{p}_{ab} \Big\{ a^+(\vec{p})\, a(\vec{p}) + b^+(\vec{p})\, b(\vec{p}) \Big\}$$

taking into account equations (1.140) and (1.141). Hence

$$\{Q_a, \bar{Q}_b\}$$

$$= \frac{1}{2} m \lambda^2 \sum_{r,s} \int d^3p\, d^+(\vec{p},r)\, d(\vec{p},s) \left(\delta_{ac}\delta_{db} - \gamma^5_{ac}\gamma^5_{db}\right) \cdot$$

$$\cdot \left(v_c(\vec{p},r)\, \bar{v}_d(\vec{p},s) + u_c(\vec{p},s)\, \bar{u}_d(\vec{p},r) \right)$$

$$+ \frac{1}{2} \lambda^2 \int d^3p\, \not{p}_{ab} \Big\{ a^+(\vec{p})\, a(\vec{p}) + b^+(\vec{p})\, b(\vec{p}) \Big\} \tag{5.70}$$

We now demonstrate in another lengthy calculation that the first integral can be rewritten in a simpler form. We start by introducing sixteen linearly independent 4x4 matrices

$$1_{4x4}, \gamma^{\mu}, \gamma^{\mu}\gamma^{5}, \sigma_4^{\mu\nu}, \gamma^{5}$$

Any 4x4 matrix can be expanded in terms of these matrices. Hence we can write

$$v_c(\vec{p},r) \otimes \bar{v}_d(\vec{p},s) + u_c(\vec{p},s) \otimes \bar{u}_d(\vec{p},r)$$
$$\equiv M_{cd}(\vec{p},r,s)$$
$$= \left(m_0 1_{4\times 4} + m_\mu \gamma^\mu + m_{\mu\nu}\sigma^{\mu\nu} + n_\mu \gamma^\mu \gamma^5 + n_5 \gamma^5\right)_{cd} \tag{5.71}$$

and we have

$$(\delta_{ac}\delta_{db} - \gamma^5_{ac}\gamma^5_{db})(v_c(\vec{p},r) \otimes \bar{v}_d(\vec{p},s)$$
$$+ u_c(\vec{p},s) \otimes \bar{u}_d(\vec{p},r))$$
$$= (\delta_{ac}\delta_{db} - \gamma^5_{ac}\gamma^5_{db}) M_{cd}(\vec{p},r,s)$$
$$= (\delta_{ac}\delta_{db} - \gamma^5_{ac}\gamma^5_{db})(m_0 1 + m_\mu \gamma^\mu + m_{\mu\nu}\sigma^{\mu\nu}$$
$$+ n_\mu \gamma^\mu \gamma^5 + n_5 \gamma^5)_{cd}$$
$$= (m_0 1 + m_\mu \gamma^\mu + m_{\mu\nu}\sigma^{\mu\nu} + n_\mu \gamma^\mu\gamma^5 + n_5\gamma^5)_{ab}$$
$$- \{\gamma^5(m_0 1 + m_\mu \gamma^\mu + m_{\mu\nu}\sigma^{\mu\nu} + n_\mu\gamma^\mu\gamma^5 + n_5\gamma^5)\gamma^5\}_{ab}$$
$$= m_0 \delta_{ab} + m_\mu \gamma^\mu_{ab} + m_{\mu\nu}(\sigma^{\mu\nu})_{ab} + n_\mu(\gamma^\mu\gamma^5)_{ab}$$
$$+ n_5\gamma^5_{ab} - (m_0\delta_{ab} - m_\mu\gamma^\mu_{ab} + m_{\mu\nu}(\sigma^{\mu\nu})_{ab}$$
$$- n_\mu(\gamma^\mu\gamma^5)_{ab} + n_5\gamma^5_{ab})$$

(using (1.140) and (1.141))

$$= 2(m_\mu \gamma^\mu + n_\mu \gamma^\mu \gamma^5)_{ab}$$

In order to obtain the coefficients m_μ and n_μ we show that

$$\text{i)} \quad m_\mu = \frac{1}{4} Tr(\gamma_\mu M) \qquad (5.72)$$

$$\text{ii)} \quad n_\mu = -\frac{1}{4} Tr(\gamma_\mu \gamma^5 M) \qquad (5.73)$$

We prove these relations as follows.

i) $\frac{1}{4} Tr(\gamma_\mu M)$

$$= \frac{1}{4} Tr\{m_0 \gamma_\mu + m_\rho \gamma_\mu \gamma^\rho + m_{\rho\sigma} \gamma_\mu \sigma^{\rho\sigma} + n_\rho \gamma_\mu \gamma^\rho \gamma^5 + n_5 \gamma_\mu \gamma^5\}$$

(with (5.71))

$$= \frac{m_0}{4} Tr\,\gamma_\mu + \frac{m_\rho}{4} Tr(\gamma_\mu \gamma^\rho) + \frac{m_{\rho\sigma}}{4} Tr(\gamma_\mu \sigma^{\rho\sigma}) + \frac{n_\rho}{4} Tr(\gamma_\mu \gamma^\rho \gamma^5) + \frac{n_\rho}{4} Tr(\gamma_\mu \gamma^5)$$

$$= \frac{1}{4} m_\rho \cdot 4\, \delta_\mu{}^\rho$$

$$= m_\mu$$

since $Tr\,\gamma_\mu = Tr(\gamma_5 \gamma^\mu) = Tr(\gamma_\mu \gamma^\rho \gamma^5) = 0$

ii) $-\frac{1}{4} Tr(\gamma_\mu \gamma^5 M)$

$$= -\frac{1}{4} Tr\{m_0 \gamma_\mu \gamma^5 + m_\rho \gamma_\mu \gamma^5 \gamma^\rho + m_{\rho\sigma} \gamma_\mu \gamma^5 \sigma^{\rho\sigma} + n_\rho \gamma_\mu \gamma^5 \gamma^\rho \gamma^5 + n_5 \gamma_\mu \gamma^5 \gamma^5\}$$

(with (5.71))

$$= \frac{1}{4} n_\rho\, 4\, \delta_\mu{}^\rho$$

$$= n_\mu$$

Then (5.70) becomes

$$\{Q_a, \bar{Q}_b\} = \frac{1}{2} m\lambda^2 \sum_{r,s} \int d^3p\, d^+(\vec{p},r)\, d(\vec{p},s) \cdot 2(m_\mu \gamma^\mu + n_\mu \gamma^\mu \gamma^5)_{ab} + \frac{1}{2}\lambda^2 \int d^3p\, \not{p}_{ab} \{a^+(\vec{p})a(\vec{p}) + b^+(\vec{p})b(\vec{p})\}$$

$$= \frac{1}{4} m\lambda^2 \sum_{r,s} \int d^3p\, d^+(\vec{p},r)d(\vec{p},s) \cdot \left[Tr(\gamma_\mu M(\vec{p},r,s))\gamma^\mu - Tr(\gamma_\mu \gamma^5 M(\vec{p},r,s))\gamma^\mu\gamma^5\right]_{ab} + \frac{1}{2}\lambda^2 \int d^3p\, \not{p}_{ab} \{a^+(\vec{p})a(\vec{p}) + b^+(\vec{p})b(\vec{p})\}$$

(5.74)

where we used (5.72) and (5.73). Evaluating the traces we have

$$Tr(\gamma_\mu M(\vec{p},r,s)) = (\gamma_\mu)_{ab} M_{ba}(\vec{p},r,s)$$

$$= (\gamma_\mu)_{ab} (v_b(\vec{p},r)\bar{v}_a(\vec{p},s) + u_b(\vec{p},r)\bar{u}_a(\vec{p},s))$$

(using (5.71))

$$= \bar{v}(\vec{p},r)\gamma_\mu v(\vec{p},s) + \bar{u}(\vec{p},r)\gamma_\mu u(\vec{p},s)$$

(5.75)

and

$$Tr(\gamma_\mu \gamma^5 M(\vec{p},r,s))$$

$$= \bar{v}(\vec{p},r)\gamma_\mu \gamma^5 v(\vec{p},s) + \bar{u}(\vec{p},r)\gamma_\mu \gamma^5 u(\vec{p},s)$$

(5.76)

Using (5.59) we have

$$\bar{u}(\vec{p},r)\gamma_\mu u(\vec{p},s) = \bar{u}(\vec{p},r)\gamma_\mu \Lambda_+(\vec{p})\, u(\vec{p},s)$$

$$= \bar{u}(\vec{p},r)\gamma_\mu\left(\frac{\not{p}+m}{2m}\right)u(\vec{p},s) \qquad \text{(with (5.56))}$$

$$= \frac{1}{2m}\bar{u}(\vec{p},r)\gamma_\mu(p_\rho\gamma^\rho+m)u(\vec{p},s)$$

$$= \frac{1}{2m}\bar{u}(\vec{p},r)(p_\rho\gamma_\mu\gamma^\rho+m\gamma_\mu)u(\vec{p},s)$$

$$= \frac{1}{2m}\bar{u}(\vec{p},r)[p_\rho 2\delta_\mu{}^\rho - p_\rho\gamma^\rho\gamma_\mu + m\gamma_\mu]u(\vec{p},s)$$

$$= \frac{1}{2m}\bar{u}(\vec{p},r)(2p_\mu - \not{p}\gamma_\mu + m\gamma_\mu)u(\vec{p},s)$$

$$= \bar{u}(\vec{p},r)\frac{p_\mu}{m}u(\vec{p},s) + \bar{u}(\vec{p},r)\Lambda_-\gamma^\mu u(\vec{p},s)$$

$$= \frac{p_\mu}{m}\bar{u}(\vec{p},r)u(\vec{p},s)$$

since

$$\bar{u}(\vec{p},r)\Lambda_- = 0$$

Using (5.30) we obtain

$$\bar{u}(\vec{p},r)\gamma_\mu u(\vec{p},s) = \frac{p_\mu}{m}\delta_{rs} \tag{5.77}$$

A corresponding relation holds for $\bar{v}(\vec{p},r)\gamma_\mu v(\vec{p},s)$,i.e.

$$\bar{v}(\vec{p},r)\gamma_\mu v(\vec{p},s) = \frac{p_\mu}{m}\delta_{rs} \tag{5.78}$$

Inserting (5,75), (5.76), (5.77) and (5.78) into (5.74) we obtain

$$\{Q_a,\bar{Q}_b\} = \frac{1}{2}\lambda^2\int d^3p\,\not{p}_{ab}\{a^+(\vec{p})a(\vec{p})+b^+(\vec{p})b(\vec{p})\}$$
$$+\frac{1}{4}m\lambda^2\sum_{r,s}\int d^3p\, d^+(\vec{p},r)d(\vec{p},s)\cdot$$
$$\cdot\{[\bar{v}(\vec{p},r)\gamma_\mu v(\vec{p},s)+\bar{u}(\vec{p},r)\gamma_\mu u(\vec{p},s)]\gamma^\mu_{ab}$$
$$-[\bar{v}(\vec{p},r)\gamma_\mu\gamma^5 v(\vec{p},s)+\bar{u}(\vec{p},r)\gamma_\mu\gamma^5 u(\vec{p},s)]\cdot$$
$$\cdot(\gamma^\mu\gamma^5)_{ab}\}$$

$$= \frac{1}{2}\lambda^2 \int d^3p \, \not{p}_{ab} \{ a^+(\vec{p}) a(\vec{p}) + b^+(\vec{p}) b(\vec{p}) \}$$

$$+ \frac{1}{4} m\lambda^2 \sum_{r,s} \int d^3p \; d^+(\vec{p},r) d(\vec{p},s) \cdot$$

$$\cdot \{ [\frac{p_\mu}{m} \delta_{rs} + \frac{p_\mu}{m} \delta_{rs}] \gamma^\mu_{ab}$$

$$- [\bar{v}(\vec{p},r) \gamma_\mu \gamma_5 v(\vec{p},s)$$

$$+ \bar{u}(\vec{p},r) \gamma_\mu \gamma^5 u(\vec{p},s)] (\gamma^\mu \gamma^5)_{ab} \}$$

$$= \frac{1}{2}\lambda^2 \int d^3p \, \not{p}_{ab} \{ a^+(\vec{p}) a(\vec{p}) + b^+(\vec{p}) b(\vec{p}) + \sum_r d^+(\vec{p},r) d(\vec{p},r) \}$$

$$- \frac{1}{4} m\lambda^2 \int d^3p \sum_{r,s} [\bar{v}(\vec{p},r) \gamma_\mu \gamma^5 v(\vec{p},s)$$

$$+ \bar{u}(\vec{p},r) \gamma_\mu \gamma_5 u(\vec{p},s)] (\gamma^\mu \gamma^5)_{ab}$$

(5.79)

We now convince ourselves that the contribution of the term proportional to $(\gamma^\mu \gamma^5)_{ab}$ vanishes. This can be shown to follow from symmetry properties of both sides. We multiply (5.79) from the right by the charge conjugation matrix C_{db} as in the calculations of the proof of (3.23) and use $C = - C^T$; then

$$\{Q_a, Q_d\} = \frac{1}{2}\lambda^2 \int d^3p \, p_\mu (\gamma^\mu C^T)_{ad} \cdot$$

$$\cdot \{ a^+(\vec{p}) a(\vec{p}) + b^+(\vec{p}) b(\vec{p}) + \sum_r d^+(\vec{p},r) d(\vec{p},r) \}$$

$$- \frac{1}{4} m\lambda^2 \int d^3p \sum_{r,s} [\bar{v}(\vec{p},r) \gamma_\mu \gamma^5 v(\vec{p},s)$$

$$+ \bar{u}(\vec{p},r) \gamma_\mu \gamma^5 u(\vec{p},s)] (\gamma^\mu \gamma^5 C^T)_{ad}$$

(5.80)

Now according to (3.14) and (3.16b) the first part, proportional to $\gamma^\mu C$, is symmetric in a and d, whereas the second part is antisymmetric in a and d. The left hand side of (5.80) is symmetric in a and d, so the coefficient of $\gamma^\mu\gamma^5 C^T$ must vanish. Thus we conclude that the anticommutator of Q_a and Q_b is

$$\{Q_a, \bar{Q}_b\} = \tfrac{1}{2}\lambda^2 \int d^3p\, \not{p}_{ab} \{a^+(\vec{p})a(\vec{p}) + b^+(\vec{p})b(\vec{p}) + \sum_r d^+(\vec{p},r)d(\vec{p},r)\} \tag{5.81}$$

In Section 5.6 we will see that the energy-momentum operator is given by

$$P^\mu := \int d^3p\, p^\mu [a^+(\vec{p})a(\vec{p}) + b^+(\vec{p})b(\vec{p}) + \sum_r d^+(\vec{p},r)d(\vec{p},r)] \tag{5.82}$$

Thus if we choose $\lambda = 2$, we obtain

$$\{Q_a, \bar{Q}_b\} = 2P^\mu(\gamma_\mu)_{ab} \tag{5.83}$$

This result shows that the Super-Poincaré algebra can indeed be represented in terms of commutators and anticommutators of linear operators in Fock space.We now demonstrate that the fourth of relations (4.1) is also satisfied.

Proposition: Given the momentum operator (5.82) and the spinor charge (5.47), then

$$[P^\mu, Q_a] = 0 \tag{5.84}$$

Proof: Using (5.47) and (5.82) we have

$$[P^\mu, Q_a]$$
$$= i(2m)^{1/2}[\int d^3p\, p^\mu (a^+(\vec{p})a(\vec{p}) + b^+(\vec{p})b(\vec{p}) + \sum_r d^+(\vec{p},r)d(\vec{p},r)),$$

$$\sum_s \int d^3k \Big(C(\vec{k}) d^+(\vec{k},s) v(\vec{k},s) - D(\vec{k}) d(\vec{k},s) u(\vec{k},s) \Big)_a \Big]$$

$$= i\,(2m)^{1/2} \int d^3p\, p^\mu \int d^3k \sum_s \cdot$$

$$\cdot \Big[\Big(a^+(\vec{p}) a(\vec{p}) + b^+(\vec{p}) b(\vec{p}) + \sum_r d^+(\vec{p},r) d(\vec{p},r) \Big), C_{ab}(\vec{k}) d^+(\vec{k},s) v_b(\vec{k},s) - D_{ab}(\vec{k}) d(\vec{k},s) u_b(\vec{k},s) \Big]$$

$$= i\,(2m)^{1/2} \int d^3p\, p^\mu \int d^3k \sum_s \cdot$$

$$\cdot \Big\{ [a^+(\vec{p}) a(\vec{p}), C_{ab}(\vec{k}) d^+(\vec{k},s)]\, v_b(\vec{k},s) - [a^+(\vec{p}) a(\vec{p}), D_{ab}(\vec{k}) d(\vec{k},s)]\, u_b(\vec{k},s) + [b^+(\vec{p}) b(\vec{p}), C_{ab}(\vec{k}) d^+(\vec{k},s)]\, v_b(\vec{k},s) - [b^+(\vec{p}) b(\vec{p}), D_{ab}(\vec{k}) d(\vec{k},s)]\, u_b(\vec{k},s) + \sum_r [d^+(\vec{p},r) d(\vec{p},r), C_{ab}(\vec{k}) d^+(\vec{k},s)]\, v_b(\vec{k},s) - \sum_r [d^+(\vec{p},r) d(\vec{p},r), D_{ab}(\vec{k}) d(\vec{k},s)]\, u_b(\vec{k},s) \Big\}$$

Consider

i) $[a^+(\vec{p})\, a(\vec{p}), C_{ab}(\vec{k})\, d^+(\vec{k},s)]$

$$= a^+(\vec{p}) [a(\vec{p}), C_{ab}(\vec{k}) d^+(\vec{k},s)] + [a^+(\vec{p}), C_{ab}(\vec{k}) d^+(\vec{k},s)]\, a(\vec{p})$$

$$= a^+(\vec{p}) [a(\vec{p}), C_{ab}(\vec{k})]\, d^+(\vec{k},s) + a^+(\vec{p}) C_{ab}(\vec{k}) [a(\vec{p}), d^+(\vec{k},s)] + C_{ab}(\vec{k}) [a^+(\vec{p}), d^+(\vec{k},s)]\, a(\vec{p}) + [a^+(\vec{p}), C_{ab}(\vec{k})]\, d^+(\vec{k},s)\, a(\vec{p})$$

$$= a^+(\vec{p}) [a(\vec{p}), a(\vec{k}) \delta_{ab} - i\gamma^5_{ab} b(\vec{k})]\, d^+(\vec{k},s)$$

$$+[a^+(\vec{p}), a(\vec{k})\,\delta_{ab} - i\gamma^5_{ab}\, b(\vec{k})]\, d^+(\vec{k},s)\, a(\vec{p})$$

(using (5.48))

where we take into account that $a(\vec{p})$ and $a^+(\vec{p})$ commute with $d^+(\vec{k})$. Furthermore

$$[a(\vec{p}), a(\vec{k})] = [a^+(\vec{p}), a^+(\vec{k})] = 0$$

$$[a(\vec{p}), a^+(\vec{k})] = \delta(\vec{p}-\vec{k})$$

and

$$[a(\vec{p}), b(\vec{k})] = [a^+(\vec{p}), b(\vec{k})] = 0$$

$$[a(\vec{p}), b^+(\vec{k})] = [a^+(\vec{p}), b^+(\vec{p})] = 0$$

so that the final expression for the first commutator is

$$[a^+(\vec{p})\, a(\vec{p}), C_{ab}(\vec{k})\, d^+(\vec{k},s)]\, v_b(\vec{k},s)$$

$$= -\delta(\vec{p}-\vec{k})\, d^+(\vec{k},s)\, a(\vec{p})\, v_a(\vec{k},s)$$

Similarly

ii) $-[a^+(\vec{p})a(\vec{p}), D_{ab}(\vec{k})d(\vec{k},s)]\, u_b(\vec{k},s)$

$$= -\delta(\vec{p}-\vec{k})\, a^+(\vec{p})\, d(\vec{k},s)\, u_a(\vec{k},s)$$

iii) $[b^+(\vec{p})b(\vec{p}), C_{ab}(\vec{k})\, d^+(\vec{k},r)]\, v_b(\vec{k},r)$

$$= i\gamma^5_{ab}\,\delta(\vec{p}-\vec{k})\, d^+(\vec{k},s)\, b(\vec{p})\, v_b(\vec{k},s)$$

iv) $-[b^+(\vec{p})b(\vec{p}), D_{ab}(\vec{k})d(\vec{k},s)]\, u_b(\vec{k},s)$

$$= i\gamma^5_{ab}\,\delta(\vec{p}-\vec{k})\, b^+(\vec{p})\, d(\vec{k},s)\, u_b(\vec{k},s)$$

v) $\sum_r [d^+(\vec{p},r)d(\vec{p},r), C_{ab}(\vec{k})\, d^+(\vec{k},s)]\, v_b(\vec{k},s)$

$$= \delta(\vec{p}-\vec{k})\, d^+(\vec{p},s)\,\{a(\vec{k})\, v_a(\vec{k},s) - i\gamma^5_{ab}\, b(\vec{k})\, v_b(\vec{k},s)\}$$

vi) $$\sum_r [d^+(\vec{p},r)d(\vec{p},r), D_{ab}(\vec{k})d(\vec{k},s)]u_b(\vec{k},s)$$
$$= \delta(\vec{p}-\vec{k})d(\vec{p},s)\{a^+(\vec{k})u_a(\vec{k},s) - i\gamma^5_{ab}b^+(\vec{k})u_b(\vec{k},s)\}$$

These relations can easily be surmised by inserting the defining expression (5.48) of C and D and observing that in each case only one type of operators does not commute with the others. Inserting these commutators in our expression for $[P_\mu, Q_a]$ we have

$$[P^\mu, Q_a] = i(2m)^{1/2}\int d^3p\, p^\mu \int d^3k \sum_s \cdot$$
$$\cdot\{-a(\vec{p})d^+(\vec{k},s)v_a(\vec{k},s)\delta(\vec{p}-\vec{k})$$
$$- a^+(\vec{p})d(\vec{k},s)u_a(\vec{k},s)\delta(\vec{p}-\vec{k})$$
$$+ i\gamma^5_{ab}b(\vec{p})d^+(\vec{k},s)v_b(\vec{k},s)\delta(\vec{p}-\vec{k})$$
$$+ i\gamma^5_{ab}b^+(\vec{p})d(\vec{k},s)u_b(\vec{k},s)\delta(\vec{p}-\vec{k})$$
$$+ a(\vec{k})d^+(\vec{p},s)v_a(\vec{k},s)\delta(\vec{p}-\vec{k})$$
$$- i\gamma^5_{ab}b(\vec{k})d^+(\vec{p},s)v_b(\vec{k},s)\delta(\vec{p}-\vec{k})$$
$$+ a^+(\vec{k})d(\vec{p},s)u_a(\vec{k},s)\delta(\vec{p}-\vec{k})$$
$$- i\gamma^5_{ab}b^+(\vec{k})d(\vec{p},s)u_b(\vec{k},s)\delta(\vec{p}-\vec{k})\}$$
$$= i(2m)^{1/2}\int d^3p\, p^\mu \sum_s \{-a(\vec{p})d^+(\vec{p},s)v_a(\vec{p},s)$$
$$- a^+(\vec{p})d(\vec{p},s)u_a(\vec{p},s) + i\gamma^5_{ab}b(\vec{p})d^+(\vec{p},s)v_b(\vec{p},s)$$
$$+ i\gamma^5_{ab}b^+(\vec{p})d(\vec{p},s)u_b(\vec{p},s) + a(\vec{p})d^+(\vec{p},s)v_a(\vec{p},s)$$
$$- i\gamma^5_{ab}b(\vec{p})d^+(\vec{p},s)v_b(\vec{p},s) + a^+(\vec{p})d(\vec{p},s)u_a(\vec{p},s)$$
$$- i\gamma^5_{ab}b^+(\vec{p})d(\vec{p},s)u_b(\vec{p},s)\}$$
$$= 0$$

5.6 The Energy-Momentum Operator of the Wess-Zumino Model

We now derive equation (5.82). In particular we will see that in the Wess-Zumino model

$$\mathcal{H} = :\mathcal{H}: \quad , \quad \mathcal{P}_i = :\mathcal{P}_i: \tag{5.85}$$

where : : denotes normal ordering (i.e. all creation operators stand to the left of all annihilation operators). This property is a reflection of the fact that supersymmetric theories possess the same number of bosonic and fermionic degrees of freedom for a given mass. Furthermore, the first of equations (5.85) implies stringent consequences for the energy of the vacuum state, as will be discussed in Chapter 9.

We recall first the following aspects of the Wess-Zumino model. The Lagrangian density of the free massive Wess-Zumino model is given by (5.10), where A(x) is a real scalar field, B(x) is a real pseudoscalar field, and ψ (x) is a spin-$\frac{1}{2}$ Majorana field. We saw earlier that these fields have the following plane-wave expansions:

$$A(x) = \frac{1}{(2\pi)^{3/2}} \int \frac{d^3p}{(2\omega_p)^{1/2}} \left[a(\vec{p})e^{-ipx} + a^+(\vec{p})e^{ipx}\right] \tag{5.86}$$

with bosonic creation and annihilation operators

$$\left.\begin{aligned} &[a(\vec{p}), a^+(\vec{p}\,')] = \delta(\vec{p} - \vec{p}\,') \\ &[a(\vec{p}), a(\vec{p}\,')] = [a^+(\vec{p}), a^+(\vec{p}\,')] = 0 \end{aligned}\right\} \tag{5.87}$$

and

$$B(x) = \frac{1}{(2\pi)^{3/2}} \int \frac{d^3p}{(2\omega_p)^{1/2}} \left[b(\vec{p})e^{-ipx} + b^+(\vec{p})e^{ipx}\right] \tag{5.88}$$

with

$$\left.\begin{aligned}&[b(\vec{p}), b^+(\vec{p}\,')] = \delta(\vec{p}-\vec{p}\,')\\&[b(\vec{p}), b(\vec{p}\,')] = [b^+(\vec{p}), b^+(\vec{p}\,')] = 0\end{aligned}\right\} \quad (5.89)$$

and

$$\psi(x) = \frac{1}{(2\pi)^{3/2}} \sum_s \int d^3p \left(\frac{m}{\omega_p}\right)^{1/2} \{ d(\vec{p},s) u(\vec{p},s) e^{-ipx} + d^+(\vec{p},s) v(\vec{p},s) e^{ipx} \} \quad (5.90)$$

with

$$\omega_p = p_0 = (\vec{p}^{\,2} + m^2)^{1/2} \quad (5.91)$$

and the anticommutation relations

$$\left.\begin{aligned}&\{d(\vec{p},r), d^+(\vec{p}\,',s)\} = \delta_{rs}\,\delta(\vec{p}-\vec{p}\,')\\&\{d(\vec{p},r), d(\vec{p}\,',s)\} = \{d^+(\vec{p},r), d^+(\vec{p}\,',s)\} = 0\end{aligned}\right\} \quad (5.92)$$

The energy-momentum operator is defined by

$$P_\mu = \int d^3x\, T_{0\mu}\,, \quad P_0 = H \quad (5.93)$$

where $T_{\nu\mu}$ is the canonical energy-momentum tensor defined by the relation

$$T_{\nu\mu} := -\eta_{\nu\mu} \mathcal{L} + \sum_i \frac{\partial \mathcal{L}}{\partial\left(\frac{\partial \phi_i}{\partial x_\nu}\right)} \cdot \frac{\partial \phi_i}{\partial x^\mu} \quad (5.94)$$

where $\mathcal{L}$ is the Lagrangian density (5.10) and the sum is to be taken over the various fields of the theory. Substituting (5.94) into (5.93) we obtain

$$P_\mu = \int d^3x\, T_{0\mu} = \int d^3x \left\{ \sum_i \frac{\partial \mathcal{L}}{\partial\left(\frac{\partial \phi_i}{\partial x_0}\right)} \frac{\partial \phi_i}{\partial x^\mu} - \eta_{0\mu} \mathcal{L} \right\} \quad (5.95)$$

For the Wess-Zumino model (5.95) is

$$P_\mu = \int d^3x \left\{ \frac{\partial \mathcal{L}}{\partial\left(\frac{\partial A(x)}{\partial x_0}\right)} \cdot \frac{\partial A(x)}{\partial x^\mu} + \frac{\partial \mathcal{L}}{\partial\left(\frac{\partial B(x)}{\partial x_0}\right)} \cdot \frac{\partial B(x)}{\partial x^\mu} \right.$$

$$+\frac{\partial \mathcal{L}}{\partial\left(\frac{\partial \psi(x)}{\partial x_0}\right)}\cdot\frac{\partial \psi(x)}{\partial x^\mu}-\eta_{0\mu}\mathcal{L}\Big\} \tag{5.96}$$

Calculating the various terms appearing in (5.96) and taking into account thereby the Lagrangian density (5.10) as well as the plane-wave expansions (5.86), (5.88) and (5. 90) we have

$$\frac{\partial \mathcal{L}}{\partial\left(\frac{\partial A(x)}{\partial x_0}\right)}=\frac{\partial A(x)}{\partial x^0}=\frac{i}{(2\pi)^{3/2}}\int\frac{d^3p}{(2/\omega_p)^{1/2}}\left[-a(\vec{p})e^{-ipx}+a^+(\vec{p})e^{ipx}\right] \tag{5.97}$$

$$\frac{\partial A(x)}{\partial x^\mu}=\frac{i}{(2\pi)^{3/2}}\int d^3p\, p_\mu\left(\frac{1}{2\omega_p}\right)^{1/2}\left[-a(\vec{p})e^{-ipx}+a^+(\vec{p})e^{ipx}\right] \tag{5.98}$$

$$\frac{\partial \mathcal{L}}{\partial\left(\frac{\partial B(x)}{\partial x_0}\right)}=\frac{\partial B(x)}{\partial x^0}=\frac{i}{(2\pi)^{3/2}}\int\frac{d^3p}{(2/\omega_p)^{1/2}}\left[-b(\vec{p})e^{-ipx}+b^+(\vec{p})e^{ipx}\right] \tag{5.99}$$

$$\frac{\partial B(x)}{\partial x^\mu}=\frac{i}{(2\pi)^{3/2}}\int d^3p\, p_\mu\left(\frac{1}{2\omega_p}\right)^{1/2}\left[-b(\vec{p})e^{-ipx}+b^+(\vec{p})e^{ipx}\right] \tag{5.100}$$

$$\frac{\partial \mathcal{L}}{\partial\left(\frac{\partial \psi(x)}{\partial x_0}\right)}=\frac{i}{2}\bar{\psi}\gamma^0=\frac{i}{2}\cdot\frac{1}{(2\pi)^{3/2}}\sum_s\int d^3p\left(\frac{m}{\omega_p}\right)^{1/2}\cdot$$

$$\cdot\left\{d^+(\vec{p},s)\bar{u}(\vec{p},s)\gamma^0 e^{ipx}+d(\vec{p},s)\bar{v}(\vec{p},s)\gamma^0 e^{-ipx}\right\} \tag{5.101}$$

$$\frac{\partial\psi(x)}{\partial x^\mu}=\frac{i}{(2\pi)^{3/2}}\sum_s\int d^3p\left(\frac{m}{\omega_p}\right)^{1/2}p_\mu\cdot$$
$$\cdot\left\{-d(\vec{p},s)u(\vec{p},s)e^{-ipx}+d^+(\vec{p},s)v(\vec{p},s)e^{ipx}\right\}$$

(5.102)

5.6.1 The Hamilton Operator

We now calculate the Hamiltonian of the Wess-Zumino model. This operator is given by

$$H\equiv P_0$$
$$=\int d^3x\left\{\frac{\partial\mathcal{L}}{\partial\left(\frac{\partial A(x)}{\partial x_0}\right)}\cdot\frac{\partial A(x)}{\partial x^0}+\frac{\partial\mathcal{L}}{\partial\left(\frac{\partial B(x)}{\partial x_0}\right)}\cdot\frac{\partial B(x)}{\partial x^0}\right.$$
$$\left.+\frac{\partial\mathcal{L}}{\partial\left(\frac{\partial\psi(x)}{\partial x^0}\right)}\cdot\frac{\partial\psi(x)}{\partial x^0}-\eta_{00}\mathcal{L}\right\}$$
$$\equiv P_0^A+P_0^B+P_0^\psi-\eta_{00}\int\mathcal{L}d^3x \qquad (5.103)$$

Here

$$P_0^A=\int d^3x\frac{\partial\mathcal{L}}{\partial\left(\frac{\partial A(x)}{\partial x_0}\right)}\cdot\frac{\partial A(x)}{\partial x^0}$$
$$=-\frac{1}{(2\pi)^3}\int d^3x\int d^3p\int d^3k\,\frac{1}{2}(\omega_p\omega_k)^{1/2}\cdot$$
$$\cdot\left[-a(\vec{p})e^{-ipx}+a^+(\vec{p})e^{ipx}\right]\cdot$$
$$\cdot\left[-a(\vec{k})e^{-ikx}+a^+(\vec{k})e^{ikx}\right]$$

(using (5.97) and (5.98))

$$=-\frac{1}{(2\pi)^3}\int d^3x\int d^3p\int d^3k\,\frac{1}{2}(\omega_p\omega_k)^{1/2}\cdot$$

$$\cdot\{a(\vec{p})a(\vec{k})e^{-i(p+k)x}+a^+(\vec{p})a^+(\vec{k})e^{i(p+k)x}$$
$$-a(\vec{p})a^+(\vec{k})e^{-i(p-k)x}-a^+(\vec{p})a(\vec{k})e^{i(p-k)x}\}$$
$$=-\int d^3p\int d^3k\,\tfrac{1}{2}(\omega_p\omega_k)^{1/2}\cdot\{a(\vec{p})a(\vec{k})e^{-2i\omega_p t}\cdot$$
$$\cdot\delta(\vec{p}+\vec{k})+a^+(\vec{p})a^+(\vec{k})e^{2i\omega_p t}\delta(\vec{p}+\vec{k})$$
$$-a(\vec{p})a^+(\vec{k})\delta(\vec{p}-\vec{k})-a^+(\vec{p})a(\vec{k})\delta(\vec{p}-\vec{k})\}$$

Thus we obtain

$$P_0^A=-\frac{1}{2}\int d^3p\;\omega_p\{a(\vec{p})a(-\vec{p})e^{-2i\omega_p t}-a(\vec{p})a^+(\vec{p})$$
$$-a^+(\vec{p})a(\vec{p})+a^+(\vec{p})a^+(-\vec{p})e^{2i\omega_p t}\}$$
(5.104)

A similar calculation leads to

$$P_0^B=-\frac{1}{2}\int d^3p\;\omega_p\{b(\vec{p})b(-\vec{p})e^{-2i\omega_p t}-b(\vec{p})b^+(\vec{p})$$
$$-b^+(\vec{p})b(\vec{p})+b^+(\vec{p})b^+(-\vec{p})e^{2i\omega_p t}\}$$
(5.105)

For the evaluation of $P_0\psi$ we require some formulae (see also (5.77) and (5.78)).

<u>Proposition</u>: The following relations can be shown to hold:

i) $\bar{u}(\vec{p},s)\gamma^0 u(\vec{p},r)=\dfrac{p_0}{m}\delta_{rs}$ (5.106a)

ii) $\bar{v}(\vec{p},s)\gamma^0 v(\vec{p},r)=\dfrac{p_0}{m}\delta_{rs}$ (5.106b)

iii) $\bar{u}(\vec{p},s)\gamma^0 v(-\vec{p},r)=0$ (5.106c)

iv) $\bar{v}(\vec{p},s)\gamma^0 u(-\vec{p},r)=0$ (5.106d)

<u>Proof</u>:

i) $\bar{u}(\vec{p},s)\gamma^0 u(\vec{p},r)$

$$= \bar{u}(\vec{p},s)\gamma^0 \Lambda_+(p) u(\vec{p},r)$$ (with (5.59))

$$= \bar{u}(\vec{p},s)\gamma^0 \left(\frac{\not{p}+m}{2m}\right) u(\vec{p},r)$$ (with (5.56))

$$= \bar{u}(\vec{p},s)\gamma^0 \frac{1}{2m}(p_0\gamma^0 - \vec{p}\cdot\vec{\gamma} + m) u(\vec{p},r)$$

$$= \bar{u}(\vec{p},s)\frac{1}{2m}(p_0\gamma^0 + \vec{p}\cdot\vec{\gamma} + m)\gamma_0 u(\vec{p},r)$$

$$= \bar{u}(\vec{p},s)\frac{1}{2m}(2p_0\gamma_0 - \not{p} + m)\gamma_0 u(\vec{p},r)$$

$$= \frac{p_0}{m}\bar{u}(\vec{p},s) u(\vec{p},r) + \bar{u}(\vec{p},s)\Lambda_-(p)\gamma_0 u(\vec{p},r)$$

$$= \frac{p_0}{m}\delta_{rs}$$ (with (5.30))

since $\bar{u}(\vec{p},s)\Lambda_-(p) = 0$.

ii) $\bar{v}(\vec{p},s)\gamma^0 v(\vec{p},r)$

$$= \bar{v}(\vec{p},s)\gamma^0 \Lambda_-(p) v(\vec{p},r)$$ (with (5.59))

$$= \bar{v}(\vec{p},s)\gamma^0 \left(\frac{-\not{p}+m}{2m}\right) v(\vec{p},r)$$

$$= \bar{v}(\vec{p},s)\gamma^0 \frac{1}{2m}(-p_0\gamma^0 + \vec{p}\cdot\vec{\gamma} + m) v(\vec{p},r)$$

$$= v(\vec{p},s)\frac{1}{2m}[-2p_0\gamma_0 + p_0\gamma_0 - \vec{p}\cdot\vec{\gamma} + m]\gamma_0 v(\vec{p},r)$$

$$= -\frac{p_0}{m}\bar{v}(\vec{p},s) v(\vec{p},r) + \bar{v}(\vec{p},s)\left(\frac{\not{p}+m}{2m}\right)\gamma_0 v(\vec{p},s)$$

$$= \frac{p_0}{m}\delta_{rs} + \bar{v}(\vec{p},s)\Lambda_+(p)\gamma_0 v(\vec{p},r)$$

(with (5.30))

$$= \frac{p_0}{m}\delta_{rs}$$

iii) $\bar{u}(\vec{p},r)\gamma_0 v(-\vec{p},r)$

$$= \bar{u}(\vec{p},s)\gamma_0 \Lambda_-(p_0,-\vec{p}) v(-\vec{p},r)$$

(using (5.59))

$$= \bar{u}(\vec{p},s)\gamma_0 \frac{1}{2m}(-p_0\gamma^0 - \vec{p}\cdot\vec{\gamma} + m)v(-\vec{p},r)$$
$$= \bar{u}(\vec{p},s)\frac{1}{2m}(-p_0\gamma_0 + \vec{p}\cdot\vec{\gamma} + m)\gamma_0 v(-\vec{p},r)$$
$$= \bar{u}(\vec{p},s)\Lambda_-(p)\gamma_0 v(-\vec{p},r)$$
$$= 0 \qquad \text{(with (5.63'))}$$

iv)
$$\bar{v}(\vec{p},r)\gamma_0 u(-\vec{p},r)$$
$$= \bar{v}(\vec{p},r)\gamma_0\Lambda_+(p_0,-\vec{p})u(-\vec{p},r) \qquad \text{(with (5.59))}$$
$$= \bar{v}(\vec{p},r)\gamma_0\frac{1}{2m}(p_0\gamma^0 + \vec{p}\cdot\vec{\gamma} + m)u(-\vec{p},r)$$
$$= \bar{v}(\vec{p},r)\frac{1}{2m}(p_0\gamma_0 - \vec{p}\cdot\vec{\gamma} + m)\gamma_0 u(-\vec{p},r)$$
$$= \bar{v}(\vec{p},r)\Lambda_+(p)\gamma_0 u(-\vec{p},r)$$
$$= 0 \qquad \text{(with (5.63'))}$$

Now we have

$$P_0^\psi = \int d^3x \frac{\partial \mathcal{L}}{\partial(\partial_0\psi)}\cdot\frac{\partial\psi(x)}{\partial x^0} = \frac{i}{2}\int d^3x\,\bar{\psi}(x)\gamma^0\partial_0\psi(x)$$
$$= -\frac{1}{2(2\pi)^3}\int d^3x\sum_{s,r}\int d^3p\int d^3k\left(\frac{m}{\omega_k}\right)^{1/2}\left(\frac{m}{\omega_p}\right)^{1/2}\omega_k\,\cdot$$
$$\cdot\left\{d^+(\vec{p},s)\bar{u}(\vec{p},s)\gamma^0 e^{ipx} + d(\vec{p},s)\bar{v}(\vec{p},s)\gamma^0 e^{-ipx}\right\}\cdot$$
$$\cdot\left\{-d(\vec{k},r)u(\vec{k},r)e^{-ikx} + d^+(\vec{k},r)v(\vec{k},r)e^{ikx}\right\}$$
$$= -\frac{1}{2(2\pi)^3}\int d^3x\sum_{s,r}\int d^3p\int d^3k\left(\frac{m}{\omega_p}\right)^{1/2}(m\omega_k)^{1/2}\,\cdot$$
$$\cdot\left\{-d^+(\vec{p},s)d(\vec{k},r)\bar{u}(\vec{p},s)\gamma^0 u(\vec{k},r)e^{i(p-k)x}\right.$$
$$+ d(\vec{p},s)d^+(\vec{k},r)\bar{v}(\vec{p},s)\gamma^0 v(\vec{k},r)e^{-i(p-k)x}$$
$$+ d^+(\vec{p},s)d^+(\vec{k},r)\bar{u}(\vec{p},s)\gamma^0 v(\vec{k},r)e^{i(p+k)x}$$
$$\left.- d(\vec{p},s)d(\vec{k},r)\bar{v}(\vec{p},s)\gamma^0 u(\vec{k},r)e^{-i(p+k)x}\right\}$$

$$= -\frac{1}{2}\sum_{s,r}\int d^3p\int d^3k\, m\left(\frac{\omega_k}{\omega_p}\right)^{\frac{1}{2}} \cdot$$
$$\cdot\{-d^+(\vec{p},s)\,d(\vec{k},r)\,\bar{u}(\vec{p},s)\gamma^0 u(\vec{k},r)\,\delta(\vec{p}-\vec{k})$$
$$+d(\vec{p},s)\,d^+(\vec{k},r)\,\bar{v}(\vec{p},s)\gamma^0 v(\vec{k},r)\,\delta(\vec{p}-\vec{k})$$
$$+d^+(\vec{p},s)\,d^+(\vec{k},r)\,\bar{u}(\vec{p},s)\gamma^0 v(\vec{k},r)\,.$$
$$.e^{2i\omega_p t}\,\delta(\vec{p}+\vec{k})$$
$$-d(\vec{p},s)\,d(\vec{k},r)\,\bar{v}(\vec{p},s)\gamma^0 u(\vec{k},r).$$
$$.e^{-2i\omega_p t}\,\delta(\vec{p}+\vec{k})\}$$
$$= -\frac{1}{2}\sum_{s,r}\int d^3p\, m\,\{-d^+(\vec{p},s)\,d(\vec{p},r)\bar{u}(\vec{p},s)\gamma^0 u(\vec{p},r)$$
$$+d(\vec{p},s)\,d^+(\vec{p},r)\bar{v}(\vec{p},s)\,\gamma^0 v(\vec{p},r)$$
$$+d^+(\vec{p},s)\,d^+(-\vec{p},r)\,\bar{u}(\vec{p},s)\,\gamma^0 v(-\vec{p},r)\,e^{2i\omega_p t}$$
$$-d(\vec{p},s)\,d(-\vec{p},r)\,\bar{v}(\vec{p},s)\gamma^0 u(-\vec{p},r)\,e^{-2i\omega_p t}\}$$
$$= -\frac{1}{2}\sum_r\int d^3p\, p_0\,\{-d^+(\vec{p},r)d(\vec{p},r)+d(\vec{p},r)\,d^+(\vec{p},r)\}$$

(using (5.106))

$$= \sum_r\int d^3p\, p_0\{d^+(\vec{p},r)\,d(\vec{p},r)-\frac{1}{2}\delta(\vec{0})\}$$

(with (5.92))

Hence

$$P_0^\Psi = \sum_r\int d^3p\, p_0\left\{d^+(\vec{p},r)d(\vec{p},r)-\frac{1}{2}\delta(\vec{0})\right\} \qquad (5.107)$$

Next we have to evaluate the Lagrangian

$$L = \int d^3x\,\mathcal{L}$$
$$= \int d^3x\,\{\frac{1}{2}(\partial_\mu A(x))(\partial^\mu A(x)) - \frac{1}{2}m^2 A^2(x)$$
$$+\frac{1}{2}(\partial_\mu B(x))(\partial^\mu B(x))-\frac{1}{2}m^2 B^2(x)$$
$$+\frac{1}{2}\bar{\Psi}(x)(i\partial\!\!\!/ - m)\Psi(x)\}$$

The contribution of the spinor field $\psi(x)$ vanishes because of (5.5) (for g = 0). Inserting the Fourier expansions of the scalar field A(x) and the pseudoscalar field B(x) we obtain

$$L = \int d^3x \Big\{ \frac{1}{2}\Big[\frac{i}{(2\pi)^{3/2}} \int d^3p \frac{p_\mu}{(2\omega_p)^{1/2}} \big(-a(\vec{p})e^{-ipx} + a^+(\vec{p})e^{ipx}\big) \cdot \frac{i}{(2\pi)^{3/2}} \int \frac{d^3k}{(2\omega_k)^{1/2}} k^\mu \cdot$$
$$\cdot \big(-a(\vec{k})e^{-ikx} + a^+(\vec{k})e^{ikx}\big)\Big]$$
$$- \frac{m^2}{2(2\pi)^3} \int \frac{d^3p}{(2\omega_p)^{1/2}} \big(a(\vec{p})e^{-ipx} + a^+(\vec{p})e^{ipx}\big) \cdot$$
$$\cdot \int \frac{d^3k}{(2\omega_k)^{1/2}} \big(a(\vec{k})e^{-ikx} + a^+(\vec{k})e^{ikx}\big)$$
$$- \frac{1}{2(2\pi)^3} \int d^3p \frac{p_\mu}{(2\omega_p)^{1/2}} \big(-b(\vec{p})e^{-ipx} + b^+(\vec{p})e^{ipx}\big) \cdot$$
$$\cdot \int \frac{d^3k}{(2\omega_k)^{1/2}} k^\mu \big(-b(\vec{k})e^{-ikx} + b^+(\vec{k})e^{ikx}\big)$$
$$- \frac{m^2}{2(2\pi)^3} \int \frac{d^3p}{(2\omega_p)^{1/2}} \big(b(\vec{p})e^{-ipx} + b^+(\vec{p})e^{ipx}\big) \cdot$$
$$\cdot \int \frac{d^3k}{(2\omega_k)^{1/2}} \big(b(\vec{k})e^{-ikx} + b^+(\vec{k})e^{ikx}\big)\Big\}$$
$$= \int d^3x \Big\{ -\frac{1}{2(2\pi)^3} \int d^3p \int d^3k \frac{p_\mu k^\mu}{2(\omega_p\omega_k)^{1/2}} \cdot$$
$$\cdot \big[a(\vec{p})a(\vec{k})e^{-i(p+k)x} + a^+(\vec{p})a^+(\vec{k})e^{i(p+k)x}$$
$$- a(\vec{p})a^+(\vec{k})e^{-i(p-k)x} - a^+(\vec{p})a(\vec{k})e^{i(p-k)x}\big]$$
$$- \frac{m^2}{2(2\pi)^3} \int d^3p \int \frac{d^3k}{2(\omega_k\omega_p)^{1/2}} \cdot \big[a(\vec{p})a(\vec{k})e^{-i(p+k)x}$$
$$+ a^+(\vec{p})a^+(\vec{k})e^{i(p+k)x} + a(\vec{p})a^+(\vec{k})e^{-i(p-k)x}$$
$$+ a^+(\vec{p})a(\vec{k})e^{i(p-k)x}\big]$$

$$- \frac{1}{2(2\pi)^3} \int d^3p \int d^3k \, \frac{p_\mu k^\mu}{2(\omega_p \omega_k)^{1/2}}$$

$$\cdot \left[b(\vec{p}) b(\vec{k}) e^{-i(p+k)x} + b^+(\vec{p}) b^+(\vec{k}) e^{i(p+k)x} \right.$$

$$\left. - b(\vec{p}) b^+(\vec{k}) e^{-i(p-k)x} - b^+(\vec{p}) b(\vec{k}) e^{i(p-k)x} \right]$$

$$- \frac{m^2}{2(2\pi)^3} \int d^3p \int d^3k \, \frac{1}{2(\omega_p \omega_k)^{1/2}} \cdot$$

$$\cdot \left[b(\vec{p}) b(\vec{k}) e^{-i(p+k)x} + b^+(\vec{p}) b^+(\vec{k}) e^{i(p+k)x} \right.$$

$$\left. + b(\vec{p}) b(\vec{k}) e^{-i(p-k)x} + b^+(\vec{p}) b(\vec{k}) e^{i(p-k)x} \right] \Big\}$$

$$= - \frac{1}{2} \int d^3p \int d^3k \, \frac{p_\mu k^\mu}{2(\omega_p \omega_k)^{1/2}} \cdot$$

$$\cdot \left[a(\vec{p}) a(\vec{k}) \delta(\vec{p} + \vec{k}) e^{-2i\omega_p t} \right.$$

$$+ a^+(\vec{p}) a^+(\vec{k}) \delta(\vec{p} + \vec{k}) e^{2i\omega_p t}$$

$$- a(\vec{p}) a^+(\vec{k}) \delta(\vec{p} - \vec{k})$$

$$\left. - a^+(\vec{k}) a(\vec{k}) \delta(\vec{p} - \vec{k}) \right]$$

$$- \frac{m^2}{2} \int d^3p \int d^3k \, \frac{1}{2(\omega_p \omega_k)^{1/2}} \cdot$$

$$\cdot \left[a(\vec{p}) a(\vec{k}) \delta(\vec{p} + \vec{k}) e^{-2i\omega_p t} \right.$$

$$+ a^+(\vec{p}) a^+(\vec{k}) \delta(\vec{p} + \vec{k}) e^{2i\omega_p t}$$

$$+ a(\vec{p}) a^+(\vec{k}) \delta(\vec{p} - \vec{k})$$

$$\left. + a^+(\vec{p}) a(\vec{k}) \delta(\vec{p} - \vec{k}) \right]$$

$$- \frac{1}{2} \int d^3p \int d^3k \, \frac{p_\mu k^\mu}{2(\omega_k \omega_p)^{1/2}} \cdot$$

$$\cdot \left[b(\vec{p}) b(\vec{k}) \delta(\vec{p} + \vec{k}) e^{-2i\omega_p t} \right.$$

$$+ b^+(\vec{p}) b^+(\vec{k}) \delta(\vec{p} + \vec{k}) e^{2i\omega_p t}$$

$$- b(\vec{p})\, b^+(\vec{k})\, \delta(\vec{p}-\vec{k})$$
$$- b^+(\vec{p})\, b(\vec{k})\, \delta(\vec{p}-\vec{k})]$$
$$- \frac{m^2}{2}\int d^3p \int d^3k \, \frac{1}{2(\omega_p \omega_k)^{1/2}} \cdot$$
$$\cdot [\, b(\vec{p})\, b(\vec{k})\, \delta(\vec{p}+\vec{k})\, e^{-2i\omega_p t}$$
$$+ b^+(\vec{p})\, b^+(\vec{k})\, \delta(\vec{p}+\vec{k})\, e^{2i\omega_p t}$$
$$+ b(\vec{p})\, b^+(\vec{k})\, \delta(\vec{p}-\vec{k})$$
$$+ b^+(\vec{p})\, b(\vec{k})\, \delta(\vec{p}-\vec{k})]$$

Hence the Lagrangian becomes

$$L = -\frac{1}{2}\int \frac{d^3p}{2\omega_p}(p_0^2 + \vec{p}^{\,2} + m^2) \cdot$$
$$\cdot [a(\vec{p})\, a(-\vec{p})\, e^{-2i\omega_p t} + a^+(\vec{p})\, a^+(-\vec{p})\, e^{2i\omega_p t}]$$
$$-\frac{1}{2}\int \frac{d^3p}{2\omega_p}(p_0^2 + \vec{p}^{\,2} + m^2) \cdot$$
$$\cdot [b(\vec{p})\, b(-\vec{p})\, e^{-2i\omega_p t} + b^+(\vec{p})\, b^+(-\vec{p})\, e^{2i\omega_p t}] \tag{5.108}$$

All other terms vanish as a result of the relation $p^2 = m^2$. Collecting all contributions we obtain the following expression for the Hamilton operator

$$H = P_0^A + P_0^B + P_0^\psi - \eta_{00} L$$
$$= \int d^3p \, \frac{\omega_p}{2} \{a(\vec{p})\, a^+(\vec{p}) + a^+(\vec{p})\, a(\vec{p})$$
$$- a(\vec{p})\, a(-\vec{p})\, e^{-2i\omega_p t} - a^+(\vec{p})\, a^+(-\vec{p})\, e^{2i\omega_p t}\}$$
$$+ \int d^3p \, \frac{\omega_p}{2} \{b(\vec{p})\, b^+(\vec{p}) + b^+(\vec{p})\, b(\vec{p})$$
$$- b(\vec{p})\, b(-\vec{p})\, e^{-2i\omega_p t} - b^+(\vec{p})\, b^+(-\vec{p})\, e^{2i\omega_p t}\}$$

$$+\sum_r \int d^3p\ \omega_p \{d^+(\vec{p},r)d(\vec{p},r) - \tfrac{1}{2}\delta(\vec{0})\}$$

$$+\frac{1}{2}\int \frac{d^3p}{2\omega_p}(p_0^2+\vec{p}^{\,2}+m^2)\cdot$$

$$\cdot\{a(\vec{p})a(-\vec{p})e^{-2i\omega_p t}+a^+(\vec{p})a^+(-\vec{p})e^{2i\omega_p t}\}$$

$$+\frac{1}{2}\int \frac{d^3p}{2\omega_p}(p_0^2+\vec{p}^{\,2}+m^2)\cdot$$

$$\cdot\{b(\vec{p})b(-\vec{p})e^{-2i\omega_p t}+b^+(\vec{p})b^+(-\vec{p})e^{2i\omega_p t}\}$$

$$= \int d^3p\ \omega_p \{a^+(\vec{p})a(\vec{p}) + \tfrac{1}{2}\delta(\vec{0}) + b^+(\vec{p})b(\vec{p})$$

$$+\tfrac{1}{2}\delta(\vec{0}) + \sum_r \left(d^+(\vec{p},r)d(\vec{p},r) - \tfrac{1}{2}\delta(\vec{0})\right)\}$$

$$+\frac{1}{2}\int d^3p\left[-\omega_p + \frac{1}{2\omega_p}(\omega_p^2+\vec{p}^{\,2}+m^2)\right]\cdot$$

$$\cdot\left[a(\vec{p})a(-\vec{p})e^{-2i\omega_p t} + a^+(\vec{p})a^+(-\vec{p})e^{2i\omega_p t}\right]$$

$$+\frac{1}{2}\int d^3p\left[-\omega_p + \frac{1}{2\omega_p}(\omega_p^2+\vec{p}^{\,2}+m^2)\right]\cdot$$

$$\cdot\left[b(\vec{p})b(-\vec{p})e^{-2i\omega_p t} + b^+(\vec{p})b^+(-\vec{p})e^{2i\omega_p t}\right]$$

Now

$$-\omega_p + \frac{1}{2\omega_p}(\omega_p^2+\vec{p}^{\,2}+m^2)$$

$$= \frac{1}{2\omega_p}(-\omega_p^2+\vec{p}^{\,2}+m^2) = \frac{1}{2\omega_p}(-p^2+m^2) = 0$$

Hence the Hamilton operator is

$$H = \int d^3p\ \omega_p \left[a^+(\vec{p})a(\vec{p}) + \tfrac{1}{2}\delta(\vec{0}) + b^+(\vec{p})b(\vec{p})\right.$$

$$\left. + \tfrac{1}{2}\delta(\vec{0}) + \sum_{r=1,2}\{d^+(\vec{p},r)d(\vec{p},r) - \tfrac{1}{2}\delta(\vec{0})\}\right.$$

Our result is therefore

$$H = \int d^3p \, \omega_p \left[a^+(\vec{p}) a(\vec{p}) + b^+(\vec{p}) b(\vec{p}) + \sum_r d^+(\vec{p},r) d(\vec{p},r) \right] \tag{5.109}$$

From (5.109) we see that the Hamilton operator is normal ordered, i.e. all terms have annihilation operators to the right of creation operators. As can be seen from the derivation of the expression, this is due to the fact that the zero-point energies of the bosonic fields cancel exactly the zero-point energies of the fermionic degrees of freedom. The crucial point is, that in the Wess-Zumino model we have an equal number of bosonic and fermionic degrees of freedom such that the cancellation of the zero-point energies can occur.

5.6.2 The Three-Momentum P_i

The three-momentum P_i of the Wess-Zumino model is defined by

$$P_i = \int d^3x \left\{ \frac{\partial \mathcal{L}}{\partial\left(\frac{\partial A(x)}{\partial x_0}\right)} \cdot \frac{\partial A(x)}{\partial x^i} + \frac{\partial \mathcal{L}}{\partial\left(\frac{\partial B(x)}{\partial x_0}\right)} \cdot \frac{\partial B(x)}{\partial x^i} + \frac{\partial \mathcal{L}}{\partial\left(\frac{\partial \psi(x)}{\partial x_0}\right)} \cdot \frac{\partial \psi(x)}{\partial x^i} \right\} \tag{5.110}$$

Evaluating this expression we have

$$P_i = \int d^3x \left\{ \frac{\partial A(x)}{\partial x_0} \cdot \frac{\partial A(x)}{\partial x^i} + \frac{\partial B(x)}{\partial x_0} \cdot \frac{\partial B(x)}{\partial x^i} + \frac{i}{2} \bar{\psi}(x) \gamma^0 \frac{\partial \psi(x)}{\partial x^i} \right\}$$

$$= \int d^3x \left\{ \frac{i}{(2\pi)^{3/2}} \int d^3p \left(\frac{\omega_p}{2}\right)^{\frac{1}{2}} \left[-a(\vec{p}) e^{-ipx} + a^+(\vec{p}) e^{ipx} \right] \right.$$
$$\left. \cdot \frac{(-i)}{(2\pi)^{3/2}} \int \frac{d^3k\, k_i}{(2\omega_k)^{1/2}} \left[a(\vec{k}) e^{-ikx} - a^+(\vec{k}) e^{ikx} \right] \right\}$$
$$+ \int d^3x \left\{ \frac{i}{(2\pi)^{3/2}} \int d^3p \left(\frac{\omega_p}{2}\right)^{\frac{1}{2}} \left[-b(\vec{p}) e^{-ipx} + b^+(\vec{p}) e^{ipx} \right] \right.$$
$$\left. \cdot \frac{(-i)}{(2\pi)^{3/2}} \int \frac{d^3k\, k_i}{(2\omega_k)^{1/2}} \left[b(\vec{k}) e^{-ikx} - b^+(\vec{k}) e^{ikx} \right] \right\}$$
$$+ \int d^3x \left\{ \frac{i}{2(2\pi)^{3/2}} \sum_r \int d^3p \left(\frac{m}{\omega_p}\right)^{\frac{1}{2}} \left[d^+(\vec{p},r) \overline{u}(\vec{p},r) \gamma^0 e^{ipx} \right. \right.$$
$$\left. + d(\vec{p},r) \overline{v}(\vec{p},r) \gamma^0 e^{-ipx} \right].$$
$$\cdot \frac{(-i)}{(2\pi)^{3/2}} \sum_s \int d^3k \left(\frac{m}{\omega_k}\right)^{\frac{1}{2}} k_i \left[d(\vec{k},s) u(\vec{k},s) e^{-ikx} \right.$$
$$\left. \left. - d^+(\vec{k},s) v(\vec{k},s) e^{ikx} \right] \right\}$$

(using (5.101) and (5.91))

$$= \frac{1}{(2\pi)^3} \int d^3x \int d^3p \int d^3k \left(\frac{\omega_p}{2}\right)^{\frac{1}{2}} \frac{k_i}{(2\omega_k)^{1/2}} \cdot$$
$$\cdot \left[-a(\vec{p}) a(\vec{k}) e^{-i(p+k)x} + a(\vec{p}) a^+(\vec{k}) e^{-i(p-k)x} \right.$$
$$+ a^+(\vec{p}) a(\vec{k}) e^{i(p-k)x} - a^+(\vec{p}) a^+(\vec{k}) e^{i(p+k)x}$$
$$- b(\vec{p}) b(\vec{k}) e^{-i(p+k)x} + b(\vec{p}) b^+(\vec{k}) e^{-i(p-k)x}$$
$$\left. + b^+(\vec{p}) b(\vec{k}) e^{i(p-k)x} - b^+(\vec{p}) b^+(\vec{k}) e^{i(p+k)x} \right]$$
$$+ \frac{1}{2(2\pi)^3} \int d^3x \sum_{r,s} \int d^3p \int d^3k \frac{m}{(\omega_p \omega_k)^{1/2}} k_i \cdot$$
$$\cdot \left[d^+(\vec{p},r) d(\vec{k},s) \overline{u}(\vec{p},r) \gamma^0 u(\vec{k},s) e^{i(p-k)x} \right.$$
$$- d(\vec{p},r) d^+(\vec{k},s) \overline{v}(\vec{p},r) \gamma^0 v(\vec{k},s) e^{-i(p-k)x}$$
$$+ d^+(\vec{p},r) d^+(\vec{k},s) \overline{u}(\vec{p},r) \gamma^0 v(\vec{k},s) e^{i(p+k)x}$$

$$+ d(\vec{p},r)\, d(\vec{k},s)\, \bar{v}(\vec{p},r)\gamma^0 u(\vec{k},s) e^{-i(p+k)x}]$$

$$= \int d^3p \int d^3k \left(\frac{\omega_p}{2}\right)^{\frac{1}{2}} \frac{k_i}{(2\omega_k)^{1/2}} \cdot$$

$$\cdot [-a(\vec{p})a(\vec{k})\, \delta(\vec{p}+\vec{k}) e^{-2i\omega_p t}$$

$$+ a(\vec{p})a^+(\vec{k})\, \delta(\vec{p}-\vec{k}) + a^+(\vec{p})a(\vec{k})\, \delta(\vec{p}-\vec{k})$$

$$- a^+(\vec{p})a^+(\vec{k})\, \delta(\vec{p}+\vec{k}) e^{2i\omega_p t}$$

$$- b(\vec{p})b(\vec{k})\, \delta(\vec{p}+\vec{k}) e^{-2i\omega_p t}$$

$$+ b(\vec{p})b^+(\vec{k})\, \delta(\vec{p}-\vec{k}) + b^+(\vec{p})b(\vec{k})\, \delta(\vec{p}-\vec{k})$$

$$- b^+(\vec{p})b^+(\vec{k})\, \delta(\vec{p}+\vec{k}) e^{2i\omega_p t}]$$

$$+ \frac{1}{2}\sum_{r,s} \int d^3p \int d^3k \frac{m}{(\omega_p\omega_k)^{1/2}} k_i \cdot$$

$$\cdot [d^+(\vec{p},r)d(\vec{k},s)\, \bar{u}(\vec{p},r)\gamma^0 u(\vec{k},s)\, \delta(\vec{p}-\vec{k})$$

$$- d(\vec{p},r)d^+(\vec{k},s)\, \bar{v}(\vec{p},r)\gamma^0 v(\vec{k},s)\, \delta(\vec{p}-\vec{k})$$

$$- d^+(\vec{p},r)d^+(\vec{k},s)\, \bar{u}(\vec{p},r)\gamma^0 v(\vec{k},s)\, \delta(\vec{p}+\vec{k}) \cdot e^{2i\omega_p t}$$

$$+ d(\vec{p},r)d(\vec{k},s)\, \bar{v}(\vec{p},r)\gamma^0 u(\vec{k},s)\, \delta(\vec{p}+\vec{k}) \cdot e^{-2i\omega_p t}]$$

$$= \int \tfrac{1}{2} d^3p\, p_i [a(\vec{p})a^+(\vec{p}) + a^+(\vec{p})a(\vec{p})$$

$$+ b(\vec{p})b^+(\vec{p}) + b^+(\vec{p})b(\vec{p})$$

$$+ a(\vec{p})a(-\vec{p}) e^{-2i\omega_p t}$$

$$+ a^+(\vec{p})a^+(-\vec{p}) e^{2i\omega_p t}$$

$$+ b(\vec{p})b(-\vec{p}) e^{-2i\omega_p t}$$

$$+ b^+(\vec{p})b^+(-\vec{p}) e^{2i\omega_p t}]$$

$$+\frac{1}{2}\sum_{r,s}\int d^3p\,\frac{m}{\omega_p}p_i\,[d^+(\vec{p},r)d(\vec{p},s)\bar{u}(\vec{p},r)\gamma^0 u(\vec{p},s)$$
$$-d(\vec{p},r)d^+(\vec{p},s)\bar{v}(\vec{p},r)\gamma^0 v(\vec{p},s)]$$
$$-\frac{1}{2}\sum_{r,s}\int d^3p\,\frac{m}{\omega_p}p_i\,[-d^+(\vec{p},r)d^+(-\vec{p},s)\cdot$$
$$\cdot\,\bar{u}(\vec{p},r)\gamma^0 v(-\vec{p},s)e^{2i\omega_p t}$$
$$+d(\vec{p},r)d(-\vec{p},s)\bar{v}(\vec{p},r)\gamma^0 u(-\vec{p},s)e^{-2i\omega_p t}]$$

We can see that the time-dependent terms all vanish since

$$\int_{-\infty}^{\infty} p_i\cdot a(\vec{p})a(-\vec{p})e^{-2i\omega_p t}d^3p$$
$$=-\int_{-\infty}^{\infty} p_i\cdot a(-\vec{p})a(\vec{p})e^{-2i\omega_p t}d^3p$$

(replacing $\vec{p}$ by $-\vec{p}$)

$$=-\int_{-\infty}^{\infty} p_i\cdot a(\vec{p})a(-\vec{p})e^{-2i\omega_p t}d^3p$$

(using (5.87))

Bringing both expressions on one side of the equation this is

$$2\int p_i\cdot a(\vec{p})a(-\vec{p})e^{-2i\omega_p t}d^3p=0$$

Furthermore, using equations (6.106a - d) we obtain

$$P_i=\int d^3p\,\frac{1}{2}p_i\,[a(\vec{p})a^+(\vec{p})+a^+(\vec{p})a(\vec{p})$$
$$+b(\vec{p})b^+(\vec{p})+b^+(\vec{p})b(\vec{p})]$$
$$+\frac{1}{2}\sum_r\int d^3p\,p_i\,[d^+(\vec{p},r)d(\vec{p},r)$$
$$-d(\vec{p},r)d^+(\vec{p},r)]$$

Using (5.87), (5.89) and (5.92) we obtain

$$P_i = \int d^3p\, p_i \left[a^+(\vec{p})\, a(\vec{p}) + \frac{1}{2}\delta(\vec{0}) + b^+(\vec{p})\, b(\vec{p}) + \frac{1}{2}\delta(\vec{0}) + \sum_r \left(d^+(\vec{p},r)\, d(\vec{p},r) - \frac{1}{2}\delta(\vec{0}) \right) \right]$$

i.e.

$$P_i = \int d^3p\, p_i \left[a^+(\vec{p})\, a(\vec{p}) + b^+(\vec{p})\, b(\vec{p}) + \sum_r d^+(\vec{p},r)\, d(\vec{p},r) \right] \tag{5.111}$$

As in the case of the Hamilton operator, the three-momentum operator of the Wess-Zumino model has the interesting property to be normal ordered, i.e.

$$P_i = :P_i: \tag{5.112}$$

Combining (5.109) and (5.111) we obtain the four-momentum operator

$$P_\mu = \int d^3p\, p_\mu \left[a^+(\vec{p})\, a(\vec{p}) + b^+(\vec{p})\, b(\vec{p}) + \sum_r d^+(\vec{p},r)\, d(\vec{p},r) \right] \tag{5.113}$$

This is the expression (5.82) referred to earlier.

5.7 Generators of Infinitesimal Supersymmetry Transformations

We now demonstrate that the spinor charges Q_a given by (5.47) (with $\lambda = 2$ as in (5.83)), generate the supersymmetry transformation (5.11) of the fields A(x), B(x) and ψ (x) of the massive free Wess-Zumino model.

We recall that in classical mechanics[c] one writes the generating function F of a canonical transformation

$$F = F_{id} + \varepsilon G$$

where F_{id} is the generating function of the identity transformation in phase space, G is the generator of the infinitesimal transformation and ε the appropriate infinitesimal parameter. It is then shown, that an arbitrary function u of the canonical variables undergoes an infinitesimal change δu under the transformation which is given by

$$\delta u = \varepsilon [u, G]$$

where in classical mechanics the bracket denotes the Poisson bracket. We now demonstrate that in the present case of the Wess-Zumino model the corresponding relations for the fields are given by the operator relations

$$\delta A(x) = -i[\bar{\varepsilon} Q, A(x)] = -i\bar{\varepsilon}_a [Q_a, A(x)] \tag{5.114a}$$

$$\delta B(x) = -i[\bar{\varepsilon} Q, B(x)] = -i\bar{\varepsilon}_a [Q_a, B(x)] \tag{5.114b}$$

$$\delta \psi(x) = -i[\bar{\varepsilon} Q, \psi(x)] = -i\bar{\varepsilon}_a \{Q_a, \psi(x)\} \tag{5.114c}$$

where $\bar{\varepsilon}$ is an infinitesimal, x independent Grassmann parameter and $\bar{\varepsilon} = \varepsilon^+ \gamma_0$. The variations

$$\delta A(x), \ \delta B(x) \ \text{and} \ \delta\psi(x)$$

are given by equations (5.11a) to (5.11d). We prove the relations (5.114) by inserting into the commutators the Fourier expansions obtained earlier for the relevant quantities.

c See e.g. H. Goldstein[48]

Proof of (5.114a): Using (5.47) and (5.45) we have

$$-i[\bar{\varepsilon}Q, A(x)] = -i\bar{\varepsilon}_a[Q_a, A(x)]$$

$$= (2m)^{\frac{1}{2}}\bar{\varepsilon}_a \sum_s \int d^3p\,(2\pi)^{-3/2}\int d^3k\,(2\omega_k)^{-1/2}\cdot$$

$$\cdot[C_{ab}(\vec{p})d^+(\vec{p},s)v_b(\vec{p},s)$$

$$- D_{ab}(\vec{p})d(\vec{p},s)u_b(\vec{p},s),$$

$$a(\vec{k})e^{-ikx} + a^+(\vec{k})e^{ikx}]$$

$$= \bar{\varepsilon}_a \sum_s (2\pi)^{-3/2}\int d^3p\int d^3k\,(m/\omega_k)^{1/2}\cdot$$

$$\cdot\{[C_{ab}(\vec{p}),a(\vec{k})]d^+(\vec{p},s)v_b(\vec{p},s)e^{-ikx}$$

$$+[C_{ab}(\vec{p}),a^+(\vec{k})]d^+(\vec{p},s)v_b(\vec{p},s)e^{ikx}$$

$$-[D_{ab}(\vec{p}),a(\vec{k})]d(\vec{p},s)u_b(\vec{p},s)e^{-ikx}$$

$$-[D_{ab}(\vec{p}),a^+(\vec{k})]d(\vec{p},s)u_b(\vec{p},s)e^{ikx}\}$$

Now, using (5.48) and (5.87) we obtain

$$[C_{ab}(\vec{p}), a(\vec{k})]$$

$$= [a(\vec{p})\delta_{ab} - i\gamma^5_{ab}b(\vec{p}), a(\vec{k})]$$

$$= [a(\vec{p}), a(\vec{k})]\delta_{ab} - i\gamma^5_{ab}[b(\vec{p}), a(\vec{k})]$$

$$= 0 \tag{5.115a}$$

$$[C_{ab}(\vec{p}), a^+(\vec{k})]$$

$$= [a(\vec{p})\delta_{ab} - i\gamma^5_{ab}b(\vec{p}), a^+(\vec{k})]$$

$$= [a(\vec{p}), a(\vec{k})]\delta_{ab} - i\gamma^5_{ab}[b(\vec{p}), a^+(\vec{k})]$$

$$= \delta(\vec{p}-\vec{k})\delta_{ab} \tag{5.115b}$$

$$[D_{ab}(\vec{p}), a(\vec{k})]$$
$$= [a^+(\vec{p})\delta_{ab} - i\gamma^5_{ab} b^+(\vec{p}), a(\vec{k})]$$
$$= [a^+(\vec{p}), a(\vec{k})]\delta_{ab} - i\gamma^5_{ab}[b^+(\vec{p}), a(\vec{k})]$$
$$= -\delta(\vec{p}-\vec{k})\delta_{ab} \qquad (5.115c)$$

$$[D_{ab}(\vec{p}), a^+(\vec{k})]$$
$$= [a^+(\vec{p})\delta_{ab} - i\gamma^5_{ab} b^+(\vec{p}), a^+(\vec{k})]$$
$$= [a^+(\vec{p}), a^+(\vec{k})]\delta_{ab} - i\gamma^5_{ab}[b^+(\vec{p}), a^+(\vec{k})]$$
$$= 0 \qquad (5.115d)$$

We obtain therefore

$$-i[\bar{\varepsilon}Q, A(x)]$$
$$= \bar{\varepsilon}_a \sum_s (2\pi)^{-3/2} \int d^3p \int d^3k \left(\frac{m}{\omega_k}\right)^{\frac{1}{2}} \cdot$$
$$\cdot \{\delta(\vec{p}-\vec{k})\delta_{ab}\, d^+(\vec{p},s)\, v_b(\vec{p},s)\, e^{ikx}$$
$$+ \delta(\vec{p}-\vec{k})\delta_{ab}\, d(\vec{p},s)\, u_b(\vec{p},s)\, e^{-ikx}\}$$
$$= \bar{\varepsilon}_a \sum_s (2\pi)^{-3/2} \int d^3p \left(\frac{m}{\omega_p}\right)^{\frac{1}{2}} \cdot$$
$$\cdot \{d^+(\vec{p},s)\, v_a(\vec{p},s)\, e^{ipx}$$
$$+ d(\vec{p},s)\, u_a(\vec{p},s)\, e^{-ipx}\}$$
$$= \bar{\varepsilon}_a \psi_a \qquad \text{(with (5.90))}$$
$$= \delta A(x) \qquad \text{(with (5.11a))}$$

which had to be shown.

Proof of (5.114b): Proceeding as before

$$-i[\bar{\varepsilon}Q_0, B(x)] = -i\bar{\varepsilon}_a[Q_a, B(x)]$$

$$= (2m)^{1/2}\bar{\varepsilon}_a\sum_s\int d^3p\,(2\pi)^{-\frac{3}{2}}\int d^3k\,(2\omega_k)^{-\frac{1}{2}}\cdot$$

$$[C_{ab}(\vec{p})\,d^+(\vec{p},s)\,v_b(\vec{p},s)$$

$$- D_{ab}(\vec{p})\,d(\vec{p},s)\,u_b(\vec{p},s),$$

$$b(\vec{k})e^{-ikx} + b^+(\vec{k})e^{ikx}]$$

$$= \bar{\varepsilon}_a(2\pi)^{-\frac{3}{2}}\sum_s\int d^3p\int d^3k\left(\frac{m}{\omega_k}\right)^{\frac{1}{2}}\cdot$$

$$\cdot\{[C_{ab}(\vec{p}), b(\vec{k})]\,d^+(\vec{p},s)\,v_b(\vec{p},s)e^{-ikx}$$

$$+ [C_{ab}(\vec{p}), b^+(\vec{k})]\,d^+(\vec{p},s)\,v_b(\vec{p},s)e^{ikx}$$

$$- [D_{ab}(\vec{p}), b(\vec{k})]\,d(\vec{p},s)\,u_b(\vec{p},s)e^{-ikx}$$

$$- [D_{ab}(\vec{p}), b^+(\vec{k})]\,d(\vec{p},s)\,u_b(\vec{p},s)e^{ikx}\}$$

$$= \bar{\varepsilon}_a(2\pi)^{-\frac{3}{2}}\sum_s\int d^3p\int d^3k\left(\frac{m}{\omega_k}\right)^{\frac{1}{2}}\cdot$$

$$\cdot\{-i\gamma^5_{ab}\,\delta(\vec{p}-\vec{k})\,d^+(\vec{p},s)\,v_b(\vec{p},s)e^{ikx}$$

$$-i\gamma^5_{ab}\,\delta(\vec{p}-\vec{k})\,d(\vec{p},s)\,u_b(\vec{p},s)e^{-ikx}\}$$

where we used the following commutation relations which may be derived in a similar way as (5.113)

$$[C_{ab}(\vec{p}), b(\vec{k})] = [D_{ab}(\vec{p}), b^+(\vec{k})] = 0$$

$$[C_{ab}(\vec{p}), b^+(\vec{k})] = -[D_{ab}(\vec{p}), b(\vec{k})]$$

$$= -i\gamma^5_{ab}\,\delta(\vec{p}-\vec{k})$$

Hence

$$-i[\bar{\varepsilon}Q, B(x)]$$

$$= -i\bar{\varepsilon}_a \gamma^5_{ab} \; (2\pi)^{-\frac{3}{2}} \sum_s \int d^3p \left(\frac{m}{\omega_p}\right)^{\frac{1}{2}} \cdot$$

$$\cdot \{ d(\vec{p},s) u_b(\vec{p},s) e^{-ipx} + d^+(\vec{p},s) v_b(\vec{p},s) e^{ipx} \}$$

$$= -i\bar{\varepsilon}\gamma^5 \psi(x) \quad \text{(with (5.90))}$$

$$= \delta B(x) \quad \text{((with (5.11b)}$$

This proves the relation (5.114b). Before we can prove the last relation (5.114c) we need some more formulae.

Proposition: The following relations hold

$$\sum_r u_a(\vec{p},r) v_b(\vec{p},r) = \left(\frac{\not{p}+m}{2m} C\right)_{ab} \qquad (5.116a)$$

$$\sum_r v_a(\vec{p},r) u_b(\vec{p},r) = \left(\frac{\not{p}-m}{2m} C\right)_{ab} \qquad (5.116b)$$

where C is the charge conjugation matrix.

Proof: Consider

$$u_a(\vec{p},r) v_b(\vec{p},r)$$

$$= v^C_a(\vec{p},r) v_b(\vec{p},r) \quad \text{(using (5.40))}$$

$$= C_{am} \bar{v}^T_m(\vec{p},r) v_b(\vec{p},r) \quad \text{(using (5.38))}$$

$$= C_{am} v_b(\vec{p},r) \bar{v}^T_m(\vec{p},r)$$

$$= C_{am} (\Lambda_-)_{bm} \quad \text{(using (5.61))}$$

$$= (\Lambda_- C^T)_{ba}$$

$$= -(\Lambda_- C)_{ba} \quad \text{(using } C^T = -C \text{)}$$

$$= -(\Lambda_- C)^T_{ab}$$

$$= (\Lambda_+ C)_{ab} \quad \text{(see below)}$$

$$=\left(\frac{\not{p}+m}{2m}C\right)_{ab}$$

since

$$\Lambda_+ C = (\not{p}+m)C/2m$$
$$= (CC^{-1}\not{p}C+mC)/2m$$
$$= (C(-\not{p}^T)+mC)/2m$$
$$= C(-\not{p}^T+m)/2m$$

and so

$$(\Lambda_+ C)^T = \frac{(-\not{p}+m)}{2m}C^T$$
$$= \Lambda_- C^T$$
$$= -\Lambda_- C$$

Hence

$$(\Lambda_- C)^T = -(\Lambda_+ C)$$

In an analogous way we have

$$v_a(\vec{p},r)\,u_b(\vec{p},r)$$

$$= u^c_a(\vec{p},r)\,u_b(\vec{p},r) \qquad \text{(using (5.39))}$$

$$= C_{am}\,\bar{u}^T_m(\vec{p},r)\,u_b(\vec{p},r) \qquad \text{(using (5.37))}$$

$$= C_{am}\,u_b(\vec{p},r)\,\bar{u}^T_m(\vec{p},r)$$

$$= C^T_{ma}\,\Lambda_{+\,bm} \qquad \text{(using (5.60))}$$

$$= (\Lambda_+ C^T)_{ba}$$

$$= -(\Lambda_+ C)_{ba}$$

$$= -(\Lambda_+ C)^T_{ab}$$

$$= -\left(\Lambda_{-} C^T\right)_{ab}$$

$$= \left(\frac{\not p - m}{2m} C\right)_{ab} \qquad \text{(using } \quad C^T = -C \text{)}$$

With these formulae we can prove (5.114c).

<u>Proof of (5.114c)</u>: Consider the commutator

$$-i\,[\bar\varepsilon Q, \psi_a(x)]$$

$$= -i\,(\bar\varepsilon_b Q_b \psi_a(x) - \psi_a(x)\,\bar\varepsilon_b Q_b)$$

$$= -i\,\bar\varepsilon_b \{Q_b, \psi_a(x)\}$$

$$= \frac{(2m)^{1/2}}{(2\pi)^{3/2}}\,\bar\varepsilon_b \sum_{r,s} \int d^3p \int d^3k \left(\frac{m}{\omega_k}\right)^{\frac{1}{2}} \cdot$$

$$\cdot \Big\{ C_{bc}(\vec p)\, d^+(\vec p, s)\, v_c(\vec p, s)$$

$$- D_{bc}(\vec p)\, d(\vec p, s)\, u_c(\vec p, s)$$

$$+ d(\vec k, r)\, u_a(\vec k, r)\, e^{-ikx}$$

$$+ d^+(\vec k, r)\, v_a(\vec k, r)\, e^{ikx} \Big\} \qquad \text{(with (5.47), (5.90))}$$

$$= \frac{(2m)^{1/2}}{(2\pi)^{3/2}}\,\bar\varepsilon_b \sum_{r,s} \int d^3p \int d^3k \left(\frac{m}{\omega_k}\right)^{\frac{1}{2}} \cdot$$

$$\cdot \Big[\{d^+(\vec p, s), d(\vec k, r)\}\, C_{bd}(\vec p)\, v_d(\vec p, s)\, u_a(\vec k, r)\, e^{-ikx}$$

$$+ \{d^+(\vec p, s), d^+(\vec k, r)\}\, C_{bd}(\vec p)\, v_d(\vec p, s)\, v_a(\vec k, r)\, e^{ikx}$$

$$- \{d(\vec p, s), d(\vec k, r)\}\, D_{bd}(\vec p)\, u_d(\vec p, s)\, u_a(\vec k, r)\, e^{-ikx}$$

$$- \{d(\vec p, s), d^+(\vec k, r)\}\, D_{bd}(\vec p)\, u_d(\vec p, s)\, v_a(\vec k, r)\, e^{ikx} \Big]$$

$$= \frac{(2m)^{1/2}}{(2\pi)^{3/2}}\,\bar\varepsilon_b \sum_{r} \int d^3p \left(\frac{m}{\omega_p}\right)^{\frac{1}{2}} \cdot$$

$$\cdot \Big[C_{bd}(\vec p)\, v_d(\vec p, r)\, u_a(\vec p, r)\, e^{-ipx}$$

$$- D_{bd}(\vec{p})\, u_d(\vec{p},r)\, v_a(\vec{p},r)\, e^{ipx} \Big]$$

(using (5.92))

$$= \frac{(2m)^{1/2}}{(2\pi)^{3/2}} \bar{\varepsilon}_b \int d^3p \left(\frac{m}{\omega_p}\right)^{\frac{1}{2}} \cdot$$

$$\cdot \Big[C_{bd}(\vec{p}) \left(\frac{\not{p}-m}{2m} C\right)_{da} e^{-ipx}$$

$$- D_{bd}(\vec{p}) \left(\frac{\not{p}+m}{2m} C\right)_{da} e^{ipx} \Big]$$

(using (5.116))

$$= \bar{\varepsilon}_b (2\pi)^{-3/2} \int d^3p\, (2\omega_p)^{-\frac{1}{2}} \cdot$$

$$\cdot \Big[\left(a(\vec{p})\, \delta_{bd} - i \gamma^5_{bd}\, b(\vec{p})\right) \left((\not{p}-m)C\right)_{da} e^{-ipx}$$

$$- \left(a^+(\vec{p})\, \delta_{bd} - i \gamma^5_{bd}\, b^+(\vec{p})\right) \left((\not{p}+m)C\right)_{da} e^{ipx} \Big]$$

$$= \bar{\varepsilon}_b \Big\{ (2\pi)^{-3/2} \int d^3p\, (2\omega_p)^{-\frac{1}{2}} \cdot \qquad \text{(with (5.48))}$$

$$\cdot \Big[a(\vec{p}) (\not{p}-m) C\, e^{-ipx}$$

$$- i b(\vec{p})\, \gamma^5 (\not{p}-m) C e^{-ipx}$$

$$- a^+(\vec{p}) (\not{p}+m) C e^{ipx}$$

$$+ i\, b^+(\vec{p})\, \gamma^5 (\not{p}+m) C e^{ipx} \Big]_{ba}$$

We now use the following equations

$$(\not{p}-m)\, e^{-ipx} = (i \not{\partial} - m)\, e^{-ipx}$$

$$(\not{p}+m)\, e^{ipx} = -(i \not{\partial} - m)\, e^{ipx}$$

Then

$$-i[\bar{\varepsilon}Q, \psi_a(x)]$$

$$= \bar{\varepsilon}_b \{(2\pi)^{-3/2} \int d^3p \, (2\omega_p)^{-\frac{1}{2}} \cdot$$

$$\cdot [a(\vec{p})(i\not{p}-m)Ce^{-ipx}$$

$$+ a^+(\vec{p})(i\not{p}-m)Ce^{ipx}$$

$$- i\, b(\vec{p})\gamma^5(i\not{p}-m)Ce^{-ipx}$$

$$- i\, b^+(\vec{p})\gamma^5(i\not{p}-m)Ce^{ipx}]\}_{ba}$$

$$= \bar{\varepsilon}_b\{(i\not{\partial}-m)C(2\pi)^{-\frac{3}{2}}\int d^3p\,(2\omega_p)^{-\frac{1}{2}} \cdot$$

$$\cdot (a(\vec{p})e^{-ipx} + a^+(\vec{p})e^{ipx})$$

$$- i\gamma^5(i\not{\partial}-m)C(2\pi)^{-\frac{3}{2}}\int d^3p\,(2\omega_p)^{-\frac{1}{2}} \cdot$$

$$(b(\vec{p})e^{-ipx} + b^+(\vec{p})e^{ipx})\}_{ba}$$

$$= \bar{\varepsilon}_b\{(i\not{\partial}-m)CA(x) - i\gamma^5(i\not{\partial}-m)CB(x)\}_{ba}$$

(using (5.86) and (5.88))

$$= \bar{\varepsilon}_b\{C(-i\not{\partial}^T-m)A(x) - i\gamma^5 C(-i\not{\partial}^T-m)B(x)\}_{ba}$$

(using (1.166))

$$= \bar{\varepsilon}_b\{C(-i\not{\partial}^T-m)A(x) - iC\gamma^{5T}(-i\not{\partial}^T-m)B(x)\}_{ba}$$

(using (1.173))

$$= \bar{\varepsilon}_b C_{bd}\{(A(x) - i\gamma^{5T}B(x))(-i\overleftarrow{\not{\partial}}^T - m)\}_{da}$$

$$= \{\varepsilon^T(A(x) - i\gamma^5 B(x))^T(-1)(i\overleftarrow{\not{\partial}}+m)^T\}_a$$

$$= (-(i\not{\partial}+m)(A(x) - i\gamma^5 B(x))\varepsilon)_a$$

$$= \delta\psi_a(x) \qquad \text{(with (5.11c)}$$

These results demonstrate that the spinor charge Q_a defined by (5.47) gives the correct supersymmetry transformation of the fields A(x), B(x) and ψ (x).

CHAPTER 6

SUPERSPACE FORMALISM AND SUPERFIELDS

6.1 Superspace

In order to be able to construct supersymmetric models one wants to have a formalism in which supersymmetry is inherently manifest like Lorentz invariance in electrodynamics. Such a formalism is the superfield formalism[49,50] introduced by Salam and Strathdee[20]. Consider, for instance, a theory formulated in three-dimensional Euclidean space. In general one cannot expect such a theory to be invariant under Lorentz transformations or to be invariant under the transformations of the Poincare group. In order to obtain such a relativistically invariant theory one is forced to extend Euclidean 3-space to a flat pseudo-Riemannian 4-space, the Minkowski space. Introducing time as the additional coordinate, it is possible to formulate the theory in a relativistically invariant form. Analogously it is not possible to have a manifestly supersymmetric theory in Minkowski space. To obtain a formalism which achieves this we have to extend Minkowski space to superspace. Elements of superspace are socalled super-coordinates which consist of the usual four Minkowski space-time coordinates and four constant (i.e. x_μ-independent), anticommuting Grassmann numbers. If we formulate the theory in terms of the two-component Weyl spinor formalism, the latter are

$$\{\theta_A\}_{A=1,2} \quad \text{and} \quad \{\bar{\theta}_{\dot{B}}\}_{\dot{B}=\dot{1},\dot{2}}$$

two-component Weyl spinors which transform under the self-representation of SL(2,C) and the complex conjugate self-representation of SL(2,C) respectively and are considered to be independent. On the other hand, if we work with four-dimensional Majorana spinors, the additional parameters are constant, anticommuting Grassmann numbers

$$\{\varepsilon_a\}_{a=1,2,3,4}$$

with ε_a satisfying the Majorana condition (1.180). The resulting "superspace" therefore has eight dimensions.

With the help of the anticommuting Grassmann parameters we can transform the graded Lie algebra (involving both commutators and anticommutators) into a regular Lie algebra (which involves only commutators) by writing the elements of the spinor sector of the algebra

$$\theta^A Q_A\,, \quad \bar{\theta}_{\dot{A}}\bar{Q}^{\dot{A}}$$

The anticommutation relations of the two-component Weyl spinors are given by

$$\left.\begin{aligned} \{\theta_A, \theta_B\} &= 0 \\ \{\bar{\theta}_{\dot{A}}, \bar{\theta}_{\dot{B}}\} &= 0 \\ \{\theta_A, \bar{\theta}_{\dot{B}}\} &= 0 \end{aligned}\right\} \qquad (6.1)$$

and an element of superspace is given by the supercoordinate $(x_\mu, \theta_A, \bar{\theta}_{\dot{A}})$. We now reformulate the supersymmetry algebra (4.1) entirely in terms of commutators. As is evident from (3.25) the relevant part of the Super-Poincare algebra in the Weyl formalism has the anticommutation relations

$$\left.\begin{aligned} &\{Q_A, Q_B\} = \{\bar{Q}_{\dot{A}}, \bar{Q}_{\dot{B}}\} = 0 \\ &\{Q_A, \bar{Q}_{\dot{B}}\} = 2\sigma^{\mu}_{A\dot{B}} P_{\mu} \end{aligned}\right\} \quad (6.2)$$

<u>Proposition</u>: The anticommutation relations (6.2) can be rewritten as commutators

$$\left.\begin{aligned} &[\theta^A Q_A, \bar{\theta}_{\dot{B}} \bar{Q}^{\dot{B}}] = 2\theta^A \sigma^{\mu}_{A\dot{B}} \bar{\theta}^{\dot{B}} P_{\mu} \\ &[\theta^A Q_A, \theta^B Q_B] = 0 \\ &[\bar{\theta}_{\dot{A}} \bar{Q}^{\dot{A}}, \bar{\theta}_{\dot{B}} \bar{Q}^{\dot{B}}] = 0 \end{aligned}\right\} \quad (6.3a)$$

where θ^A and $\bar{\theta}_{\dot{A}}$ are anticommuting Grassmann numbers which also anticommute with the spinor charges Q_A and $\bar{Q}_{\dot{A}}$. In (6.3a) we use the summation convention for two-component Weyl spinors as in (1.77) and (1.78).

<u>Proof</u>: We start with the anticommutator

$$\{Q_A, \bar{Q}_{\dot{B}}\} = 2\sigma^{\mu}_{A\dot{B}} P_{\mu}$$

i.e.

$$Q_A \bar{Q}_{\dot{B}} + \bar{Q}_{\dot{B}} Q_A = 2\sigma^{\mu}_{A\dot{B}} P_{\mu}$$

Multiplying this equation from the left by θ^A and from the right by $\bar{\theta}^{\dot{B}}$ we obtain

$$\theta^A Q_A \bar{Q}_{\dot{B}} \bar{\theta}^{\dot{B}} + \theta^A \bar{Q}_{\dot{B}} Q_A \bar{\theta}^{\dot{B}} = 2\theta^A \sigma^{\mu}_{A\dot{B}} \bar{\theta}^{\dot{B}} P_{\mu}$$

We now assume that Q_A , $\bar{Q}_{\dot{A}}$ are linear operators with the appropriate transformation properties under SL(2,C).

Then

$$Q_A \sim \frac{\partial}{\partial\theta^A} \quad , \quad \bar{Q}_{\dot{A}} \sim \frac{\partial}{\partial\bar{\theta}^{\dot{A}}}$$

and so

$$(Q_A \bar{\theta}_{\dot{B}}) = 0 \, , \quad (\bar{Q}_{\dot{B}} \, \theta_A) = 0$$

The left hand side of the above equation can then be written (with quantities in brackets representing c-numbers)

$$\begin{aligned}
& -\theta^A Q_A \bar{\theta}^{\dot{B}} \bar{Q}_{\dot{B}} + \theta^A Q_A (\bar{Q}_{\dot{B}} \bar{\theta}^{\dot{B}}) \\
& -\bar{Q}_{\dot{B}} \theta^A Q_A \bar{\theta}^{\dot{B}} + (\bar{Q}_{\dot{B}} \theta^A) Q_A \bar{\theta}^{\dot{B}} \\
= & -\theta^A Q_A \bar{\theta}^{\dot{B}} \bar{Q}_{\dot{B}} + \theta^A Q_A (\bar{Q}_{\dot{B}} \bar{\theta}^{\dot{B}}) \\
& + \bar{Q}_{\dot{B}} \theta^A \bar{\theta}^{\dot{B}} Q_A - \bar{Q}_{\dot{B}} \theta^A (Q_A \bar{\theta}^{\dot{B}}) \\
& \qquad + (\bar{Q}_{\dot{B}} \theta^A) Q_A \bar{\theta}^{\dot{B}} \\
= & -\theta^A Q_A \bar{\theta}^{\dot{B}} \bar{Q}_{\dot{B}} + \theta^A Q_A (\bar{Q}_{\dot{B}} \bar{\theta}^{\dot{B}}) \\
& - \bar{Q}_{\dot{B}} \bar{\theta}^{\dot{B}} \theta^A Q_A \\
= & -\theta^A Q_A \bar{\theta}^{\dot{B}} \bar{Q}_{\dot{B}} + \theta^A Q_A (\bar{Q}_{\dot{B}} \bar{\theta}^{\dot{B}}) \\
& + \bar{\theta}^{\dot{B}} \bar{Q}_{\dot{B}} \theta^A Q_A - (\bar{Q}_{\dot{B}} \bar{\theta}^{\dot{B}}) \theta^A Q_A \\
= & + \theta^A Q_A \, \bar{\theta}_{\dot{B}} \bar{Q}^{\dot{B}} - \bar{\theta}_{\dot{B}} \bar{Q}^{\dot{B}} \theta^A Q_A
\end{aligned}$$

since (using (1.76c) and (1.76d))

$$\begin{aligned}
\bar{\theta}^{\dot{B}} \bar{Q}_{\dot{B}} & = -\bar{\theta}_{\dot{A}} \varepsilon^{\dot{A}\dot{B}} \varepsilon_{\dot{B}\dot{C}} \bar{Q}^{\dot{C}} \\
& = -\bar{\theta}_{\dot{A}} \, \delta^{\dot{A}}{}_{\dot{C}} \bar{Q}^{\dot{C}} \\
& = -\bar{\theta}_{\dot{A}} \, \bar{Q}^{\dot{A}} \\
& = -(\bar{\theta}\bar{Q})
\end{aligned}$$

Hence returning to the original equation we obtain

$$(\theta Q)(\bar{\theta}\bar{Q}) - (\bar{\theta}\bar{Q})(\theta Q) = 2\,(\theta\sigma^{\mu}\bar{\theta})P_{\mu}$$

and so the first of equations (6.3a). The remaining two relations can be shown to hold in a similar way.
For later calculations we need some more commutators which are readily verified. We state them here without proof.

Proposition: The following relations can be shown to hold

$$\left.\begin{array}{l} [P_{\mu},(\theta Q)] = 0, \quad [P_{\mu},(\bar{\theta}\bar{Q})] = 0 \\ [(\theta Q),(\theta\sigma_{\mu}\bar{\theta})] = 0 \\ [(\bar{\theta}\bar{Q}),(\theta\sigma_{\mu}\bar{\theta})] = 0 \end{array}\right\} \qquad (6.3b)$$

We have seen earlier that one way to find irreducible representations of the supersymmetry algebra is to proceed as in the case of the Poincaré algebra and to find Casimir operators and then to construct the appropriate Fock space. The superspace formalism is an alternative method to find irreducible representations of the supersymmetry algebra, and is particularly useful in performing calculations. In order to be able to write down explicit representations of the supersymmetry algebra in terms of linear operators acting on functions in superspace, it is necessary to have at our disposal a calculus for differentiating with respect to Grassmann numbers. We therefore introduce this first.

6.2 Differentiation with respect to Grassmann Numbers[51]

We consider a set of N discrete numbers $\{a_1, a_2, \ldots a_N\}$ obeying

$$\{a_i, a_j\} = 0, \qquad \forall\, i, j = 1,2,\ldots, N$$

We can construct a differential calculus for these Grassmann variables, and similarly (later) an integral calculus. However, the numbers a_i, i = 1,2,..., N, are discrete objects; for this reason the derivative is defined formally as

$$\frac{\partial a_i}{\partial a_j} := \delta_{ij} \tag{6.4a}$$

This is simply a definition of the symbol on the left hand side and is not to be looked at as the ratio of two infinitesimal increments.

To obtain a product rule we must take into account the anticommutative character of the variables a_i. The product rule is then given by the relation

$$\begin{aligned}\frac{\partial}{\partial a_p}(a_{i_1} a_{i_2} \cdots a_{i_r}) &= \delta_{p i_1} a_{i_2} \cdots a_{i_r} \\ &\quad - \delta_{p i_2} a_{i_1} a_{i_3} \cdots a_{i_r} + \cdots \\ &\quad + (-1)^{r-1} \delta_{p i_r} a_{i_1} a_{i_2} \cdots a_{i_{r-1}}\end{aligned} \tag{6.4b}$$

Equation (6.4b) implies that $\partial/\partial a_i$ is an "antiderivative". We also have the following relation

$$\begin{aligned}\left\{\frac{\partial}{\partial a_p}, a_r\right\} &= \frac{\partial}{\partial a_p} a_r + a_r \frac{\partial}{\partial a_p} \\ &= \frac{\partial a_r}{\partial a_p} - a_r \frac{\partial}{\partial a_p} + a_r \frac{\partial}{\partial a_p} \quad \text{(with (6.4b))} \\ &= \frac{\partial a_r}{\partial a_p} = \delta_{rp} \quad \text{(with (6.4a))}\end{aligned}$$

Hence

$$\left\{\frac{\partial}{\partial a_p}, a_r\right\} = \delta_{pr} \tag{6.4c}$$

Given any function $f(a_1,\ldots,a_N)$ we have

$$\frac{\partial}{\partial a_i}\frac{\partial}{\partial a_j} f(a_1,\ldots,a_N) = -\frac{\partial}{\partial a_j}\frac{\partial}{\partial a_i} f(a_1,\ldots,a_N)$$

so that

$$\left\{\frac{\partial}{\partial a_i}, \frac{\partial}{\partial a_j}\right\} f(a_1,\ldots,a_N) = 0 \tag{6.4d}$$

As an example we consider

$$f(a_1,\ldots,a_N) = a_r a_p$$

Then

$$\frac{\partial}{\partial a_i}\left(\frac{\partial}{\partial a_j} a_r a_p\right)$$

$$= \frac{\partial}{\partial a_i}\left(\delta_{jr} a_p - a_r \delta_{jp}\right) \qquad \text{(with (6.4a,b))}$$

$$= \delta_{jr}\delta_{ip} - \delta_{ir}\delta_{jp} \qquad \text{(with (6.4a))}$$

On the other hand

$$\frac{\partial}{\partial a_j}\left(\frac{\partial}{\partial a_i} a_r a_p\right)$$

$$= \frac{\partial}{\partial a_j}\left(\delta_{ir} a_p - a_r \delta_{ip}\right)$$

$$= \delta_{ir}\delta_{jp} - \delta_{jr}\delta_{ip}$$

$$= -\frac{\partial}{\partial a_i}\left(\frac{\partial}{\partial a_j} a_r a_p\right)$$

Hence

$$\left\{\frac{\partial}{\partial a_i}, \frac{\partial}{\partial a_j}\right\} a_r a_p = 0$$

We now consider the two-component Weyl spinors θ^A and $\bar{\theta}^{\dot{A}}$ and define

$$\partial_A := \frac{\partial}{\partial\theta^A}, \qquad \partial^A := \frac{\partial}{\partial\theta_A}$$
$$\bar{\partial}_{\dot{A}} := \frac{\partial}{\partial\bar{\theta}^{\dot{A}}}, \qquad \bar{\partial}^{\dot{A}} := \frac{\partial}{\partial\bar{\theta}_{\dot{A}}} \tag{6.4e}$$

According to (6.4a) differentiation with respect to θ is defined as

$$\partial_A\theta^B = \frac{\partial}{\partial\theta^A}\theta^B = \delta_A{}^B$$

$$\partial^A\theta_B = \frac{\partial}{\partial\theta_A}\theta_B = \delta^A{}_B$$

$$\partial_A\theta_B = \frac{\partial}{\partial\theta^A}(\varepsilon_{BC}\theta^C) = \varepsilon_{BC}\delta_A{}^C$$
$$= \varepsilon_{BA}$$

$$\bar{\partial}_{\dot{A}}\bar{\theta}^{\dot{B}} = \frac{\partial}{\partial\bar{\theta}^{\dot{A}}}\bar{\theta}^{\dot{B}} = \delta_{\dot{A}}{}^{\dot{B}}$$

$$\bar{\partial}^{\dot{A}}\bar{\theta}_{\dot{B}} = \frac{\partial}{\partial\bar{\theta}_{\dot{A}}}\bar{\theta}_{\dot{B}} = \delta^{\dot{A}}{}_{\dot{B}} \tag{6.4f}$$

We can use the metric tensor to raise or lower indices of derivatives.

Proposition: Undotted indices of derivatives can be raised or lowered with the formulae

$$\varepsilon^{AB}\partial_B = -\partial^A , \quad \partial_A = -\varepsilon_{AB}\partial^B \tag{6.4g}$$

Proof: We have

$$\begin{aligned}\varepsilon^{AB}\partial_B\theta^C &= \varepsilon^{AB}\frac{\partial}{\partial\theta^B}\theta^C \\ &= \varepsilon^{AB}\delta_B{}^C \\ &= \varepsilon^{AC}\end{aligned}$$

and

$$\begin{aligned}-\partial^A\theta^C &= -\partial^A\varepsilon^{CD}\theta_D \\ &= -\varepsilon^{CD}\frac{\partial}{\partial\theta_A}\theta_D \\ &= -\varepsilon^{CD}\delta_D{}^A \\ &= -\varepsilon^{CA} \\ &= \varepsilon^{AC} \\ &= \varepsilon^{AB}\partial_B\theta^C\end{aligned}$$

This implies for the operators

$$\varepsilon^{AB}\partial_B = -\partial^A$$

and the other relation follows similarly.

We conclude therefore that

$$\left.\begin{aligned}\partial^A\theta^C &= -\varepsilon^{AC} \\ \text{and similarly}\qquad \partial_A\theta_C &= -\varepsilon_{AC}\end{aligned}\right\} \tag{6.4h}$$

Equation (6.4d) implies for our special case

$$\{\partial_A, \partial_B\} = 0 \tag{6.4i}$$

Analogous formulae hold for $\bar{\partial}$. Thus

$$\left.\begin{aligned} &\bar{\partial}^{\dot{A}} = \bar{\partial}_{\dot{B}}\varepsilon^{\dot{B}\dot{A}} \quad \text{or} \quad \bar{\partial}^{\dot{A}} = -\varepsilon^{\dot{A}\dot{B}}\bar{\partial}_{\dot{B}} \\ &\text{and} \\ &\bar{\partial}_{\dot{A}} = -\varepsilon_{\dot{A}\dot{B}}\bar{\partial}^{\dot{B}} \end{aligned}\right\} \tag{6.4j}$$

implying

$$\left.\begin{aligned} &\bar{\partial}^{\dot{A}}\bar{\theta}^{\dot{B}} = -\varepsilon^{\dot{A}\dot{B}} \\ &\text{and similarly} \\ &\bar{\partial}_{\dot{A}}\bar{\theta}_{\dot{B}} = -\varepsilon_{\dot{A}\dot{B}} \end{aligned}\right\} \tag{6.4k}$$

since

$$\begin{aligned} \bar{\partial}_{\dot{A}}\bar{\theta}_{\dot{B}} &= \bar{\partial}_{\dot{A}}\,\varepsilon_{\dot{B}\dot{C}}\bar{\theta}^{\dot{C}} \qquad \text{(using (1.76d))} \\ &= -\varepsilon_{\dot{B}\dot{C}}\,\delta_{\dot{A}}^{\ \dot{C}} \\ &= \varepsilon_{\dot{B}\dot{A}} = -\varepsilon_{\dot{A}\dot{B}} \end{aligned} \tag{6.4l}$$

and as a consequence of (6.4d)

$$\{\bar{\partial}_{\dot{A}}, \bar{\partial}_{\dot{B}}\} = 0$$

In addition, since θ and $\bar{\theta}$ are considered to be independent, we demand

$$\{\bar{\partial}_{\dot{A}}, \theta^B\} = 0 = \{\partial_A, \bar{\theta}^{\dot{B}}\} \tag{6.4m}$$

and hence

$$\bar{\partial}_{\dot{A}}\theta^B = \partial_A\bar{\theta}^{\dot{B}} = 0 \tag{6.4n}$$

<u>Proposition</u>: We have

$$\partial_A\theta^2 = 2\theta_A \ , \quad \bar{\partial}_{\dot{A}}\bar{\theta}^2 = -2\bar{\theta}_{\dot{A}} \tag{6.4p}$$

<u>Proof</u>: Consider

$$\begin{aligned} \partial_A\theta^2 &= \partial_A(\theta^B\theta_B) \\ &= (\partial_A\theta^B)\theta_B - \theta^B(\partial_A\theta_B) \end{aligned}$$

$$= \delta_A{}^B \theta_B - \theta^B[-\varepsilon_{AD}\partial^D\theta_B]$$

$$= \theta_A + \theta^B \varepsilon_{AB}$$

$$= 2\theta_A \qquad \text{(with (1.76b))}$$

Similarly

$$\bar{\partial}_{\dot{A}}(\bar{\theta}^2) = \bar{\partial}_{\dot{A}}(\bar{\theta}_{\dot{B}}\bar{\theta}^{\dot{B}}) = -2\bar{\theta}_{\dot{A}}$$

<u>Proposition</u>: The following relations hold

$$\left.\begin{aligned} \varepsilon^{AB}\partial_A\partial_B(\theta\theta) &= 4 = \partial^B\partial_B(\theta\theta) \\ \varepsilon_{\dot{A}\dot{B}}\bar{\partial}^{\dot{A}}\bar{\partial}^{\dot{B}}(\bar{\theta}\bar{\theta}) &= 4 = \bar{\partial}_{\dot{B}}\bar{\partial}^{\dot{B}}(\bar{\theta}\bar{\theta}) \end{aligned}\right\} \qquad (6.4q)$$

<u>Proof</u>: Consider

$$\varepsilon^{AB}\partial_A\partial_B(\theta\theta) = 2\varepsilon^{AB}\partial_A\theta_B \qquad \text{(with (6.4p))}$$

$$= 2\varepsilon^{AB}(-\varepsilon_{AD}\partial^D\theta_B)$$

$$= 2\varepsilon^{AB}(-\varepsilon_{AD}\delta^D{}_B)$$

$$= 2\varepsilon^{AB}(-\varepsilon_{AB})$$

$$= 2\varepsilon^{AB}\varepsilon_{BA}$$

$$= 2\delta^A{}_A$$

$$= 4$$

Similarly

$$\varepsilon_{\dot{A}\dot{B}}\,\bar{\partial}^{\dot{A}}\bar{\partial}^{\dot{B}}\bar{\theta}^2 = 2\varepsilon_{\dot{A}\dot{B}}\,\bar{\partial}^{\dot{A}}\bar{\theta}^{\dot{B}}$$

$$= -2\varepsilon_{\dot{A}\dot{B}}\varepsilon^{\dot{A}\dot{B}} = 2\varepsilon_{\dot{A}\dot{B}}\varepsilon^{\dot{B}\dot{A}} \qquad \text{(with (6.4k))}$$

$$= 2\delta_{\dot{A}}{}^{\dot{A}}$$

$$= 4$$

6.3 Supersymmetry Transformations in the Weyl Formalism

6.3.1 Finite Supersymmetry Transformations

In Chapter 5 we discussed infinitesimal variations of the fields appearing in the Wess-Zumino model. Such changes of the fields correspond to infinitesimal supersymmetry transformations. The consideration can be extended to finite supersymmetry transformations in a natural way in the context of the superspace formalism. Following Salam and Strathdee[49] we consider the action of the supersymmetry group on the space of left cosets with respect to the subgroup of homogeneous Lorentz transformations (i.e. there are no terms in $M_{\mu\nu}$ in the arguments of the exponentials of (6.5) below; we consider only the subgroup of transformations with the homogeneous Lorentz transformations factorized out). This space is a homogeneous space on which the factor group acts transitively, and as described above, it is an eight-dimensional space which is parametrized in terms of Minkowski space-time coordinates x_μ and Weyl spinors θ_A, $\bar\theta_{\dot A}$, $A = 1,2$, $\dot A = \dot 1, \dot 2$. It is this space which is called superspace. Since we factorize out the homogeneous Lorentz transformations, we construct and define the following operators:

$$L(x_\mu, \theta_A, \bar\theta^{\dot A}) := \exp(-ix_\mu P^\mu + i\theta Q + i\bar\theta\bar Q)$$

$$L_1(x_\mu, \theta_A, \bar\theta^{\dot A})$$
$$:= \exp(-ix_\mu P^\mu + i\theta Q) . \exp(i\bar\theta\bar Q)$$

$$L_2(x_\mu, \theta_A, \bar\theta^{\dot A})$$
$$:= \exp(-ix_\mu P^\mu + i\bar\theta\bar Q) . \exp(i\theta Q)$$

(6.5)

$L(x, \theta, \bar\theta)$ is a unitary operator. L_1 and L_2 are also unitary and related to $L(x, \theta, \bar{\theta})$ as in (6.8) below. $L(x, \theta, \bar{\theta})$, $L_1(x, \theta, \bar{\theta})$ and $L_2(x, \theta, \bar{\theta})$ are operators which describe three different but equivalent actions of the supersymmetry group on functions defined on superspace. Hence (6.5) leads to three different but equivalent realizations of one and the same supersymmetry transformation, and hence, as we shall see, to three different definitions of superfields. In (6.5) P_μ, Q_A and $\bar{Q}^{\dot{A}}$ denote hermitian operators which act on functions in superspace. These operators correspond to the basic elements of the Super-Poincaré algebra. It should be noted that we use the same symbols for these operators and their corresponding Lie algebra elements. It should also be noted from (6.5) that θ and $\bar{\theta}$ have dimension $+\frac{1}{2}$ in length since Q, $\bar{Q}$ have dimension $-\frac{1}{2}$ as follows e.g. from (6.2).

A general (Lorentz scalar or pseudoscalar) superfield Φ is an operator-valued function defined on superspace which is to be understood in terms of its power series expansion in θ and $\bar{\theta}$. Since θ and $\bar{\theta}$ are anti-commuting Grassmann numbers, this power series expansion is finite, i.e.

$$\begin{aligned}
&\Phi(x, \theta, \bar{\theta}) \\
&= f(x) + \theta^A \phi_A(x) + \bar{\theta}_{\dot{A}} \bar{\chi}^{\dot{A}}(x) \\
&\quad + (\theta\theta) m(x) + (\bar{\theta}\bar{\theta}) n(x) \\
&\quad + (\theta \sigma^\mu \bar{\theta}) V_\mu(x) + (\theta\theta) \bar{\theta}_{\dot{A}} \bar{\lambda}^{\dot{A}}(x) \\
&\quad + (\bar{\theta}\bar{\theta}) \theta^A \psi_A(x) + (\theta\theta)(\bar{\theta}\bar{\theta}) d(x)
\end{aligned} \tag{6.6}$$

In constructing this expression one recalls, of course, that $(\theta\theta) \equiv \theta^A \theta_A$, $(\bar{\theta}\bar{\theta}) \equiv \bar{\theta}_{\dot{A}} \bar{\theta}^{\dot{A}}$

are Lorentz scalars (cf. (1.77)), $\theta\sigma^\mu\bar\theta$ is a Lorentz vector (cf. (1.117)) and so on. One can easily convince oneself that the sum of terms of (6.6) exhausts all possibilities of nonvanishing combinations of powers of θ, $\bar\theta$; e.g. with (1.115) the term $\bar\theta\bar\sigma^\mu\theta$ which does not appear in (6.6) can be reduced to $-\theta\sigma^\mu\bar\theta$. All higher powers of Grassmann numbers θ, $\bar\theta$ vanish as explained earlier. The quantities $f(x)$, $\phi_A(x)$, $\bar\chi^{\dot A}(x)$, etc are called component fields. The power series expansion (6.6) will be discussed in detail later.

The three different operators (6.5) lead to three different types of superfields $\Phi(x,\theta,\bar\theta)$, $\Phi_1(x,\theta,\bar\theta)$ and $\Phi_2(x,\theta,\bar\theta)$ respectively. They are related by the following identity

$$\begin{aligned}\Phi(x,\theta,\bar\theta) &= \Phi_1(x_\mu + i\theta\sigma_\mu\bar\theta,\theta,\bar\theta)\\ &= \Phi_2(x_\mu - i\theta\sigma_\mu\bar\theta,\theta,\bar\theta)\end{aligned} \tag{6.7}$$

Before we demonstrate the validity of this relation we show that the following connection holds between the operators (6.5)

$$\begin{aligned}L(x,\theta,\bar\theta) &= L_1(x+i\theta\sigma\bar\theta,\theta,\bar\theta)\\ &= L_2(x-i\theta\sigma\bar\theta,\theta,\bar\theta)\end{aligned} \tag{6.8}$$

In order to prove (6.8) we need the Baker-Campbell-Hausdorff formula[44]

$$\begin{aligned}&\exp(A).\exp(B)\\ &= \exp\Big(A+B+\tfrac{1}{2}[A,B]+\tfrac{1}{12}[A,[A,B]]\\ &\qquad -\tfrac{1}{12}[B,[B,A]]+\cdots\Big)\end{aligned} \tag{6.9}$$

Hence

$$L_1(x+i\theta\sigma\bar{\theta},\theta,\bar{\theta})$$
$$=\exp\{-i(x_\mu+i\theta\sigma_\mu\bar{\theta})P^\mu+i\theta Q\}\exp\{i\bar{\theta}\bar{Q}\}$$

(using (6.5))

$$=\exp\{-i(x_\mu+i\theta\sigma_\mu\bar{\theta})P^\mu+i\theta Q+i\bar{\theta}\bar{Q}$$
$$+\frac{1}{2}[-ix_\mu P^\mu+\theta\sigma_\mu\bar{\theta}P^\mu+i\theta Q, i\bar{\theta}\bar{Q}]$$
$$+0\}$$
$$=\exp\{-i(x_\mu+i\theta\sigma_\mu\bar{\theta})P^\mu+i\theta Q+i\bar{\theta}\bar{Q}$$
$$+\frac{1}{2}x_\mu[P^\mu,\bar{\theta}\bar{Q}]-\frac{1}{2}[\theta Q,\bar{\theta}\bar{Q}]$$
$$+\frac{i}{2}\theta\sigma_\mu\bar{\theta}[P^\mu,\bar{\theta}\bar{Q}]\}$$
$$=\exp\{-i(x_\mu+i\theta\sigma_\mu\bar{\theta})P^\mu+i\theta Q+i\bar{\theta}\bar{Q}$$
$$-\theta\sigma_\mu\bar{\theta}P^\mu\}$$

(using (6.3a), (6.3b))

$$=\exp\{-ix_\mu P^\mu+i\theta Q+i\bar{\theta}\bar{Q}\}$$
$$=L(x,\theta,\bar{\theta})$$ (with (6.5))

It should be noted that all higher order commutators, e.g. $[A, [A, B]]$, in the expression of the Baker-Campbell-Hausdorff formula vanish, because the commutators close into P_μ, e.g. formally (using (6.3a) and (6.3b))

$$[P+\theta Q,[P+\theta Q,\bar{\theta}\bar{Q}]]$$
$$=[P+\theta Q,[P,\bar{\theta}\bar{Q}]+[\theta Q,\bar{\theta}\bar{Q}]]$$
$$=[P+\theta Q,P]=[P,P]+[\theta Q,P]$$
$$=0$$

Consider in the same way

$$L_2(x-i\theta\sigma\bar{\theta},\theta,\bar{\theta})$$

$$=\exp\{-i(x_\mu-i\theta\sigma_\mu\bar{\theta})P^\mu+i\bar{\theta}\bar{Q}\}\exp\{i\theta Q\}$$

$$=\exp\{-i(x_\mu-i\theta\sigma_\mu\bar{\theta})P^\mu+i\bar{\theta}\bar{Q}+i\theta Q$$
$$+\frac{1}{2}[-ix_\mu P^\mu-\theta\sigma_\mu\bar{\theta}P^\mu+i\bar{\theta}\bar{Q},i\theta Q]\}$$

(with (6.9))

$$=\exp\{-ix_\mu P_\mu-\theta\sigma_\mu\bar{\theta}P^\mu+i\bar{\theta}\bar{Q}+i\theta Q$$
$$+\frac{1}{2}x_\mu[P^\mu,\theta Q]-\frac{1}{2}[\bar{\theta}\bar{Q},\theta Q]$$
$$-\frac{i}{2}\theta\sigma_\mu\bar{\theta}[P^\mu,\theta Q]\}$$

$$=\exp\{-ix_\mu P^\mu-\theta\sigma_\mu\bar{\theta}P^\mu+i\bar{\theta}\bar{Q}+i\theta Q$$
$$+\theta\sigma_\mu\bar{\theta}P^\mu\}$$

$$=\exp\{-ix_\mu P^\mu+i\bar{\theta}\bar{Q}+i\theta Q\}$$

$$=L(x,\theta,\bar{\theta}) \qquad \text{(with (6.5))}$$

We can now demonstrate that the relation (6.7) holds. For a given field configuration

$$\Phi_0\equiv\Phi(x_0,\theta_0,\bar{\theta}_0)$$

we have

$$\Phi(x,\theta,\bar{\theta}):=L(x,\theta,\bar{\theta})\Phi_0L^{-1}(x,\theta,\bar{\theta})$$
$$=L_1(x+i\theta\sigma\bar{\theta},\theta,\bar{\theta})\Phi_0L_1^{-1}(x+i\theta\sigma\bar{\theta},\theta,\bar{\theta})$$

(using (6.8))

$$=:\Phi_1(x+i\theta\sigma\bar{\theta},\theta,\bar{\theta}) \tag{6.10}$$

and

$$\Phi(x,\theta,\bar\theta) := L(x,\theta,\bar\theta)\Phi_0 L^{-1}(x,\theta,\bar\theta)$$
$$= L_2(x-i\theta\sigma\bar\theta,\theta,\bar\theta)\Phi_0 L_2^{-1}(x-i\theta\sigma\bar\theta,\theta,\bar\theta)$$
$$=: \Phi_2(x-i\theta\sigma\bar\theta,\theta,\bar\theta) \tag{6.11}$$

Under a finite supersymmetry transformation denoted by T_α the superfields Φ , Φ_1 and Φ_2 undergo the following transformations

$$T_\alpha\Phi(x,\theta,\bar\theta) = \Phi(x+i\theta\sigma\bar\alpha-i\alpha\sigma\bar\theta,\theta+\alpha,\bar\theta+\bar\alpha) \tag{6.12}$$

$$T_\alpha\Phi_1(x,\theta,\bar\theta) = \Phi_1(x+2i\theta\sigma\bar\alpha+i\alpha\sigma\bar\alpha,\theta+\alpha,\bar\theta+\bar\alpha) \tag{6.13}$$

$$T_\alpha\Phi_2(x,\theta,\bar\theta) = \Phi_2(x-2i\alpha\sigma\bar\theta-i\alpha\sigma\bar\alpha,\theta+\alpha,\bar\theta+\bar\alpha) \tag{6.14}$$

Proof: Considering the left hand side of (6.12) we have

$$T_\alpha\Phi(x,\theta,\bar\theta)$$
$$= L(0,\alpha,\bar\alpha)\Phi(x,\theta,\bar\theta)L^{-1}(0,\alpha,\bar\alpha)$$
$$= L(0,\alpha,\bar\alpha)[L(x,\theta,\bar\theta)\Phi_0 L^{-1}(x,\theta,\bar\theta)]L^{-1}(0,\alpha,\bar\alpha)$$
$$= [L(0,\alpha,\bar\alpha)L(x,\theta,\bar\theta)]\Phi_0[L(0,\alpha,\bar\alpha)L(x,\theta,\bar\theta)]^{-1} \tag{6.15}$$

Now (using (6.5) and then (6.9))

$$L(0,\alpha,\bar\alpha)L(x,\theta,\bar\theta)$$
$$= \exp\{i\alpha Q+i\bar\alpha\bar Q\}.\exp\{-ix_\mu P^\mu+i\theta Q+i\bar\theta\bar Q\}$$
$$= \exp\{-ix_\mu P_\mu+i(\theta+\alpha)Q+i(\bar\theta+\bar\alpha)\bar Q$$
$$+\tfrac{1}{2}[i\alpha Q+i\bar\alpha\bar Q,-ix_\mu P^\mu+i\theta Q+i\bar\theta\bar Q]\}$$

$$= \exp\{-ix_\mu P^\mu + i(\theta+\alpha)Q + i(\bar\theta+\bar\alpha)\bar Q$$
$$+ \tfrac{1}{2}x_\mu[\alpha Q, P^\mu] - \tfrac{1}{2}[\alpha Q, \theta Q]$$
$$- \tfrac{1}{2}[\alpha Q, \bar\theta\bar Q] + \tfrac{1}{2}x_\mu[\bar\alpha\bar Q, P^\mu]$$
$$- \tfrac{1}{2}[\bar\alpha\bar Q, \theta Q] - \tfrac{1}{2}[\bar\alpha\bar Q, \bar\theta\bar Q]\}$$
$$= \exp\{-ix_\mu P^\mu + i(\theta+\alpha)Q + i(\bar\theta+\bar\alpha)\bar Q$$
$$- \alpha\sigma_\mu\bar\theta P^\mu + \theta\sigma_\mu\bar\alpha P^\mu\}$$

(using (6.3) and (6.4))

$$= \exp\{-i[x_\mu - i\alpha\sigma_\mu\bar\theta + i\theta\sigma_\mu\bar\alpha]P^\mu$$
$$+ i(\theta+\alpha)Q + i(\bar\theta+\bar\alpha)\bar Q\}$$
$$= L(x - i\alpha\sigma\bar\theta + i\theta\sigma\bar\alpha,\ \theta+\alpha,\ \bar\theta+\bar\alpha)$$

(with (6.5)) (6.16)

Inserting (6.16) into (6.15) we obtain

$$T_\alpha \Phi(x,\theta,\bar\theta)$$
$$= L(x - i\alpha\sigma\bar\theta + i\theta\sigma\bar\alpha,\ \theta+\alpha,\ \bar\theta+\bar\alpha)\Phi_0 \cdot$$
$$\cdot L^{-1}(x - i\alpha\sigma\bar\theta + i\theta\sigma\bar\alpha,\ \theta+\alpha,\ \bar\theta+\bar\alpha)$$
$$= \Phi(x - i\alpha\sigma\bar\theta + i\theta\sigma\bar\alpha,\ \theta+\alpha,\ \bar\theta+\bar\alpha)$$

This proves (6.12). In order to prove (6.13) we consider

$$T_\alpha \Phi_1(x,\theta,\bar\theta)$$
$$= L_1(i\alpha\sigma\bar\alpha, \alpha, \bar\alpha)\Phi_1(x,\theta,\bar\theta)L_1^{-1}(i\alpha\sigma\bar\alpha, \alpha, \bar\alpha)$$
$$= L_1(i\alpha\sigma\bar\alpha, \alpha, \bar\alpha)[L_1(x,\theta,\bar\theta)\Phi_0 L_1^{-1}(x,\theta,\bar\theta)] \cdot$$
$$\cdot L_1^{-1}(i\alpha\sigma\bar\alpha, \alpha, \bar\alpha)$$

$$= [L_1(i\alpha\sigma\bar{\alpha}, \alpha, \bar{\alpha}) L_1(x, \theta, \bar{\theta})] \Phi_0 \cdot$$
$$\cdot [L_1(i\alpha\sigma\bar{\alpha}, \alpha, \bar{\alpha}) L_1(x, \theta, \bar{\theta})]^{-1}$$

Using (6.5) we have

$$L_1(i\alpha\sigma\bar{\alpha}, \alpha, \bar{\alpha}) L_1(x, \theta, \bar{\theta})$$
$$= exp\{-i(i\alpha\sigma_\mu\bar{\alpha})P^\mu + i\alpha Q\} exp\{i\bar{\alpha}\bar{Q}\} \cdot$$
$$\cdot exp\{-ix_\mu P^\mu + i\theta Q\} exp\{i\bar{\theta}\bar{Q}\}$$

In the next step we use the Baker-Campbell-Hausdorff formula (6.9) in the form of the following expression

$$exp\{A\} exp\{B\}$$
$$= exp\{A + B + \tfrac{1}{2}[A, B]\}$$
$$= exp\{A + B - \tfrac{1}{2}[A, B] + [A, B]\}$$
$$= exp\{B + A + \tfrac{1}{2}[B, A] + [A, B]\}$$
$$= exp\{B + [A, B]\} exp\{A\} \qquad (6.17)$$

Setting

$$e^A = e^{i\bar{\alpha}\bar{Q}} \quad \text{and} \quad e^B = e^{-ix_\mu P^\mu + i\theta Q}$$

we obtain

$$exp\{\alpha\sigma_\mu\bar{\alpha}P^\mu + i\alpha Q\} exp\{i\bar{\alpha}\bar{Q}\} \cdot$$
$$\cdot exp\{-ix_\mu P^\mu + i\theta Q\} exp\{i\bar{\theta}\bar{Q}\}$$
$$= exp\{\alpha\sigma_\mu\bar{\alpha}P^\mu + i\alpha Q\} \cdot$$
$$\cdot exp\{-ix_\mu P^\mu + i\theta Q + [i\bar{\alpha}\bar{Q}, -ix_\mu P^\mu + i\theta Q]\} \cdot$$
$$\cdot exp\{i\bar{\alpha}\bar{Q}\} exp\{i\bar{\theta}\bar{Q}\} \quad \text{(with (6.17))}$$

$$= \exp\{\alpha\sigma_\mu\bar{\alpha}P^\mu + i\alpha Q\}\cdot\exp\{-ix_\mu P^\mu + i\theta Q - [\bar{\alpha}\bar{Q}, \theta Q]\}\cdot\exp\{i(\bar{\theta}+\bar{\alpha})\bar{Q}\}$$

(using (6.4))

$$= \exp\{\alpha\sigma_\mu\bar{\alpha}P^\mu + i\alpha Q\}\exp\{-ix_\mu P^\mu + i\theta Q + 2\theta\sigma_\mu\bar{\alpha}P^\mu\}\cdot\exp\{i(\bar{\theta}+\bar{\alpha})\bar{Q}\}$$

(using (6.3))

$$= \exp\{\alpha\sigma_\mu\bar{\alpha}P^\mu + i\alpha Q\}\exp\{-i(x_\mu + 2i\theta\sigma_\mu\bar{\alpha})P^\mu + i\theta Q\}.\exp\{i(\bar{\theta}+\bar{\alpha})\bar{Q}\}$$

$$= \exp\{-i(x_\mu + 2i\theta\sigma_\mu\bar{\alpha} + i\alpha\sigma_\mu\bar{\alpha})P^\mu + i(\theta+\alpha)Q\}\cdot \cdot\exp\{i(\bar{\theta}+\bar{\alpha})\bar{Q}\}$$

(using (6.9))

In the last step we took into account that the commutator of the two exponents vanishes. Hence finally

$$L_1(i\alpha\sigma\bar{\alpha}, \alpha, \bar{\alpha})\, L_1(x, \theta, \bar{\theta}) = L_1(x + 2i\theta\sigma\bar{\alpha} + i\alpha\sigma\bar{\alpha}, \theta+\alpha, \bar{\theta}+\bar{\alpha}) \tag{6.18}$$

and we have

$$T_\alpha\Phi_1(x, \theta, \bar{\theta})$$
$$= [L_1(i\alpha\sigma\bar{\alpha}, \alpha, \bar{\alpha})\, L_1(x, \theta, \bar{\theta})]\,\Phi_0\cdot \cdot[L_1(i\alpha\sigma\bar{\alpha}, \alpha, \bar{\alpha})\, L_1(x, \theta, \bar{\theta})]^{-1}$$
$$= L_1(x + 2i\theta\sigma\bar{\alpha} + i\alpha\sigma\bar{\alpha}, \theta+\alpha, \bar{\theta}+\bar{\alpha})\,\Phi_0\cdot \cdot L_1^{-1}(x + 2i\theta\sigma\bar{\alpha} + i\alpha\sigma\bar{\alpha}, \theta+\alpha, \bar{\theta}+\bar{\alpha})$$
$$= \Phi(x_\mu + 2i\theta\sigma_\mu\bar{\alpha} + i\alpha\sigma_\mu\bar{\alpha}, \theta+\alpha, \bar{\theta}+\bar{\alpha})$$

In equations (6.12) to (6.14) the symbol T_α refers to a particular supersymmetry transformation. Thus T_α acting on superfields Φ corresponds to the application of the operator $L(0,\alpha,\bar{\alpha})$ whereas T_α acting on a superfield of type 1, i.e. Φ_1, requires the operator $L_1(i\alpha\sigma\bar{\alpha},\alpha,\bar{\alpha})$. This is due to the fact that according to (6.8) $L(0,\alpha,\bar{\alpha})$ and $L_1(i\alpha\sigma\bar{\alpha},\alpha,\bar{\alpha})$ generate the same supersymmetry transformation in different representations. Correspondingly T_α acting on superfields of type 2, i.e. Φ_2, requires operators $L_2(-i\alpha\sigma\bar{\alpha},\alpha,\bar{\alpha})$. Consider

$$
\begin{aligned}
& T_\alpha \Phi_2(x,\theta,\bar{\theta}) \\
&= L_2(-i\alpha\sigma\bar{\alpha},\alpha,\bar{\alpha})\Phi_2(x,\theta,\bar{\theta})L_2^{-1}(-i\alpha\sigma\bar{\alpha},\alpha,\bar{\alpha}) \\
&= L_2(-i\alpha\sigma\bar{\alpha},\alpha,\bar{\alpha})[L_2(x,\theta,\bar{\theta})\Phi_0 L_2^{-1}(x,\theta,\bar{\theta})]\cdot \\
&\qquad \cdot L_2^{-1}(-i\alpha\sigma\bar{\alpha},\alpha,\bar{\alpha}) \\
&= [L_2(-i\alpha\sigma\bar{\alpha},\alpha,\bar{\alpha})L_2(x,\theta,\bar{\theta})]\Phi_0\cdot \\
&\qquad \cdot[L_2(-i\alpha\sigma\bar{\alpha},\alpha,\bar{\alpha})L_2(x,\theta,\bar{\theta})]^{-1} \\
&= \text{(see below)}\ L_2(x_\mu-2i\alpha\sigma_\mu\bar{\theta}-i\alpha\sigma_\mu\bar{\alpha},\theta+\alpha,\bar{\theta}+\bar{\alpha})\cdot \\
&\qquad \cdot\Phi_0 L_2^{-1}(x_\mu-2i\alpha\sigma_\mu\bar{\theta}-i\alpha\sigma_\mu\bar{\alpha},\theta+\alpha,\bar{\theta}+\bar{\alpha}) \\
&= \Phi_2(x-2i\alpha\sigma\bar{\theta}-i\alpha\sigma\bar{\alpha},\theta+\alpha,\bar{\theta}+\bar{\alpha})
\end{aligned}
$$

where

$$
\begin{aligned}
& L_2(-i\alpha\sigma\bar{\alpha},\alpha,\bar{\alpha})L_2(x,\theta,\bar{\theta}) \\
&= \exp\{-i(-i\alpha\sigma_\mu\bar{\alpha})P^\mu+i\bar{\alpha}\bar{Q}\}\exp\{i\alpha Q\}\cdot \\
&\quad \exp\{-ix_\mu P^\mu+i\bar{\theta}\bar{Q}\}\exp\{i\theta Q\}
\end{aligned}
$$

(using (6.5))

$$= \exp\{-\alpha\sigma_\mu\bar{\alpha}P^\mu + i\bar{\alpha}\bar{Q}\}\cdot$$
$$\cdot\exp\{-ix_\mu P^\mu + i\bar{\theta}\bar{Q} + [i\alpha Q, i\bar{\theta}\bar{Q}]\}\cdot$$
$$\cdot\exp\{i(\theta+\alpha)Q\} \quad \text{(with (6.17))}$$
$$= \exp\{-\alpha\sigma_\mu\bar{\alpha}P^\mu + i\bar{\alpha}\bar{Q}\}\cdot$$
$$\cdot\exp\{-ix_\mu P^\mu + i\bar{\theta}\bar{Q} - 2\alpha\sigma_\mu\bar{\theta}P^\mu\}\cdot$$
$$\cdot\exp\{i(\theta+\alpha)Q\}$$
$$= \exp\{-\alpha\sigma_\mu\bar{\alpha}P^\mu + i\bar{\alpha}\bar{Q}\}\cdot$$
$$\cdot\exp\{-i(x_\mu - 2i\alpha\sigma_\mu\bar{\theta})P^\mu + i\bar{\theta}\bar{Q}\}\cdot$$
$$\cdot\exp\{i(\theta+\alpha)Q\}$$
$$= \exp\{-i(x_\mu - 2i\alpha\sigma_\mu\bar{\theta} - i\alpha\sigma_\mu\bar{\alpha})P^\mu$$
$$+ i(\bar{\theta}+\bar{\alpha})\bar{Q}\}\cdot\exp\{i(\theta+\alpha)Q\}$$
$$= L_2(x_\mu - 2i\alpha\sigma_\mu\bar{\theta} - i\alpha\sigma_\mu\bar{\alpha}, \theta+\alpha, \bar{\theta}+\bar{\alpha})$$

This completes the proof of the last of relations (6.12) to (6.14).

6.3.2 Infinitesimal Supersymmetry Transformations and Differential Operator Representations of the Generators

From equations (6.12) to (6.14) we can derive the transformation properties of the three types of superfields in infinitesimal form, i.e. we now consider T_α of (6.12) to (6.14) for infinitesimal α. Starting with the superfield Φ we have

$$
\begin{aligned}
\delta_S \Phi &= T_\alpha \Phi(x,\theta,\bar{\theta}) - \Phi(x,\theta,\bar{\theta}) \\
&= \Phi(x + i\theta\sigma\bar{\alpha} - i\alpha\sigma\bar{\theta}, \theta+\alpha, \bar{\theta}+\bar{\alpha}) \\
&\quad - \Phi(x,\theta,\bar{\theta})
\end{aligned}
$$

(using (6.12))

We now derive the differential operator representation of the generators Q , $\overline{Q}$ in much the same way as the differential operator representation of the momentum operator P is obtained. Thus, considering the translation of the variable x of some function f(x) we have

$$
\begin{aligned}
\delta_t f(x) &= f(x+a) - f(x) \\
&= a\frac{df}{dx} + \cdots \\
&= e^{iPa} f(x) e^{-iPa} - f(x) \\
&= ia(Pf - fP) + \cdots \\
&= ia(Pf)_c + \cdots
\end{aligned}
$$

where $(Pf)_c$ is a c-number. Equating the expression in the last line and that in the second line yields the operator representation $P = -i\, d/dx$. Proceeding analogously in the present case we obtain (with the appropriate Taylor expansions)

$$
\begin{aligned}
&\delta_S \Phi \\
&= \Phi(x,\theta,\bar{\theta}) + i(\theta\sigma^\mu\bar{\alpha} - \alpha\sigma^\mu\bar{\theta})\partial_\mu \Phi(x,\theta,\bar{\theta}) \\
&\quad + \alpha\frac{\partial}{\partial\theta}\Phi(x,\theta,\bar{\theta}) + \bar{\alpha}\frac{\partial}{\partial\bar{\theta}}\Phi(x,\theta,\bar{\theta}) \\
&\quad + \cdots - \Phi(x,\theta,\bar{\theta})
\end{aligned}
$$

$$= \left\{ \alpha \frac{\partial}{\partial \theta} + \bar{\alpha} \frac{\partial}{\partial \bar{\theta}} + i(\theta \sigma^{\mu} \bar{\alpha} - \alpha \sigma^{\mu} \bar{\theta}) \partial_{\mu} + \cdots \right\} \Phi(x, \theta, \bar{\theta}) \tag{6.19}$$

But we have also (using (6.5))

$$\begin{aligned}
&\delta_S \Phi \\
&= L(0, \alpha, \bar{\alpha}) \Phi(x, \theta, \bar{\theta}) L^{-1}(0, \alpha, \bar{\alpha}) - \Phi(x, \theta, \bar{\theta}) \\
&= \exp\{i\alpha Q + i\bar{\alpha}\bar{Q}\} \Phi(x, \theta, \bar{\theta}) . \\
&\quad . \exp\{-i\alpha Q - i\bar{\alpha}\bar{Q}\} - \Phi(x, \theta, \bar{\theta}) \\
&= (1 + i\alpha Q + i\bar{\alpha}\bar{Q} + \cdots) \Phi(x, \theta, \bar{\theta}) \cdot \\
&\quad \cdot (1 - i\alpha Q - i\bar{\alpha}\bar{Q} + \cdots) - \Phi(x, \theta, \bar{\theta}) \\
&= i\alpha Q \Phi(x, \theta, \bar{\theta}) + i\bar{\alpha}\bar{Q} \Phi(x, \theta, \bar{\theta}) \\
&\quad - i\Phi(x, \theta, \bar{\theta}) \alpha Q - i\Phi(x, \theta, \bar{\theta}) \bar{\alpha}\bar{Q} + \cdots \\
&= i[\alpha Q, \Phi(x, \theta, \bar{\theta})] + i[\bar{\alpha}\bar{Q}, \Phi(x, \theta, \bar{\theta})] + \cdots
\end{aligned} \tag{6.20}$$

where the derivatives with respect to Grassmann numbers are to be understood in the sense of the definition given in Section 6.2 . Considering any function $F(x, \theta, \bar{\theta})$ on superspace we have

$$\begin{aligned}
&i[\alpha Q, \Phi(x, \theta, \bar{\theta})] F(x, \theta, \bar{\theta}) \\
&= i\{\alpha Q \Phi(x, \theta, \bar{\theta}) - \Phi(x, \theta, \bar{\theta}) \alpha Q\} F(x, \theta, \bar{\theta}) \\
&= i\alpha Q (\Phi(x, \theta, \bar{\theta}) F(x, \theta, \bar{\theta})) \\
&\quad - \Phi(x, \theta, \bar{\theta}) (i\alpha Q F(x, \theta, \bar{\theta}))
\end{aligned}$$

$$= i\alpha[Q\Phi(x,\theta,\bar\theta)]F(x,\theta,\bar\theta)$$
$$+ i\Phi(x,\theta,\bar\theta)\alpha[QF(x,\theta,\bar\theta)]$$
$$- i\Phi(x,\theta,\bar\theta)\alpha[QF(x,\theta,\bar\theta)]$$
$$= i\alpha[Q\Phi(x,\theta,\bar\theta)]F(x,\theta,\bar\theta) \qquad (6.21)$$

Hence in the linear approximation we can rewrite (6.20)

$$\delta_S\Phi = i\alpha Q\Phi(x,\theta,\bar\theta) + i\bar\alpha\bar Q\Phi(x,\theta,\bar\theta)$$
$$= [i\alpha^A Q_A + i\bar\alpha_{\dot A}\bar Q^{\dot A}]\Phi(x,\theta,\bar\theta) \qquad (6.22)$$

Comparing equations (6.19) and (6.22) we obtain the following differential operator representation of the group generators

$$Q_A = -i(\partial_A - i\sigma^\mu_{A\dot B}\bar\theta^{\dot B}\partial_\mu)\,, \quad \partial_A \equiv \frac{\partial}{\partial\theta^A} \qquad (6.23)$$

$$\bar Q^{\dot A} = -i(\bar\partial^{\dot A} - i\theta^A\sigma^\mu_{A\dot B}\varepsilon^{\dot B\dot A}\partial_\mu)$$
$$= -i(\bar\partial^{\dot A} - i(\bar\sigma^\mu\theta)^{\dot A}\partial_\mu)\,, \quad \bar\partial^{\dot A} \equiv \frac{\partial}{\partial\bar\theta_{\dot A}} \qquad (6.24)$$

since, using (1.88) and (1.76),

$$(\bar\sigma^\mu\theta)^{\dot A} = \bar\sigma^{\mu\dot A A}\theta_A$$
$$= \varepsilon^{AB}\varepsilon^{\dot A\dot B}\sigma^\mu_{B\dot B}\theta_A$$
$$= \varepsilon^{BA}\theta_A\sigma^\mu_{B\dot B}\varepsilon^{\dot B\dot A}$$
$$= \theta^B\sigma^\mu_{B\dot B}\varepsilon^{\dot B\dot A}$$

We can proceed in a similar way for type 1 superfields Φ_1. Then, using (6.13),

$$\delta_S \Phi_1 = T_\alpha \Phi_1(x,\theta,\bar\theta) - \Phi(x,\theta,\bar\theta)$$

$$= \Phi_1(x+2i\theta\sigma\bar\alpha + i\alpha\sigma\bar\alpha, \theta+\alpha, \bar\theta+\bar\alpha)$$

$$- \Phi_1(x,\theta,\bar\theta)$$

$$= 2i\theta\sigma_\mu\bar\alpha\partial^\mu\Phi_1(x,\theta,\bar\theta) + \alpha^A \frac{\partial}{\partial\theta^A}\Phi_1(x,\theta,\bar\theta)$$

$$+ \bar\alpha_{\dot A}\frac{\partial}{\partial\bar\theta_{\dot A}}\Phi_1(x,\theta,\bar\theta) + \cdots$$

$$= \alpha^A \frac{\partial}{\partial\theta^A}\Phi_1(x,\theta,\bar\theta)$$

$$+ \bar\alpha_{\dot A}\left(\frac{\partial}{\partial\bar\theta_{\dot A}} - 2i\theta^A\sigma^\mu_{A\dot B}\varepsilon^{\dot B\dot A}\partial_\mu\right)\Phi_1(x,\theta,\bar\theta)$$

$$+ \cdots \tag{6.25}$$

On the other hand we also have

$$\delta_S\Phi_1$$

$$= L_1(i\alpha\sigma\bar\alpha,\alpha,\bar\alpha)\Phi_1(x,\theta,\bar\theta)L_1^{-1}(i\alpha\sigma\bar\alpha,\alpha,\bar\alpha)$$

$$- \Phi_1(x,\theta,\bar\theta)$$

$$= \exp\{\alpha\sigma_\mu\bar\alpha P^\mu + i\alpha Q^{(1)}\}\cdot\exp\{i\bar\alpha\bar Q^{(1)}\}\cdot$$

$$\cdot\Phi_1(x,\theta,\bar\theta)\cdot\exp\{-\alpha\sigma_\mu\bar\alpha P^\mu + i\alpha Q^{(1)}\}\cdot$$

$$\cdot\exp\{-i\bar\alpha\bar Q^{(1)}\} - \Phi_1(x,\theta,\bar\theta)$$

(with (6.5))

$$= (1+i\alpha Q^{(1)}+\cdots)(1+i\bar\alpha\bar Q^{(1)})\Phi_1(x,\theta,\bar\theta)\cdot$$

$$\cdot(1-i\alpha Q^{(1)}+\cdots)(1-i\bar\alpha\bar Q^{(1)}+\cdots) - \Phi_1(x,\theta,\bar\theta)$$

(using (6.9) and (6.3b))

$$= i[\alpha Q^{(1)}, \Phi_1(x,\theta,\bar\theta)] + i[\bar\alpha \bar Q^{(1)}, \Phi_1(x,\theta,\bar\theta)] + \cdots$$

$$= i\alpha^A (Q_A^{(1)} \Phi_1(x,\theta,\bar\theta)) + i\bar\alpha_{\dot A} (\bar Q^{(1)\dot A} \Phi_1(x,\theta,\bar\theta)) + \cdots \tag{6.26}$$

Comparing (6.25) and (6.26) we obtain the following representations of the group generators

$$Q_A^{(1)} = -i \frac{\partial}{\partial \theta^A} \equiv -i \partial_A \tag{6.27}$$

$$\bar Q^{(1)\dot A} = -i\left(\frac{\partial}{\partial \bar\theta_{\dot A}} - 2i\,\theta^A \sigma^\mu_{A\dot B} \varepsilon^{\dot B \dot A} \partial_\mu\right) \tag{6.28}$$

For convenience we add the following relation which is obtained with (6.4j)

$$\bar Q^{(1)}_{\dot A} = \varepsilon_{\dot A \dot B} \bar Q^{(1)\dot B}$$
$$= -i(-\bar\partial_{\dot A} + 2i\,\theta^A \sigma^\mu_{A\dot A} \partial_\mu)$$

Finally, for superfields of type 2, i.e. Φ_2, the variation under infinitesimal supersymmetry transformations is given by

$$\delta_S \Phi_2 = T_\alpha \Phi_2(x,\theta,\bar\theta) - \Phi_2(x,\theta,\bar\theta)$$
$$= \Phi_2(x - 2i\alpha\sigma\bar\theta - i\alpha\sigma\bar\alpha, \theta+\alpha, \bar\theta+\bar\alpha) - \Phi_2(x,\theta,\bar\theta)$$

(with (6.14))

$$= \Phi_2(x,\theta,\bar\theta) - 2i\alpha\sigma^\mu\bar\theta\,\partial_\mu \Phi_2(x,\theta,\bar\theta)$$
$$+ \alpha\frac{\partial}{\partial\theta}\Phi_2(x,\theta,\bar\theta) + \bar\alpha\frac{\partial}{\partial\bar\theta}\Phi_2(x,\theta,\bar\theta) + \cdots - \Phi_2(x,\theta,\bar\theta)$$

$$= \alpha^A \left[\frac{\partial}{\partial\theta^A} - 2i\sigma^\mu_{A\dot{B}}\bar{\theta}^{\dot{B}}\partial_\mu\right]\Phi_2(x,\theta,\bar{\theta})$$

$$+ \bar{\alpha}_{\dot{A}} \frac{\partial}{\partial\bar{\theta}_{\dot{A}}}\Phi_2(x,\theta,\bar{\theta}) + \cdots \qquad (6.29)$$

On the other hand

$$\delta_S\Phi_2(x,\theta,\bar{\theta})$$

$$= L_2(-i\alpha\sigma\bar{\alpha},\alpha,\bar{\alpha})\Phi_2(x,\theta,\bar{\theta})L_2^{-1}(-i\alpha\sigma\bar{\alpha},\alpha,\bar{\alpha}) - \Phi_2(x,\theta,\bar{\theta})$$

$$= \exp\{-\alpha\sigma_\mu\bar{\alpha}P^\mu + i\bar{\alpha}\bar{Q}^{(2)}\}\exp\{i\alpha Q^{(2)}\}\cdot \Phi_2(x,\theta,\bar{\theta})\cdot\exp\{\alpha\sigma_\mu\bar{\alpha}P^\mu - i\bar{\alpha}\bar{Q}^{(2)}\}\cdot \exp\{-i\alpha Q^{(2)}\} - \Phi_2(x,\theta,\bar{\theta})$$

(using (6.5))

$$= (1 + i\bar{\alpha}\bar{Q}^{(2)} + \cdots)(1 + i\alpha Q^{(2)} + \cdots)\Phi_2(x,\theta,\bar{\theta})\cdot (1 - i\bar{\alpha}\bar{Q}^{(2)} + \cdots)(1 - i\alpha Q^{(2)} + \cdots) - \Phi_2(x,\theta,\bar{\theta})$$

$$= i[\bar{\alpha}\bar{Q}^{(2)}, \Phi_2(x,\theta,\bar{\theta})] + i[\alpha Q^{(2)}, \Phi_2(x,\theta,\bar{\theta})] + \cdots$$

$$= i\bar{\alpha}_{\dot{A}}\left(\bar{Q}^{(2)\dot{A}}\Phi_2(x,\theta,\bar{\theta})\right) + i\left(Q_A^{(2)}\Phi_2(x,\theta,\bar{\theta})\right) + \cdots$$

Comparing this expression with (6.29) we obtain the following representation of the generators

$$Q_A^{(2)} = -i\left(\frac{\partial}{\partial\theta^A} - 2i\sigma^\mu_{A\dot{B}}\bar{\theta}^{\dot{B}}\partial_\mu\right) \qquad (6.30)$$

$$\left.\begin{aligned} \bar{Q}^{(2)\dot{A}} &= -i\frac{\partial}{\partial\bar{\theta}_{\dot{A}}} \equiv -i\bar{\partial}^{\dot{A}} \\ \bar{Q}^{(2)}_{\dot{A}} &= i\bar{\partial}_{\dot{A}} \end{aligned}\right\} \tag{6.31}$$

Equations (6.23), (6.24), (6.27) and (6.28), (6.30), (6.31) are three different representations of the spinor charges of the Super-Poincaré algebra as differential operators acting in superspace. For reasons of consistency these operators must obey the same anticommutation relations as the spinor charges of the algebra. It should be noted that the construction of the differential operator representations of the spinor charges was possible only as a result of the spinor extension of space-time. We now verify the following proposition.

<u>Proposition</u>: Given Q_A, $\bar{Q}_{\dot{A}}$ as in (6.23), (6.24), we have

$$\{Q_A, Q_B\} = \{\bar{Q}_{\dot{A}}, \bar{Q}_{\dot{B}}\} = 0 \tag{6.32}$$

$$\{Q_A, \bar{Q}_{\dot{B}}\} = -2i\sigma^{\mu}_{A\dot{B}}\partial_{\mu} = 2\sigma^{\mu}_{A\dot{B}}P_{\mu} \tag{6.33}$$

<u>Proof</u>: Consider

$$\begin{aligned} &-\{Q_A, Q_B\} \\ &= \{\partial_A - i\sigma^{\mu}_{A\dot{B}}\bar{\theta}^{\dot{B}}\partial_{\mu}, \partial_B - i\sigma^{\nu}_{B\dot{C}}\bar{\theta}^{\dot{C}}\partial_{\nu}\} \\ &= \{\partial_A, \partial_B\} - i\{\partial_A, \sigma^{\nu}_{B\dot{C}}\bar{\theta}^{\dot{C}}\partial_{\nu}\} \\ &\quad - i\{\sigma^{\mu}_{A\dot{B}}\bar{\theta}^{\dot{B}}\partial_{\mu}, \partial_B\} \\ &\quad - \{\sigma^{\mu}_{A\dot{B}}\bar{\theta}^{\dot{B}}\partial_{\mu}, \sigma^{\nu}_{B\dot{C}}\bar{\theta}^{\dot{C}}\partial_{\nu}\} \\ &= 0 \end{aligned}$$

since (see (6.4i))

$$\{\partial_A, \partial_B\} = 0, \quad \partial_A\bar{\theta}^{\dot{C}} = 0, \quad \{\bar{\theta}^{\dot{B}}, \bar{\theta}^{\dot{C}}\} = 0$$

This establishes the first of relations (6.32). Before we proceed we lower the index of $\bar{Q}^{\dot{A}}$. Thus, using (1.76d) we have (using (6.4j))

$$\begin{aligned}
\bar{Q}_{\dot{A}} &= \varepsilon_{\dot{A}\dot{B}}\bar{Q}^{\dot{B}} \\
&= \varepsilon_{\dot{A}\dot{B}}\frac{1}{i}\{\bar{\partial}^{\dot{B}} - i\theta^A\sigma^\mu_{A\dot{C}}\varepsilon^{\dot{C}\dot{B}}\partial_\mu\} \\
&= \frac{1}{i}\{-\bar{\partial}_{\dot{A}} - i\theta^A\sigma^\mu_{A\dot{C}}\varepsilon^{\dot{C}\dot{B}}\varepsilon_{\dot{A}\dot{B}}\partial_\mu\} \\
&= \frac{1}{i}\{-\bar{\partial}_{\dot{A}} + i\theta^A\sigma^\mu_{A\dot{C}}\delta^{\dot{C}}{}_{\dot{A}}\partial_\mu\}
\end{aligned}$$

Hence

$$\bar{Q}_{\dot{A}} = i(\bar{\partial}_{\dot{A}} - i\theta^A\sigma^\mu_{A\dot{A}}\partial_\mu) \tag{6.34}$$

We now have

$$\begin{aligned}
&-\{\bar{Q}_{\dot{A}}, \bar{Q}_{\dot{B}}\} \\
&= \{-\bar{\partial}_{\dot{A}} + i\theta^A\sigma^\mu_{A\dot{A}}\partial_\mu, -\bar{\partial}_{\dot{B}} + i\theta^B\sigma^\nu_{B\dot{B}}\partial_\nu\} \\
&= \{\bar{\partial}_{\dot{A}}, \bar{\partial}_{\dot{B}}\} - i\{\bar{\partial}_{\dot{A}}, \theta^B\sigma^\nu_{B\dot{B}}\partial_\nu\} \\
&\quad - i\{\theta^A\sigma^\mu_{A\dot{A}}\partial_\mu, \bar{\partial}_{\dot{B}}\} - \{\theta^A\sigma^\mu_{A\dot{A}}\partial_\mu, \theta^B\sigma^\nu_{B\dot{B}}\partial_\nu\} \\
&= 0
\end{aligned}$$

using (6.4n).

Finally consider

$$\begin{aligned}
&-\{Q_A, \bar{Q}_{\dot{B}}\} \\
&= \{\partial_A - i\sigma^\mu_{A\dot{C}}\bar{\theta}^{\dot{C}}\partial_\mu, -\bar{\partial}_{\dot{B}} + i\theta^C\sigma^\nu_{C\dot{B}}\partial_\nu\} \\
&= -\{\partial_A, \bar{\partial}_{\dot{B}}\} + i\{\partial_A, \theta^C\sigma^\nu_{C\dot{B}}\partial_\nu\} \\
&\qquad + i\{\sigma^\mu_{A\dot{C}}\bar{\theta}^{\dot{C}}\partial_\mu, \bar{\partial}_{\dot{B}}\} \\
&\qquad + \{\sigma^\mu_{A\dot{C}}\bar{\theta}^{\dot{C}}\partial_\mu, \theta^C\sigma^\nu_{C\dot{B}}\partial_\nu\}
\end{aligned}$$

$$= i\left(\frac{\partial}{\partial\theta^A}\theta^C\right)\sigma^\nu_{C\dot{B}}\partial_\nu + i\left(\frac{\partial}{\partial\bar{\theta}^{\dot{B}}}\sigma^\mu_{A\dot{C}}\bar{\theta}^{\dot{C}}\right)\partial_\mu$$

$$= i\,\delta_A{}^C\sigma^\nu_{C\dot{B}}\partial_\nu + i\,\sigma^\mu_{A\dot{C}}\,\delta^{\dot{C}}{}_{\dot{B}}\partial_\mu$$

$$= i\,\sigma^\mu_{A\dot{B}}\partial_\mu + i\,\sigma^\mu_{A\dot{B}}\partial_\mu$$

$$= 2i\,\sigma^\mu_{A\dot{B}}\partial_\mu = -2\,\sigma^\mu_{A\dot{B}}P_\mu \qquad (6.35)$$

Similarly we can establish the following result.

Proposition: The differential operators defined by (6.27), (6.28) and (6.30), (6.31) satisfy the following relations

$$\{Q^{(1)}_A, Q^{(1)}_B\} = \{\bar{Q}^{(1)}_{\dot{A}}, \bar{Q}^{(1)}_{\dot{B}}\} = 0 \qquad (6.36)$$

$$\{Q^{(1)}_A, \bar{Q}^{(1)}_{\dot{B}}\} = -2i\,\sigma^\mu_{A\dot{B}}\partial_\mu = 2\,\sigma^\mu_{A\dot{B}}P_\mu \qquad (6.37)$$

and

$$\{Q^{(2)}_A, Q^{(2)}_B\} = \{\bar{Q}^{(2)}_{\dot{A}}, \bar{Q}^{(2)}_{\dot{B}}\} = 0 \qquad (6.38)$$

$$\{Q^{(2)}_A, \bar{Q}^{(2)}_{\dot{B}}\} = -2i\,\sigma^\mu_{A\dot{B}}\partial_\mu = 2\,\sigma^\mu_{A\dot{B}}P_\mu \qquad (6.39)$$

Proof: Equations (6.36) are readily verified. Consider (using (6.27) and (6.28))

$$\{Q^{(1)}_A, \bar{Q}^{(1)}_{\dot{A}}\} = -\{\partial_A, -\bar{\partial}_{\dot{A}} + 2i\theta^B\sigma^\mu_{B\dot{A}}\partial_\mu\}$$

$$= \{\partial_A, \bar{\partial}_{\dot{A}}\} - 2i\{\partial_A, \theta^B\sigma^\mu_{B\dot{A}}\partial_\mu\}$$

$$= -2i\frac{\partial}{\partial\theta^A}(\theta^B\sigma^\mu_{B\dot{A}}\partial_\mu) = -2i\,\delta_A{}^B\sigma^\mu_{B\dot{A}}\partial_\mu$$

$$= -2i\,\sigma^\mu_{A\dot{A}}\partial_\mu = 2\,\sigma^\mu_{A\dot{A}}P_\mu$$

Equations (6.38) are also readily seen to hold. Consider therefore (6.39). We have (using (6.30), (6.31))

$$\{Q^{(2)}_A, \bar{Q}_{\dot{B}}{}^{(2)}\} = -\{\partial_A - 2i\,\sigma^\mu_{A\dot{C}}\bar{\theta}^{\dot{C}}\partial_\mu, -\bar{\partial}_{\dot{B}}\}$$

$$= \{\partial_A, \bar{\partial}_{\dot{B}}\} - 2i\{\sigma^\mu_{A\dot{C}}\bar{\theta}^{\dot{C}}\partial_\mu, \bar{\partial}_{\dot{B}}\}$$
$$= -2i\bar{\partial}_{\dot{B}}(\sigma^\mu_{A\dot{C}}\bar{\theta}^{\dot{C}}\partial_\mu)$$
$$= -2i\sigma^\mu_{A\dot{C}}\delta^{\dot{C}}{}_{\dot{B}}\partial_\mu$$
$$= -2i\sigma^\mu_{A\dot{B}}\partial_\mu = 2\sigma^\mu_{A\dot{B}}P_\mu$$

6.4 Consistency with the Majorana Formalism

We can verify the consistency of the anticommutators obtained above with those of the Majorana formulation, i.e. (3.23). In order to see this, we need in addition to (6.23), (6.24) and (6.34) the derivative representation of Q^A. This may be obtained from (6.23). Thus

$$Q^A = \varepsilon^{AB}Q_B$$
$$= -i\varepsilon^{AB}(\partial_B - i\sigma^\mu_{B\dot{B}}\bar{\theta}^{\dot{B}}\partial_\mu)$$
$$= i(\partial^A + i\varepsilon^{AB}\sigma^\mu_{B\dot{B}}\bar{\theta}^{\dot{B}}\partial_\mu)$$

(using (6.4g)

$$= i(\partial^A + i\varepsilon^{AB}(\varepsilon_{BD}\varepsilon_{\dot{B}\dot{D}}\bar{\sigma}^{\mu\dot{D}D})\bar{\theta}^{\dot{B}}\partial_\mu)$$
$$= i(\partial^A - i\delta^A{}_D\varepsilon_{\dot{D}\dot{B}}\bar{\theta}^{\dot{B}}\bar{\sigma}^{\mu\dot{D}D}\partial_\mu)$$

(using (1.88))

$$= i(\partial^A - i\bar{\theta}_{\dot{D}}\bar{\sigma}^{\mu\dot{D}A}\partial_\mu) \qquad \text{(with (1.76d))}$$

This result together with (6.23), (6.24) and (6.34) allows us to write

$$\begin{pmatrix} Q_A \\ \bar{Q}^{\dot{A}} \end{pmatrix} = -i\left[\begin{pmatrix} \frac{\partial}{\partial\theta^A} \\ \frac{\partial}{\partial\bar{\theta}_{\dot{A}}} \end{pmatrix} - i\begin{pmatrix} 0 & \sigma^\mu_{A\dot{B}} \\ \bar{\sigma}^{\mu\dot{A}B} & 0 \end{pmatrix}\begin{pmatrix} \theta_B \\ \bar{\theta}^{\dot{B}} \end{pmatrix}\partial_\mu\right]$$

and

$$(Q^A, \bar{Q}_{\dot{A}})$$

$$= -i\left[\left(-\frac{\partial}{\partial\theta_A}, -\frac{\partial}{\partial\bar{\theta}^{\dot{A}}}\right)\right.$$

$$\left. + i(\theta^B, \bar{\theta}_{\dot{B}})\begin{pmatrix} 0 & \sigma^\mu_{B\dot{A}} \\ \bar{\sigma}^{\mu\dot{B}A} & 0 \end{pmatrix}\partial_\mu\right]$$

The Majorana character of these spinors is evident since the same Q appears in both Weyl spinors. Defining as usual

$$(\theta)_a = \begin{pmatrix} \theta_B \\ \bar{\theta}^{\dot{B}} \end{pmatrix} \quad \text{so that} \quad (\bar{\theta})_a = (\theta^B, \bar{\theta}_{\dot{B}})$$

we have

$$\left(\frac{\partial}{\partial\theta}\right)_a = \left(\frac{\partial}{\partial\theta_A}, \frac{\partial}{\partial\bar{\theta}^{\dot{A}}}\right), \quad \left(\frac{\partial}{\partial\bar{\theta}}\right)_a = \begin{pmatrix} \frac{\partial}{\partial\theta^A} \\ \frac{\partial}{\partial\bar{\theta}_{\dot{A}}} \end{pmatrix}$$

Hence we can rewrite the above relations

$$(Q) = -i\left(\frac{\partial}{\partial\bar{\theta}} - i\gamma^\mu\theta\,\partial_\mu\right)$$

$$(\bar{Q}) = -i\left(-\frac{\partial}{\partial\theta} + i\,\bar{\theta}\gamma^\mu\partial_\mu\right)$$

and the anticommutator (3.23), i.e.

$$\{Q_a, \bar{Q}_b\} = 2\gamma^\mu_{ab} P_\mu$$

is readily verified. Together with this anticommutator we also have

$$\{Q_a, Q_b\} = -2(\gamma^\mu C)_{ab} P_\mu$$

and

$$\{\bar{Q}_a, \bar{Q}_b\} = 2(C^{-1}\gamma^\mu)_{ab} P_\mu$$

The first of these relations has been obtained before (see (3.19) , where a = - 2 as stated in the corollary following (3.23)). We can verify the last relation by considering (since for Majorana spinors $\psi = \psi^c = C\bar{\Psi}^T$, $\psi^T = \bar{\Psi} C^T = \bar{\Psi} C^{-1}$)

$$\{Q_a, Q_b\} = \{\bar{Q}_{a'}, \bar{Q}_{b'}\} C^T_{a'a} C^T_{b'b}$$

$$= 2 C_{aa'} (C^{-1}\gamma^\mu)_{a'b'} C^T_{b'b} P_\mu$$

$$= 2 (\gamma^\mu C^T)_{ab} P_\mu$$

$$= -2 (\gamma^\mu C)_{ab} P_\mu \qquad \text{(since } C^T = - C)$$

We can verify that the derivative representations of $Q_a, \bar{Q}_a$ indeed satisfy the above anticommutation relations among themselves. Thus

$$\{Q_a, Q_b\} = -2 \frac{\partial}{\partial \bar{\theta}_a} [-i(\gamma^\mu \theta)_b \partial_\mu]$$

Using (3.24b) this becomes

$$\{Q_a, Q_b\} = -2(\gamma^\mu)_{bc} P_\mu \frac{\partial}{\partial \bar{\theta}_a} \theta_c$$

$$= -2(\gamma^\mu)_{bc} P_\mu C_{ca}$$

$$= -2(\gamma^\mu C)_{ba} P_\mu$$

$$= -2(\gamma^\mu C)_{ab} P_\mu$$

since $\gamma^\mu C$ is symmetric. Correspondingly

$$\{\bar{Q}_a, \bar{Q}_b\}$$

$$= -2\left[-\frac{\partial}{\partial\theta_a} i(\bar{\theta}\gamma^\mu)_b \partial_\mu\right]$$

$$= -2\frac{\partial}{\partial\theta_a}(\bar{\theta}\gamma^\mu)_b P_\mu$$

$$= -2(C\gamma^\mu)_{ab} P_\mu \qquad \text{(with (3.24c))}$$

$$= 2(C^{-1}\gamma^\mu)_{ab} P_\mu \qquad (\text{since } C = -C^{-1})$$

6.5 Covariant Derivatives

Covariant derivatives are derivatives which are useful in the construction of manifestly supersymmetric Lagrangians. We can define three types of covariant derivatives corresponding to the three types of superfields introduced previously.

We define as covariant derivatives for superfields $\Phi(x,\theta,\bar{\theta})$ the operators

$$\left.\begin{aligned} D_A &:= \partial_A + i\sigma^\mu_{A\dot{B}}\bar{\theta}^{\dot{B}}\partial_\mu \\ D^A &:= \varepsilon^{AB}D_B = -\partial^A - i\bar{\theta}_{\dot{C}}\bar{\sigma}^{\mu\dot{C}A}\partial_\mu \end{aligned}\right\} \qquad (6.40)$$

$$\left.\begin{aligned} \bar{D}_{\dot{A}} &:= -\bar{\partial}_{\dot{A}} - i\theta^B\sigma^\mu_{B\dot{A}}\partial_\mu \\ \bar{D}^{\dot{A}} &:= \varepsilon^{\dot{A}\dot{B}}\bar{D}_{\dot{B}} = \bar{\partial}^{\dot{A}} + i\bar{\sigma}^{\mu\dot{A}C}\theta_C\partial_\mu \end{aligned}\right\} \qquad (6.41)$$

Proposition: The operators D_A and $\overline{D}_{\dot{A}}$ are covariant derivatives, i.e. D_A and $\overline{D}_{\dot{A}}$ are invariant under supersymmetry transformations in the sense that

$$[D_A, \delta_S] = 0 \tag{6.42a}$$

$$[\overline{D}_{\dot{A}}, \delta_S] = 0 \tag{6.43}$$

Proof: Consider (using (6.22))

$$\begin{aligned}
&[D_A, \delta_S]\Phi(x,\theta,\overline{\theta}) \\
&= [D_A, i\alpha Q + i\overline{\alpha}\overline{Q}]\Phi(x,\theta,\overline{\theta}) \\
&= i[D_A, \alpha^B Q_B]\Phi(x,\theta,\overline{\theta}) \\
&\qquad + i[D_A, \overline{\alpha}_{\dot{B}}\overline{Q}^{\dot{B}}]\Phi(x,\theta,\overline{\theta}) \\
&= i(D_A\alpha^B Q_B - \alpha^B Q_B D_A)\Phi(x,\theta,\overline{\theta}) \\
&\quad + i(D_A\overline{\alpha}_{\dot{B}}\overline{Q}^{\dot{B}} - \overline{\alpha}_{\dot{B}}\overline{Q}^{\dot{B}}D_A)\Phi(x,\theta,\overline{\theta}) \\
&= -i\alpha^B(D_A Q_B + Q_B D_A)\Phi(x,\theta,\overline{\theta}) \\
&\quad - i\overline{\alpha}_{\dot{B}}(D_A\overline{Q}^{\dot{B}} + \overline{Q}^{\dot{B}}D_A)\Phi(x,\theta,\overline{\theta}) \\
&= -i\alpha^B\{D_A, Q_B\}\Phi(x,\theta,\overline{\theta}) \\
&\quad - i\overline{\alpha}_{\dot{B}}\{D_A, \overline{Q}^{\dot{B}}\}\Phi(x,\theta,\overline{\theta})
\end{aligned}$$

Thus in order to prove (6.42a) we have to demonstrate that

$$\{D_A, Q_B\} = 0, \quad \{D_A, \overline{Q}^{\dot{B}}\} = 0 \tag{6.42b}$$

Consider

$$\begin{aligned}
&\{D_A, Q_B\} \\
&= -i\{\partial_A + i\sigma^\mu_{A\dot{B}}\overline{\theta}^{\dot{B}}\partial_\mu, \partial_B - i\sigma^\nu_{B\dot{C}}\overline{\theta}^{\dot{C}}\partial_\nu\} \\
&= -i\{\partial_A, \partial_B\} - \{\partial_A, \sigma^\nu_{B\dot{C}}\overline{\theta}^{\dot{C}}\partial_\nu\}
\end{aligned}$$

$$+\{\sigma^{\mu}_{A\dot{B}}\bar{\theta}^{\dot{B}}\partial_\mu, \partial_B\}$$
$$-i\{\sigma^{\mu}_{A\dot{B}}\bar{\theta}^{\dot{B}}\partial_\mu, \sigma^{\nu}_{B\dot{C}}\bar{\theta}^{\dot{C}}\partial_\nu\}$$
$$= 0 \tag{6.44}$$

since

$$\{\partial_A, \partial_B\} = \{\partial_A, \bar{\theta}^{\dot{B}}\} = \{\bar{\theta}^{\dot{B}}, \bar{\theta}^{\dot{C}}\} = 0$$

Now consider

$$i\{D_A, \bar{Q}^{\dot{B}}\}$$
$$= \{\partial_A + i\sigma^{\mu}_{A\dot{C}}\bar{\theta}^{\dot{C}}\partial_\mu, \bar{\partial}^{\dot{B}} - i\theta^C\sigma^{\nu}_{C\dot{D}}\varepsilon^{\dot{D}\dot{B}}\partial_\nu\}$$
$$= \{\partial_A, \bar{\partial}^{\dot{B}}\} - i\{\partial_A, \theta^C\sigma^{\nu}_{C\dot{D}}\varepsilon^{\dot{D}\dot{B}}\partial_\nu\}$$
$$+ i\{\sigma^{\mu}_{A\dot{C}}\bar{\theta}^{\dot{C}}\partial_\mu, \bar{\partial}^{\dot{B}}\}$$
$$+ \{\sigma^{\mu}_{A\dot{C}}\bar{\theta}^{\dot{C}}\partial_\mu, \theta^C\sigma^{\nu}_{C\dot{D}}\varepsilon^{\dot{D}\dot{B}}\partial_\nu\}$$
$$= -i\sigma^{\nu}_{A\dot{D}}\varepsilon^{\dot{D}\dot{B}}\partial_\nu + i\sigma^{\mu}_{A\dot{C}}(\bar{\partial}^{\dot{B}}\bar{\theta}^{\dot{C}})\partial_\mu$$

Using (6.4j) we have

$$\bar{\partial}^{\dot{B}}\bar{\theta}^{\dot{C}} = -\varepsilon^{\dot{B}\dot{D}}\bar{\partial}_{\dot{D}}\bar{\theta}^{\dot{C}} = -\varepsilon^{\dot{B}\dot{D}}\delta_{\dot{D}}^{\dot{C}}$$
$$= -\varepsilon^{\dot{B}\dot{C}} \tag{6.45}$$

Hence

$$i\{D_A, \bar{Q}^{\dot{B}}\}$$
$$= -i\sigma^{\nu}_{A\dot{D}}\varepsilon^{\dot{D}\dot{B}}\partial_\nu + i\sigma^{\mu}_{A\dot{C}}(-\varepsilon^{\dot{B}\dot{C}})\partial_\mu$$
$$= -i\sigma^{\nu}_{A\dot{D}}\varepsilon^{\dot{D}\dot{B}}\partial_\nu + i\sigma^{\nu}_{A\dot{C}}\varepsilon^{\dot{C}\dot{B}}\partial_\nu$$
$$= 0 \tag{6.46}$$

With (6.44) and (6.46) we therefore obtain (6.42a).

Now

$$[\bar{D}_{\dot{A}}, \delta_S]\,\Phi(x,\theta,\bar{\theta})$$

$$= [\bar{D}_{\dot{A}}, i\alpha^B Q_B + i\bar{\alpha}_{\dot{B}}\bar{Q}^{\dot{B}}]\,\Phi(x,\theta,\bar{\theta})$$

$$= \left([\bar{D}_{\dot{A}}, i\alpha^B Q_B] + [\bar{D}_{\dot{A}}, i\bar{\alpha}_{\dot{B}}\bar{Q}^{\dot{B}}]\right)\Phi(x,\theta,\bar{\theta})$$

$$= \left(-i\alpha^B\{\bar{D}_{\dot{A}}, Q_B\} - i\bar{\alpha}_{\dot{B}}\{\bar{D}_{\dot{A}}, \bar{Q}^{\dot{B}}\}\right)\Phi(x,\theta,\bar{\theta})$$

Hence in order to demonstrate (6.43) we have to show that

$$\{\bar{D}_{\dot{A}}, Q_B\} = \{\bar{D}_{\dot{A}}, \bar{Q}^{\dot{B}}\} = 0$$

Consider

$$i\{\bar{D}_{\dot{A}}, Q_B\}$$

$$= \{-\bar{\partial}_{\dot{A}} - i\theta^B\sigma^\mu_{B\dot{A}}\partial_\mu, \partial_B - i\sigma^\mu_{B\dot{C}}\bar{\theta}^{\dot{C}}\partial_\mu\}$$

$$= -\{\bar{\partial}_{\dot{A}}, \partial_B\} + i\{\bar{\partial}_{\dot{A}}, \sigma^\mu_{B\dot{C}}\bar{\theta}^{\dot{C}}\partial_\mu\}$$

$$-i\{\theta^C\sigma^\mu_{C\dot{A}}\partial_\mu, \partial_B\} - \{\theta^C\sigma^\mu_{C\dot{A}}\partial_\mu, \sigma^\mu_{B\dot{C}}\bar{\theta}^{\dot{C}}\partial_\mu\}$$

$$= -i\sigma^\mu_{B\dot{C}}(\bar{\partial}_{\dot{A}}\bar{\theta}^{\dot{C}})\partial_\mu - i\partial_B\theta^C\sigma^\mu_{C\dot{A}}\partial_\mu$$

$$= i\sigma^\mu_{B\dot{C}}\delta^{\dot{C}}{}_{\dot{A}}\partial_\mu - i\delta_B{}^C\sigma^\mu_{C\dot{A}}\partial_\mu$$

$$= i\sigma^\mu_{B\dot{A}}\partial_\mu - i\sigma^\mu_{B\dot{A}}\partial_\mu$$

$$= 0$$

and

$$i\{\bar{D}_{\dot{A}}, \bar{Q}^{\dot{B}}\}$$

$$= \{-\bar{\partial}_{\dot{A}} - i\theta^B\sigma^\mu_{B\dot{A}}\partial_\mu, \bar{\partial}^{\dot{B}} - i\theta^C\sigma^\mu_{C\dot{D}}\varepsilon^{\dot{D}\dot{B}}\partial_\mu\}$$

$$= -\{\bar{\partial}_{\dot{A}}, \bar{\partial}^{\dot{B}}\} + i\{\bar{\partial}_{\dot{A}}, \theta^C\sigma^\mu_{C\dot{D}}\varepsilon^{\dot{D}\dot{B}}\partial_\mu\}$$

$$-i\{\theta^B\sigma^\mu_{B\dot{A}}\partial_\mu, \bar{\partial}^{\dot{B}}\}$$
$$-\{\theta^B\sigma^\mu_{B\dot{A}}\partial_\mu, \theta^C\sigma^\mu_{C\dot{D}}\varepsilon^{\dot{D}\dot{B}}\partial_\mu\}$$

$$= 0$$

since
$$\{\bar{\partial}_{\dot{A}}, \bar{\partial}^{\dot{B}}\} = 0 \ , \ \bar{\partial}_{\dot{A}}\theta^C = 0$$
and
$$\{\theta^B, \theta^C\} = 0$$

Hence we obtain the result
$$[\bar{D}_{\dot{A}}, \delta_S]\Phi(x,\theta,\bar{\theta}) = 0$$
This completes the proof of (6.42a) and (6.43).

In a similar way we can find covariant derivatives for superfields of types 1 and 2.

<u>Proposition</u>: The operators
$$D_A^{(1)} := \partial_A + 2i\sigma^\mu_{A\dot{B}}\bar{\theta}^{\dot{B}}\partial_\mu \tag{6.47}$$
$$\bar{D}_{\dot{A}}^{(1)} := -\bar{\partial}_{\dot{A}} \tag{6.48}$$
and
$$D_A^{(2)} := \partial_A \tag{6.49}$$
$$\bar{D}_{\dot{A}}^{(2)} := -\bar{\partial}_{\dot{A}} - 2i\theta^B\sigma^\mu_{B\dot{A}}\partial_\mu \tag{6.50}$$
are covariant derivatives, i.e.
$$\left.\begin{aligned} [D_A^{(1)}, \delta_S]\Phi_1(x,\theta,\bar{\theta}) &= 0 \\ [\bar{D}_{\dot{A}}^{(1)}, \delta_S]\Phi_1(x,\theta,\bar{\theta}) &= 0 \end{aligned}\right\} \tag{6.51}$$

and

$$\left.\begin{array}{l}[\mathcal{D}_A^{(2)}, \delta_S]\, \Phi_2(x,\theta,\bar{\theta}) = 0 \\ [\bar{\mathcal{D}}_{\dot{A}}^{(2)}, \delta_S]\, \Phi_2(x,\theta,\bar{\theta}) = 0\end{array}\right\} \qquad (6.52)$$

Proof: Consider (using (6.26))

$$[\mathcal{D}_A^{(1)}, \delta_S]\, \Phi_1(x,\theta,\bar{\theta})$$

$$= [\mathcal{D}_A^{(1)}, i\alpha Q^{(1)} + i\bar{\alpha}\bar{Q}^{(1)}]\, \Phi_1(x,\theta,\bar{\theta})$$

$$= -i\alpha^B \{\mathcal{D}_A^{(1)}, Q_B^{(1)}\}\, \Phi_1(x,\theta,\bar{\theta})$$

$$- i\bar{\alpha}_{\dot{A}} \{\mathcal{D}_A^{(1)}, \bar{Q}^{(1)\dot{A}}\}\, \Phi_1(x,\theta,\bar{\theta})$$

But

$$\{\mathcal{D}_A^{(1)}, Q_B^{(1)}\}$$

$$= -i\{\partial_A + 2i\sigma^\mu_{A\dot{B}}\bar{\theta}^{\dot{B}}\partial_\mu, \partial_B\}$$

(with (6.47) and (6.27))

$$= -i\{\partial_A, \partial_B\} + \{\sigma^\mu_{A\dot{B}}\bar{\theta}^{\dot{B}}\partial_\mu, \partial_B\}$$

and

$$i\{\mathcal{D}_A^{(1)}, \bar{Q}^{(1)\dot{A}}\}$$

$$= \{\partial_A + 2i\sigma^\mu_{A\dot{B}}\bar{\theta}^{\dot{B}}\partial_\mu, \bar{\partial}^{\dot{A}} - 2i\theta^C\sigma^\mu_{C\dot{B}}\varepsilon^{\dot{B}\dot{A}}\partial_\mu\}$$

(with (6.47) and (6.28))

$$= \{\partial_A, \bar{\partial}^{\dot{A}}\} - 2i\{\partial_A, \theta^C\sigma^\mu_{C\dot{B}}\varepsilon^{\dot{B}\dot{A}}\partial_\mu\}$$

$$+ 2i\{\sigma^\mu_{A\dot{B}}\bar{\theta}^{\dot{B}}\partial_\mu, \bar{\partial}^{\dot{A}}\}$$

$$+ 4\{\sigma^\mu_{A\dot{B}}\bar{\theta}^{\dot{B}}\partial_\mu, \theta^C\sigma^\mu_{C\dot{B}}\varepsilon^{\dot{B}\dot{A}}\partial_\mu\}$$

$$= -2i(\partial_A\theta^C)\sigma^\mu_{C\dot{B}}\varepsilon^{\dot{B}\dot{A}}\partial_\mu$$

$$+ 2i\sigma^{\mu}_{A\dot{B}}(\bar{\partial}^{\dot{A}}\bar{\theta}^{\dot{B}})\partial_{\mu}$$

$$= - 2i\delta_{A}{}^{C}\sigma^{\mu}_{C\dot{B}}\varepsilon^{\dot{B}\dot{A}}\partial_{\mu} + 2i\sigma^{\mu}_{A\dot{B}}\varepsilon^{\dot{B}\dot{A}}\partial_{\mu}$$

(with (6.46))

$$= 0$$

This demonstrates that $D_A^{(1)}$ is a covariant derivative for type 1 superfields. Now consider (using (6.26))

$$[\bar{D}_{\dot{A}}^{(1)}, \delta_S]\Phi_1(x,\theta,\bar{\theta})$$

$$= [\bar{D}_{\dot{A}}^{(1)}, i\alpha Q^{(1)} + i\bar{\alpha}\bar{Q}^{(1)}]\Phi_1(x,\theta,\bar{\theta})$$

$$= -i\alpha^A\{\bar{D}_{\dot{A}}^{(1)}, Q_A^{(1)}\}\Phi_1(x,\theta,\bar{\theta})$$

$$-i\bar{\alpha}_{\dot{B}}\{\bar{D}_{\dot{A}}^{(1)}, \bar{Q}^{(1)\dot{B}}\}\Phi_1(x,\theta,\bar{\theta})$$

Now

$$\{\bar{D}_{\dot{A}}^{(1)}, Q_A^{(1)}\}$$

$$= -i\{-\bar{\partial}_{\dot{A}}, \partial_A\} \quad \text{(using (6.27), (6.48))}$$

$$= 0$$

and

$$\{\bar{D}_{\dot{A}}^{(1)}, \bar{Q}^{(1)\dot{B}}\}$$

$$= -i\{-\bar{\partial}_{\dot{A}}, \bar{\partial}^{\dot{B}} - 2i\theta^A\sigma^{\mu}_{A\dot{C}}\varepsilon^{\dot{C}\dot{B}}\partial_{\mu}\}$$

(using (6.48) and (6.28))

$$= 0$$

Hence

$$[\bar{D}_{\dot{A}}^{(1)}, \delta_S]\Phi_1(x,\theta,\bar{\theta}) = 0$$

and (6.51) is verified. In order to verify (6.52) we

consider

$$[D_A^{(2)}, \delta_S]\Phi_2(x,\theta,\bar\theta)$$
$$= [D_A^{(2)}, i\alpha Q^{(2)} + i\bar\alpha\bar Q^{(2)}]\Phi_2(x,\theta,\bar\theta)$$
$$= -i\alpha^B\{D_A^{(2)}, Q_B^{(2)}\}\Phi_2(x,\theta,\bar\theta)$$
$$-i\bar\alpha_{\dot B}\{D_A^{(2)}, \bar Q^{(2)\dot B}\}\Phi_2(x,\theta,\bar\theta)$$

Again we have to evaluate anticommutators. Thus, using (6.49) and (6.30) we have

$$\{D_A^{(2)}, Q_B^{(2)}\}$$
$$= -i\{\partial_A, \partial_B - 2i\sigma^\mu_{B\dot C}\bar\theta^{\dot C}\partial_\mu\}$$
$$= -i\{\partial_A, \partial_B\} - 2\{\partial_A, \sigma^\mu_{B\dot C}\bar\theta^{\dot C}\partial_\mu\}$$
$$= 0$$

Analogously, using (6.49) and (6.31), we have

$$\{D_A^{(2)}, \bar Q^{(2)\dot B}\}$$
$$= \{\partial_A, \bar\partial^{\dot B}\}$$
$$= 0$$

Hence

$$[D_A^{(2)}, \delta_S]\Phi_2(x,\theta,\bar\theta) = 0$$

Finally consider

$$[\bar D_{\dot A}^{(2)}, \delta_S]\Phi_2(x,\theta,\bar\theta)$$
$$= [\bar D_{\dot A}^{(2)}, i\alpha Q^{(2)} + i\bar\alpha\bar Q^{(2)}]\Phi_2(x,\theta,\bar\theta)$$
$$= -i\alpha^A\{\bar D_{\dot A}^{(2)}, Q_A^{(2)}\}\Phi_2(x,\theta,\theta)$$

$$-i\bar{\alpha}_{\dot{B}}\{\bar{D}^{(2)}_{\dot{A}}, \bar{Q}^{(2)\dot{B}}\}\Phi_2(x,\theta,\bar{\theta})$$

Using (6.30) and (6.50) we have

$$\begin{aligned}
& i\{\bar{D}^{(2)}_{\dot{A}}, Q^{(2)}_{A}\} \\
&= \{-\bar{\partial}_{\dot{A}} - 2i\theta^B\sigma^\mu_{B\dot{A}}\partial_\mu, \partial_A - 2i\sigma^\mu_{A\dot{C}}\bar{\theta}^{\dot{C}}\partial_\mu\} \\
&= -\{\bar{\partial}_{\dot{A}}, \partial_A\} + 2i\{\bar{\partial}_{\dot{A}}, \sigma^\mu_{A\dot{C}}\bar{\theta}^{\dot{C}}\partial_\mu\} \\
&\quad - 2i\{\theta^B\sigma^\mu_{B\dot{A}}\partial_\mu, \partial_A\} \\
&\quad - 4\{\theta^B\sigma^\mu_{B\dot{A}}\partial_\mu, \sigma^\nu_{A\dot{C}}\bar{\theta}^{\dot{C}}\partial_\nu\} \\
&= 2i\sigma^\mu_{A\dot{C}}(\bar{\partial}_{\dot{A}}\bar{\theta}^{\dot{C}})\partial_\mu - 2i(\partial_A\theta^B)\sigma^\mu_{B\dot{A}}\partial_\mu \\
&= 2i\sigma^\mu_{A\dot{A}}\partial_\mu - 2i\sigma^\mu_{A\dot{A}}\partial_\mu \\
&= 0
\end{aligned}$$

The covariant derivatives have important properties which we summarize in the following proposition.

<u>Proposition</u>: The covariant derivatives obey the following algebra

a)
$$\left.\begin{aligned}
&\{D_A, D_B\} = \{\bar{D}_{\dot{A}}, \bar{D}_{\dot{B}}\} = 0 \\
&D^3 = \bar{D}^3 = 0 \\
&\{D_A, \bar{D}_{\dot{B}}\} = -2i\sigma^\mu_{A\dot{B}}\partial_\mu \\
&\qquad\qquad = 2\sigma^\mu_{A\dot{B}}P_\mu \\
&\{D^A, \bar{D}^{\dot{B}}\} = -2i\bar{\sigma}^{\mu\dot{B}A}\partial_\mu
\end{aligned}\right\} \qquad (6.53)$$

b)
$$\left.\begin{aligned} \{D_A^{(1)}, D_B^{(1)}\} &= \{\bar{D}_{\dot{A}}^{(1)}, \bar{D}_{\dot{B}}^{(1)}\} = 0 \\ \{D_A^{(1)}, \bar{D}_{\dot{B}}^{(1)}\} &= -2i\sigma^{\mu}_{A\dot{B}}\partial_\mu \\ &= 2\sigma^{\mu}_{A\dot{B}}P_\mu \end{aligned}\right\} \qquad (6.54)$$

c)
$$\left.\begin{aligned} \{D_A^{(2)}, D_B^{(2)}\} &= \{\bar{D}_{\dot{A}}^{(2)}, \bar{D}_{\dot{B}}^{(2)}\} = 0 \\ \{D_A^{(2)}, \bar{D}_{\dot{B}}^{(2)}\} &= -2i\sigma^{\mu}_{A\dot{B}}\partial_\mu \\ &= 2\sigma^{\mu}_{A\dot{B}}P_\mu \end{aligned}\right\} \qquad (6.55)$$

<u>Proof</u>:

a) $\{D_A, D_B\}$

$$= \{\partial_A + i\sigma^{\mu}_{A\dot{B}}\bar{\theta}^{\dot{B}}\partial_\mu, \partial_B + i\sigma^{\nu}_{B\dot{C}}\bar{\theta}^{\dot{C}}\partial_\nu\}$$
$$= \{\partial_A, \partial_B\} + i\{\partial_A, \sigma^{\nu}_{B\dot{C}}\bar{\theta}^{\dot{C}}\partial_\nu\}$$
$$+ i\{\sigma^{\mu}_{A\dot{B}}\bar{\theta}^{\dot{B}}\partial_\mu, \partial_B\}$$
$$- \{\sigma^{\mu}_{A\dot{B}}\bar{\theta}^{\dot{B}}\partial_\mu, \sigma^{\nu}_{B\dot{C}}\bar{\theta}^{\dot{C}}\partial_\nu\}$$
$$= 0$$

Also

$$\{\bar{D}_{\dot{A}}, \bar{D}_{\dot{B}}\}$$
$$= \{-\bar{\partial}_{\dot{A}} - i\theta^B\sigma^{\mu}_{B\dot{A}}\partial_\mu, -\bar{\partial}_{\dot{B}} - i\theta^C\sigma^{\nu}_{C\dot{B}}\partial_\nu\}$$
$$= \{\bar{\partial}_{\dot{A}}, \bar{\partial}_{\dot{B}}\} + i\{\bar{\partial}_{\dot{A}}, \theta^C\}\sigma^{\nu}_{C\dot{B}}\partial_\nu$$
$$+ i\{\theta^B, \bar{\partial}_{\dot{B}}\}\sigma^{\mu}_{B\dot{A}}\partial_\mu$$

$$-\{\theta^B, \theta^C\}\sigma^\mu_{B\dot{A}}\partial_\mu \sigma^\nu_{C\dot{B}}\partial_\nu$$

$$= 0$$

and

$$\{D_A, \bar{D}_{\dot{B}}\}$$

$$= \{\partial_A + i\sigma^\mu_{A\dot{C}}\bar{\theta}^{\dot{C}}\partial_\mu, -\bar{\partial}_{\dot{B}} - i\theta^B\sigma^\nu_{B\dot{B}}\partial_\nu\}$$

$$= -\{\partial_A, \bar{\partial}_{\dot{B}}\} - i\{\partial_A, \theta^B\sigma^\nu_{B\dot{B}}\partial_\nu\}$$

$$-i\{\sigma^\mu_{A\dot{C}}\bar{\theta}^{\dot{C}}\partial_\mu, \bar{\partial}_{\dot{B}}\}$$

$$+\sigma^\mu_{A\dot{C}}\partial_\mu\{\bar{\theta}^{\dot{C}}, \theta^B\}\sigma^\nu_{B\dot{B}}\partial_\nu$$

$$= -i(\partial_A\theta^B)\sigma^\nu_{B\dot{B}}\partial_\nu - i\sigma^\mu_{A\dot{C}}(\bar{\partial}_{\dot{B}}\bar{\theta}^{\dot{C}})\partial_\mu$$

$$= -i\sigma^\nu_{A\dot{B}}\partial_\nu - i\sigma^\mu_{A\dot{B}}\partial_\mu$$

$$= -2i\sigma^\nu_{A\dot{B}}\partial_\nu$$

and

$$\{D^A, \bar{D}^{\dot{B}}\}$$

$$= \varepsilon^{AC}\varepsilon^{\dot{B}\dot{E}}(D_C\bar{D}_{\dot{E}} + \bar{D}_{\dot{E}}D_C)$$

$$= \varepsilon^{AC}\varepsilon^{\dot{B}\dot{E}}(-2i\sigma^\mu_{C\dot{E}}\partial_\mu)$$

$$= -2i\bar{\sigma}^{\mu\dot{B}A}\partial_\mu$$

b)

$$\{D^{(1)}_A, D^{(1)}_B\}$$

$$= \{\partial_A + 2i\sigma^\mu_{A\dot{B}}\bar{\theta}^{\dot{B}}\partial_\mu, \partial_B + 2i\sigma^\nu_{B\dot{C}}\bar{\theta}^{\dot{C}}\partial_\nu\}$$

$$= \{\partial_A, \partial_B\} + 2i\sigma^\nu_{B\dot{C}}\partial_\nu\{\partial_A, \bar{\theta}^{\dot{C}}\}$$

$$+2i\sigma^\mu_{A\dot{B}}\partial_\mu\{\bar{\theta}^{\dot{B}}, \partial_B\}$$

$$-4\sigma^{\mu}_{A\dot{B}}\partial_\mu \sigma^{\nu}_{B\dot{C}}\partial_\nu\{\bar{\theta}^{\dot{B}},\bar{\theta}^{\dot{C}}\}$$
$$=0$$

Also

$$\{\bar{D}^{(1)}_{\dot{A}},\bar{D}^{(1)}_{\dot{B}}\}=\{\bar{\partial}_{\dot{A}},\bar{\partial}_{\dot{B}}\}=0$$

and

$$\{D^{(1)}_{A},\bar{D}^{(1)}_{\dot{B}}\}$$
$$=\{\partial_A+2i\sigma^{\mu}_{A\dot{C}}\bar{\theta}^{\dot{C}}\partial_\mu,-\bar{\partial}_{\dot{B}}\}$$
$$=-\{\partial_A,\bar{\partial}_{\dot{B}}\}-2i\sigma^{\mu}_{A\dot{C}}\partial_\mu\{\bar{\theta}^{\dot{C}},\bar{\partial}_{\dot{B}}\}$$
$$=-2i\sigma^{\mu}_{A\dot{C}}\partial_\mu\delta^{\dot{C}}_{\dot{B}}$$
$$=-2i\sigma^{\mu}_{A\dot{B}}\partial_\mu$$

c) $$\{D^{(2)}_{A},D^{(2)}_{B}\}=\{\partial_A,\partial_B\}=0$$

Also

$$\{\bar{D}^{(2)}_{\dot{A}},\bar{D}^{(2)}_{\dot{B}}\}$$
$$=\{-\bar{\partial}_{\dot{A}}-2i\theta^{B}\sigma^{\mu}_{B\dot{A}}\partial_\mu,-\bar{\partial}_{\dot{B}}-2i\theta^{C}\sigma^{\nu}_{C\dot{B}}\partial_\nu\}$$
$$=\{\bar{\partial}_{\dot{A}},\bar{\partial}_{\dot{B}}\}+2i\{\bar{\partial}_{\dot{A}},\theta^{C}\}\sigma^{\nu}_{C\dot{B}}\partial_\nu$$
$$+2i\{\theta^{B},\bar{\partial}_{\dot{B}}\}\sigma^{\mu}_{B\dot{A}}\partial_\mu$$
$$-4\{\theta^{B},\theta^{C}\}\sigma^{\mu}_{B\dot{A}}\partial_\mu\sigma^{\nu}_{C\dot{B}}\partial_\nu$$
$$=0$$

and

$$\{D_A^{(2)}, \bar{D}_{\dot{A}}^{(2)}\}$$
$$= \{\partial_A, -\bar{\partial}_{\dot{A}} - 2i\theta^B \sigma^\mu_{B\dot{A}}\partial_\mu\}$$
$$= -\{\partial_A, \bar{\partial}_{\dot{A}}\} - 2i\{\partial_A, \theta^B\}\sigma^\mu_{B\dot{A}}\partial_\mu$$
$$= -2i\delta_A{}^B\sigma^\mu_{B\dot{A}}\partial_\mu$$
$$= -2i\sigma^\mu_{A\dot{A}}\partial_\mu$$

6.6 Projection Operators

In later sections we will be concerned with superfields which satisfy certain constraints. These special types of fields can also be obtained from the most general form of a superfield by the application of projection operators. It is convenient to define these projection operators here after the introduction of the covariant derivatives. We begin by proving a set of relations which are of considerable use in calculations.

Proposition: The follwing relations hold

a)
$$[D_A, \bar{D}^2] = -4i\sigma^\mu_{A\dot{A}}\bar{D}^{\dot{A}}\partial_\mu \qquad (6.56)$$

b)
$$[D^A, \bar{D}^2] = 4i\bar{D}_{\dot{C}}\bar{\sigma}^{\mu\dot{C}A}\partial_\mu \qquad (6.57)$$

c)
$$[\bar{D}_{\dot{A}}, D^2] = 4iD^A\sigma^\mu_{A\dot{A}}\partial_\mu \qquad (6.58)$$

d)

$$[\bar{D}^{\dot{A}}, D^2] = -4i\bar{\sigma}^{\mu\dot{A}A} D_A \partial_\mu \tag{6.59}$$

e)

$$\bar{D}\bar{\sigma}^\mu D = -D\sigma^\mu\bar{D} - 4i\partial^\mu \tag{6.60}$$

f)

$$[D^2, \bar{D}^2] = -8i(D\sigma^\mu\bar{D})\partial_\mu + 16\Box \tag{6.61}$$

g)

$$[\bar{D}^2, D^2] = -8i(\bar{D}\bar{\sigma}^\mu D)\partial_\mu + 16\Box \tag{6.62}$$

h)

$$\begin{aligned} &\sigma^\mu_{A\dot{A}}\sigma^\nu_{B\dot{B}} + \sigma^\nu_{A\dot{A}}\sigma^\mu_{B\dot{B}} \\ &= \eta^{\mu\nu}\varepsilon_{AB}\varepsilon_{\dot{A}\dot{B}} \\ &\quad + 4(\sigma^{\mu\rho}\varepsilon^T)_{AB}(\varepsilon\bar{\sigma}_\rho{}^\nu)_{\dot{A}\dot{B}} \end{aligned} \tag{6.63}$$

i)

$$\left.\begin{aligned} D\sigma^{\mu\nu}\varepsilon^T D &= 0 \\ \bar{D}\varepsilon\bar{\sigma}^{\mu\nu}\bar{D} &= 0 \end{aligned}\right\} \tag{6.64}$$

j)

$$D^A\bar{D}^2 D_A = \bar{D}_{\dot{A}} D^2 \bar{D}^{\dot{A}} \tag{6.65}$$

<u>Proof</u>:

a) We have (6.53), i.e.

$$\{D_A, \bar{D}_{\dot{A}}\} = -2i\sigma^\mu_{A\dot{A}}\partial_\mu$$

Multiplying this equation from the right by $\bar{D}^{\dot{A}}$ we obtain

$$D_A\bar{D}^2 + \bar{D}_{\dot{A}} D_A\bar{D}^{\dot{A}} = -2i\sigma^\mu_{A\dot{A}}\bar{D}^{\dot{A}}\partial_\mu \tag{6.66}$$

Alternatively, multiplying (6.53) by $\varepsilon^{\dot{A}\dot{B}}$ we obtain

$$(D_A\bar{D}_{\dot{A}}+\bar{D}_{\dot{A}}D_A)\varepsilon^{\dot{A}\dot{B}}=-2i\sigma^{\mu}_{A\dot{A}}\partial_{\mu}\varepsilon^{\dot{A}\dot{B}}$$

Using (1.76), i.e.

$$\bar{D}^{\dot{B}}=-\bar{D}_{\dot{A}}\varepsilon^{\dot{A}\dot{B}}$$

this becomes

$$-(D_A\bar{D}^{\dot{B}}+\bar{D}^{\dot{B}}D_A)=-2i\sigma^{\mu}_{A\dot{A}}\partial_{\mu}\varepsilon^{\dot{A}\dot{B}}$$

For $\dot{B}=\dot{A}$ this yields

$$D_A\bar{D}^{\dot{A}}=-\bar{D}^{\dot{A}}D_A+2i\sigma^{\mu}_{A\dot{D}}\partial_{\mu}\varepsilon^{\dot{D}\dot{A}}$$

Inserting this into (6.66) we obtain

$$D_A\bar{D}^2+\bar{D}_{\dot{A}}(-\bar{D}^{\dot{A}}D_A+2i\sigma^{\mu}_{A\dot{D}}\partial_{\mu}\varepsilon^{\dot{D}\dot{A}})$$
$$=-2i\sigma^{\mu}_{A\dot{A}}\bar{D}^{\dot{A}}\partial_{\mu}$$

i.e.

$$[D_A,\bar{D}^2]$$
$$=-2i(\sigma^{\mu}_{A\dot{A}}\bar{D}^{\dot{A}}+\bar{D}_{\dot{A}}\sigma^{\mu}_{A\dot{D}}\varepsilon^{\dot{D}\dot{A}})\partial_{\mu}$$
$$=-4i\sigma^{\mu}_{A\dot{A}}\bar{D}^{\dot{A}}\partial_{\mu}\qquad\text{(with (1.76))}$$

This had to be shown.

b) The result b) follows by multiplying (6.56) by ε^{BA} and using (1.76) and (1.88).

c) We proceed as under a) with

$$\{D_A,\bar{D}_{\dot{A}}\}=-2i\sigma^{\mu}{}_{A\dot{A}}\partial_{\mu}$$

Hence

$$D^2\bar{D}_{\dot{A}}+D^A\bar{D}_{\dot{A}}D_A=-2iD^A\sigma^{\mu}_{A\dot{A}}\partial_{\mu}\qquad(6.67)$$

Also

$$\varepsilon^{AB}(D_B\bar{D}_{\dot{A}}+\bar{D}_{\dot{A}}D_B)=\varepsilon^{AB}(-2i\sigma^{\mu}_{B\dot{A}}\partial_{\mu})$$

Using (1.76) this gives

$$D^A \bar{D}_{\dot{A}} + \bar{D}_{\dot{A}} D^A = -2i\,\varepsilon^{AB} \sigma^\mu_{B\dot{A}} \partial_\mu$$

Inserting this into (6.67) we obtain

$$\begin{aligned}[D^2, \bar{D}_{\dot{A}}] &= 2i\varepsilon^{AB} \sigma^\mu_{B\dot{A}} \partial_\mu D_A \\ &\quad - 2i\, D^A \sigma^\mu_{A\dot{A}} \partial_\mu \\ &= -4i\, D^A \sigma^\mu_{A\dot{A}} \partial_\mu\end{aligned}$$

with (1.76).

d) The result (6.59) follows by multiplying (6.58) by $\varepsilon^{\dot{B}\dot{A}}$ and using (1.76) and (1.88).

e) Consider

$$\bar{D}_{\dot{A}} \bar{\sigma}^{\mu\dot{A}A} D_A$$

$$= (-2i\, \sigma^\rho_{A\dot{A}} \partial_\rho - D_A \bar{D}_{\dot{A}}) \bar{\sigma}^{\mu\dot{A}A} \quad \text{(with (6.53))}$$

$$= -D_A \bar{D}_{\dot{A}} \varepsilon^{AB} \varepsilon^{\dot{A}\dot{B}} \sigma^\mu_{B\dot{B}} - 2i\sigma^\rho_{A\dot{A}} \bar{\sigma}^{\mu\dot{A}A} \partial_\rho$$

(using (1.88))

$$= -(D\sigma^\mu \bar{D}) - 2i\, \mathrm{Tr}(\sigma^\rho \bar{\sigma}^\mu) \partial_\rho$$

(using (1.76))

$$= -(D\sigma^\mu \bar{D}) - 4i\partial^\mu \quad \text{(using (1.91))}$$

f) Consider

$$[D^2, \bar{D}^2]$$

$$= D^A [D_A, \bar{D}^2] + [D^A, \bar{D}^2] D_A$$

$$= D^A [D_A, \bar{D}^2] + \varepsilon^{AB} [D_B, \bar{D}^2] \varepsilon_{AC} D^C$$

(with (1.76))

$$= D^A [D_A, \bar{D}^2] - [D_A, \bar{D}^2] D^A$$

$$= D^A[-4i\sigma^\mu_{A\dot{A}}\bar{D}^{\dot{A}}\partial_\mu]$$

$$+[4i\sigma^\mu_{A\dot{A}}\bar{D}^{\dot{A}}\partial_\mu]D^A \quad \text{(with (6.56))}$$

$$= 4i\sigma^\mu_{A\dot{A}}\partial_\mu[\bar{D}^{\dot{A}}D^A - D^A\bar{D}^{\dot{A}}]$$

$$= 4i\sigma^\mu_{A\dot{A}}\partial_\mu[-2D^A\bar{D}^{\dot{A}} - 2i\bar{\sigma}^{\rho\dot{A}A}\partial_\rho]$$

(with (6.53))

$$= -8i(D\sigma^\mu\bar{D})\partial_\mu + 8(\sigma^\mu_{A\dot{A}}\bar{\sigma}^{\rho\dot{A}A})\partial_\mu\partial_\rho$$

$$= -8i(D\sigma^\mu\bar{D})\partial_\mu + 16\Box \quad \text{(with (1.91))}$$

g) The result (6.62) follows immediately from (6.60) and (6.61).

h) Consider

$$\eta^{\mu\nu}\varepsilon_{AB}\varepsilon_{\dot{A}\dot{B}}$$

$$= \tfrac{1}{2}\mathrm{Tr}(\sigma^\mu\bar{\sigma}^\nu)\varepsilon_{AB}\varepsilon_{\dot{A}\dot{B}} \quad \text{(using (1.91))}$$

$$= \tfrac{1}{2}\sigma^\mu_{D\dot{C}}\bar{\sigma}^{\nu\dot{C}D}\varepsilon_{AB}\varepsilon_{\dot{A}\dot{B}}$$

$$= \tfrac{1}{2}\sigma^\mu_{D\dot{E}}(\varepsilon^{\dot{E}\dot{F}}\varepsilon_{\dot{F}\dot{C}})\bar{\sigma}^{\nu\dot{C}K}(\varepsilon_{KL}\varepsilon^{LD})\varepsilon_{AB}\varepsilon_{\dot{A}\dot{B}}$$

(using (1.71))

$$= -\tfrac{1}{2}\sigma^\mu_{D\dot{E}}\varepsilon^{\dot{E}\dot{F}}\sigma^\nu_{L\dot{F}}\varepsilon^{LD}\varepsilon_{AB}\varepsilon_{\dot{A}\dot{B}} \quad \text{(with (1.88))}$$

Now from the explicit form of the ε matrices we deduce

$$\varepsilon^{AB}\varepsilon_{CD} = \delta^A{}_D\delta^B{}_C - \delta^A{}_C\delta^B{}_D$$

$$\varepsilon^{\dot{A}\dot{B}}\varepsilon_{\dot{C}\dot{D}} = \delta^{\dot{A}}{}_{\dot{D}}\delta^{\dot{B}}{}_{\dot{C}} - \delta^{\dot{A}}{}_{\dot{C}}\delta^{\dot{B}}{}_{\dot{D}}$$

Hence

$$\eta^{\mu\nu}\varepsilon_{AB}\varepsilon_{\dot{A}\dot{B}}$$

$$= -\frac{1}{2}\sigma^{\mu}_{D\dot{E}}\sigma^{\nu}_{L\dot{F}}(\delta^{\dot{E}}{}_{\dot{B}}\delta^{\dot{F}}{}_{\dot{A}} - \delta^{\dot{E}}{}_{\dot{A}}\delta^{\dot{F}}{}_{\dot{B}}) \cdot$$

$$\cdot (\delta^{L}{}_{B}\delta^{D}{}_{A} - \delta^{L}{}_{A}\delta^{D}{}_{B})$$

$$= -\frac{1}{2}(\sigma^{\mu}_{A\dot{B}}\sigma^{\nu}_{B\dot{A}} - \sigma^{\mu}_{B\dot{B}}\sigma^{\nu}_{A\dot{A}}$$

$$- \sigma^{\mu}_{A\dot{A}}\sigma^{\nu}_{B\dot{B}} + \sigma^{\mu}_{B\dot{A}}\sigma^{\nu}_{A\dot{B}})$$

On the other hand

$$4(\sigma^{\mu\rho}\varepsilon^{T})_{AB}(\varepsilon\bar{\sigma}_{\rho}{}^{\nu})_{\dot{A}\dot{B}}$$

$$= -\frac{1}{4}[(\sigma^{\mu}_{A\dot{C}}\bar{\sigma}^{\rho\dot{C}C} - \sigma_{A\dot{C}}{}^{\rho}\bar{\sigma}^{\mu\dot{C}C})\varepsilon^{T}_{CB}] \cdot$$

$$\cdot [\varepsilon_{\dot{A}\dot{B}}(\bar{\sigma}_{\rho}{}^{\dot{E}D}\sigma^{\nu}_{D\dot{B}} - \bar{\sigma}^{\nu\dot{E}D}\sigma_{\rho D\dot{B}})]$$

(with (1.119))

$$= -\frac{1}{4}[T_1 + T_2 + T_3 + T_4]$$

where

$$T_1 = \sigma^{\mu}_{A\dot{C}}\bar{\sigma}^{\rho\dot{C}C}\varepsilon^{T}_{CB}\varepsilon_{\dot{A}\dot{E}}\bar{\sigma}_{\rho}{}^{\dot{E}F}\delta_{F}{}^{D}\sigma^{\nu}_{D\dot{B}}$$

$$= \sigma^{\mu}_{A\dot{C}}\bar{\sigma}^{\rho\dot{C}C}\varepsilon^{T}_{CB}\varepsilon_{\dot{A}\dot{E}}\bar{\sigma}_{\rho}{}^{\dot{E}F}\varepsilon_{FK}\varepsilon^{KD}\sigma^{\nu}_{D\dot{B}}$$

$$= \sigma^{\mu}_{A\dot{C}}\bar{\sigma}^{\rho\dot{C}C}\varepsilon^{T}_{CB}(-\sigma_{\rho K\dot{A}})\varepsilon^{KD}\sigma^{\nu}_{D\dot{B}}$$

(with (1.88))

$$= \sigma^{\mu}_{A\dot{C}}\varepsilon^{T}_{CB}\varepsilon^{DK}\sigma^{\nu}_{D\dot{B}} . 2\delta_{K}{}^{C}\delta_{\dot{A}}{}^{\dot{C}}$$

(with (1.92))

$$= -2\sigma^{\mu}_{A\dot{A}}\varepsilon_{BC}\varepsilon^{CD}\sigma^{\nu}_{D\dot{B}}$$

$$= -2\sigma^{\mu}_{A\dot{A}}\,\delta_B^{\ D}\,\sigma^{\nu}_{D\dot{B}}$$

$$= -2\sigma^{\mu}_{A\dot{A}}\,\sigma^{\nu}_{B\dot{B}}$$

$$T_2 = -\sigma^{\rho}_{A\dot{C}}\,\bar{\sigma}^{\mu\dot{C}C}\,\varepsilon^{T}_{CB}\,\varepsilon_{\dot{A}\dot{E}}\,\bar{\sigma}^{\dot{E}D}_{\rho}\,\sigma^{\nu}_{D\dot{B}}$$

$$= -2\,\delta_A^{\ D}\,\delta_{\dot{C}}^{\ \dot{E}}\,\bar{\sigma}^{\mu\dot{C}C}\,\varepsilon^{T}_{CB}\,\varepsilon_{\dot{A}\dot{E}}\,\sigma^{\nu}_{D\dot{B}}$$

(with (1.92))

$$= -2\bar{\sigma}^{\mu\dot{C}C}\,\varepsilon^{T}_{CB}\,\varepsilon_{\dot{A}\dot{C}}\,\sigma^{\nu}_{A\dot{B}}$$

$$= -2\sigma^{\mu}_{B\dot{A}}\,\sigma^{\nu}_{A\dot{B}}$$

$$T_3 = -\sigma^{\mu}_{A\dot{G}}\,\bar{\sigma}^{\rho\dot{G}C}\,\varepsilon^{T}_{CB}\,\varepsilon_{\dot{A}\dot{E}}\,\bar{\sigma}^{\nu\dot{E}D}\,\sigma_{\rho D\dot{B}}$$

$$= -\sigma^{\mu}_{A\dot{G}}\,\varepsilon^{T}_{CB}\,\varepsilon_{\dot{A}\dot{E}}\,\bar{\sigma}^{\nu\dot{E}D}\,(2\,\delta_D^{\ C}\,\delta^{\dot{G}}_{\ \dot{B}})$$

(using (1.92))

$$= -2\sigma^{\mu}_{A\dot{B}}\,\varepsilon^{T}_{CB}\,\varepsilon_{\dot{A}\dot{E}}\,\bar{\sigma}^{\nu\dot{E}C}$$

$$= -2\sigma^{\mu}_{A\dot{B}}\,\sigma^{\nu}_{B\dot{A}} \qquad \text{(with (1.88))}$$

$$T_4 = \sigma^{\rho}_{A\dot{G}}\,\bar{\sigma}^{\mu\dot{G}C}\,\varepsilon^{T}_{CB}\,\varepsilon_{\dot{A}\dot{E}}\,\bar{\sigma}^{\nu\dot{E}D}\,\sigma_{\rho D\dot{B}}$$

$$= \sigma^{\rho}_{A\dot{G}}\,\bar{\sigma}^{\mu\dot{G}C}\,\varepsilon^{T}_{CB}\,\varepsilon_{\dot{A}\dot{E}}\,\bar{\sigma}^{\nu\dot{E}D}\,\varepsilon_{DK}\,\varepsilon_{\dot{B}\dot{K}}\,\bar{\sigma}^{\dot{K}K}_{\rho}$$

(with (1.88))

$$= \bar{\sigma}^{\mu\dot{G}C}\,\varepsilon^{T}_{CB}\,\varepsilon_{\dot{A}\dot{E}}\,\bar{\sigma}^{\nu\dot{E}D}\,\varepsilon_{DK}\,\varepsilon_{\dot{B}\dot{K}}\cdot 2\,\delta_A^{\ K}\,\delta_{\dot{G}}^{\ \dot{K}}$$

(using (1.92))

$$= 2\bar{\sigma}^{\mu\dot{K}C}\,\varepsilon^{T}_{CB}\,\varepsilon_{\dot{A}\dot{E}}\,\bar{\sigma}^{\nu\dot{E}D}\,\varepsilon_{DA}\,\varepsilon_{\dot{B}\dot{K}}$$

$$= -2\sigma^{\mu}_{B\dot{B}}\,\sigma^{\nu}_{A\dot{A}} \qquad \text{(with (1.88))}$$

Hence

$$4(\sigma^{\mu\rho}\varepsilon^T)_{AB}(\varepsilon\bar{\sigma}_\rho{}^\nu)_{\dot{A}\dot{B}}$$

$$= \frac{1}{2}\Big[\sigma^\mu_{A\dot{A}}\sigma^\nu_{B\dot{B}} + \sigma^\mu_{B\dot{A}}\sigma^\nu_{A\dot{B}} + \sigma^\mu_{A\dot{B}}\sigma^\nu_{B\dot{A}} + \sigma^\mu_{B\dot{B}}\sigma^\nu_{A\dot{A}}\Big]$$

and so

$$\eta^{\mu\nu}\varepsilon_{AB}\varepsilon_{\dot{A}\dot{B}} + 4(\sigma^{\mu\rho}\varepsilon^T)_{AB}(\varepsilon\bar{\sigma}_\rho{}^\nu)_{\dot{A}\dot{B}}$$

$$= \sigma^\mu_{A\dot{A}}\sigma^\nu_{B\dot{B}} + \sigma^\mu_{B\dot{B}}\sigma^\nu_{A\dot{A}}$$

as had to be shown.

i) Consider

$$D\sigma^{\mu\rho}\varepsilon^T D$$

$$= D^C(\sigma^{\mu\rho}\varepsilon^T)_{CE}D^E$$

$$= \frac{1}{2}D^C(\sigma^{\mu\rho}\varepsilon^T)_{CE}D^E + \frac{1}{2}D^C(\sigma^{\mu\rho}\varepsilon^T)_{EC}D^E$$

(using (1.127))

$$= \frac{1}{2}D^C(\sigma^{\mu\rho}\varepsilon^T)_{CE}D^E - \frac{1}{2}D^E(\sigma^{\mu\rho}\varepsilon^T)_{EC}D^C$$

(using (6.53))

$$= 0$$

Analogously

$$\bar{D}(\varepsilon\bar{\sigma}^{\mu\rho})\bar{D}$$

$$= \frac{1}{2}\bar{D}^{\dot{A}}(\varepsilon\bar{\sigma}^{\mu\rho})_{\dot{A}\dot{C}}\bar{D}^{\dot{C}} + \frac{1}{2}\bar{D}^{\dot{A}}(\varepsilon\bar{\sigma}^{\mu\rho})_{\dot{C}\dot{A}}\bar{D}^{\dot{C}}$$

(using (1.127))

$$= \frac{1}{2}\bar{D}^{\dot{A}}(\varepsilon\bar{\sigma}^{\mu\rho})_{\dot{A}\dot{C}}\bar{D}^{\dot{C}} - \frac{1}{2}\bar{D}^{\dot{C}}(\varepsilon\bar{\sigma}^{\mu\rho})_{\dot{C}\dot{A}}\bar{D}^{\dot{A}}$$

(using (6.53))

$$= 0$$

j) From (6.56) and (6.58) we obtain

$$D^A(D_A\bar{D}^2 - \bar{D}^2 D_A) = -4i D^A \sigma^\mu_{A\dot{A}} \bar{D}^{\dot{A}} \partial_\mu$$

$$(\bar{D}_{\dot{A}} D^2 - D^2 \bar{D}_{\dot{A}}) \bar{D}^{\dot{A}} = 4i D^A \sigma^\mu_{A\dot{A}} \bar{D}^{\dot{A}} \partial_\mu$$

Adding we obtain

$$D^A \bar{D}^2 D_A = \bar{D}_{\dot{A}} D^2 \bar{D}^{\dot{A}}$$

This completes the proof of the relations (6.56) to (6.65).

We now define a set of operators and then verify that they are projection operators, i.e. that each is idempotent and their sum is the identity operator.

Definition: We define the operators Π_+, Π_-, Π_T by

$$\Pi_+ := -\frac{1}{16\Box} \bar{D}^2 D^2 \tag{6.68}$$

$$\Pi_- := -\frac{1}{16\Box} D^2 \bar{D}^2 \tag{6.69}$$

$$\Pi_T := \frac{1}{8\Box} \bar{D}_{\dot{A}} D^2 \bar{D}^{\dot{A}} \tag{6.70}$$

$$= \frac{1}{8\Box} D^A \bar{D}^2 D_A$$

$$\Pi_0 := \Pi_+ + \Pi_- \tag{6.71}$$

Proposition: The operators (6.68), (6.69) and (6.70) are projection operators.

Proof: Consider (6.68). The square of this operator is

$$\Pi_+\Pi_+ = \frac{1}{16\Box} \bar{D}^2 D^2 \frac{1}{16\Box} \bar{D}^2 D^2$$

$$= \left(\frac{1}{16\Box}\right)^2 \bar{D}^2 D^2 \bar{D}^2 D^2$$

$$= \left(\frac{1}{16\Box}\right)^2 \bar{D}^2 D^2 \left(D^2 \bar{D}^2 + 8i D \sigma^\mu \bar{D} \partial_\mu - 16\Box\right)$$

(with (6.61))

$$= \left(\frac{1}{16\Box}\right)^2 \bar{D}^2 D^2 (-16\Box)$$

(since by (6.53) $D^3 = D^4 = 0$)

$$= \Pi_+$$

Similarly

$$\Pi_- \Pi_- = \left(\frac{1}{16\Box}\right)^2 D^2 \bar{D}^2 D^2 \bar{D}^2$$

$$= \left(\frac{1}{16\Box}\right)^2 D^2 \bar{D}^2 \left(\bar{D}^2 D^2 + 8i \bar{D} \bar{\sigma}^\mu D \partial_\mu - 16\Box\right)$$

(with (6.62))

$$= \Pi_- \quad \text{(using (6.53))}$$

Also

$$\Pi_+ \Pi_- = 0 = \Pi_- \Pi_+$$

since $D^4 = 0 = \bar{D}^4$,

and similarly

$$\Pi_\pm \Pi_T = \Pi_T \Pi_\pm = 0$$

Next consider

$$\Pi_T \Pi_T = \left(\frac{1}{8\Box}\right)^2 \bar{D}_{\dot{A}} D^2 \bar{D}^{\dot{A}} \bar{D}_{\dot{B}} D^2 \bar{D}^{\dot{B}}$$

$$= \left(\frac{1}{8\Box}\right)^2 \bar{D}_{\dot{A}} \left(\bar{D}^{\dot{A}} D^2 + 4i \bar{\sigma}^{\mu \dot{A} A} D_A \partial_\mu\right) \cdot$$

$$\cdot \left(D^2 \bar{D}_{\dot{B}} + 4i D^E \sigma^\rho{}_{E\dot{B}} \partial_\rho\right) \bar{D}^{\dot{B}}$$

(using (6.58) and (6.59))

$$=\left(\frac{1}{8\Box}\right)^2 \bar{D}_{\dot{A}}\,(4i\,\bar{\sigma}^{\mu\dot{A}A}D_A\partial_\mu)(4i\,D^E\sigma^\rho_{E\dot{B}}\partial_\rho)\bar{D}^{\dot{B}}$$

(since $D^3 = D^4 = 0$)

$$=-16\left(\frac{1}{8\Box}\right)^2(\bar{D}_{\dot{A}}\bar{\sigma}^{\mu\dot{A}A}D_AD^E\sigma^\rho_{E\dot{B}}\bar{D}^{\dot{B}})\partial_\mu\partial_\rho$$

$$=-\left(\frac{1}{2\Box}\right)^2(\bar{D}_{\dot{A}}\varepsilon^{AC}\varepsilon^{\dot{A}\dot{C}}\sigma^\mu_{C\dot{C}}D_AD^E\sigma^\rho_{E\dot{B}}\bar{D}^{\dot{B}})\partial_\mu\partial_\rho$$

(with (1.88))

$$=-\left(\frac{1}{2\Box}\right)^2[\bar{D}^{\dot{C}}\sigma^\mu_{C\dot{C}}D^CD^E\sigma^\rho_{E\dot{B}}\bar{D}^{\dot{B}}$$

$$+\bar{D}^{\dot{C}}\sigma^\rho_{C\dot{C}}D^CD^E\sigma^\mu_{E\dot{B}}\bar{D}^{\dot{B}}]\partial_\mu\partial_\rho$$

(with (1.76))

$$=-\frac{1}{2}\left(\frac{1}{2\Box}\right)^2[\eta^{\mu\rho}\varepsilon_{CE}\varepsilon_{\dot{C}\dot{B}}\bar{D}^{\dot{C}}D^CD^E\bar{D}^{\dot{B}}$$

$$+4(\sigma^{\mu\rho}\varepsilon^T)_{CE}(\varepsilon\bar{\sigma}_\rho{}^\nu)_{\dot{C}\dot{B}}\bar{D}^{\dot{C}}D^CD^E\bar{D}^{\dot{B}}]\partial_\mu\partial_\rho$$

(with (6.63))

$$=-\frac{1}{2}\left(\frac{1}{2\Box}\right)^2[-\eta^{\mu\rho}\bar{D}_{\dot{B}}D^2\bar{D}^{\dot{B}}$$

$$+4\bar{D}^{\dot{C}}(D\sigma^{\mu\rho}\varepsilon^TD)(\varepsilon\bar{\sigma}_\rho{}^\nu)_{\dot{C}\dot{B}}\bar{D}^{\dot{B}}]\partial_\mu\partial_\rho$$

(using (1.76))

$$=\frac{1}{8\Box}\bar{D}_{\dot{B}}D^2\bar{D}^{\dot{B}} \qquad \text{(with (6.64))}$$

$$=\pi_T$$

Finally we consider the sum of the three operators, i.e.

$$\pi_+ + \pi_- + \pi_T$$

$$=\frac{1}{16\Box}[2\bar{D}_{\dot{A}}D^2\bar{D}^{\dot{A}}-\bar{D}^2D^2-D^2\bar{D}^2]$$

$$= \frac{1}{16\Box}\left[\bar{D}_{\dot{A}} D^2 \bar{D}^{\dot{A}} + D^A \bar{D}^2 D_A - \bar{D}^2 D^2 - D^2 \bar{D}^2\right]$$

(using (6.65))

$$= \frac{1}{16\Box}\left[(\bar{D}_{\dot{A}} D^2 - D^2 \bar{D}_{\dot{A}})\bar{D}^{\dot{A}} + (D^A \bar{D}^2 - \bar{D}^2 D^A) D_A\right]$$

$$= \frac{1}{16\Box}\left[4i D^A \sigma^{\mu}_{A\dot{A}} \partial_\mu \bar{D}^{\dot{A}} + 4i \bar{D}_{\dot{C}} \bar{\sigma}^{\mu \dot{C} A} \partial_\mu D_A\right]$$

(using (6.57) and (6.58))

$$= \frac{1}{16\Box} 4i (D\sigma^\mu \bar{D} + \bar{D}\bar{\sigma}^\mu D)\partial_\mu$$

$$= \frac{1}{16\Box} 4i(-4i\partial^\mu)\partial_\mu$$

$$= 1$$

Hence the sum of the operators is the identity operator.

6.7 Constraints

From the definition of the projection operators we obtain the following constraints

$$\bar{D}\Pi_+ = 0 \qquad \text{(since } \bar{D}^3 = 0) \qquad (6.72)$$

$$D\Pi_- = 0 \qquad \text{(since } D^3 = 0) \qquad (6.73)$$

Superfields

$$\Phi_\mp = \Pi_\pm \Phi \qquad (6.74)$$

satisfying these constraints are called respectively left-handed and right-handed chiral superfields in analogy to left-handed and right-handed fermionic fields

$$\psi_{L,R} = d_{\mp}\psi, \quad d_{\mp} = \tfrac{1}{2}(1 \mp \gamma_5)$$

constructed from the fermionic field ψ since $\psi_{L,R}$ obey the constraints

$$d_{\pm}\psi_{L,R} = 0$$

Superfields satisfying constraints, in particular chiral superfields, will be discussed in detail in Chapter 7.

6.8 Transformation Properties of Component Fields

We introduced superfields as operator-valued functions, defined on superspace, which are to be understood in terms of their power series expansions in the Grassmann variables θ and $\bar{\theta}$ (see (6.6)), i.e.

$$\begin{aligned}\Phi(x,\theta,\bar{\theta}) = f(x) + \theta\phi(x) + \bar{\theta}\bar{\chi}(x) \\ + (\theta\theta)m(x) + (\bar{\theta}\bar{\theta})n(x) \\ + (\theta\sigma^{\mu}\bar{\theta})V_{\mu}(x) + (\theta\theta)\bar{\theta}\bar{\lambda}(x) \\ + (\bar{\theta}\bar{\theta})(\theta\psi(x)) + (\theta\theta)(\bar{\theta}\bar{\theta})d(x)\end{aligned} \tag{6.75}$$

The quantities $f(x), \phi(x), \bar{\chi}(x), m(x), n(x), V_{\mu}(x)$ $\bar{\lambda}(x), \psi(x)$ and $d(x)$ are called component fields. Their geometric character is determined by their transformation properties under the Lorentz group and the additional requirement that $\Phi(x,\theta,\bar{\theta})$ be a Lorentz scalar or pseudoscalar. From these conditions we deduce that

f(x), m(x), n(x)	are complex scalar or pseudoscalar fields (see below),
$\psi(x)$, $\phi(x)$	are left-handed Weyl spinor fields,
$\bar{\chi}(x)$, $\bar{\lambda}(x)$	are right-handed Weyl spinor fields,
$V_\mu(x)$	is a Lorentz four-vector field,
$d(x)$	is a scalar field

Thus a superfield is a short way to denote a finite multiplet of fields.

In an expansion such as (6.75) in the 2x2 Weyl formulation, f(x) need not have a well-defined parity. We have seen before that the Dirac equation is invariant under parity transformations whereas the individual Weyl equations are not. The latter is reflected in the expansion (6.75). Thus f(x) may be the sum of scalar and pseudoscalar contributions and correspondingly m(x) may be the sum of pseudoscalar and scalar contributions which are such that if we rewrite the superfield in the 4x4 Dirac formulation it will have a well-defined property under parity transformations , i.e. as a scalar or pseudoscalar.

Our next task is the computation of the transformation laws of the component fields with respect to supersymmetry transformations. This will enable us to obtain the Weyl representation of the transformations (5.11) of the Wess-Zumino model.

The transformation law for superfields is defined as

$$\delta_S \Phi(x,\theta,\bar{\theta}) = \delta_S' f(x) + \theta^A \delta_S' \phi_A(x) + \bar{\theta}_{\dot{A}}\, \delta_S' \bar{\chi}^{\dot{A}}(x) + (\theta\theta)\, \delta_S' m(x)$$

$$+(\bar{\theta}\bar{\theta})\,\delta'_S n(x)+(\theta\sigma^{\mu}\bar{\theta})\,\delta'_S V_{\mu}(x)$$
$$+(\theta\theta)\bar{\theta}_{\dot{A}}\,\delta'_S\bar{\lambda}^{\dot{A}}(x)+(\bar{\theta}\bar{\theta})\theta^{A}\delta'_S\psi_A(x)$$
$$+(\theta\theta)(\bar{\theta}\bar{\theta})\,\delta'_S d(x) \qquad (6.77)$$

On the other hand, according to (6.19), we have the infinitesimal supersymmetry variation

$$\delta_S\Phi(x,\theta,\bar{\theta})$$
$$=[\alpha^{A}\partial_A+\bar{\alpha}_{\dot{A}}\bar{\partial}^{\dot{A}}+i\theta\sigma^{\mu}\bar{\alpha}\partial_{\mu}$$
$$-i\alpha\sigma^{\mu}\bar{\theta}\partial_{\mu}]\Phi(x,\theta,\bar{\theta})$$
$$=[\alpha^{A}\partial_A+\bar{\alpha}_{\dot{A}}\bar{\partial}^{\dot{A}}+i\theta\sigma^{\mu}\bar{\alpha}\partial_{\mu}-i\alpha\sigma^{\mu}\bar{\theta}\partial_{\mu}]\cdot$$
$$\cdot\{f(x)+\theta^{B}\phi_B(x)+\bar{\theta}_{\dot{B}}\bar{\chi}^{\dot{B}}(x)+(\theta\theta)m(x)$$
$$+(\bar{\theta}\bar{\theta})n(x)+\theta\sigma^{\nu}\bar{\theta}\,V_{\nu}(x)+(\theta\theta)\bar{\theta}_{\dot{B}}\bar{\lambda}^{\dot{B}}(x)$$
$$+(\bar{\theta}\bar{\theta})\theta^{B}\psi_B(x)+(\theta\theta)(\bar{\theta}\bar{\theta})d(x)\}$$

(using (6.75))

$$=\alpha^{A}\phi_A(x)+2\alpha^{A}\theta_A m(x)+\alpha^{A}\sigma^{\nu}_{A\dot{B}}\bar{\theta}^{\dot{B}}V_{\nu}(x)$$
$$+2\alpha^{A}\theta_A\bar{\theta}_{\dot{B}}\bar{\lambda}^{\dot{B}}(x)+(\bar{\theta}\bar{\theta})\alpha^{A}\psi_A(x)$$
$$+2\alpha^{A}\theta_A(\bar{\theta}\bar{\theta})d(x)+\bar{\alpha}_{\dot{A}}\bar{\chi}^{\dot{A}}(x)$$
$$+2\bar{\alpha}_{\dot{A}}\bar{\theta}^{\dot{A}}n(x)-\bar{\alpha}_{\dot{A}}\theta^{A}\sigma^{\nu}_{A\dot{B}}\bar{\partial}^{\dot{A}}\bar{\theta}^{\dot{B}}V_{\nu}(x)$$
$$+(\theta\theta)\bar{\alpha}_{\dot{A}}\bar{\lambda}^{\dot{A}}(x)+2\bar{\alpha}_{\dot{A}}\bar{\theta}^{\dot{A}}\theta^{B}\psi_B(x)$$
$$+2(\theta\theta)\bar{\alpha}_{\dot{A}}\bar{\theta}^{\dot{A}}d(x)$$
$$+i[(\theta\sigma^{\mu}\bar{\alpha})\partial_{\mu}f(x)+(\theta\sigma^{\mu}\bar{\alpha})\theta^{A}\partial_{\mu}\phi_A(x)$$

$$+(\theta\sigma^\mu\bar{\alpha})\bar{\theta}_{\dot{A}}\partial_\mu\bar{\chi}^{\dot{A}}(x)$$
$$+(\theta\sigma^\mu\bar{\alpha})(\bar{\theta}\bar{\theta})\partial_\mu n(x)$$
$$+(\theta\sigma^\mu\bar{\alpha})(\theta\sigma^\nu\bar{\theta})\partial_\mu V_\nu(x)$$
$$+(\theta\sigma^\mu\bar{\alpha})(\bar{\theta}\bar{\theta})\theta^A\partial_\mu\psi_A(x)\Big]$$
$$-i\Big[(\alpha\sigma^\mu\bar{\theta})\partial_\mu f(x)+(\alpha\sigma^\mu\bar{\theta})\bar{\theta}_{\dot{A}}\partial_\mu\bar{\chi}^{\dot{A}}(x)$$
$$+(\alpha\sigma^\mu\bar{\theta})\theta^B\partial_\mu\phi_B(x)$$
$$+(\alpha\sigma^\mu\bar{\theta})(\theta\theta)\partial_\mu m(x)$$
$$+(\alpha\sigma^\mu\bar{\theta})(\theta\sigma^\nu\bar{\theta})\partial_\mu V_\nu(x)$$
$$+(\alpha\sigma^\mu\bar{\theta})(\theta\theta)\bar{\theta}_{\dot{A}}\partial_\mu\bar{\lambda}^{\dot{A}}(x)\Big]$$

where we used the fact that the third power of any Grassmann number in the Weyl representation vanishes. Hence (using (1.80) and (6.4k))

$$\delta_S\Phi(x,\theta,\bar{\theta})$$
$$=\alpha^A\phi_A(x)+\bar{\alpha}_{\dot{A}}\bar{\chi}^{\dot{A}}(x)$$
$$+\theta^A\Big\{2\alpha_A m(x)+i\sigma^\mu_{A\dot{B}}\bar{\alpha}^{\dot{B}}\partial_\mu f(x)$$
$$+\bar{\alpha}_{\dot{A}}\sigma^\nu_{A\dot{B}}\bar{\partial}^{\dot{A}}\bar{\theta}^{\dot{B}}V_\nu(x)\Big\}$$
$$+\bar{\theta}_{\dot{A}}\Big\{2\bar{\alpha}^{\dot{A}}n(x)+i(\alpha\sigma^\mu)_{\dot{B}}\varepsilon^{\dot{B}\dot{A}}\partial_\mu f(x)$$
$$-(\alpha\sigma^\mu)_{\dot{B}}\varepsilon^{\dot{B}\dot{A}}V_\mu(x)\Big\}$$
$$+(\theta\theta)\bar{\alpha}_{\dot{A}}\bar{\lambda}^{\dot{A}}(x)+i(\theta\sigma^\mu\bar{\alpha})\theta^A\partial_\mu\phi_A(x)$$
$$+(\bar{\theta}\bar{\theta})\alpha^A\psi_A(x)-i(\alpha\sigma^\mu\bar{\theta})\bar{\theta}_{\dot{A}}\partial_\mu\bar{\chi}^{\dot{A}}(x)$$
$$+(\theta\theta)\bar{\theta}_{\dot{A}}2\bar{\alpha}^{\dot{A}}d(x)$$

$$+i(\theta\sigma^{\mu}\bar{\alpha})(\theta\sigma^{\nu}\bar{\theta})\partial_{\mu}V_{\nu}(x)$$
$$-i(\alpha\sigma^{\mu}\bar{\theta})(\theta\theta)\partial_{\mu}m(x)$$
$$+(\bar{\theta}\bar{\theta})\theta^{A}\cdot 2\alpha_{A}d(x)$$
$$+i(\bar{\theta}\bar{\theta})\theta^{A}\sigma^{\mu}_{A\dot{B}}\bar{\alpha}^{\dot{B}}\partial_{\mu}n(x)$$
$$-i(\alpha\sigma^{\mu}\bar{\theta})(\theta\sigma^{\nu}\bar{\theta})\partial_{\mu}V_{\nu}(x)$$
$$+i(\theta\sigma^{\mu}\bar{\alpha})(\bar{\theta}\bar{\theta})\theta^{A}\partial_{\mu}\psi_{A}(x)$$
$$-i(\alpha\sigma^{\mu}\bar{\theta})(\theta\theta)\bar{\theta}_{\dot{A}}\partial_{\mu}\bar{\lambda}^{\dot{A}}(x)$$
$$+2\alpha^{A}\theta_{A}\bar{\theta}_{\dot{B}}\bar{\lambda}^{\dot{B}}(x)+2\bar{\alpha}^{\dot{A}}\bar{\theta}_{\dot{A}}\theta^{B}\psi_{B}(x)$$
$$+i(\theta\sigma^{\mu}\bar{\alpha})\bar{\theta}_{\dot{A}}\partial_{\mu}\bar{\chi}^{\dot{A}}(x)$$
$$-i(\alpha\sigma^{\mu}\bar{\theta})\theta^{A}\partial_{\mu}\phi_{A}(x) \qquad (6.78)$$

Now

$$i\theta^{A}\sigma^{\mu}_{A\dot{A}}\bar{\alpha}^{\dot{A}}\theta^{B}\partial_{\mu}\phi_{B}(x)$$
$$=-i\theta^{A}\theta^{B}\sigma^{\mu}_{A\dot{A}}\bar{\alpha}^{\dot{A}}\partial_{\mu}\phi_{B}(x)$$

(since $\{\bar{\alpha}^{\dot{A}},\theta^{B}\}=0$)

$$=\frac{1}{2}i\varepsilon^{AB}(\theta\theta)\sigma^{\mu}_{A\dot{A}}\bar{\alpha}^{\dot{A}}\partial_{\mu}\phi_{B}(x)$$

(using (1.83a))

$$=(\theta\theta)\left[-\frac{i}{2}\partial_{\mu}\phi^{A}(x)\sigma^{\mu}_{A\dot{A}}\bar{\alpha}^{\dot{A}}\right]$$

(since $\{\bar{\alpha}^{\dot{A}},\phi_{B}\}=0$)

$$=(\theta\theta)\left[-\frac{i}{2}\partial_{\mu}\phi(x)\sigma^{\mu}\bar{\alpha}\right] \qquad (6.79)$$

Also

$$-i(\alpha\sigma^{\mu}\bar{\theta})\bar{\theta}_{\dot{A}}\partial_{\mu}\bar{\chi}^{\dot{A}}(x)$$
$$=-i\alpha^{A}\sigma^{\mu}_{A\dot{B}}\bar{\theta}^{\dot{B}}\bar{\theta}_{\dot{A}}\partial_{\mu}\bar{\chi}^{\dot{A}}(x)$$

$$= -i\alpha^A \sigma^\mu_{A\dot{B}} \varepsilon^{\dot{B}\dot{C}} \bar{\theta}_{\dot{C}} \bar{\theta}_{\dot{A}} \partial_\mu \bar{\chi}^{\dot{A}}(x)$$

(with (1.76c))

$$= \frac{i}{2} \alpha^A \sigma^\mu_{A\dot{B}} \varepsilon^{\dot{B}\dot{C}} \varepsilon_{\dot{C}\dot{A}} (\bar{\theta}\bar{\theta}) \partial_\mu \bar{\chi}^{\dot{A}}(x)$$

(with (1.83d))

$$= (\bar{\theta}\bar{\theta}) \left\{ \frac{i}{2} \alpha^A \sigma^\mu_{A\dot{A}} \partial_\mu \bar{\chi}^{\dot{A}}(x) \right\}$$

(since $\{\alpha^A, \bar{\theta}_{\dot{A}}\} = 0$)

$$= (\bar{\theta}\bar{\theta}) \left\{ \frac{i}{2} \alpha \sigma^\mu \partial_\mu \bar{\chi}(x) \right\} \tag{6.80}$$

Using (1.80) and (1.118) we have

$$i(\theta\sigma^\mu\bar{\alpha})(\theta\sigma^\nu\bar{\theta})\partial_\mu V_\nu(x)$$
$$= \frac{i}{2}(\theta\theta)(\bar{\theta}\bar{\alpha})\partial^\mu V_\mu(x) \tag{6.81}$$

Also

$$-i(\alpha\sigma^\mu\bar{\theta})(\theta\theta)\partial_\mu m(x)$$
$$= -i(\theta\theta)(\alpha^A \sigma^\mu_{A\dot{B}} \bar{\theta}^{\dot{B}})\partial_\mu m(x)$$
$$= -i(\theta\theta)(\alpha^A \sigma^\mu_{A\dot{C}} \varepsilon^{\dot{C}\dot{B}} \bar{\theta}_{\dot{B}})\partial_\mu m(x)$$

(with (1.76c))

$$= i(\theta\theta)\bar{\theta}_{\dot{A}}(\alpha\sigma^\mu\varepsilon)^{\dot{A}}\partial_\mu m(x) \tag{6.82}$$

Again from (1.118)

$$-i(\alpha\sigma^\mu\bar{\theta})(\theta\sigma^\nu\bar{\theta})\partial_\mu V_\nu(x)$$
$$= -\frac{i}{2}(\bar{\theta}\bar{\theta})(\theta\alpha)\partial^\mu V_\mu(x) \tag{6.83}$$

Furthermore

$$i(\theta\sigma^\mu\bar{\alpha})(\bar{\theta}\bar{\theta})\theta^A\partial_\mu\psi_A(x)$$
$$= i\theta^B\sigma^\mu_{B\dot{C}}\bar{\alpha}^{\dot{C}}(\bar{\theta}\bar{\theta})\theta^A\partial_\mu\psi_A(x)$$

$$= -i\theta^B\theta^A(\bar\theta\bar\theta)\sigma^\mu_{B\dot C}\bar\alpha^{\dot C}\partial_\mu\psi_A(x)$$

$$= \frac{i}{2}(\theta\theta)(\bar\theta\bar\theta)\varepsilon^{BA}\sigma^\mu_{B\dot C}\bar\alpha^{\dot C}\partial_\mu\psi_A(x)$$

(with (1.83a))

$$= \frac{i}{2}(\theta\theta)(\bar\theta\bar\theta)\partial_\mu\psi_A(x)\varepsilon^{AB}\sigma^\mu_{B\dot C}\bar\alpha^{\dot C}$$

(since $\varepsilon^{BA} = -\varepsilon^{AB}$,

$\{\bar\alpha, \psi\} = 0$)

$$= \frac{i}{2}(\theta\theta)(\bar\theta\bar\theta)\partial_\mu\psi^A(x)\sigma^\mu_{A\dot C}\bar\alpha^{\dot C}$$

$$= (\theta\theta)(\bar\theta\bar\theta)\left\{\frac{i}{2}(\partial_\mu\psi(x)\sigma^\mu\bar\alpha)\right\} \qquad (6.84)$$

and

$$i(\alpha\sigma^\mu\bar\theta)(\theta\theta)\bar\theta_{\dot A}\partial_\mu\bar\lambda^{\dot A}(x)$$

$$= i(\theta\theta)\alpha^A\sigma^\mu_{A\dot B}\bar\theta^{\dot B}\bar\theta_{\dot A}\partial_\mu\bar\lambda^{\dot A}(x)$$

$$= i(\theta\theta)\alpha^A\sigma^\mu_{A\dot B}\varepsilon^{\dot B\dot C}\bar\theta_{\dot C}\bar\theta_{\dot A}\partial_\mu\bar\lambda^{\dot A}(x)$$

$$= -\frac{i}{2}(\theta\theta)\alpha^A\sigma^\mu_{A\dot B}\varepsilon^{\dot B\dot C}\varepsilon_{\dot C\dot A}(\bar\theta\bar\theta)\partial_\mu\bar\lambda^{\dot A}(x)$$

(using (1.83d))

$$= -\frac{i}{2}(\theta\theta)(\bar\theta\bar\theta)\alpha^A\sigma^\mu_{A\dot A}\partial_\mu\bar\lambda^{\dot A}(x)$$

$$= (\theta\theta)(\bar\theta\bar\theta)\left\{-\frac{i}{2}\alpha\sigma^\mu\partial_\mu\bar\lambda(x)\right\} \qquad (6.85)$$

Substituting expressions (6.79) to (6.85) into (6.78) we obtain

$$\delta_S\Phi(x,\theta,\bar\theta) = \alpha^A\phi_A(x) + \bar\alpha_{\dot A}\bar\chi^{\dot A}(x)$$
$$+ \theta^A\{2\alpha_A m(x) + i(\sigma^\mu\bar\alpha)_A\partial_\mu f(x)$$
$$+ (\sigma^\nu\bar\alpha)_A V_\nu(x)\}$$

$$+\bar{\theta}_{\dot{A}}\{2\bar{\alpha}^{\dot{A}}n(x)+i(\alpha\sigma^{\mu}\varepsilon)^{\dot{A}}\partial_{\mu}f(x)$$
$$-(\alpha\sigma^{\nu}\varepsilon)^{\dot{A}}V_{\nu}(x)\}$$
$$+(\theta\theta)\{\bar{\alpha}\bar{\lambda}(x)-\frac{i}{2}\partial_{\mu}\phi(x)\sigma^{\mu}\bar{\alpha}\}$$
$$+(\bar{\theta}\bar{\theta})\{\alpha\psi(x)+\frac{i}{2}\alpha\sigma^{\mu}\partial_{\mu}\bar{\chi}(x)\}$$
$$+(\theta\theta)\bar{\theta}_{\dot{A}}\{2\bar{\alpha}^{\dot{A}}d(x)+\frac{i}{2}\bar{\alpha}^{\dot{A}}\partial^{\mu}V_{\mu}(x)$$
$$+i(\alpha\sigma^{\mu}\varepsilon)^{\dot{A}}\partial_{\mu}m(x)\}$$
$$+(\bar{\theta}\bar{\theta})\theta^{A}\{2\alpha_{A}d(x)+i(\sigma^{\mu}\bar{\alpha})_{A}\partial_{\mu}n(x)$$
$$-\frac{i}{2}\alpha_{A}\partial^{\mu}V_{\mu}(x)\}$$
$$+(\theta\theta)(\bar{\theta}\bar{\theta})\frac{i}{2}[\partial_{\mu}\psi(x)\sigma^{\mu}\bar{\alpha}$$
$$-\alpha\sigma^{\mu}\partial_{\mu}\bar{\lambda}(x)]$$
$$+2(\alpha\theta)\bar{\theta}\bar{\lambda}(x)+2(\bar{\alpha}\bar{\theta})\theta\psi(x)$$
$$+i(\theta\sigma^{\mu}\bar{\alpha})\bar{\theta}\partial_{\mu}\bar{\chi}(x)$$
$$-i(\alpha\sigma^{\mu}\bar{\theta})\theta\partial_{\mu}\phi(x) \qquad (6.86)$$

We now Fierz transform the last four terms of (6.86). Thus

$$2(\alpha\theta)(\bar{\theta}\bar{\lambda}(x))$$
$$=2(\alpha\theta)\bar{\theta}_{\dot{A}}\bar{\lambda}^{\dot{A}}(x)$$
$$=2(\theta\alpha)\bar{\theta}_{\dot{A}}\bar{\lambda}^{\dot{A}}(x) \qquad \text{(using (1.80))}$$
$$=(\theta\sigma^{\mu}\bar{\theta})(\alpha\sigma_{\mu}\bar{\lambda}(x)) \qquad \text{(using (1.114))}$$
$$(6.87)$$

In order to Fierz transform the term

$$(\bar{\alpha}\bar{\theta})(\theta\psi(x))$$

we require a formula analogous to (1.114). Let $\bar{\Phi}$, $\bar{\Psi}$ and χ be two-component Weyl spinors. Then

$$(\bar{\Phi}\bar{\Psi})\chi^B = (\bar{\Phi}_{\dot{A}}\bar{\Psi}^{\dot{A}})\chi^B$$

$$= \bar{\Phi}_{\dot{A}}\varepsilon^{\dot{A}\dot{C}}\bar{\Psi}_{\dot{C}}\varepsilon^{BD}\chi_D$$

$$= \bar{\Phi}_{\dot{A}}\varepsilon^{\dot{A}\dot{C}}\delta_{\dot{C}}{}^{\dot{D}}\bar{\Psi}_{\dot{D}}\varepsilon^{BD}\delta_D{}^F\chi_F$$

$$= \bar{\Phi}_{\dot{A}}\varepsilon^{\dot{A}\dot{C}}\varepsilon^{BD}\bar{\Psi}_{\dot{D}}\delta_{\dot{C}}{}^{\dot{D}}\delta_D{}^F\chi_F$$

$$= \frac{1}{2}\bar{\Phi}_{\dot{A}}\varepsilon^{\dot{A}\dot{C}}\varepsilon^{BD}\bar{\Psi}_{\dot{D}}\sigma^\mu_{D\dot{C}}\bar{\sigma}_\mu{}^{\dot{D}F}\chi_F$$

(using (1.92))

$$= \frac{1}{2}\bar{\Psi}_{\dot{D}}\bar{\sigma}_\mu{}^{\dot{D}F}\chi_F\bar{\Phi}_{\dot{A}}\varepsilon^{\dot{A}\dot{C}}\varepsilon^{BD}\sigma^\mu_{D\dot{C}}$$

$$= \frac{1}{2}(\bar{\Psi}\bar{\sigma}_\mu\chi)\bar{\Phi}_{\dot{A}}\bar{\sigma}^{\mu\dot{A}B}$$

(using (1.88))

$$= -\frac{1}{2}(\chi\sigma_\mu\bar{\Psi})(\bar{\Phi}\bar{\sigma}^\mu)^B \tag{6.88}$$

where we used (1.115) in the last step. Hence (using (6.88) and (1.115))

$$2(\bar{\alpha}\bar{\theta})(\theta\psi(x))$$

$$= -(\theta\sigma^\mu\bar{\theta})(\bar{\alpha}\bar{\sigma}_\mu\psi(x))$$

$$= (\theta\sigma^\mu\bar{\theta})(\psi(x)\sigma_\mu\bar{\alpha}) \tag{6.89}$$

The third term we Fierz transform is

$$-i(\alpha\sigma^\mu\bar{\theta})(\theta\partial_\mu\phi(x))$$

$$= -i\alpha^A \sigma^\mu_{A\dot{B}} \bar{\theta}^{\dot{B}} \theta^C \partial_\mu \phi_C(x)$$

$$= -i\alpha^A \sigma^\mu_{A\dot{B}} \theta^C \partial_\mu \phi_C(x) \bar{\theta}^{\dot{B}}$$

$$= -i\alpha^A \sigma^\mu_{A\dot{B}} \varepsilon^{\dot{B}\dot{C}} \theta^C \partial_\mu \phi_C(x) \bar{\theta}_{\dot{C}}$$

$$= -\frac{i}{2} \alpha^A \sigma^\mu_{A\dot{B}} \varepsilon^{\dot{B}\dot{C}} (\theta \sigma_\rho \bar{\theta})(\partial_\mu \phi(x) \sigma^\rho)_{\dot{C}}$$

(using (1.114))

$$= -\frac{i}{2} (\theta \sigma_\rho \bar{\theta}) \alpha^A \sigma^\mu_{A\dot{B}} \varepsilon^{\dot{B}\dot{C}} \partial_\mu \phi^B(x) \sigma^\rho_{B\dot{C}}$$

$$= -\frac{i}{2} (\theta \sigma_\rho \bar{\theta}) \alpha^A \partial_\mu \phi^B(x) \sigma^\mu_{A\dot{B}} \varepsilon^{\dot{B}\dot{C}} \sigma^\rho_{B\dot{C}}$$

$$= \frac{i}{4} (\theta \sigma_\rho \bar{\theta})(\alpha \partial_\mu \phi(x)) \varepsilon^{AB} \sigma^\mu_{A\dot{B}} \varepsilon^{\dot{B}\dot{C}} \sigma^\rho_{B\dot{C}}$$

(with (1.83a))

$$= -\frac{i}{4} (\theta \sigma_\rho \bar{\theta})(\alpha \partial_\mu \phi(x)) \varepsilon^{BA} \sigma^\mu_{A\dot{B}} \varepsilon^{\dot{B}\dot{C}} \sigma^\rho_{B\dot{C}}$$

$$= \frac{i}{4} (\theta \sigma_\rho \bar{\theta})(\alpha \partial_\mu \phi(x))(\bar{\sigma}^\mu)^{\dot{C}B} (\sigma^\rho)_{B\dot{C}}$$

(with (1.89))

$$= \frac{i}{4} (\theta \sigma_\rho \bar{\theta})(\alpha \partial_\mu \phi(x)) \mathrm{Tr}[\bar{\sigma}^\mu \sigma^\rho]$$

$$= \frac{i}{2} (\theta \sigma_\rho \bar{\theta})(\alpha \partial_\mu \phi(x)) \eta^{\mu\rho}$$

(using (1.91))

$$= \frac{i}{2} (\theta \sigma^\mu \bar{\theta})(\alpha \partial_\mu \phi(x)) \qquad (6.90)$$

The last contribution we want to Fierz transform is

$$i(\theta \sigma^\mu \bar{\alpha})(\bar{\theta} \partial_\mu \bar{\chi}(x))$$

$$= i\theta^A \sigma^\mu_{A\dot{B}} \bar{\alpha}^{\dot{B}} (\bar{\theta} \partial_\mu \bar{\chi}(x))$$

$$= -i \sigma^\mu_{A\dot{B}} \bar{\alpha}^{\dot{B}} (\partial_\mu \bar{\chi}(x) \bar{\theta}) \theta^A \qquad \text{(with (1.81))}$$

$$= \frac{i}{2} \sigma^{\mu}_{A\dot{B}} \bar{\alpha}^{\dot{B}} (\theta \sigma_\nu \bar{\theta})(\partial_\mu \bar{\chi}(x) \bar{\sigma}^\nu)^A$$

(with (6.88))

$$= -\frac{i}{2} (\theta \sigma_\nu \bar{\theta}) \partial_\mu \bar{\chi}_{\dot{A}}(x) \bar{\sigma}^{\nu \dot{A} A} \sigma^{\mu}_{A\dot{B}} \bar{\alpha}^{\dot{B}}$$

$$= -\frac{i}{2} (\theta \sigma_\nu \bar{\theta}) \partial_\mu \bar{\chi}_{\dot{A}}(x) \bar{\alpha}_{\dot{C}} \bar{\sigma}^{\nu \dot{A} A} \sigma^{\mu}_{A\dot{B}} \epsilon^{\dot{B}\dot{C}}$$

$$= \frac{i}{4} (\theta \sigma_\nu \bar{\theta})(\partial_\mu \bar{\chi}(x) \bar{\alpha}) \varepsilon_{\dot{A}\dot{C}} \bar{\sigma}^{\nu \dot{A} A} \sigma^{\mu}_{A\dot{B}} \varepsilon^{\dot{B}\dot{C}}$$

(with (1.83d))

$$= -\frac{i}{4} (\theta \sigma_\nu \bar{\theta})(\partial_\mu \bar{\chi}(x) \bar{\alpha}) \bar{\sigma}^{\nu \dot{A} A} \sigma^{\mu}_{A\dot{B}} \delta^{\dot{B}}{}_{\dot{A}}$$

$$= -\frac{i}{4} (\theta \sigma_\nu \bar{\theta})(\partial_\mu \bar{\chi}(x) \bar{\alpha}) \mathrm{Tr}[\bar{\sigma}^\nu \sigma^\mu]$$

$$= -\frac{i}{2} (\theta \sigma_\nu \bar{\theta})(\partial_\mu \bar{\chi}(x) \bar{\alpha}) \, \eta^{\nu\mu}$$

(using (1.91))

$$= -\frac{i}{2} (\theta \sigma_\mu \bar{\theta})(\partial^\mu \bar{\chi}(x) \bar{\alpha}) \tag{6.91}$$

This result could have been guessed from the Fierz formula (1.84). Hence the final expression for $\delta_S \Phi$ is given by

$$\begin{aligned}
\delta_S \Phi(x, \theta, \bar{\theta}) = {} & \alpha \phi(x) + \bar{\alpha} \bar{\chi}(x) \\
& + \theta \{ 2\alpha m(x) + i (\sigma^\mu \bar{\alpha}) \partial_\mu f(x) \\
& \qquad + (\sigma^\mu \bar{\alpha}) V_\mu(x) \} \\
& + \bar{\theta} \{ 2\bar{\alpha} n(x) + i (\alpha \sigma^\mu \varepsilon) \partial_\mu f(x) \\
& \qquad - (\alpha \sigma^\mu \varepsilon) V_\mu(x) \} \\
& + (\theta\theta) \{ \bar{\alpha} \bar{\lambda}(x) - \frac{i}{2} \partial_\mu \phi(x) \sigma^\mu \bar{\alpha} \}
\end{aligned}$$

$$+(\bar{\theta}\bar{\theta})\{\alpha\psi(x)+\frac{i}{2}\alpha\sigma^{\mu}\partial_{\mu}\bar{\chi}(x)\}$$
$$+(\theta\sigma^{\mu}\bar{\theta})\{\alpha\sigma_{\mu}\bar{\lambda}(x)+\psi(x)\sigma_{\mu}\bar{\alpha}$$
$$+\frac{i}{2}\alpha\partial_{\mu}\phi(x)$$
$$-\frac{i}{2}\partial_{\mu}\bar{\chi}(x)\bar{\alpha}\}$$
$$+(\theta\theta)\bar{\theta}\{2\bar{\alpha}d(x)+\frac{i}{2}\bar{\alpha}\partial^{\mu}V_{\mu}(x)$$
$$+i(\alpha\sigma^{\mu}\varepsilon)\partial_{\mu}m(x)\}$$
$$+(\bar{\theta}\bar{\theta})\theta\{2\alpha\, d(x)-\frac{i}{2}\alpha\partial^{\mu}V_{\mu}(x)$$
$$+i(\sigma^{\mu}\bar{\alpha})\partial_{\mu}n(x)\}$$
$$+(\theta\theta)(\bar{\theta}\bar{\theta})\frac{i}{2}\{\partial_{\mu}\psi(x)\sigma^{\mu}\bar{\alpha}$$
$$+\alpha\sigma^{\mu}\partial_{\mu}\bar{\lambda}(x)\} \tag{6.92}$$

Comparing the coefficients of the same powers of θ and $\bar{\theta}$ in (6.77) and (6.92) we obtain the transformation properties of the component fields. Thus

$$\delta_S' f(x) = \alpha\phi(x)+\bar{\alpha}\bar{\chi}(x)$$
$$\delta_S'\phi_A(x)= 2\alpha_A m(x)$$
$$+(\sigma^{\mu}\bar{\alpha})_A\{i\partial_{\mu}f(x)+V_{\mu}(x)\}$$
$$\delta_S'\bar{\chi}^{\dot{A}}(x)=2\bar{\alpha}^{\dot{A}}n(x)$$
$$+(\alpha\sigma^{\mu}\varepsilon)^{\dot{A}}\{i\partial_{\mu}f(x)-V_{\mu}(x)\}$$

$$\delta_S' m(x) = \bar{\alpha}\bar{\lambda}(x) - \frac{i}{2}\partial_\mu \phi(x)\sigma^\mu\bar{\alpha}$$

$$\delta_S' n(x) = \alpha\psi(x) + \frac{i}{2}\alpha\sigma^\mu\partial_\mu\bar{\chi}(x)$$

$$\delta_S' V_\mu(x) = \alpha\sigma_\mu\bar{\lambda}(x) + \psi(x)\sigma_\mu\bar{\alpha} + \frac{i}{2}\alpha\partial_\mu\phi(x) - \frac{i}{2}\partial_\mu\bar{\chi}(x)\bar{\alpha}$$

$$\delta_S'\bar{\lambda}^{\dot{A}}(x) = 2\bar{\alpha}^{\dot{A}}d(x) + \frac{i}{2}\bar{\alpha}^{\dot{A}}\partial^\mu V_\mu(x) + i(\alpha\sigma^\mu\varepsilon)^{\dot{A}}\partial_\mu m(x)$$

$$\delta_S'\psi_A(x) = 2\alpha_A d(x) - \frac{i}{2}\alpha_A\partial^\mu V_\mu(x) + i(\sigma^\mu\bar{\alpha})_A\partial_\mu n(x)$$

$$\begin{aligned}\delta_S' d(x) &= \frac{i}{2}\partial_\mu\psi(x)\sigma^\mu\bar{\alpha} + \frac{i}{2}\alpha\sigma^\mu\partial_\mu\bar{\lambda}(x)\\ &= \frac{i}{2}\partial_\mu\psi(x)\sigma^\mu\bar{\alpha} - \frac{i}{2}\partial_\mu\bar{\lambda}(x)\bar{\sigma}^\mu\alpha\end{aligned} \qquad (6.93)$$

(using (1.115))

We make the very important observation that $\delta_S' d(x)$ is a total derivative (the significance of this observation will be seen later).

The set of equations (6.93) gives the general transformation laws of the component fields under supersymmetry transformations. We shall see for special examples that linear combinations of superfields are again superfields because the generators Q and $\overline{Q}$ are linear differential operators. This means that superfields form linear representations of the supersymmetry algebra. In general the representations are highly reducible. However, one can eliminate the extra component fields by imposing covariant constraints. In this way superfields replace the problem of finding supersymmetry representations via the Casimir invariants by that of finding appropriate constraints. Such covariant constraints are, for example,

$$\overline{D}_{\dot{A}} \Phi(x,\theta,\overline{\theta}) = 0 \tag{6.94}$$

$$D_A \Phi^+(x,\theta,\overline{\theta}) = 0 \tag{6.95}$$

$$\Phi(x,\theta,\overline{\theta}) = \Phi^+(x,\theta,\overline{\theta}) \tag{6.96}$$

Superfields satisfying the constraint (6.94) or (6.95) are called chiral or scalar superfields; superfields obeying the reality condition (6.96) are called vector superfields. The constraints (6.94), (6.95) can be expressed in terms of projection operators (6.72), (6.73).

Of course, a general superfield Φ can also be expressed in terms of four-component Grassmann variables θ_a which obey the Majorana condition, i.e. we can write

$$\begin{aligned}\Phi(x,\theta) = {} & A(x) + \overline{\theta}\psi(x) + \overline{\theta}\theta F(x) \\ & + i\overline{\theta}\gamma^5\theta\, G(x) \\ & + \overline{\theta}\gamma^\mu\gamma^5\theta\, A_\mu(x) + \overline{\theta}\theta\, \overline{\theta}\chi(x) \\ & + (\overline{\theta}\theta)(\overline{\theta}\theta) D(x)\end{aligned}$$

Now, of course, powers of θ higher than the fourth are zero, i.e.

$$\theta^n = 0, \qquad n \geqslant 5$$

Terms such as

$$\bar{\theta}\gamma^\mu\theta, \; \bar{\theta}\sigma^{\mu\nu}\theta$$

do not appear in the expansion. They vanish on account of relations (1.186) and (1.115), (1.128).

We have seen that the covariant derivatives D and $\overline{D}$ satisfy an algebra similar to that of the generators Q and $\overline{Q}$, and we have also seen that the latter act as Fock space annihilation and creation operators with respect to a Clifford vacuum. Equation (6.94) shows that $\overline{D}$ mimics this behaviour with respect to the field Φ. The somewhat ad hoc way of introducing the chirality constraints (6.94), (6.95) and the constraint (6.96) has been criticized[52] in view of their unclear connection with well defined particle representations, and a direct derivation of the fields from known irreducible particle representations has been given in the literature.[52]

CHAPTER 7

CONSTRAINED SUPERFIELDS AND SUPERMULTIPLETS

7.1 Chiral Superfields

Superfields which satisfy either of the constraints (6.94), (6.95) are called chiral or scalar superfields. A superfield $\Phi(x,\theta,\bar{\theta})$ which satisfies (6.94), i.e. $\bar{D}\Phi = 0$, is called a left-handed chiral superfield, and a superfield $\Phi(x,\theta,\bar{\theta})$ which satisfies (6.95), i.e. $D\Phi^+ = 0$, is called a right-handed chiral superfield. The origin of this terminology has been mentioned earlier, but will also become clear in the following.

We consider first left-handed chiral superfields. The constraint

$$\bar{D}_{\dot{A}}\Phi = 0, \quad \bar{D}_{\dot{A}} = -\bar{\partial}_{\dot{A}} - i\theta^B \sigma^\mu_{B\dot{A}}\partial_\mu \tag{7.1}$$

(recall (6.41)) is easily solved in terms of new variables given by

$$\left.\begin{aligned} y^\mu &:= x^\mu + i\theta\sigma^\mu\bar{\theta} \\ \theta'_A &:= \theta_A \ , \quad \bar{\theta}'_{\dot{A}} := \bar{\theta}_{\dot{A}} \end{aligned}\right\} \tag{7.2}$$

Proposition: The new variables (7.2) satisfy the following conditions

$$\text{a)}\quad \bar{D}_{\dot{A}}\, y^{\mu} = 0 \qquad\qquad (7.3)$$
$$\text{b)}\quad \bar{D}_{\dot{A}}\, \theta_{A} = 0$$

Proof:

a) We have

$$\bar{D}_{\dot{A}}\, y^{\mu} = \{-\bar{\partial}_{\dot{A}} - i\theta^{A}\sigma^{\nu}_{A\dot{A}}\frac{\partial}{\partial x^{\nu}}\}(x^{\mu} + i\theta\sigma^{\mu}\bar{\theta})$$
$$= i\theta^{B}\sigma^{\mu}_{B\dot{A}} - i\,\theta^{A}\sigma^{\nu}_{A\dot{A}}\delta_{\nu}{}^{\mu}$$
$$= 0$$

since

$$\{\bar{\partial}_{\dot{A}}, \theta\} = 0$$

b) This result is obvious.

Hence, in view of (7.3) any function of y^{μ} and θ satisfies the constraint (7.1). We now change the variables of the covariant derivatives.

Proposition: In terms of the new set of variables (7.2) the covariant derivatives D_{A}, $\bar{D}_{\dot{A}}$ of (6.40), (6.41) become

$$D_{A} \longrightarrow D_{A}^{(1)} = \partial_{A} + 2i\sigma^{\mu}_{A\dot{A}}\bar{\theta}^{\dot{A}}\frac{\partial}{\partial y^{\mu}}$$
$$\bar{D}_{\dot{A}} \longrightarrow \bar{D}^{(1)}_{\dot{A}} = -\bar{\partial}_{\dot{A}}$$

Verification: The covariant derivatives in the variables x, θ, $\bar{\theta}$ are given by (6.40), (6.41), i.e.

$$D_{A}(x,\theta,\bar{\theta}) = \frac{\partial}{\partial\theta^{A}} + i\sigma^{\mu}_{A\dot{A}}\bar{\theta}^{\dot{A}}\frac{\partial}{\partial x^{\mu}}$$
$$\bar{D}_{\dot{A}}(x,\theta,\bar{\theta}) = -\frac{\partial}{\partial\bar{\theta}^{\dot{A}}} - i\theta^{A}\sigma^{\mu}_{A\dot{A}}\frac{\partial}{\partial x^{\mu}}$$

Transforming these operators according to (7.2) we have

$$\frac{\partial}{\partial x^\mu} = \frac{\partial y^\nu}{\partial x^\mu}\frac{\partial}{\partial y^\nu} = \delta_\mu{}^\nu \frac{\partial}{\partial y^\nu} = \frac{\partial}{\partial y^\mu}$$

$$\frac{\partial}{\partial \theta^A} = \frac{\partial \theta'^B}{\partial \theta^A}\frac{\partial}{\partial \theta'^B} + \frac{\partial y^\mu}{\partial \theta^A}\frac{\partial}{\partial y^\mu}$$

$$= \delta_A{}^B \frac{\partial}{\partial \theta'^B} + i\sigma^\mu_{A\dot{A}}\bar{\theta}^{\dot{A}}\frac{\partial}{\partial y^\mu}$$

$$= \frac{\partial}{\partial \theta'^A} + i\sigma^\mu_{A\dot{A}}\bar{\theta}'^{\dot{A}}\frac{\partial}{\partial y^\mu} \quad \text{(since } \bar{\theta}' = \bar{\theta}\text{)}$$

$$\frac{\partial}{\partial \bar{\theta}^{\dot{A}}} = \frac{\partial \theta'^{\dot{B}}}{\partial \bar{\theta}^{\dot{A}}}\frac{\partial}{\partial \bar{\theta}'^{\dot{B}}} + \frac{\partial y^\mu}{\partial \bar{\theta}^{\dot{A}}}\frac{\partial}{\partial y^\mu}$$

$$= \frac{\partial}{\partial \bar{\theta}'^{\dot{A}}} - i\theta'^A\sigma^\mu_{A\dot{A}}\frac{\partial}{\partial y^\mu}$$

Hence

$$D_A(x,\theta,\bar{\theta}) = \frac{\partial}{\partial \theta^A} + i\sigma^\mu_{A\dot{A}}\bar{\theta}^{\dot{A}}\frac{\partial}{\partial x^\mu}$$

$$= \frac{\partial}{\partial \theta'^A} + i\sigma^\mu_{A\dot{A}}\bar{\theta}'^{\dot{A}}\frac{\partial}{\partial y^\mu} + i\sigma^\mu_{A\dot{A}}\bar{\theta}'^{\dot{A}}\frac{\partial}{\partial y^\mu}$$

$$= \frac{\partial}{\partial \theta'^A} + 2i\sigma^\mu_{A\dot{A}}\bar{\theta}'^{\dot{A}}\frac{\partial}{\partial y^\mu}$$

$$= D_A^{(1)}(y,\theta',\bar{\theta}') \quad \text{(cf. (6.47))} \qquad (7.4a)$$

and

$$\bar{D}_{\dot{A}}(x,\theta,\bar{\theta}) = -\frac{\partial}{\partial \bar{\theta}^{\dot{A}}} - i\theta^A\sigma^\mu_{A\dot{A}}\frac{\partial}{\partial x^\mu}$$

$$= -\frac{\partial}{\partial \bar{\theta}'^{\dot{A}}} + i\theta'^A\sigma^\mu_{A\dot{A}}\frac{\partial}{\partial y^\mu} - i\theta'^A\sigma^\mu_{A\dot{A}}\frac{\partial}{\partial y^\mu}$$

$$= -\frac{\partial}{\partial \bar{\theta}'^{\dot{A}}}$$

$$= \bar{D}^{(1)}_{\dot{A}}(y, \theta', \bar{\theta}') \tag{7.4b}$$

In this representation the constraint (7.1) has the simple implication that

$$\Phi(x, \theta, \bar{\theta}) = \Phi(y - i\theta\sigma\bar{\theta}, \theta, \bar{\theta})$$
$$\equiv \Phi_1(y, \theta, \bar{\theta})$$

(see (6.7)) does not possess any explicit dependence on $\bar{\theta}$, i.e.

$$\bar{D}^{(1)}_{\dot{A}} \Phi_1(y, \theta, \bar{\theta}) = -\frac{\partial}{\partial \bar{\theta}^{\dot{A}}} \Phi_1(y, \theta, \bar{\theta}) = 0$$

by (7.1), so that $\Phi_1(y, \theta, \bar{\theta}) \equiv \Phi_1(y, \theta)$, implying that Φ_1 has the power series expansion in θ:

$$\Phi_1(y, \theta) = A(y) + 2^{\frac{1}{2}}\theta\psi(y) + (\theta\theta)F(y) \tag{7.5}$$

and is independent of $\bar{\theta}$. From (7.5) we see that a superfield Φ_1 satisfying the constraint (7.1) depends on the two-component Weyl spinor $\psi(y)$ which, according to (1.53), is a left-handed Weyl spinor. This dependence of the superfield on the left-handed Weyl spinor $\psi(y)$ is the origin for the name "left-handed chiral superfield".

In (7.5) A(y) and F(y) describe complex scalar fields (A and F have no well defined parity, the change of parity being given by complex conjugation $A \rightarrow A^*$, $F \rightarrow F^*$ as explained earlier). From (7.5) we regain the original field $\Phi(x, \theta, \bar{\theta})$ satisfying the constraint (7.1) in the variables x, θ, $\bar{\theta}$ by expanding the component fields in the following way. We have

$$
\begin{aligned}
&\Phi_1(y,\theta)\\
&= A(y) + 2^{\frac{1}{2}}\theta\psi(y) + (\theta\theta)F(y)\\
&= A(x+i\theta\sigma\bar{\theta}) + 2^{\frac{1}{2}}\theta\psi(x+i\theta\sigma\bar{\theta})\\
&\qquad + (\theta\theta)F(x+i\theta\sigma\bar{\theta})\\
&= A(x) + i\theta\sigma^{\mu}\bar{\theta}\,\partial_{\mu}A(x)\\
&\quad -\frac{1}{2}(\theta\sigma_{\mu}\bar{\theta})(\theta\sigma_{\nu}\bar{\theta})\,\partial^{\mu}\partial^{\nu}A(x)\\
&\quad + 2^{\frac{1}{2}}\theta^{A}\psi_{A}(x) + 2^{\frac{1}{2}}i\,\theta^{A}(\theta\sigma^{\mu}\bar{\theta})\,\partial_{\mu}\psi_{A}(x)\\
&\quad + (\theta\theta)F(x)\\
&= A(x) + i(\theta\sigma^{\mu}\bar{\theta})\,\partial_{\mu}A(x)\\
&\quad -\frac{1}{4}(\theta\theta)(\bar{\theta}\bar{\theta})\,\Box A(x) + 2^{\frac{1}{2}}\theta^{A}\psi_{A}(x)\\
&\quad + \frac{i}{2^{1/2}}(\theta\theta)\,\bar{\theta}_{\dot{A}}\,\partial_{\mu}\psi^{A}(x)\,\sigma^{\mu}_{A\dot{B}}\,\varepsilon^{\dot{B}\dot{A}}\\
&\quad + (\theta\theta)F(x) \qquad\qquad (7.6)
\end{aligned}
$$

where we used (1.118), (1.73) and (1.83a), and the term containing $\partial_{\mu}\psi_{A}$ was rewritten in the following way

$$
\begin{aligned}
&i\theta^{A}(\theta\sigma^{\mu}\bar{\theta})\,\partial_{\mu}\psi_{A}\\
&= i\theta^{A}(\theta^{B}\sigma^{\mu}_{B\dot{B}}\,\bar{\theta}^{\dot{B}})\,\partial_{\mu}\psi_{A}\\
&= i\left\{-\frac{1}{2}\varepsilon^{AB}(\theta\theta)\right\}\sigma^{\mu}_{B\dot{B}}\,\bar{\theta}^{\dot{B}}\,\partial_{\mu}\psi_{A} \quad \text{(with (1.83a))}\\
&= -\frac{i}{2}(\theta\theta)\,\bar{\theta}^{\dot{B}}\,\partial_{\mu}\psi_{A}\,\varepsilon^{AB}\sigma^{\mu}_{B\dot{B}}\\
&= \frac{i}{2}(\theta\theta)\,\bar{\theta}_{\dot{C}}\,\partial_{\mu}\psi^{B}\sigma^{\mu}_{B\dot{B}}\,\varepsilon^{\dot{B}\dot{C}}\\
&= \frac{i}{2}(\theta\theta)\,\bar{\theta}_{\dot{A}}\,\partial_{\mu}\psi^{A}(x)\,\sigma^{\mu}_{A\dot{B}}\,\varepsilon^{\dot{B}\dot{A}}
\end{aligned}
$$

As a consistency check we demonstrate the vanishing of

$$\bar{D}_{\dot{A}}\,\Phi(x,\theta,\bar{\theta})$$
$$= \{-\bar{\partial}_{\dot{A}} - i\theta^{A}\sigma^{\mu}_{A\dot{A}}\partial_{\mu}\}\{A(x) + i(\theta\sigma^{\mu}\bar{\theta})\partial_{\mu}A(x)$$
$$+ 2^{\frac{1}{2}}\theta^{A}\psi_{A}(x) - \tfrac{1}{4}(\theta\theta)(\bar{\theta}\bar{\theta})\Box A(x)$$
$$+ (\theta\theta)F(x) + \frac{i}{2^{1/2}}(\theta\theta)\partial_{\rho}\psi(x)\sigma^{\rho}\bar{\theta}\}$$

(with (6.41))

$$= i\theta^{A}\sigma^{\rho}_{A\dot{A}}\partial_{\rho}A(x) - \tfrac{1}{2}(\theta\theta)\bar{\theta}_{\dot{A}}\Box A(x)$$
$$- \frac{i}{2^{\frac{1}{2}}}(\theta\theta)\partial_{\rho}\psi^{A}(x)\sigma^{\rho}_{A\dot{A}} - i\theta^{A}\sigma^{\mu}_{A\dot{A}}\partial_{\mu}A(x)$$
$$+ \theta^{A}\sigma^{\mu}_{A\dot{A}}(\theta\sigma^{\rho}\bar{\theta})\partial_{\mu}\partial_{\rho}A(x)$$
$$- 2^{\frac{1}{2}}i\,\theta^{A}\sigma^{\mu}_{A\dot{A}}\theta^{B}\partial_{\mu}\psi_{B}(x) \qquad (7.7)$$

where we used $\theta^{3}=0$ and where

$$-\bar{\partial}_{\dot{A}}(\theta\theta)(\bar{\theta}\bar{\theta})$$
$$= -(\theta\theta)\bar{\partial}_{\dot{A}}(\bar{\theta}_{\dot{B}}\bar{\theta}^{\dot{B}}) \qquad \text{(using (6.4m))}$$
$$= -(\theta\theta)\{(\bar{\partial}_{\dot{A}}\bar{\theta}_{\dot{B}})\bar{\theta}^{\dot{B}} - \bar{\theta}_{\dot{B}}(\bar{\partial}_{\dot{A}}\bar{\theta}^{\dot{B}})\}$$

(using (6.4b))

$$= -(\theta\theta)\{-\varepsilon_{\dot{A}\dot{B}}\bar{\theta}^{\dot{B}} - \bar{\theta}_{\dot{B}}\delta_{\dot{A}}{}^{\dot{B}}\}$$

(using (6.4k))

$$= 2(\theta\theta)\bar{\theta}_{\dot{A}}$$

and therefore

$$-\bar{\partial}_{\dot{A}}\{-\tfrac{1}{4}(\theta\theta)(\bar{\theta}\bar{\theta})\Box A(x)\}$$
$$= -\tfrac{1}{2}(\theta\theta)\bar{\theta}_{\dot{A}}\Box A(x)$$

Using (1.118) in the form

$$(\theta\sigma^{\rho}\bar{\theta})(\theta\sigma^{\mu})_{\dot{A}}=\frac{1}{2}\eta^{\rho\mu}(\theta\theta)\bar{\theta}_{\dot{A}} \tag{7.8}$$

and (1.83a), we may rewrite (7.7)

$$\begin{aligned}
&\bar{D}_{\dot{A}}\,\Phi(x,\theta,\bar{\theta})\\
&=-\frac{1}{2}(\theta\theta)\,\bar{\theta}_{\dot{A}}\,\Box A(x)-\frac{i}{2^{\frac{1}{2}}}(\theta\theta)(\partial_{\rho}\psi(x)\sigma^{\rho})_{\dot{A}}\\
&\quad+\frac{1}{2}(\theta\theta)\,\bar{\theta}_{\dot{A}}\,\Box A(x)+\frac{i}{2^{\frac{1}{2}}}(\theta\theta)(\partial_{\mu}\psi(x)\sigma^{\mu})_{\dot{A}}\\
&=0
\end{aligned}$$

This result states that the superfield (7.6) (with variables $x,\theta,\bar{\theta}$) is the most general solution of (7.1). It should be noted that the superfield (7.6) obtained in this way depends only on the fields A(x), F(x) and $\psi(x)$, a dependence which is characteristic for a left-handed chiral superfield.

The superfield $\Phi^{+}(x,\theta,\bar{\theta})$ satisfies the constraint

$$D_{A}\Phi^{+}(x,\theta,\bar{\theta})=0 \tag{7.9}$$

and is called a right-handed chiral superfield. Again we make a shift of variables to solve the constraint equation (7.9) by introducing the variables

$$\left.\begin{aligned}
z^{\mu}&:=x^{\mu}-i\,\theta\sigma^{\mu}\bar{\theta}\\
\theta'_{A}&:=\theta_{A}\,,\quad \bar{\theta}'_{\dot{A}}:=\bar{\theta}_{\dot{A}}
\end{aligned}\right\} \tag{7.10}$$

<u>Proposition</u>: The new variables (7.10) satisfy the following conditions

$$\begin{aligned}
&\text{a)}\quad D_{A}z^{\mu}=0\\
&\text{b)}\quad D_{A}\,\bar{\theta}_{\dot{A}}=0
\end{aligned} \tag{7.11}$$

where (cf. (6.40))

$$D_A \doteq \partial_A + i\sigma^\mu_{A\dot{A}}\,\bar{\theta}^{\dot{A}}\frac{\partial}{\partial x^\mu}$$

Proof:

a) We have

$$\begin{aligned} D_A z^\mu &= (\partial_A + i\sigma^\mu_{A\dot{A}}\bar{\theta}^{\dot{A}}\partial_\rho)(x^\mu - i\theta\sigma^\mu\bar{\theta}) \\ &= -i\sigma^\mu_{A\dot{A}}\bar{\theta}^{\dot{A}} + i\sigma^\rho_{A\dot{A}}\bar{\theta}^{\dot{A}}\frac{\partial x^\mu}{\partial x^\rho} \\ &= 0 \end{aligned}$$

b) This result is obvious.

Hence, in view of (7.11) any function of z^μ and $\bar{\theta}$ satisfies the constraint (7.9). We now change the variables of the covariant derivatives.

Proposition: In terms of the new set of variables (7.10) the covariant derivatives $D_A, \bar{D}_{\dot{A}}$ of (6.40), (6.41) become

$$\begin{aligned} D_A &\longrightarrow D_A^{(2)} = \partial_A \\ \bar{D}_{\dot{A}} &\longrightarrow \bar{D}_{\dot{A}}^{(2)} = -\bar{\partial}_{\dot{A}} - 2i\theta^A\sigma^\mu_{A\dot{A}}\frac{\partial}{\partial z^\mu} \end{aligned} \qquad (7.12)$$

$$D_A^{(2)} \equiv D_A^{(2)}(z,\theta,\bar{\theta})$$

$$\bar{D}_{\dot{A}}^{(2)} \equiv \bar{D}_{\dot{A}}^{(2)}(z,\theta,\bar{\theta})$$

Verification: The expressions (7.12) are easily verified in analogy to (7.4a) and (7.4b).

In view of (7.11) we can set

$$\Phi^+(z,\bar{\theta}) = A^*(z) + 2^{\frac{1}{2}}\bar{\theta}\bar{\Psi}(z) + (\bar{\theta}\bar{\theta})F^*(z) \qquad (7.13)$$

It should be observed that Φ^+ which is the hermitian conjugate of the left-handed chiral superfield Φ is a type 2 superfield (cf.(6.11)). In this representation we have, of course,

$$D_A^{(2)} \Phi^+(x,\bar{\theta}) = 0 = D_A^{(2)} \Phi_2(x,\bar{\theta})$$

expressing the fact that $\Phi^+(x,\bar{\theta})$ does not depend explicitly on θ .Expanding (7.13) we obtain

$$\begin{aligned}
&\Phi_2(x,\bar{\theta}) \\
&= A^*(x) + 2^{\frac{1}{2}}\bar{\theta}\bar{\Psi}(x) + (\bar{\theta}\bar{\theta})F^*(x) \\
&= A^*(x - i\theta\sigma\bar{\theta}) + 2^{\frac{1}{2}}\bar{\theta}\bar{\Psi}(x - i\theta\sigma\bar{\theta}) \\
&\qquad + (\bar{\theta}\bar{\theta})F^*(x - i\theta\sigma\bar{\theta}) \\
&= A^*(x) - i\,\theta\sigma^\mu\bar{\theta}\,\partial_\mu A^*(x) \\
&\quad - \frac{1}{2}(\theta\sigma^\mu\bar{\theta})(\theta\sigma^\nu\bar{\theta})\partial_\mu\partial_\nu A^*(x) \\
&\quad + 2^{\frac{1}{2}}\bar{\theta}\bar{\Psi}(x) + 2^{\frac{1}{2}} i\,\theta\sigma^\mu\bar{\theta}\,\bar{\theta}_{\dot{A}}\partial_\mu\bar{\Psi}^{\dot{A}}(x) \\
&\quad + (\bar{\theta}\bar{\theta})F^*(x)
\end{aligned}$$

Rearranging the third and the fifth terms with the help of (1.118) and (1.83) we finally obtain

$$\begin{aligned}
&\Phi_2(x,\bar{\theta}) \\
&= A^*(x) - i(\theta\sigma^\mu\bar{\theta})\partial_\mu A^*(x) \\
&\quad - \frac{1}{4}(\theta\theta)(\bar{\theta}\bar{\theta})\Box A^*(x) + 2^{\frac{1}{2}}\bar{\theta}\bar{\Psi}(x) \\
&\quad + (\bar{\theta}\bar{\theta})F^*(x) \\
&\quad - \frac{i}{2^{1/2}}(\bar{\theta}\bar{\theta})(\theta\sigma^\mu\partial_\mu\bar{\Psi}(x)) \qquad (7.14) \\
&= \Phi^+(x,\theta,\bar{\theta})
\end{aligned}$$

The result is seen to be the hermitian conjugate of (7.6). The sixth term of this expression follows from the fifth term of (7.6) upon using (1.116).

As a consistency check we verify that

$$\mathcal{D}_A \Phi^+(x,\theta,\bar\theta) = 0 \qquad (7.15)$$

where

$$\mathcal{D}_A = \partial_A + i\,\sigma^\mu_{A\dot A}\,\bar\theta^{\dot A}\,\partial_\mu$$

Consider

$$\mathcal{D}_A \Phi^+(x,\theta,\bar\theta)$$

$$= \{\partial_A + i\sigma^\mu_{A\dot A}\bar\theta^{\dot A}\partial_\mu\}\{A^*(x) - i(\theta\sigma^\mu\bar\theta)\partial_\mu A^*(x)$$

$$-\tfrac{1}{4}(\theta\theta)(\bar\theta\bar\theta)\Box A^*(x) + 2^{\frac{1}{2}}\bar\theta\bar\Psi(x)$$

$$-\frac{i}{2^{1/2}}(\bar\theta\bar\theta)\,\theta\sigma^\mu\partial_\mu\bar\Psi(x) + (\bar\theta\bar\theta)F^*(x)\}$$

$$= -i\sigma^\mu_{A\dot A}\bar\theta^{\dot A}\partial_\mu A^*(x) - \tfrac{1}{2}\theta_A(\bar\theta\bar\theta)\Box A^*(x)$$

$$-\frac{i}{2^{1/2}}(\bar\theta\bar\theta)\,\sigma^\mu_{A\dot B}\partial_\mu\bar\Psi^{\dot B}(x)$$

$$+ i\sigma^\mu_{A\dot A}\bar\theta^{\dot A}\partial_\mu A^*(x)$$

$$+ \sigma^\mu_{A\dot A}\bar\theta^{\dot A}(\theta\sigma^\mu\bar\theta)\partial_\mu\partial_\nu A^*(x)$$

$$+ 2^{\frac{1}{2}} i\,\sigma^\mu_{A\dot A}\bar\theta^{\dot A}\bar\theta_{\dot B}\,\partial^\mu\bar\Psi^{\dot B}(x)$$

$$= -\tfrac{1}{2}\theta_A(\bar\theta\bar\theta)\Box A^*(x) - \frac{i}{2^{1/2}}(\bar\theta\bar\theta)\sigma^\mu_{A\dot B}\partial_\mu\bar\Psi^{\dot B}(x)$$

$$+ \tfrac{1}{2}\theta_A(\bar\theta\bar\theta)\Box A^*(x) + \frac{i}{2^{1/2}}(\bar\theta\bar\theta)\sigma^\mu_{A\dot B}\partial_\mu\bar\Psi^{\dot B}(x)$$

$$= 0$$

Equation (7.15) states that (7.14) is the most general solution of the constraint equation $\mathcal{D}_A\Phi^+ = 0$. It should be observed that Φ^+ depends only on the complex scalar fields A^*, F^* and a right-handed chiral Weyl spinor

$\Psi(x)$. This is characteristic for right-handed chiral superfields.

Products of scalar superfields are again scalar superfields. In order to see this let Φ_i and Φ_k be two scalar superfields with component expansions

$$\Phi_i = A_i(y) + 2^{\frac{1}{2}}\theta\psi_i(y) + (\theta\theta)F_i(y)$$
$$\Phi_k = A_k(y) + 2^{\frac{1}{2}}\theta\psi_k(y) + (\theta\theta)F_k(y)$$

Then

$$\begin{aligned}
&\Phi_i\Phi_k \\
&= [A_i(y) + 2^{\frac{1}{2}}\theta\psi_i(y) + (\theta\theta)F_i(y)]\cdot \\
&\quad\cdot[A_k(y) + 2^{\frac{1}{2}}\theta\psi_k(y) + (\theta\theta)F_k(y)] \\
&= A_i(y)A_k(y) + 2(\theta\psi_i(y))(\theta\psi_k(y)) \\
&\quad + 2^{\frac{1}{2}}\theta A_i(y)\psi_k(y) + (\theta\theta)A_i(y)F_k(y) \\
&\quad + 2^{\frac{1}{2}}\theta\psi_i(y)A_k(y) + (\theta\theta)F_i(y)A_k(y) \\
&= A_i(y)A_k(y) \\
&\quad + 2^{\frac{1}{2}}\theta[A_i(y)\psi_k(y) + \psi_i(y)A_k(y)] \\
&\quad + (\theta\theta)[A_i(y)F_k(y) + F_i(y)A_k(y) \\
&\qquad - \psi_i(y)\psi_k(y)]
\end{aligned} \tag{7.16}$$

From the expansion (7.16) we see that $\Phi_i\cdot\Phi_k$ is again a function of y and θ, and therefore

$$\bar{D}_{\dot{A}}(\Phi_i\cdot\Phi_k) = 0$$

Of course, this result can also be obtained by taking into account the fact that $\bar{D}_{\dot{A}}$ is a linear operator. Hence $\Phi_i \Phi_k$ is again a left-handed chiral superfield. Now consider the product of three left-handed chiral superfields, i.e.

$$
\begin{aligned}
&\Phi_l \Phi_i \Phi_k \\
&= A_l(y) A_i(y) A_k(y) \\
&\quad + 2^{\frac{1}{2}} \theta \{ A_l(y) A_i(y) \psi_k(y) + A_l(y) \psi_i(y) A_k(y) \\
&\qquad\qquad + \psi_l(y) A_i(y) A_k(y) \} \\
&\quad + (\theta\theta) \{ A_l(y) A_i(y) F_k(y) + A_l(y) F_i(y) A_k(y) \\
&\qquad\qquad + F_l(y) A_i(y) A_k(y) - \psi_l(y) \psi_k(y) A_i(y) \\
&\qquad\qquad - \psi_l(y) \psi_i(y) A_k(y) - A_l(y) \psi_i(y) \psi_k(y) \}
\end{aligned}
\tag{7.17}
$$

and again $\Phi_l \Phi_i \Phi_k$ is a left-handed chiral superfield satisfying the condition

$$\bar{D}_{\dot{A}} (\Phi_l \Phi_i \Phi_k) = 0 \tag{7.18}$$

Similar results hold, of course, for the conjugate fields.

An important point in the construction of invariant actions (which will be considered later) is the transformation property of the highest component of a superfield. From the general expression (6.93) we obtain the transformation properties of the component fields of the left-handed chiral superfield (7.5) as (see below)

$$
\begin{aligned}
&\delta_S' A(y) = 2^{\frac{1}{2}} \alpha \psi(y) \\
&\delta_S' \psi_A(y) = 2^{\frac{1}{2}} \alpha_A F(y) + 2^{\frac{1}{2}} i \, \sigma^{\mu}_{A\dot{A}} \bar{\alpha}^{\dot{A}} \partial_{\mu} A(y)
\end{aligned}
$$

$$\delta_S' F(y) = -2^{\frac{1}{2}} i\, \partial_\mu \psi(y) \sigma^\mu \bar{\alpha} \qquad (7.19)$$

These relations are obtained by first comparing (6.75) with (7.6) and making the following identifications

$$f(x) \longrightarrow A(x)$$
$$\phi(x) \longrightarrow 2^{\frac{1}{2}} \psi(x)$$
$$\bar{\chi}(x) \longrightarrow 0$$
$$m(x) \longrightarrow F(x)$$
$$n(x) \longrightarrow 0$$
$$V_\mu(x) \longrightarrow i \partial_\mu A(x)$$
$$\bar{\lambda}^{\dot{A}}(x) \longrightarrow -\frac{i}{2^{1/2}} \partial_\mu \psi^A(x) \sigma^\mu_{A\dot{B}} \varepsilon^{\dot{B}\dot{A}}$$
$$\psi(x) \longrightarrow 0$$
$$d(x) \longrightarrow -\frac{1}{4} \Box A(x)$$

The appropriate substitutions in (6.93) then yield (7.19), now, of course, in conformity with the variable of (7.5), in terms of y. It is evident from (7.19) that the fields A, ψ, F constitute an irreducible representation of the supersymmetry algebra since their multiplet transforms under the supersymmetry transformation into itself. From (7.19) one can also see the significance of the auxiliary field F: the supersymmetry algebra is closed linearly in the off-shell case (i.e. in the presence of F), whereas one has a nonlinear representation of the supersymmetry algebra in the on-shell case.

An important aspect of the transformation properties of superfields is that the highest component field of Φ, i.e. F, transforms into a total space-time derivative.

A space-time integral $\int d^4x$ of this quantity is thus invariant under supersymmetry transformations, because the supersymmetric variation of this component field (which is the total derivative) can be transformed into a surface integral which vanishes, provided the fields fall off sufficiently fast at infinity.

In order to be able to construct supersymmetric Lagrangians in terms of superfields we need the product of a right-handed chiral superfield and a left-handed chiral superfield. However, this product is neither chiral nor antichiral. Consider

$$\Phi_i^\dagger = A_i^*(z) + 2^{\frac{1}{2}}\,\bar\theta\,\bar\Psi_i(z) + (\bar\theta\bar\theta)F_i^*(z)$$

$$\Phi_j = A_j(y) + 2^{\frac{1}{2}}\,\theta\psi_j(y) + (\theta\theta)F_j(y)$$

Then

$$\Phi_i^\dagger\,\Phi_j$$

$$= \{A_i^*(z) + 2^{\frac{1}{2}}\,\bar\theta\,\bar\Psi_i(z) + (\bar\theta\,\bar\theta)F_i^*(z)\}\cdot$$
$$\{A_j(y) + 2^{\frac{1}{2}}\,\theta\psi_j(y) + (\theta\theta)F_j(y)\}$$

$$= A_i^*(z)A_j(y) + 2^{\frac{1}{2}}\,\theta A_i^*(z)\psi_j(y)$$
$$+ 2^{\frac{1}{2}}\,\bar\theta\bar\Psi_i(z)A_j(y) + (\theta\theta)A_i^*(z)F_j(y)$$
$$+ (\bar\theta\bar\theta)F_i^*(z)A_j(y) + 2(\bar\theta\bar\Psi_i(z))(\theta\psi_j(z))$$
$$+ 2^{\frac{1}{2}}(\bar\theta\bar\theta)\theta\psi_j(y)F_i^*(z)$$
$$+ 2^{\frac{1}{2}}(\theta\theta)\bar\theta\bar\Psi_i(z)F_j(y)$$
$$+ (\theta\theta)(\bar\theta\bar\theta)F_i^*(z)F_j(y)$$

$$= A_i^*(x - i\theta\sigma\bar\theta)A_j(x + i\theta\sigma\bar\theta)$$
$$+ 2^{\frac{1}{2}}\theta A_i^*(x - i\theta\sigma\bar\theta)\psi_j(x + i\theta\sigma\bar\theta)$$

$$
\begin{aligned}
&+ 2^{\frac{1}{2}}\, \bar\theta \bar\psi_{\dot i}(x - i\theta\sigma\bar\theta) A_{\dot j}(x + i\theta\sigma\bar\theta) \\
&+ (\theta\theta) A^*_{\dot i}(x - i\theta\sigma\bar\theta) F_{\dot j}(x + i\theta\sigma\bar\theta) \\
&+ (\bar\theta\bar\theta) F^*_{\dot i}(x - i\theta\sigma\bar\theta) A_{\dot j}(x + i\theta\sigma\bar\theta) \\
&+ 2[\bar\theta\bar\psi_{\dot i}(x - i\theta\sigma\bar\theta)][\theta\psi_{\dot j}(x + i\theta\sigma\bar\theta)] \\
&+ 2^{\frac{1}{2}} (\bar\theta\bar\theta)\, \theta\psi_{\dot j}(x + i\theta\sigma\bar\theta) F^*_{\dot i}(x - i\theta\sigma\bar\theta) \\
&+ 2^{\frac{1}{2}} (\theta\theta)\, \bar\theta\psi^*_{\dot i}(x - i\theta\sigma\bar\theta) F_{\dot j}(x + i\theta\sigma\bar\theta) \\
&+ (\theta\theta)(\bar\theta\bar\theta) F^*_{\dot i}(x - i\theta\sigma\bar\theta) F_{\dot j}(x + i\theta\sigma\bar\theta)
\end{aligned}
$$

(7.20)

We consider various terms separately . Thus

$$
\begin{aligned}
&A^*_{\dot i}(x - i\theta\sigma\bar\theta) A_{\dot j}(x + i\theta\sigma\bar\theta) \\
&= \{A^*_{\dot i}(x) - i\,\theta\sigma^\mu\bar\theta\, \partial_\mu A^*_{\dot i}(x) - \tfrac{1}{4}(\theta\theta)(\bar\theta\bar\theta)\Box A^*_{\dot i}(x)\} \\
&\quad \cdot \{A_{\dot j}(x) - i\theta\sigma^\rho\bar\theta\, \partial_\rho A_{\dot j}(x) - \tfrac{1}{4}(\theta\theta)(\bar\theta\bar\theta)\Box A_{\dot j}(x)\}
\end{aligned}
$$

(using (1.118))

$$
\begin{aligned}
&= A^*_{\dot i}(x) A_{\dot j}(x) + i\,\theta\sigma^\rho\bar\theta\, \partial_\rho A_{\dot j}(x) A^*_{\dot i}(x) \\
&\quad - \tfrac{1}{4}(\theta\theta)(\bar\theta\bar\theta)\{A^*_{\dot i}(x)\,\Box A_{\dot j}(x) \\
&\qquad\qquad + \Box A^*_{\dot i}(x) A_{\dot j}(x)\} \\
&\quad - i\,\theta\sigma^\mu\bar\theta\, \partial_\mu A^*_{\dot i}(x) A_{\dot j}(x) \\
&\quad + (\theta\sigma^\mu\bar\theta)(\theta\sigma^\rho\bar\theta)\, \partial_\mu A^*_{\dot i}(x)\, \partial_\rho A_{\dot j}(x) \\
&= A^*_{\dot i}(x) A_{\dot j}(x) \\
&\quad + i(\theta\sigma^\rho\bar\theta)[(\partial_\rho A_{\dot j}(x)) A^*_{\dot i}(x) \\
&\qquad\qquad - (\partial_\rho A^*_{\dot i}(x)) A_{\dot j}(x)]
\end{aligned}
$$

$$+(\theta\theta)(\bar\theta\bar\theta)\Big\{-\frac{1}{4}A^*_{i\cdot}(x)\,\Box A_{j\cdot}(x)$$
$$-\frac{1}{4}\Box A^*_{i\cdot}(x)A_{j\cdot}(x)$$
$$+\frac{1}{2}\partial_\mu A^*_{i\cdot}(x)\,\partial^\mu A_{j\cdot}(x)\Big\}$$

(7.21)

where we used again (1.118). Next consider

$$2^{\frac{1}{2}}\theta^A A^*_{i\cdot}(x-i\theta\sigma\bar\theta)\psi_{j\cdot A}(x+i\theta\sigma\bar\theta)$$
$$=2^{\frac{1}{2}}\theta^A\{(A^*_{i\cdot}(x)-i\theta\sigma^\mu\bar\theta\partial_\mu A^*_{i\cdot}(x))\cdot$$
$$\cdot(\psi_{j\cdot A}(x)+i\theta\sigma^\nu\bar\theta\partial_\nu\psi_{j\cdot A}(x))\}$$
$$=2^{\frac{1}{2}}\theta^A\psi_{j\cdot A}(x)A^*_{i\cdot}(x)$$
$$-2^{\frac{1}{2}}i\,\theta^A\theta^B\sigma^\mu_{B\dot A}\bar\theta^{\dot A}\partial_\mu A^*_{i\cdot}(x)\psi_{j\cdot A}(x)$$
$$+2^{\frac{1}{2}}i\,\theta^A A^*_{i\cdot}(x)\theta^B\sigma^\mu_{B\dot A}\bar\theta^{\dot A}\partial_\mu\psi_{j\cdot A}(x)$$
$$=2^{\frac{1}{2}}(\theta\psi_{j\cdot}(x))A^*_{i\cdot}(x)$$
$$+\frac{i}{2^{1/2}}(\theta\theta)\varepsilon^{AB}\sigma^\mu_{B\dot A}\bar\theta^{\dot A}\partial_\mu A^*_{i\cdot}(x)\psi_{j\cdot A}(x)$$
$$-\frac{i}{2^{1/2}}(\theta\theta)A^*_{i\cdot}(x)\varepsilon^{AB}\sigma^\mu_{B\dot C}\bar\theta^{\dot C}\partial_\mu\psi_{j\cdot A}(x)$$

(using (1.83a))

$$=2^{\frac{1}{2}}(\theta\psi_{j\cdot}(x))A^*_{i\cdot}(x)$$
$$-\frac{i}{2^{1/2}}(\theta\theta)\bar\theta^{\dot A}\psi_{j\cdot}^{A}(x)\sigma^\mu_{A\dot A}\partial_\mu A^*_{i\cdot}(x)$$
$$+\frac{i}{2^{1/2}}(\theta\theta)\bar\theta^{\dot A}A^*_{i\cdot}(x)\partial_\mu\psi^A_{j\cdot}(x)\sigma^\mu_{A\dot A}$$

(using (1.76a))

$$= 2^{\frac{1}{2}} \left(\theta \psi_{j}(x) \right) A_{i}^{*}(x)$$
$$- (\theta\theta) \bar{\theta}^{\dot{A}} \frac{i}{2^{1/2}} \sigma^{\mu}_{A\dot{A}} \left(\psi_{j}^{A}(x) \partial_{\mu} A_{i}^{*}(x) - A_{i}^{*}(x) \partial_{\mu} \psi_{j}^{A}(x) \right)$$

(7.22)

Now consider

$$2^{\frac{1}{2}} \bar{\theta} \bar{\Psi}_{i} (x - i\theta\sigma\bar{\theta}) A_{j} (x + i\theta\sigma\bar{\theta})$$
$$= 2^{\frac{1}{2}} \bar{\theta}_{\dot{A}} \left(\bar{\Psi}_{i}^{\dot{A}}(x) - i\theta\sigma^{\mu}\bar{\theta}\, \partial_{\mu} \bar{\Psi}_{i}^{\dot{A}}(x) \right) \cdot$$
$$\cdot \left(A_{j}(x) + i\theta\sigma^{\mu}\bar{\theta}\, \partial_{\mu} A_{j}(x) \right)$$
$$= 2^{\frac{1}{2}} \left(\bar{\theta} \bar{\Psi}_{i}(x) \right) A_{j}(x)$$
$$+ 2^{\frac{1}{2}} i\, \bar{\theta}_{\dot{A}} \bar{\Psi}_{i}^{\dot{A}}(x)\, \theta\sigma^{\mu}\bar{\theta}\, \partial_{\mu} A_{j}(x)$$
$$- 2^{\frac{1}{2}} i\, \bar{\theta}_{\dot{A}} (\theta\sigma^{\mu}\bar{\theta})\, \partial_{\mu} \bar{\Psi}_{i}^{\dot{A}}(x) A_{j}(x)$$
$$= 2^{\frac{1}{2}} \left(\bar{\theta} \bar{\Psi}_{i}(x) \right) A_{j}(x)$$
$$- 2^{\frac{1}{2}} i\, \bar{\Psi}_{i\dot{A}}(x)\, \bar{\theta}^{\dot{A}} \bar{\theta}^{\dot{B}} (\theta\sigma^{\mu})_{\dot{B}}\, \partial_{\mu} A_{j}(x)$$
$$+ 2^{\frac{1}{2}} i\, \bar{\theta}_{\dot{A}} \bar{\theta}_{\dot{B}} (\theta\sigma^{\mu})_{\dot{C}}\, \varepsilon^{\dot{C}\dot{B}}\, \partial_{\mu} \bar{\Psi}_{i}^{\dot{A}}(x) A_{j}(x)$$

(using (1.81) and (1.76))

$$= 2^{\frac{1}{2}} \left(\bar{\theta} \bar{\Psi}_{i}(x) \right) A_{j}(x) - \frac{i}{2^{1/2}} (\bar{\theta}\bar{\theta}) \left(\theta\sigma^{\mu} \bar{\Psi}_{i}(x) \right) \partial_{\mu} A_{j}(x)$$
$$+ \frac{i}{2^{1/2}} (\bar{\theta}\bar{\theta})\, \theta\sigma^{\mu} \partial_{\mu} \bar{\Psi}_{i}(x) A_{j}(x)$$

(with (1.83))

$$= 2^{\frac{1}{2}} \left(\bar{\theta} \bar{\Psi}_{i}(x) \right) A_{j}(x)$$
$$+ (\bar{\theta}\bar{\theta}) \theta^{A} \left[-\frac{i}{4} \sigma^{\mu}_{A\dot{A}} \left(\bar{\Psi}_{i}^{\dot{A}}(x) \partial_{\mu} A_{j}(x) - \partial_{\mu} \bar{\Psi}_{i}^{\dot{A}}(x) A_{j}(x) \right] \right.$$

(7.23)

where the second term has been reexpressed in the following way:

$$2^{\frac{1}{2}} i\, \bar{\theta}_{\dot{A}} \bar{\Psi}_{i}^{\dot{A}}(x)\, \theta^{A} \sigma^{\mu}_{A\dot{B}}\, \bar{\theta}^{\dot{B}} \partial_{\mu} A_{j}(x)$$

$$= 2^{\frac{1}{2}} i\, \bar{\Psi}_{i\dot{A}}(x)\, \bar{\theta}^{\dot{A}} \theta^{A} \sigma^{\mu}_{A\dot{B}}\, \bar{\theta}^{\dot{B}} \partial_{\mu} A_{j}(x)$$

(using (1.81))

$$= -2^{\frac{1}{2}} i\, \bar{\Psi}_{i\dot{A}}(x)\, \bar{\theta}^{\dot{A}} \bar{\theta}^{\dot{B}} \theta^{A} \sigma^{\mu}_{A\dot{B}} \partial_{\mu} A_{j}(x)$$

$$= -\frac{i}{2^{1/2}} \bar{\Psi}_{i\dot{A}}(x)\, \varepsilon^{\dot{A}\dot{B}} (\bar{\theta}\bar{\theta})\, \theta^{A} \sigma^{\mu}_{A\dot{B}} \partial_{\mu} A_{j}(x)$$

(using (1.83c))

$$= \frac{i}{2^{1/2}} \bar{\Psi}_{i\dot{A}}(x) (\bar{\theta}\bar{\theta})\, \theta^{A} \sigma^{\mu}_{A\dot{B}}\, \varepsilon^{\dot{B}\dot{A}} \partial_{\mu} A_{j}(x)$$

$$= -\frac{i}{2^{1/2}} (\bar{\theta}\bar{\theta})\, \theta^{A} \sigma^{\mu}_{A\dot{B}}\, \varepsilon^{\dot{B}\dot{A}} \bar{\Psi}_{i\dot{A}}(x) \partial_{\mu} A_{j}(x)$$

$$= -\frac{i}{2^{1/2}} (\bar{\theta}\bar{\theta}) (\theta \sigma^{\mu} \bar{\Psi}_{i}(x)) \partial_{\mu} A_{j}(x)$$

Also

$$(\theta\theta) A_{i}^{*}(x - i\theta\sigma\bar{\theta}) F_{j}(x + i\theta\sigma\bar{\theta})$$
$$= (\theta\theta) A_{i}^{*}(x) F_{j}(x), \tag{7.24}$$

$$(\bar{\theta}\bar{\theta}) A_{j}(x + i\theta\sigma\bar{\theta}) F_{i}^{*}(x - i\theta\sigma\bar{\theta})$$
$$= (\bar{\theta}\bar{\theta}) A_{j}(x) F_{i}^{*}(x) \tag{7.25}$$

and

$$2\bar{\theta}\bar{\Psi}_{i}(x - i\theta\sigma\bar{\theta})\, \theta\psi_{j}(x + i\theta\sigma\bar{\theta})$$

$$= -2\bar{\theta}_{\dot{A}} \theta^{A} \bar{\Psi}_{i}^{\dot{A}}(x - i\theta\sigma\bar{\theta})\, \psi_{jA}(x + i\theta\sigma\bar{\theta})$$

$$= -2\bar{\theta}_{\dot{A}} \theta^{A} \{ \bar{\Psi}_{i}^{\dot{A}}(x) \psi_{jA}(x)$$
$$+ i\bar{\Psi}_{i}^{\dot{A}}(x)\, \theta\sigma^{\mu}\bar{\theta}\, \partial_{\mu} \bar{\Psi}_{jA}(x)$$

$$-i\theta\sigma^{\mu}\bar{\theta}(\partial_{\mu}\bar{\Psi}_{i}^{\dot{A}}(x))\psi_{jA}(x)\}$$
$$= 2(\bar{\theta}\bar{\Psi}_{i}(x))(\theta\psi_{j}(x))$$
$$-2(\theta\theta)(\bar{\theta}\bar{\theta})\frac{i}{4}\{\psi_{j}(x)\sigma^{\mu}\partial_{\mu}\bar{\Psi}_{i}(x)$$
$$-\partial_{\mu}\psi_{j}(x)\sigma^{\mu}\bar{\Psi}_{i}(x)\} \tag{7.26}$$

since e.g.

$$-2\bar{\theta}_{\dot{A}}\theta^{A}i\bar{\Psi}_{i}^{\dot{A}}\theta^{B}\sigma^{\mu}_{B\dot{B}}\bar{\theta}^{\dot{B}}\partial_{\mu}\psi_{jA}$$
$$= -2i\bar{\theta}_{\dot{A}}\bar{\theta}^{\dot{B}}\theta^{A}\theta^{B}\bar{\Psi}_{i}^{\dot{A}}\sigma^{\mu}_{B\dot{B}}\partial_{\mu}\psi_{jA}$$
$$= -2i\left(\frac{1}{2}\delta^{\dot{B}}{}_{\dot{A}}(\bar{\theta}\bar{\theta})\right)\left(-\frac{1}{2}\varepsilon^{AB}(\theta\theta)\right)\bar{\Psi}_{i}^{\dot{A}}\sigma^{\mu}_{B\dot{B}}\partial_{\mu}\psi_{jA}$$

(using (1.83a) and (1.83e))

$$= \frac{i}{2}(\bar{\theta}\bar{\theta})(\theta\theta)\bar{\Psi}_{i}^{\dot{B}}\sigma^{\mu}_{B\dot{B}}\varepsilon^{AB}\partial_{\mu}\psi_{jA}$$
$$= -\frac{i}{2}(\bar{\theta}\bar{\theta})(\theta\theta)(\partial_{\mu}\psi_{jA})\varepsilon^{AB}\sigma^{\mu}_{B\dot{B}}\bar{\Psi}_{i}^{\dot{B}}$$
$$= \frac{i}{2}(\bar{\theta}\bar{\theta})(\theta\theta)(\partial_{\mu}\varepsilon^{BA}\psi_{jA})\sigma^{\mu}_{B\dot{B}}\bar{\Psi}_{i}^{\dot{B}}$$
$$= \frac{i}{2}(\bar{\theta}\bar{\theta})(\theta\theta)(\partial_{\mu}\psi_{j}^{B})\sigma^{\mu}_{B\dot{B}}\bar{\Psi}_{i}^{\dot{B}}$$

(using (1.65))

$$= \frac{i}{2}(\theta\theta)(\bar{\theta}\bar{\theta})((\partial_{\mu}\psi_{j})\sigma^{\mu}\bar{\Psi}_{i})$$

and

$$2i\bar{\theta}_{\dot{A}}\theta^{A}(\theta\sigma^{\mu}\bar{\theta})(\partial_{\mu}\bar{\Psi}_{i}^{\dot{A}}(x))\psi_{jA}(x)$$
$$= 2i\bar{\theta}_{\dot{A}}\theta^{A}\theta^{B}\sigma^{\mu}_{B\dot{B}}\bar{\theta}^{\dot{B}}(\partial_{\mu}\bar{\Psi}_{i}^{\dot{A}}(x))\psi_{jA}(x)$$
$$= 2i(\bar{\theta}_{\dot{A}}\bar{\theta}^{\dot{B}})(\theta^{A}\theta^{B})\sigma^{\mu}_{B\dot{B}}(\partial_{\mu}\bar{\Psi}_{i}^{\dot{A}}(x))\psi_{jA}(x)$$

$$= -\frac{i}{2}\,\delta^{\dot{B}}{}_{\dot{A}}\,(\bar{\theta}\bar{\theta})\,\varepsilon^{AB}(\theta\theta)\cdot$$

$$\cdot\,\sigma^{\mu}_{B\dot{B}}\,(\partial_{\mu}\bar{\Psi}_{i}^{\dot{A}}(x))\,\psi_{jA}(x)$$

(using (1.83e) and (1.83a))

$$= -\frac{i}{2}(\theta\theta)(\bar{\theta}\bar{\theta})(\varepsilon^{BA}\psi_{jA}(x))\cdot$$

$$\cdot\,\delta^{\dot{B}}{}_{\dot{A}}\,\sigma^{\mu}_{B\dot{B}}\,(\partial_{\mu}\bar{\Psi}_{i}^{\dot{A}}(x))$$

$$= -\frac{i}{2}(\theta\theta)(\bar{\theta}\bar{\theta})\,\psi_{j}^{B}(x)\,\sigma^{\mu}_{B\dot{B}}\,\partial_{\mu}\bar{\Psi}_{i}^{\dot{B}}(x)$$

$$= -\frac{i}{2}(\theta\theta)(\bar{\theta}\bar{\theta})\,(\psi_{j}(x)\sigma^{\mu}\partial_{\mu}\bar{\Psi}_{i}(x))$$

Inserting (7.21) to (7.26) into (7.20) we finally obtain

$$\begin{aligned}
\Phi_i^{+}\Phi_j = {} & A_i^{*}(x)\,A_j(x) + 2^{\frac{1}{2}}\,\theta\psi_j(x)\,A_i^{*}(x) \\
& + 2^{\frac{1}{2}}\,\bar{\theta}\bar{\Psi}_i(x)\,A_j(x) + (\theta\theta)A_i^{*}(x)F_j(x) \\
& + (\bar{\theta}\bar{\theta})F_i^{*}(x)\,A_j(x) + 2\bar{\theta}\bar{\Psi}_i(x)\,\theta\psi_j(x) \\
& + \theta\sigma^{\mu}\bar{\theta}\,i\,[(\partial_{\mu}A_j(x))A_i^{*}(x) \\
& \qquad - (\partial_{\mu}A_i^{*}(x))\,A_j(x)] \\
& - 2^{\frac{1}{2}}(\theta\theta)\bar{\theta}_{\dot{A}}\{\tfrac{i}{2}\sigma^{\mu}_{A\dot{B}}\,\varepsilon^{\dot{B}\dot{A}}\cdot \\
& \quad \cdot(\psi_j^{A}(x)\,\partial_{\mu}A_i^{*}(x) - A_i^{*}(x)\,\partial_{\mu}\psi_j^{A}(x)) \\
& \qquad + \bar{\Psi}_i^{\dot{A}}(x)F_j(x)\} \\
& + 2^{\frac{1}{2}}(\bar{\theta}\bar{\theta})\theta^{A}\{-\tfrac{i}{2}\sigma^{\mu}_{A\dot{A}}\cdot \\
& \quad \cdot(\bar{\Psi}_i^{\dot{A}}(x)\,\partial_{\mu}A_j(x) - A_j(x)\,\partial_{\mu}\bar{\Psi}_i^{\dot{A}}(x)) \\
& \qquad + \psi_{jA}(x)F_i^{*}(x)\}
\end{aligned}$$

$$+ (\theta\theta)(\bar{\theta}\bar{\theta}) \{ \frac{1}{2} \partial_\mu A_i^*(x) \partial^\mu A_j(x)$$
$$- \frac{1}{4} A_i^*(x) \Box A_j(x) - \frac{1}{4} A_j(x) \Box A_i^*(x)$$
$$+ \frac{i}{2} \partial_\mu \psi_j(x) \sigma^\mu \bar{\psi}_i(x)$$
$$- \frac{i}{2} \psi_j(x) \sigma^\mu \partial_\mu \bar{\psi}_i(x)$$
$$+ F_i^*(x) F_j(x) \} \tag{7.27a}$$

We will show later that for i = j the $(\theta\theta)(\bar{\theta}\bar{\theta})$ -component of this expression can be rewritten (cf. (8.28))

$$(\theta\theta)(\bar{\theta}\bar{\theta}) \{ - A_i^*(x) \Box A_i(x) + |F_i(x)|^2$$
$$+ i (\partial_\mu \bar{\psi}_i(x)) \bar{\sigma}^\mu \psi_i(x)$$
$$+ \text{total derivatives} \} \tag{7.27b}$$

This expression will be required later at various points (in action integral (8.30); see also (9.71)).

In the above product the $(\theta\theta)(\bar{\theta}\bar{\theta})$-component transforms under supersymmetry transformations into a space-time derivative (i.e. like d(x) of (6.75) and $\delta_S' d(x)$ of (6.93)). We also observe that the product $\Phi_i^\dagger \Phi_j$ of antichiral and chiral superfields generates spin 1 component fields (i.e. the term proportional to $\theta\sigma^\mu\bar{\theta}$) in much the same way as the product of elementary Weyl spinors generates higher spin fields (see for example (1.129a)).

7.2 Vector Superfields and Generalized Gauge Transformations

Vector superfields satisfy the reality condition

$$V(x,\theta,\bar{\theta}) = V^{+}(x,\theta,\bar{\theta}) \tag{7.28}$$

(see (6.96)). Like any other superfield the vector superfield V is defined in terms of its power series expansion in θ and $\bar{\theta}$, i.e.

$$\begin{aligned} V(x,\theta,\bar{\theta}) = {} & C(x) + \theta\phi(x) + \bar{\theta}\bar{\chi}(x) \\ & + \theta\theta M(x) + \bar{\theta}\bar{\theta} N(x) \\ & + \theta\sigma^{\mu}\bar{\theta} V_{\mu}(x) + (\theta\theta)\bar{\theta}\bar{\lambda}(x) \\ & + (\bar{\theta}\bar{\theta})\theta\psi(x) \\ & + (\theta\theta)(\bar{\theta}\bar{\theta}) D(x) \end{aligned}$$

The hermitian conjugate is (verified with the help of (1.116) and (1.116'))

$$\begin{aligned} V^{+}(x,\theta,\bar{\theta}) = {} & C^{*}(x) + \bar{\theta}\bar{\phi}(x) + \theta\chi(x) \\ & + \bar{\theta}\bar{\theta} M^{*}(x) + \theta\theta N^{*}(x) \\ & + \theta\sigma^{\mu}\bar{\theta} V_{\mu}^{*}(x) + (\bar{\theta}\bar{\theta})\theta\lambda(x) \\ & + (\theta\theta)(\bar{\theta}\bar{\psi}(x)) \\ & + (\theta\theta)(\bar{\theta}\bar{\theta}) D^{*}(x) \end{aligned}$$

The reality condition (7.28) is satisfied if and only if

$C(x) = C^{*}(x) \longrightarrow C(x)$: a real scalar field

$\phi(x) = \chi(x)$

$M(x) = N^{*}(x)$

$V_{\mu}(x) = V_{\mu}^{*}(x) \longrightarrow V_{\mu}(x)$: a real vector field

$$\lambda(x) = \psi(x)$$

$$D(x) = D^*(x) \longrightarrow D(x): \text{ a real scalar field}$$

Hence a vector superfield obeying the constraint (7.28) has the general expansion

$$\begin{aligned} V(x,\theta,\bar{\theta}) = C(x) &+ \theta\phi(x) + \bar{\theta}\bar{\phi}(x) \\ &+ (\theta\theta)M(x) + (\bar{\theta}\bar{\theta})M^*(x) \\ &+ \theta\sigma_\mu\bar{\theta}\,V^\mu(x) + (\theta\theta)\bar{\theta}\bar{\lambda}(x) \\ &+ (\bar{\theta}\bar{\theta})\theta\lambda(x) \\ &+ (\theta\theta)(\bar{\theta}\bar{\theta})D(x) \end{aligned} \tag{7.29}$$

where

$C(x)$, $V_\mu(x)$ and $D(x)$	are real fields, and
$M(x)$, $D(x)$, $C(x)$	are scalar fields,
$\lambda(x)$, $\phi(x)$	are spinor fields,
$V_\mu(x)$	is a vector field

The vector field $V_\mu(x)$ lends its name to the entire multiplet $V(x,\theta,\bar{\theta})$. From $\delta_S' d(x)$ of (6.93) we know that under a supersymmetry transformation the $(\theta\theta)(\bar{\theta}\bar{\theta})$-component of $V(x,\theta,\bar{\theta})$ transforms into a space-time derivative of $\lambda(x)$, i.e.

$$\delta_S' D(x) = \frac{i}{2}\left(\partial_\mu\lambda(x)\sigma^\mu\bar{\alpha} - \partial_\mu\bar{\lambda}(x)\bar{\sigma}^\mu\alpha\right) \tag{7.30}$$

As explained earlier, this indicates that the $(\theta\theta)(\bar{\theta}\bar{\theta})$-component of the vector superfield is a candidate for a supersymmetric Lagrangian.

A particular example of a vector superfield is the product of a right-handed chiral superfield and a left-handed chiral superfield $\Phi^+\Phi$, as given by (7.27), since in this case

$$(\Phi^+\Phi)^+ = \Phi^+(\Phi^+)^+ = \Phi^+\Phi \qquad (7.31)$$

and the reality condition (7.28) is satisfied.

Another important example of a vector superfield is the sum of a left-handed chiral superfield and a right-handed chiral superfield, since

$$(\Phi + \Phi^+)^+ = \Phi^+ + \Phi = \Phi + \Phi^+ \qquad (7.32)$$

and again the reality condition (7.28) is satisfied. Expressing this sum in terms of component fields, we have (adding (7.6) and (7.14))

$$\begin{aligned}\Phi + \Phi^+ = {} & A(x) + A^*(x) + 2^{\frac{1}{2}}\,\theta\psi(x) \\ & + 2^{\frac{1}{2}}\,\bar\theta\bar\psi(x) + \theta\theta\,F(x) + \bar\theta\bar\theta\,F^*(x) \\ & + i\,\theta\sigma^\mu\bar\theta\,\partial_\mu[A(x) - A^*(x)] \\ & - \frac{i}{2^{1/2}}(\theta\theta)\,\bar\theta\bar\sigma^\mu\partial_\mu\psi(x) \\ & - \frac{i}{2^{1/2}}(\bar\theta\bar\theta)\,\theta\sigma^\mu\partial_\mu\bar\psi(x) \\ & - \frac{1}{4}(\theta\theta)(\bar\theta\bar\theta)\,\Box\,[A(x) + A^*(x)]\end{aligned} \qquad (7.33)$$

It is important to observe, that this combination of scalar superfields has the gradient $i\partial_\mu[A(x) - A^*(x)]$ as coefficient of $\theta\sigma^\mu\bar\theta$. We now consider a special choice of V which is such that certain components of V are invariant under the gauge transformations to be defined below. This is achieved by making in (7.29) the replacements

$$\lambda(x) \longrightarrow \lambda(x) - \frac{i}{2}\sigma^\mu\partial_\mu\bar\Phi(x)$$

$$D(x) \longrightarrow D(x) - \frac{1}{4}\Box C(x)$$

Then

$$V(x,\theta,\bar\theta) = C(x) + \theta\phi(x) + \bar\theta\bar\phi(x) + (\theta\theta)M(x) + (\bar\theta\bar\theta)M^*(x) + \theta\sigma^\mu\bar\theta V_\mu(x) + (\theta\theta)\bar\theta\left(\bar\lambda(x) - \frac{i}{2}\bar\sigma^\mu\partial_\mu\phi(x)\right) + (\bar\theta\bar\theta)\theta\left(\lambda(x) - \frac{i}{2}\sigma^\mu\partial_\mu\bar\phi(x)\right) + (\theta\theta)(\bar\theta\bar\theta)\left(D(x) - \frac{1}{4}\Box C(x)\right) \qquad (7.34)$$

This field again satisfies the condition (7.28) as can be verified with (1.115). The significance of the choice of components in (7.34) will become clear below. The following transformation of vector superfields (with V given by (7.34)) is the supersymmetric generalization of a gauge transformation

$$V(x,\theta,\bar\theta) \longrightarrow V'(x,\theta,\bar\theta) = V(x,\theta,\bar\theta) + \Phi(x,\theta,\bar\theta) + \Phi^+(x,\theta,\bar\theta) \equiv V(x,\theta,\bar\theta) + i\left(\Lambda(x,\theta,\bar\theta) - \Lambda^+(x,\theta,\bar\theta)\right) \qquad (7.35)$$

In this transformation $\Phi \equiv i\Lambda$ is any chiral superfield. The component expansion of the transformed vector superfield $V'(x,\theta,\bar\theta)$ is seen to be (using (7.33))

$$V'(x,\theta,\bar\theta) = C(x) + A(x) + A^*(x)$$

$$
\begin{aligned}
&+ \theta[\phi(x) + 2^{\frac{1}{2}}\psi(x)] \\
&+ \bar\theta[\bar\Phi(x) + 2^{\frac{1}{2}}\bar\Psi(x)] \\
&+ \theta\theta[M(x) + F(x)] + \bar\theta\bar\theta[M^*(x) + F^*(x)] \\
&+ \theta\sigma^\mu\bar\theta[V_\mu(x) + i\partial_\mu(A(x) - A^*(x))] \\
&+ \theta\theta\bar\theta[\bar\lambda(x) - \frac{i}{2}\bar\sigma^\mu\partial_\mu(\phi(x) + 2^{\frac{1}{2}}\psi(x))] \\
&+ \bar\theta\bar\theta\theta[\lambda(x) - \frac{i}{2}\sigma^\mu\partial_\mu(\bar\Phi(x) + 2^{\frac{1}{2}}\bar\Psi(x))] \\
&+ (\theta\theta)(\bar\theta\bar\theta)[D(x) \\
&\qquad - \frac{1}{4}\Box(C(x) + A(x) + A^*(x))]
\end{aligned} \tag{7.36}
$$

Hence the transformation (7.35) leads to the following transformation of the component fields

$$
\begin{aligned}
&C(x) \to C'(x) = C(x) + A(x) + A^*(x) \\
&\phi(x) \to \phi'(x) = \phi(x) + 2^{1/2}\psi(x) \\
&M(x) \to M'(x) = M(x) + F(x) \\
&V_\mu(x) \to V'_\mu(x) = V_\mu(x) + i\,\partial_\mu(A(x) - A^*(x)) \\
&\lambda(x) \to \lambda'(x) = \lambda(x) \\
&D(x) \to D'(x) = D(x)
\end{aligned} \tag{7.37}
$$

From (7.37) we see that the special choice of V $(x, \theta, \bar\theta)$ given by (7.34) implies that the λ and D component fields are invariant under the transformation (7.35). We also observe that the field $V_\mu(x)$ transforms as

$$V_\mu(x) \to V'_\mu(x) = V_\mu(x) + i\partial_\mu(A(x) - A^*(x)) \tag{7.38}$$

which corresponds to an abelian gauge transformation. We also see that

$$F_{\mu\nu} = \partial_\mu V_\nu - \partial_\nu V_\mu$$

is super-gauge invariant. Following Wess and Zumino[26,27] one calls the transformation (7.35) the supersymmetric extension of a gauge transformation. Since, as stated above, the $(\theta\theta)(\bar{\theta}\bar{\theta})$-component, i.e. D(x), is a good candidate for a supersymmetric Lagrangian (it transforms into a space-time derivative under supersymmetric transformations), we see that the invariance of D(x) under (7.35) as demonstrated by (7.37) implies the invariance of this Lagrangian under supersymmetric gauge transformations.

From the set of equations (7.37) we see that one can choose a particular scalar field Φ, i.e. choose a particular gauge, such that in the gauge transformed vector field $V'(x,\theta,\bar{\theta})$ the component fields C', ϕ' and M' vanish. This gauge is called the Wess-Zumino gauge. Taking

$$\begin{aligned} 2^{\frac{1}{2}}\psi(x) &= -\phi(x) \\ F(x) &= -M(x) \\ 2\,\mathrm{Re}\,A(x) = A(x)+A^*(x) &= -C(x) \end{aligned} \tag{7.39}$$

in (7.33), the transformed vector field $V_{WZ}(x,\theta,\bar{\theta})$ assumes the form (the index WZ indicating the Wess-Zumino gauge)

$$\begin{aligned} &V_{WZ}(x,\theta,\bar{\theta}) \\ &= V(x,\theta,\bar{\theta}) + \Phi(x,\theta,\bar{\theta}) + \Phi^+(x,\theta,\bar{\theta}) \\ &= C(x) - C(x) + \theta[\phi(x) - \phi(x)] \\ &\quad + \bar{\theta}[\bar{\phi}(x) - \bar{\phi}(x)] + \theta\theta[M(x) - M(x)] \end{aligned}$$

$$+ \bar{\theta}\bar{\theta}[M^*(x) - M^*(x)]$$
$$+ \theta\sigma^\mu\bar{\theta}[V_\mu(x) + i\partial_\mu(A(x) - A^*(x))]$$
$$+ (\bar{\theta}\bar{\theta})\theta[\lambda(x) - \frac{i}{2}\sigma^\mu\partial_\mu(\bar{\Phi}(x) - \bar{\Phi}(x))]$$
$$+ (\theta\theta)\bar{\theta}[\bar{\lambda}(x) - \frac{i}{2}\bar{\sigma}^\mu\partial_\mu(\phi(x) - \phi(x))]$$
$$+ (\theta\theta)(\bar{\theta}\bar{\theta})[D(x) - \frac{1}{4}\Box(C(x) - C(x))]$$

Hence

$$\begin{aligned} &V_{WZ}(x,\theta,\bar{\theta}) \\ &= \theta\sigma^\mu\bar{\theta}[V_\mu(x) + i\partial_\mu(A(x) - A^*(x))] \\ &\quad + (\theta\theta)\bar{\theta}\bar{\lambda}(x) + (\bar{\theta}\bar{\theta})\theta\lambda(x) \\ &\quad + (\theta\theta)(\bar{\theta}\bar{\theta})D(x) \end{aligned} \tag{7.40}$$

Here $V_\mu(x)$ is the gauge field and λ is its supersymmetric partner. D(x) is the socalled auxiliary field, the significance of which will become clear later.

It is important to observe that in (7.39) we have not fixed the imaginary part of the component scalar field A(x) which causes the shift of the vector field $V_\mu(x)$. Hence the Wess-Zumino gauge does not fix the gauge freedom completely; one still has the gauge degree of freedom of conventional gauge theories. However, the Wess-Zumino gauge breaks supersymmetry in the sense that the supersymmetry variation of $\phi_A(x)$ and M(x) violates the gauge condition

$$C(x) = \phi_A(x) = M(x) = 0$$

For example, comparing (6.75) with (7.29) and using

(6.93) and (7.39) we have

$$\delta_S' \phi_A(x) = 2\alpha_A M(x) + (\sigma^\mu \bar{\alpha})_A \{ i\partial_\mu C(x) + V_\mu(x) \}$$

and

$$\delta_S' M(x) = \bar{\alpha}\bar{\lambda}(x) - \frac{i}{2}\partial_\mu \phi(x) \sigma^\mu \bar{\alpha}$$

Hence in the Wess-Zumino gauge (ϕ_A = C = M = 0)

$$\left(\delta_S' \phi_A(x)\right)_{WZ} = \sigma^\mu_{A\dot{A}}\, \bar{\alpha}^{\dot{A}}\, V_\mu(x)$$

$$\left(\delta_S' M(x)\right)_{WZ} = \bar{\alpha}\bar{\lambda}(x)$$

Thus the supersymmetry variations of $\phi_A(x)$ and M(x) in this gauge do not vanish. From this we deduce that the Wess-Zumino gauge is noncovariant.

It is easy to compute powers of V_{WZ} by taking into account the anticommuting character of the Grassmann variables θ. With Im A = 0 we have (using (1.118) and $\theta^3 = 0$)

$$\begin{aligned} V_{WZ}(x,\theta,\bar{\theta}) &= \theta\sigma^\mu\bar{\theta}\, V_\mu + \theta\theta\, \bar{\theta}\bar{\lambda}(x) \\ &\quad + \bar{\theta}\bar{\theta}\, \theta\lambda(x) + \theta\theta\, \bar{\theta}\bar{\theta}\, D(x) \\ V^2_{WZ}(x,\theta,\bar{\theta}) &= \frac{1}{2}\theta\theta\,\bar{\theta}\bar{\theta}\, V_\mu(x) V^\mu(x) \\ V^3_{WZ}(x,\theta,\bar{\theta}) &= 0 \end{aligned} \tag{7.41}$$

It is the last of these properties which makes the Wess-Zumino gauge a particularly convenient gauge to work in. Then we have, for instance,

$$\exp\{V\} = 1 + V + \frac{1}{2}V^2$$

$$= 1 + \theta\sigma^\mu\bar{\theta}\, V_\mu(x) + (\theta\theta)\,\bar{\theta}\bar{\lambda}(x) + (\bar{\theta}\bar{\theta})\,\theta\lambda(x) + (\theta\theta)(\bar{\theta}\bar{\theta})\left\{D(x) + \frac{1}{4}V_\mu(x)V^\mu(x)\right\} \tag{7.42}$$

This exponential of V will be used later in the construction of supersymmetric gauge theories (see Section 10.1).

Counting the number of bosonic and fermionic degrees of freedom of the vector super-multiplet before and after Wess-Zumino gauge fixing we have

a) in the case of the general vector superfield:

$C(x)$	: real scalar field:	1 bosonic degree
$\phi(x)$	: complex two-spinor field:	4 fermionic degrees
$M(x)$	: complex scalar field:	2 bosonic degrees
$V_\mu(x)$	: real vector field:	4 bosonic degrees
$\lambda(x)$	: complex two-spinor field:	4 fermionic degrees
$D(x)$	: real scalar field:	1 bosonic degree

thus altogether the multiplet has sixteen degrees of freedom, eight bosonic degrees and eight fermionic degrees of freedom;

b) in the case of the vector superfield in the Wess-Zumino gauge:

$V_\mu(x)$	: real vector field:	3 bosonic degrees (since we are still free to choose the gauge of a conventional abelian gauge theory, e.g. $V_0 = 0$)
$\lambda(x)$	: complex two-spinor field:	4 fermionic degrees
$D(x)$	: real scalar field:	1 bosonic degree

thus altogether the multiplet has eight degrees of freedom, four bosonic degrees of freedom and four fermionic degrees of freedom. We observe that the number of fermionic degrees of freedom is the same as the number of bosonic degrees of freedom in either case, as expected from the general considerations following (4.34).

7.3 The Supersymmetric Field Strength

The supersymmetric field strength for an arbitrary vector superfield V $(x, \theta, \bar{\theta})$ (not in the special Wess-Zumino gauge) is defined by the components

$$W_A := -\frac{1}{4}(\bar{D}\bar{D})D_A V(x,\theta,\bar{\theta}) \tag{7.43}$$

$$\bar{W}_{\dot{A}} := -\frac{1}{4}(DD)\bar{D}_{\dot{A}} V(x,\theta,\bar{\theta}) \tag{7.44}$$

W_A and $\bar{W}_{\dot{A}}$ are examples of spinor superfields. We first show that W_A and $\bar{W}_{\dot{A}}$ are,in fact, chiral superfields and so represent irreducible representations of the Super-Poincaré algebra. The fact that superfields with spinor indices A, $\dot{A}$ can be chiral in the sense that they satisfy one of the chirality constraints (6.94), (6.95) like the superfields Φ, Φ^+ without spinor indices, is a reflection of the property of these constraints to select irreducible representations of the Super-Poincaré algebra. It may be recalled that in Section 4.2 we discussed the corresponding cases of the lowest-dimensional representations for bosonic and fermionic Clifford vacuum states.

<u>Proposition</u>: W_A is a left-handed chiral superfield, and $\bar{W}_{\dot{A}}$ is a right-handed chiral superfield. Moreover, both superfields are invariant under the supersymmetric

gauge transformation (7.35).

Proof:

i) We first establish the chirality of W_A, $\overline{W}_{\dot{A}}$. This is easily shown since

$$\overline{D}_{\dot{A}} W_A = -\frac{1}{4}\overline{D}_{\dot{A}}(\overline{D}\overline{D})D_A V$$

$$= 0 \qquad (7.45a)$$

since $\overline{D}^3 = 0$ (see (6.53)). Similarly

$$D_A \overline{W}_{\dot{A}} = -\frac{1}{4} D_A (DD)\overline{D}_{\dot{A}} V$$

$$= 0$$

since $D^3 = 0$.

ii) In order to establish the gauge invariance of W_A we apply the transformation (7.35). Then

$$W_A \longrightarrow W_A' = -\frac{1}{4}(\overline{D}\overline{D})D_A V'$$

$$= -\frac{1}{4}(\overline{D}\overline{D})D_A(V + \Phi + \Phi^+)$$

Since D and $\overline{D}$ are linear operators we have

$$W_A' = -\frac{1}{4}(\overline{D}\overline{D})D_A V - \frac{1}{4}(\overline{D}\overline{D})D_A\Phi - \frac{1}{4}(\overline{D}\overline{D})D_A\Phi^+$$

$$= W_A - \frac{1}{4}\overline{D}\overline{D}D_A\Phi \qquad \text{(with (7.43) and (7.15))}$$

$$= W_A - \frac{1}{4}\overline{D}_{\dot{A}}\overline{D}^{\dot{A}}D_A\Phi - \frac{1}{4}\overline{D}_{\dot{A}}D_A\overline{D}^{\dot{A}}\Phi$$

$$\text{(using (7.1), i.e. } \overline{D}_{\dot{A}}\Phi = 0 \text{)}$$

$$= W_A - \frac{1}{4}\overline{D}_{\dot{A}}\{\overline{D}^{\dot{A}}, D_A\}\Phi$$

From (6.53) we know that $\{D, \overline{D}\}$ closes into P_μ. The latter, however, commutes with $\overline{D}_{\dot{A}}$ so that the last contribution vanishes on account of $\overline{D}_{\dot{A}}\Phi = 0$. Hence

$$W_A' = W_A \qquad \text{q.e.d.}$$

In a similar way one demonstrates the invariance of W_A . Thus, with (7.35),

$$\bar{W}_{\dot{A}} \longrightarrow \bar{W}_{\dot{A}}' = -\frac{1}{4} DD\bar{D}_{\dot{A}} V'$$

$$= -\frac{1}{4} DD\bar{D}_{\dot{A}} (V + \Phi + \Phi^+)$$

$$= -\frac{1}{4} DD\bar{D}_{\dot{A}} V - \frac{1}{4} DD\bar{D}_{\dot{A}} \Phi$$

$$- \frac{1}{4} DD\bar{D}_{\dot{A}} \Phi^+$$

$$= \bar{W}_{\dot{A}} - \frac{1}{4} DD\bar{D}_{\dot{A}} \Phi^+$$

(with (7.1))

$$= \bar{W}_{\dot{A}} - \frac{1}{4} D^A D_A \bar{D}_{\dot{A}} \Phi^+$$

$$- \frac{1}{4} D^A \bar{D}_{\dot{A}} D_A \Phi^+$$

(with (7.9))

$$= \bar{W}_{\dot{A}} - \frac{1}{4} D^A \{ D_A , \bar{D}_{\dot{A}} \} \Phi^+$$

Using (6.53) and again (7.9) this is

$$\bar{W}_{\dot{A}}' = \bar{W}_{\dot{A}} + \frac{i}{2} D^A \sigma^\mu_{A\dot{A}} \partial_\mu \Phi^+$$

$$= \bar{W}_{\dot{A}} + \frac{i}{2} \sigma^\mu_{A\dot{A}} \partial_\mu D^A \Phi^+$$

$$= \bar{W}_{\dot{A}}$$

Our next task is the calculation of the component expansion of W_A in the Wess-Zumino gauge. This calculation is, of course, simplified if we use the variable

$$y^\mu := x^\mu + i\theta\sigma^\mu\bar{\theta}$$

(cf. (7.2)) and correspondingly

$$z^\mu = x^\mu - i\theta\sigma^\mu\bar{\theta}$$

(cf. (7.10)) for the calculation of the component expansion of $\overline{W}_{\dot{A}}$. Now

$$V_{WZ}(x,\theta,\bar{\theta}) = \theta\sigma^\mu\bar{\theta}V_\mu(x) + \theta\theta\,\bar{\theta}\bar{\lambda}(x) + \bar{\theta}\bar{\theta}\,\theta\lambda(x) + \theta\theta\,\bar{\theta}\bar{\theta}D(x)$$

Setting

$$x^\mu = y^\mu - i\theta\sigma^\mu\bar{\theta} = z^\mu + i\theta\sigma^\mu\bar{\theta}$$

we obtain the vector superfields in the coordinates y and z respectively, i.e.

$$\begin{aligned}
V_{WZ}(x,\theta,\bar{\theta}) &= V_{WZ}(y - i\theta\sigma\bar{\theta}, \theta, \bar{\theta}) \\
&= \theta\sigma^\mu\bar{\theta}V_\mu(y - i\theta\sigma\bar{\theta}) + \theta\theta\,\bar{\theta}\bar{\lambda}(y - i\theta\sigma\bar{\theta}) \\
&\quad + \bar{\theta}\bar{\theta}\,\theta\lambda(y - i\theta\sigma\bar{\theta}) + \theta\theta\,\bar{\theta}\bar{\theta}D(y - i\theta\sigma\bar{\theta}) \\
&= \theta\sigma^\mu\bar{\theta}\,V_\mu(y) - i(\theta\sigma^\mu\bar{\theta})(\theta\sigma^\rho\bar{\theta})\partial_\rho V_\mu(y) \\
&\quad + (\theta\theta)(\bar{\theta}\bar{\lambda}(y)) + (\bar{\theta}\bar{\theta})(\theta\lambda(y)) + (\theta\theta)(\bar{\theta}\bar{\theta})D(y) \\
&= (\theta\sigma^\mu\bar{\theta})V_\mu(y) + (\theta\theta)(\bar{\theta}\bar{\lambda}(y)) + (\bar{\theta}\bar{\theta})(\theta\lambda(y)) \\
&\quad + (\theta\theta)(\bar{\theta}\bar{\theta})\left[D(y) - \frac{i}{2}\partial_\mu V^\mu(y)\right] \\
&\equiv V^{(1)}_{WZ}(y,\theta,\bar{\theta})
\end{aligned} \tag{7.46}$$

where we used (1.118), and similarly

$$\begin{aligned}
V_{WZ}(x,\theta,\bar{\theta}) &= V_{WZ}(z + i\theta\sigma\bar{\theta}, \theta, \bar{\theta}) \\
&= \theta\sigma^\mu\bar{\theta}V_\mu(z + i\theta\sigma\bar{\theta}) + \theta\theta\,\bar{\theta}\bar{\lambda}(z + i\theta\sigma\bar{\theta}) \\
&\quad + \bar{\theta}\bar{\theta}\,\theta\lambda(z + i\theta\sigma\bar{\theta}) + \theta\theta\,\bar{\theta}\bar{\theta}D(z + i\theta\sigma\bar{\theta}) \\
&= \theta\sigma^\mu\bar{\theta}\,V_\mu(z) + i(\theta\sigma^\mu\bar{\theta})(\theta\sigma^\nu\bar{\theta})\partial_\nu V_\mu(z)
\end{aligned}$$

$$+ (\theta\theta)\,\bar{\theta}\bar{\lambda}(z) + (\bar{\theta}\bar{\theta})\,\theta\lambda(z) + (\theta\theta)(\bar{\theta}\bar{\theta})D(z)$$

$$= \theta\sigma^{\mu}\bar{\theta}\,V_{\mu}(z) + (\theta\theta)\,\bar{\theta}\bar{\lambda}(z) + (\bar{\theta}\bar{\theta})\,\theta\lambda(z)$$

$$+ (\theta\theta)(\bar{\theta}\bar{\theta})\,[D(z) + i\,\partial_{\mu}V^{\mu}(z)]$$

$$\equiv V^{(2)}_{WZ}(z,\theta,\bar{\theta}) \tag{7.47}$$

again using (1.118). Thus we have

$$V_{WZ}(x,\theta,\bar{\theta}) = V^{(1)}_{WZ}(y,\theta,\bar{\theta}) = V^{(2)}_{WZ}(z,\theta,\bar{\theta})$$

A suitable set of coordinates to calculate the component expansion of W_A in is given by $(y, \theta, \bar{\theta})$. Then

$$W_A = -\frac{1}{4}\,\bar{D}\bar{D}\,D_A\,V_{WZ}(x,\theta,\bar{\theta})$$

$$= -\frac{1}{4}\,\bar{D}^{(1)}\bar{D}^{(1)}D_A^{(1)}\,V^{(1)}_{WZ}(y,\theta,\bar{\theta})$$

since in terms of y the covariant derivatives D and $\bar{D}$ are given by $D^{(1)}$ and $\bar{D}^{(1)}$ of (7.4a,b). Considering the first derivative we have

$$D_A^{(1)}\,V^{(1)}_{WZ}(y,\theta,\bar{\theta})$$

$$= (\partial_A + 2i\,\sigma^{\mu}_{A\dot{B}}\,\bar{\theta}^{\dot{B}}\partial_{\mu})\,V^{(1)}_{WZ}(y,\theta,\bar{\theta})$$

$$\text{(with (6.47))}$$

$$= \partial_A\{\theta\sigma^{\mu}\bar{\theta}\,V_{\mu}(y) + (\theta\theta)\,\bar{\theta}\bar{\lambda}(y) + (\bar{\theta}\bar{\theta})\,\theta\lambda(y)$$

$$+ (\theta\theta)(\bar{\theta}\bar{\theta})\,[D(y) - \tfrac{i}{2}\,\partial_{\mu}V^{\mu}(y)]\}$$

$$+ 2i\,\sigma^{\rho}_{A\dot{B}}\,\bar{\theta}^{\dot{B}}\partial_{\rho}\{\theta\sigma^{\mu}\bar{\theta}\,V_{\mu}(y) + (\theta\theta)\,\bar{\theta}\bar{\lambda}(y)$$

$$+ (\bar{\theta}\bar{\theta})\,\theta\lambda(y) + (\theta\theta)(\bar{\theta}\bar{\theta})\,[D(y) - \tfrac{i}{2}\,\partial_{\mu}V^{\mu}(y)]\}$$

$$\text{(with (7.46))}$$

$$= (\partial_A\theta^B)\,\sigma^{\mu}_{B\dot{C}}\,\bar{\theta}^{\dot{C}}V_{\mu}(y) + \partial_A(\theta\theta)\,\bar{\theta}\bar{\lambda}(y)$$

$$+(\bar\theta\bar\theta)\partial_A\theta^B\lambda_B(y)+\partial_A(\theta\theta)(\bar\theta\bar\theta)[D(y)-\tfrac{i}{2}\partial_\mu V^\mu(y)]$$
$$+2i\sigma^\rho_{A\dot B}\bar\theta^{\dot B}\theta^C\sigma^\mu_{C\dot D}\bar\theta^{\dot D}\partial_\rho V_\mu(y)$$
$$+2i\sigma^\rho_{A\dot B}\bar\theta^{\dot B}\theta\theta\,\bar\theta\partial_\rho\bar\lambda(y) \quad \text{(since } \bar\theta\bar\theta\bar\theta = 0)$$
$$=\sigma^\mu_{A\dot C}\bar\theta^{\dot C}V_\mu(y)+2\theta_A\bar\theta\bar\lambda(y)+\bar\theta\bar\theta\lambda_A(y)$$
$$+2\theta_A\bar\theta\bar\theta[D(y)-\tfrac{i}{2}\partial_\mu V^\mu(y)]$$
$$+2i\sigma^\rho_{A\dot B}\bar\theta^{\dot B}\theta^C\sigma^\mu_{C\dot D}\bar\theta^{\dot D}\partial_\rho V_\mu(y)$$
$$+2i\sigma^\rho_{A\dot B}\bar\theta^{\dot B}\theta\theta\,\bar\theta_{\dot C}\partial_\rho\bar\lambda^{\dot C}(y) \tag{7.48}$$

The second last term can be rewritten as follows

$$2i\sigma^\mu_{A\dot B}\bar\theta^{\dot B}\theta^C\sigma^\rho_{C\dot D}\bar\theta^{\dot D}\partial_\mu V_\rho(y)$$
$$=-2i\sigma^\mu_{A\dot B}\bar\theta^{\dot B}\bar\theta^{\dot D}\theta^C\sigma^\rho_{C\dot D}\partial_\mu V_\rho(y)$$
$$=-i\sigma^\mu_{A\dot B}\bar\theta\bar\theta\varepsilon^{\dot B\dot D}\theta^C\sigma^\rho_{C\dot D}\partial_\mu V_\rho(y)$$

(using (1.83c))

$$=-i\sigma^\mu_{A\dot B}\bar\theta\bar\theta\varepsilon^{\dot B\dot D}\varepsilon^{CD}\sigma^\rho_{C\dot D}\partial_\mu V_\rho(y)\theta_D$$

(using (1.65))

$$=i\sigma^\mu_{A\dot B}\bar\theta\bar\theta\,\varepsilon^{DC}\varepsilon^{\dot B\dot D}\sigma^\rho_{C\dot D}\partial_\mu V_\rho(y)\theta_D$$
$$=i(\bar\theta\bar\theta)\sigma^\mu_{A\dot B}\bar\sigma^{\rho\,\dot BD}\partial_\mu V_\rho(y)\theta_D$$

(using (1.88))

$$=i(\bar\theta\bar\theta)(\sigma^\mu\bar\sigma^\rho)_A{}^D\partial_\mu V_\rho(y)\theta_D$$

The last term of (7.48) can be written (using $\bar\theta_{\dot C}\bar\lambda^{\dot C} = -\bar\theta^{\dot C}\bar\lambda_{\dot C}$)

$$2i\sigma^\mu_{A\dot B}\bar\theta^{\dot B}(\theta\theta)\bar\theta_{\dot C}\partial_\mu\bar\lambda^{\dot C}(y)$$
$$=-2i\sigma^\mu_{A\dot B}\bar\theta^{\dot B}\bar\theta^{\dot C}(\theta\theta)\partial_\mu\bar\lambda_{\dot C}(y)$$

$$= -i\sigma^{\mu}_{A\dot{B}}(\bar{\theta}\bar{\theta})\varepsilon^{\dot{B}\dot{C}}(\theta\theta)\partial_\mu\bar{\lambda}_{\dot{C}}(y)$$

(with (1.83c))

$$= -i(\theta\theta)(\bar{\theta}\bar{\theta})\sigma^{\mu}_{A\dot{B}}\partial_\mu\bar{\lambda}^{\dot{B}}(y)$$

Hence

$$D_A^{(1)}V_{WZ}^{(1)}(y,\theta,\bar{\theta})$$

$$= \sigma^{\mu}_{A\dot{B}}\bar{\theta}^{\dot{B}}V_\mu(y) + 2\theta_A\bar{\theta}\bar{\lambda}(y) + \bar{\theta}\bar{\theta}\lambda_A(y)$$
$$+ \bar{\theta}\bar{\theta}\{2\delta_A{}^B D(y) + i(\sigma^\mu\bar{\sigma}^\nu)_A{}^B\partial_\mu V_\nu(y)$$
$$- i\delta_A{}^B\eta^{\mu\nu}\partial_\mu V_\nu(y)\}\theta_B$$
$$- i(\theta\theta)(\bar{\theta}\bar{\theta})(\sigma^\mu\partial_\mu\bar{\lambda}(y))_A \qquad (7.49)$$

In rewriting this expression we make use of the following result.

Proposition: If $\sigma^{\mu\nu}$ is given by (1.119a) it can be expressed in the following way

$$2\sigma^{\mu\nu}{}_A{}^B = i\left[-\delta_A{}^B\eta^{\mu\nu} + (\sigma^\mu\bar{\sigma}^\nu)_A{}^B\right] \qquad (7.50)$$

Proof: From (1.122a) we know that

$$(\sigma^\mu\bar{\sigma}^\nu)_A{}^B + (\sigma^\nu\bar{\sigma}^\mu)_A{}^B = 2\eta^{\mu\nu}\delta_A{}^B$$

Subtracting from both sides $2(\sigma^\mu\bar{\sigma}^\nu)_A{}^B$ we obtain

$$-(\sigma^\mu\bar{\sigma}^\nu)_A{}^B + (\sigma^\nu\bar{\sigma}^\mu)_A{}^B = 2\left(\eta^{\mu\nu}\delta_A{}^B - \sigma^\mu\bar{\sigma}^\nu{}_A{}^B\right)$$

But (see (1.119a))

$$\frac{i}{4}(\sigma^\nu\bar{\sigma}^\mu - \sigma^\mu\bar{\sigma}^\nu)_A{}^B = (\sigma^{\nu\mu})_A{}^B$$

Hence

$$(-\sigma^{\mu}\bar{\sigma}^{\nu}+\sigma^{\nu}\bar{\sigma}^{\mu})_A{}^B=-2[-\delta_A{}^B\eta^{\mu\nu}+(\sigma^{\mu}\bar{\sigma}^{\nu})_A{}^B]$$

and so

$$-4i(\sigma^{\nu\mu})_A{}^B=-2[-\delta_A{}^B\eta^{\mu\nu}+(\sigma^{\mu}\bar{\sigma}^{\nu})_A{}^B]$$

i.e.

$$2(\sigma^{\mu\nu})_A{}^B=i[-\delta_A{}^B\eta^{\mu\nu}+(\sigma^{\mu}\bar{\sigma}^{\nu})_A{}^B]$$

which had to be shown. Using (7.50) we can rewrite the coefficient of the term in $\bar{\theta}\bar{\theta}$ of (7.49), i.e.

$$2\delta_A{}^B D(y)+i(\sigma^{\mu}\bar{\sigma}^{\nu})_A{}^B\partial_\mu V_\nu(y)-i\delta_A{}^B\eta^{\mu\nu}\cdot\partial_\mu V_\nu(y)$$

$$=2\delta_A{}^B D(y)+i[-\delta_A{}^B\eta^{\mu\nu}+(\sigma^{\mu}\bar{\sigma}^{\nu})_A{}^B]\partial_\mu V_\nu(y)$$

$$=2\delta_A{}^B D(y)+2\sigma^{\mu\nu}{}_A{}^B\partial_\mu V_\nu(y)$$

$$=2\delta_A{}^B D(y)+\sigma^{\mu\nu}{}_A{}^B\partial_\mu V_\nu(y)+\sigma^{\nu\mu}{}_A{}^B\partial_\nu V_\mu(y)$$

$$=2\delta_A{}^B D(y)+\sigma^{\mu\nu}{}_A{}^B(\partial_\mu V_\nu(y)-\partial_\nu V_\mu(y))$$

$$=2\delta_A{}^B D(y)+\sigma^{\mu\nu}{}_A{}^B F_{\mu\nu}(y)$$

where we made use of the antisymmetry of $\sigma^{\mu\nu}$ in μ and ν and we set

$$F_{\mu\nu}(y):=\partial_\mu V_\nu(y)-\partial_\nu V_\mu(y)$$

as the usual expression for the field strength tensor. Hence we obtain

$$D_A^{(1)}V_{WZ}^{(1)}(y,\theta,\bar{\theta})$$

$$=\sigma^{\mu}{}_{A\dot{B}}\bar{\theta}^{\dot{B}}V_\mu(y)+2\theta_A\bar{\theta}\bar{\lambda}(y)+\bar{\theta}\bar{\theta}\lambda_A(y)$$

$$+\bar{\theta}\bar{\theta}[2\delta_A{}^B D(y)+(\sigma^{\mu\nu})_A{}^B F_{\mu\nu}(y)]\theta_B$$
$$-i(\theta\theta)(\bar{\theta}\bar{\theta})\sigma^{\mu}_{A\dot{B}}\partial_\mu \bar{\lambda}^{\dot{B}}(y) \qquad (7.51)$$

Then

$$W_A = -\frac{1}{4}\bar{D}^{(1)}\bar{D}^{(1)}D_A^{(1)}V_{WZ}^{(1)}(y,\theta,\bar{\theta})$$
$$= -\frac{1}{4}\bar{\partial}_{\dot{A}}\bar{\partial}^{\dot{A}}\left(D_A^{(1)}W_{WZ}^{(1)}(y,\theta,\bar{\theta})\right)$$

(with (6.48))

$$= -\frac{1}{4}\varepsilon_{\dot{A}\dot{B}}\bar{\partial}^{\dot{B}}\bar{\partial}^{\dot{A}}\left(D_A^{(1)}V_{WZ}^{(1)}(y,\theta,\bar{\theta})\right)$$
$$= \frac{1}{4}\varepsilon_{\dot{A}\dot{B}}\bar{\partial}^{\dot{A}}\bar{\partial}^{\dot{B}}\left(D_A^{(1)}V_{WZ}^{(1)}(y,\theta,\bar{\theta})\right)$$
$$= \frac{1}{4}\varepsilon_{\dot{A}\dot{B}}\bar{\partial}^{\dot{A}}\bar{\partial}^{\dot{B}}\{\sigma^{\mu}_{A\dot{C}}\bar{\theta}^{\dot{C}}V_\mu(y)+2\theta_A\bar{\theta}\bar{\lambda}(y)$$
$$+\bar{\theta}\bar{\theta}\lambda_A(y)$$
$$+\bar{\theta}\bar{\theta}[2\delta_A{}^B D(y)+\sigma^{\mu\nu}{}_A{}^B F_{\mu\nu}(y)]\theta_B$$
$$-i(\theta\theta)(\bar{\theta}\bar{\theta})\sigma^{\mu}_{A\dot{B}}\partial_\mu\bar{\lambda}^{\dot{B}}(y)\}$$
$$= \lambda_A(y)+2\theta_A D(y)+\sigma^{\mu\nu}{}_A{}^B\theta_B F_{\mu\nu}$$
$$-i(\theta\theta)\sigma^{\mu}_{A\dot{B}}\partial_\mu\bar{\lambda}^{\dot{B}}(y)$$

Hence the component expansion of W_A is given by

$$W_A = \lambda_A(y)+2D(y)\theta_A+(\sigma^{\mu\nu}\theta)_A F_{\mu\nu}(y)$$
$$-i(\theta\theta)\sigma^{\mu}_{A\dot{B}}\partial_\mu\bar{\lambda}^{\dot{B}}(y) \qquad (7.52)$$

In a similar way we can find that the component expansion of $\overline{W}_{\dot{A}}$ is given by

$$\overline{W}_{\dot{A}} = \overline{\lambda}_{\dot{A}}(z) + 2D(z)\overline{\theta}_{\dot{A}} - \varepsilon_{\dot{A}\dot{B}}(\overline{\sigma}^{\mu\nu}\overline{\theta})^{\dot{B}} F_{\mu\nu}(z)$$
$$+ i(\overline{\theta}\overline{\theta})(\partial_\mu \lambda(z)\sigma^\mu)_{\dot{A}} \qquad (7.53)$$

In order to verify this result we recall that $(z^\mu, \theta, \overline{\theta})$ is a convenient set of coordinates in which we can compute the component expansion of $\overline{W}_{\dot{A}}$. Thus

$$\overline{W}_{\dot{A}} = -\frac{1}{4} DD\overline{D}_{\dot{A}} V_{WZ}(x,\theta,\overline{\theta})$$
$$= -\frac{1}{4} D^{(2)}D^{(2)}\overline{D}^{(2)}_{\dot{A}} V^{(2)}_{WZ}(z,\theta,\overline{\theta})$$

We first calculate the expansion of $\overline{D}^{(2)}_{\dot{A}} V^{(2)}_{WZ}$, i.e.

$$\overline{D}^{(2)}_{\dot{A}} V^{(2)}_{WZ}(z,\theta,\overline{\theta})$$
$$= \{-\overline{\partial}_{\dot{A}} - 2i\theta^A \sigma^\mu_{A\dot{A}}\partial_\mu\} V^{(2)}_{WZ}(z,\theta,\overline{\theta})$$

(with (6.50))

$$= -\overline{\partial}_{\dot{A}} V^{(2)}_{WZ}(z,\theta,\overline{\theta}) - 2i(\theta\sigma^\mu)_{\dot{A}}\partial_\mu V^{(2)}_{WZ}(z,\theta,\overline{\theta})$$
$$= -\overline{\partial}_{\dot{A}}\{\theta\sigma^\mu\overline{\theta}V_\mu(z) + (\theta\theta)\overline{\theta}\overline{\lambda}(z) + (\overline{\theta}\overline{\theta})\theta\lambda(z)$$
$$+ (\theta\theta)(\overline{\theta}\overline{\theta})[D(z) + \frac{i}{2}\partial_\mu V^\mu(z)]\}$$
$$-2i(\theta\sigma^\mu)_{\dot{A}}\partial_\mu\{\theta\sigma^\rho\overline{\theta}V_\rho(z) + (\theta\theta)\overline{\theta}\overline{\lambda}(z)$$
$$+ (\overline{\theta}\overline{\theta})\theta\lambda(z) + (\theta\theta)(\overline{\theta}\overline{\theta})[D(z)$$
$$+ \frac{i}{2}\partial_\mu V^\mu(z)]\}$$

(using (7.47))

$$= \theta^A\sigma^\mu_{A\dot{B}}\overline{\partial}_{\dot{A}}\overline{\theta}^{\dot{B}} V_\mu(z) - (\theta\theta)\overline{\partial}_{\dot{A}}\overline{\theta}_{\dot{B}}\overline{\lambda}^{\dot{B}}(z)$$

$$- \bar{\partial}_{\dot{A}} (\bar{\theta}\bar{\theta})\, \theta\lambda(x)$$
$$- (\theta\theta)\bar{\partial}_{\dot{A}} (\bar{\theta}\bar{\theta}) [D(x) + \frac{i}{2} \partial_\mu V^\mu (x)]$$
$$- 2i (\theta\sigma^\mu)_{\dot{A}} (\theta\sigma^\rho\bar{\theta}) \partial_\mu V_\rho (x)$$
$$- 2i (\theta\sigma^\mu)_{\dot{A}} (\bar{\theta}\bar{\theta})\, \theta\, \partial_\mu \lambda(x)$$

(plus terms which vanish because they contain θ^3)

$$= (\theta\sigma^\mu)_{\dot{A}} V_\mu(x) + (\theta\theta)\bar{\lambda}_{\dot{A}}(x) + 2\bar{\theta}_{\dot{A}}\theta\lambda(x)$$
$$+ 2(\theta\theta)\bar{\theta}_{\dot{A}} [D(x) + \frac{i}{2} \partial_\mu V^\mu (x)]$$
$$- i (\theta\theta) \varepsilon_{\dot{A}\dot{B}} (\bar{\sigma}^\mu\sigma^\nu)^{\dot{B}}{}_{\dot{C}}\, \partial_\mu V_\nu (x) \bar{\theta}^{\dot{C}}$$
$$+ i (\theta\theta)(\bar{\theta}\bar{\theta}) (\partial_\mu \lambda(x) \sigma^\mu)_{\dot{A}}$$

where we used the following relations

i) $\bar{\partial}_{\dot{A}} \bar{\theta}_{\dot{B}} = - \varepsilon_{\dot{A}\dot{B}}$ (see (6.41))

ii) $\bar{\partial}_{\dot{A}} (\bar{\theta}\bar{\theta}) = - 2\bar{\theta}_{\dot{A}}$ (as in (6.4p))

iii)
$$- 2i\, \theta^B \sigma^\mu_{B\dot{A}}\, \theta^C \sigma^\nu_{C\dot{D}}\, \bar{\theta}^{\dot{D}} \partial_\mu V_\nu (x)$$
$$= - 2i\, \theta^B \theta^C \sigma^\mu_{B\dot{A}} \sigma^\nu_{C\dot{D}}\, \bar{\theta}^{\dot{D}} \partial_\mu V_\nu (x)$$
$$= i (\theta\theta) \varepsilon^{BC} \sigma^\mu_{B\dot{A}} \sigma^\nu_{C\dot{D}}\, \bar{\theta}^{\dot{D}} \partial_\mu V_\nu (x)$$

(using (1.83a))

$$= - i (\theta\theta) \varepsilon^{BC} \sigma^\mu_{B\dot{B}}\, \delta^{\dot{B}}{}_{\dot{A}}\, \sigma^\nu_{C\dot{D}}\, \bar{\theta}^{\dot{D}} \partial_\mu V_\nu (x)$$
$$= i (\theta\theta) \varepsilon^{BC} \sigma^\mu_{B\dot{B}}\, \varepsilon^{\dot{B}\dot{C}} \varepsilon_{\dot{C}\dot{A}}\, \sigma^\nu_{C\dot{D}}\, \bar{\theta}^{\dot{D}} \partial_\mu V_\nu (x)$$

(using (1.71))

$$= -i(\theta\theta)\varepsilon_{\dot{A}\dot{C}}\varepsilon^{\dot{C}\dot{B}}\varepsilon^{CB}\sigma^{\mu}_{B\dot{B}}\sigma^{\nu}_{C\dot{D}}\bar{\theta}^{\dot{D}}\partial_\mu V_\nu(z)$$

$$= -i(\theta\theta)\varepsilon_{\dot{A}\dot{C}}(\bar{\sigma}^{\mu})^{\dot{C}C}(\sigma^{\nu})_{C\dot{D}}\bar{\theta}^{\dot{D}}\partial_\mu V_\nu(z)$$

(using (1.88))

$$= -i(\theta\theta)\varepsilon_{\dot{A}\dot{C}}(\bar{\sigma}^{\mu}\sigma^{\nu})^{\dot{C}}{}_{\dot{D}}\bar{\theta}^{\dot{D}}\partial_\mu V_\nu(z)$$

and

iv) $$-2i\,\theta^B\sigma^{\mu}_{B\dot{A}}(\bar{\theta}\bar{\theta})\theta^C\partial_\mu\lambda_C(z)$$

$$= -2i\,\theta^B\theta^C\sigma^{\mu}_{B\dot{A}}(\bar{\theta}\bar{\theta})\partial_\mu\lambda_C(z)$$

$$= -i(\theta\theta)(\bar{\theta}\bar{\theta})\varepsilon^{BC}\sigma^{\mu}_{B\dot{A}}\partial_\mu\lambda_C(z)$$

(using (1.83a))

$$= -i(\theta\theta)(\bar{\theta}\bar{\theta})\partial_\mu\lambda_C(z)\varepsilon^{CB}\sigma^{\mu}_{B\dot{A}}$$

$$= i(\theta\theta)(\bar{\theta}\bar{\theta})\varepsilon^{BC}\partial_\mu\lambda_C(z)\sigma^{\mu}_{B\dot{A}}$$

$$= i(\theta\theta)(\bar{\theta}\bar{\theta})(\partial_\mu\lambda(z)\sigma^{\mu})_{\dot{A}}$$

Hence

$$\bar{D}^{(2)}_{\dot{A}} V^{(2)}_{WZ}(z,\theta,\bar{\theta})$$

$$\begin{aligned}
&= (\theta\sigma^{\mu})_{\dot{A}}V_\mu(z) + (\theta\theta)\bar{\lambda}_{\dot{A}}(z) + 2\bar{\theta}_{\dot{A}}(\theta\lambda(z)) \\
&\quad + (\theta\theta)\{2\varepsilon_{\dot{A}\dot{C}}D(z) + i\,\varepsilon_{\dot{A}\dot{C}}\eta^{\mu\nu}\partial_\mu V_\nu(z) \\
&\qquad - i\,\varepsilon_{\dot{A}\dot{B}}(\bar{\sigma}^{\mu}\sigma^{\nu})^{\dot{B}}{}_{\dot{C}}\partial_\mu V_\nu(z)\}\bar{\theta}^{\dot{C}} \\
&\quad + i(\theta\theta)(\bar{\theta}\bar{\theta})(\partial_\mu\lambda(z)\sigma^{\mu})_{\dot{A}}
\end{aligned}$$

$$= (\theta\sigma^{\mu})_{\dot{A}}V_\mu(z) + (\theta\theta)\bar{\lambda}_{\dot{A}}(z) + 2\bar{\theta}_{\dot{A}}(\theta\lambda(z))$$

$$+(\theta\theta)\{2\varepsilon_{\dot{A}\dot{C}}D(x) - i\varepsilon_{\dot{A}\dot{B}}[-\delta^{\dot{B}}{}_{\dot{C}}\eta^{\mu\nu}$$
$$+(\bar{\sigma}^{\mu}\sigma^{\nu})^{\dot{B}}{}_{\dot{C}}]\partial_{\mu}V_{\nu}(x)\}\bar{\theta}^{\dot{C}}$$
$$+i(\theta\theta)(\bar{\theta}\bar{\theta})(\partial_{\mu}\lambda(x)\sigma^{\mu})_{\dot{A}} \tag{7.54}$$

Next we need the following result.

<u>Proposition</u>: The following relation holds

$$2\bar{\sigma}^{\mu\nu\,\dot{B}}{}_{\dot{C}} = i[-\delta^{\dot{B}}{}_{\dot{C}}\eta^{\mu\nu} + (\bar{\sigma}^{\mu}\sigma^{\nu})^{\dot{B}}{}_{\dot{C}}] \tag{7.55}$$

<u>Proof</u>: From (1.122b) we have

$$(\bar{\sigma}^{\mu}\sigma^{\nu} + \bar{\sigma}^{\nu}\sigma^{\mu})^{\dot{B}}{}_{\dot{C}} = 2\eta^{\mu\nu}\delta^{\dot{B}}{}_{\dot{C}}$$

Subtracting from both sides $2(\bar{\sigma}^{\mu}\sigma^{\nu})^{\dot{B}}{}_{\dot{C}}$ we obtain

$$-(\bar{\sigma}^{\mu}\sigma^{\nu} - \bar{\sigma}^{\nu}\sigma^{\mu})^{\dot{B}}{}_{\dot{C}} = 2\eta^{\mu\nu}\delta^{\dot{B}}{}_{\dot{C}} - 2(\bar{\sigma}^{\mu}\sigma^{\nu})^{\dot{B}}{}_{\dot{C}}$$

Using (1.119b),

$$4i(\bar{\sigma}^{\mu\nu})^{\dot{B}}{}_{\dot{C}} = 2\eta^{\mu\nu}\delta^{\dot{B}}{}_{\dot{C}} - 2(\bar{\sigma}^{\mu}\sigma^{\nu})^{\dot{B}}{}_{\dot{C}}$$

i.e.

$$2\bar{\sigma}^{\mu\nu\,\dot{B}}{}_{\dot{C}} = i[-\delta^{\dot{B}}{}_{\dot{C}}\eta^{\mu\nu} + (\bar{\sigma}^{\mu}\sigma^{\nu})^{\dot{B}}{}_{\dot{C}}]$$

which is (7.55).

Using (7.55) we can rewrite (7.54)

$$\bar{D}^{(2)}_{\dot{A}}V^{(2)}_{WZ}(x,\theta,\bar{\theta})$$
$$=(\theta\sigma^{\mu})_{\dot{A}}V_{\mu}(x) + (\theta\theta)\bar{\lambda}_{\dot{A}}(x) + 2\bar{\theta}_{\dot{A}}(\theta\lambda(x))$$
$$+(\theta\theta)\{2\varepsilon_{\dot{A}\dot{C}}D(x) - 2\varepsilon_{\dot{A}\dot{B}}(\bar{\sigma}^{\mu\nu})^{\dot{B}}{}_{\dot{C}}\partial_{\mu}V_{\nu}(x)\}\bar{\theta}^{\dot{C}}$$
$$+i(\theta\theta)(\bar{\theta}\bar{\theta})(\partial_{\mu}\lambda(x)\sigma^{\mu})_{\dot{A}}$$

Using the antisymmetry of $\bar{\sigma}^{\mu\nu}$ in μ and ν, we obtain

$$\bar{\sigma}^{\mu\nu}\partial_\mu V_\nu(z)$$
$$=\frac{1}{2}\bar{\sigma}^{\mu\nu}\partial_\mu V_\nu(z)+\frac{1}{2}\bar{\sigma}^{\nu\mu}\partial_\nu V_\mu(z)$$
$$=\frac{1}{2}\bar{\sigma}^{\mu\nu}[\partial_\mu V_\nu(z)-\partial_\nu V_\mu(z)]$$
$$=\frac{1}{2}\bar{\sigma}^{\mu\nu}F_{\mu\nu}(z)$$

where as before

$$F_{\mu\nu}(z)=\partial_\mu V_\nu(z)-\partial_\nu V_\mu(z)$$

Hence

$$\bar{D}^{(2)}_{\dot{A}}V^{(2)}_{WZ}(z,\theta,\bar{\theta})$$
$$=(\theta\sigma^\mu)_{\dot{A}}V_\mu(z)+(\theta\theta)\bar{\lambda}_{\dot{A}}(z)+2\bar{\theta}_{\dot{A}}\,\theta\lambda(z)$$
$$+(\theta\theta)\{2\varepsilon_{\dot{A}\dot{C}}D(z)-\varepsilon_{\dot{A}\dot{B}}(\bar{\sigma}^{\mu\nu})^{\dot{B}}{}_{\dot{C}}F_{\mu\nu}(z)\}\bar{\theta}^{\dot{C}}$$
$$+i(\theta\theta)(\bar{\theta}\bar{\theta})(\partial_\mu\lambda(z)\sigma^\mu)_{\dot{A}} \quad (7.56)$$

Now we can easily obtain $\bar{W}_{\dot{A}}$, i.e.

$$\bar{W}_{\dot{A}}=-\frac{1}{4}D^{(2)}D^{(2)}\bar{D}^{(2)}_{\dot{A}}V^{(2)}_{WZ}(z,\theta,\bar{\theta})$$
$$=-\frac{1}{4}\partial^A\partial_A\bar{D}^{(2)}_{\dot{A}}V^{(2)}_{WZ}(z,\theta,\bar{\theta})$$

(with (6.49))

$$=\frac{1}{4}\varepsilon^{AB}\partial_A\partial_B\{\bar{D}^{(2)}_{\dot{A}}V^{(2)}_{WZ}(z,\theta,\bar{\theta})\}$$
$$=\bar{\lambda}_{\dot{A}}(z)+2\varepsilon_{\dot{A}\dot{C}}D(z)\bar{\theta}^{\dot{C}}$$
$$-\varepsilon_{\dot{A}\dot{B}}(\bar{\sigma}^{\mu\nu})^{\dot{B}}{}_{\dot{C}}F_{\mu\nu}(z)\bar{\theta}^{\dot{C}}$$
$$+i(\bar{\theta}\bar{\theta})(\partial_\mu\lambda(z)\sigma^\mu)_{\dot{A}}$$

(using (6.4q))

We have thus established (7.53).

We see from (7.52) and (7.53) that the superfields W_A and $\overline{W}_{\dot{A}}$ contain only the gauge invariant fields D, λ and $F_{\mu\nu} = \partial_\mu V_\nu - \partial_\nu V_\mu$. Furthermore, as shown with (7.45), these fields are chiral. We can also prove the following result.

Proposition: The fields W_A, $\overline{W}^{\dot{A}}$ obey the following relation

$$\overline{D}_{\dot{A}} \overline{W}^{\dot{A}} = D^A W_A \tag{7.57}$$

Proof: We have

$$\begin{aligned} \overline{D}_{\dot{A}} \overline{W}^{\dot{A}} &= \varepsilon^{\dot{A}\dot{B}} \overline{D}_{\dot{A}} \overline{W}_{\dot{B}} \\ &= \varepsilon^{\dot{A}\dot{B}} \left\{ -\frac{1}{4} \overline{D}_{\dot{A}} (DD) \overline{D}_{\dot{B}} V(x,\theta,\bar{\theta}) \right\} \end{aligned}$$

(with (7.44))

$$\begin{aligned} &= -\frac{1}{4} \overline{D}_{\dot{A}} (DD) \overline{D}^{\dot{A}} V(x,\theta,\bar{\theta}) \\ &= -\frac{1}{4} D^A (\overline{D}\overline{D}) D_A V(x,\theta,\bar{\theta}) \end{aligned}$$

(using (6.70))

$$= D^A W_A \qquad \text{(with (7.43))}$$

CHAPTER 8

SUPERSYMMETRIC LAGRANGIANS

8.1 Integration with respect to Grassmann Numbers[51]

As in the case of differentiation with respect to a Grassmann variable in Section 6.2, we begin by defining the relevant symbol, i.e. the integral

$$\int da\, f(a) = I[f]$$

where "a" denotes a single Grassmann number. Since Grassmann variables are discrete objects, the integral does not represent the area under a curve f(a), nor is any meaning attached to upper and lower limits of the integral. Rather we define a functional, which associates a c-number $I[f]$ with every element f(a) $\in$ G (G being the Grassmann algebra with one element). Furthermore we demand

i)

$$\int da\, f(a+b) = \int da\, f(a) \tag{8.1}$$

implying translation invariance of the integral,

and

ii)

$$\int da \{\alpha f(a) + \beta g(a)\}$$
$$= \alpha \int da\, f(a) + \beta \int da\, g_{\circ}(a) \tag{8.2}$$

implying complex linearity, for every $\alpha, \beta \in C$.
If"a"denotes a single Grassmann number, any function f(a) can be written

$$f(a) = f(0) + f^{(1)} . a$$

since $a^2 = 0$. We now define $I[f]$ to be equal to $f^{(1)}$,i.e.

$$\begin{aligned} I[f] &= \int da\, f(a) \\ &= \int da\, [f(0) + f^{(1)}.a] \\ &= f(0) \int da\, 1 + f^{(1)} \int da\, a \\ &:= f^{(1)} \end{aligned} \tag{8.3}$$

where we used (8.2). Thus our definition implies

$$\int da\, 1 := 0 , \quad \int da\, a := 1 \tag{8.4}$$

It should be observed that (8.4) satisfies (8.1) and (8.2). We also observe that there is no difference between differentiation and integration with respect to a Grassmann variable, i.e.

$$\frac{\partial}{\partial a} f(a) = f^{(1)} = \int da\, f(a)$$

Next we consider the Grassmann algebra G_2 generated by two elements θ_1 and θ_2. G_2 has four independent elements, i.e.

$$1_{G_2} , \quad \theta_1 , \quad \theta_2 , \quad \theta_1 \theta_2$$

We wish to define the integral

$$\int d\theta_1 d\theta_2\, f(\theta_1, \theta_2).$$

First we demand that the $d\theta_A$'s also anticommute, i.e.

$$\{d\theta_A, d\theta_B\} = \{d\theta_A, \theta_B\} = 0 \qquad (8.5)$$

In order to preserve consistency with (8.4) we define

i) $$\int d\theta_1 \int d\theta_2\, 1 = \int d\theta_1 \left[\int d\theta_2\, 1\right] = 0 \qquad (8.6a)$$

ii)
$$\int d\theta_1 \int d\theta_2\, \theta_1 = -\int d\theta_2 \left[\int d\theta_1\, \theta_1\right]$$

(using (8.5))

$$= -\int d\theta_2\, 1$$

(using (8.4))

$$= 0 \qquad (8.6b)$$

(again using (8.4))

iii)
$$\int d\theta_1 \int d\theta_2\, \theta_2 = \int d\theta_1\, 1$$

$$= 0 \qquad (8.6c)$$

(using (8.4))

iv) $$\int d\theta_1 \int d\theta_2\, \theta_1\theta_2 = -\int d\theta_1 \left[\int d\theta_2\, \theta_2\right]\theta_1$$

$$= -\int d\theta_1\, \theta_1 = -1 \qquad (8.6d)$$

again using (8.4). The integral of an arbitrary function $f(\theta_1, \theta_2) \in G_2$ is then obtained by linearity, i.e.

$$\int d\theta_1 \int d\theta_2\, f(\theta_1, \theta_2)$$

$$= \int d\theta_1 \int d\theta_2 \left\{ f^{(0)} + \theta_1 f^{(1)} + \theta_2 f^{(2)} + \theta_1\theta_2 f^{(3)} \right\}$$

$$= f^{(0)}\int d\theta_1 \int d\theta_2 + f^{(1)}\int d\theta_1 \int d\theta_2\, \theta_1$$
$$+ f^{(2)}\int d\theta_1 \int d\theta_2\, \theta_2 + f^{(3)}\int d\theta_1 \int d\theta_2\, \theta_1\theta_2$$
(using (8.2))

$$= -f^{(3)}$$ (using (8.6a,b,c,d)) (8.7)

We see from this result that an integral $\int d\theta_1 \int d\theta_2 f(\theta_1,\theta_2)$ corresponds to a projection such that the highest order component of the expansion of $f(\theta_1,\theta_2)$ is projected out.

Delta functions are defined by the relation

$$\int da\, f(a)\delta(a) = f(0)\,, \quad f(a) \in G_1 \tag{8.8}$$

Since

$$f(a) = f(0) + f^{(1)}a$$

and

$$\int da\, f(a)a = \int da\,[f(0) + f^{(1)}a]\,a$$
$$= \int da\,[f(0)a + f^{(1)}a^2]$$
$$= f(0)\int da\, a$$
$$= f(0)$$ (using (8.4))

we conclude that

$$\delta(a) = a \tag{8.9}$$

We now define volume elements of superspace by the relations

$$d^2\theta := -\frac{1}{4}\, d\theta^A d\theta^B \varepsilon_{AB} \tag{8.10}$$

$$d^2\bar{\theta} := -\frac{1}{4}\, d\bar{\theta}_{\dot{A}}\, d\bar{\theta}_{\dot{B}}\, \varepsilon^{\dot{A}\dot{B}} \tag{8.11}$$

$$d^4\theta := d^2\theta\, d^2\bar{\theta} \tag{8.12}$$

<u>Proposition</u>: With the definitions (8.10) to (8.12) we have

$$\int d^2\theta\,(\theta\theta) = 1 \tag{8.13a}$$

$$\int d^2\bar{\theta}\,(\bar{\theta}\bar{\theta}) = 1 \tag{8.13b}$$

and, of course,

$$\int d^2\theta = 0 = \int d^2\bar{\theta}$$

and

$$\int d^2\theta\,\theta_A = 0 = \int d^2\bar{\theta}\,\bar{\theta}_{\dot{A}} \tag{8.13c}$$

<u>Proof</u>:

i) Consider (using the definition (8.10))

$$\int d^2\theta\,(\theta\theta) = -\frac{1}{4}\int d\theta^A d\theta^B \varepsilon_{AB}(\theta\theta)$$

$$= \frac{1}{2}\int d\theta^A d\theta^B \varepsilon_{AB}\,\theta^1\theta^2 \quad \text{(using (1.82))}$$

$$= \frac{1}{2}\int (\varepsilon_{12}\,d\theta^1 d\theta^2 + \varepsilon_{21}\,d\theta^2 d\theta^1)\,\theta^1\theta^2$$

$$= \frac{1}{2}\int (\varepsilon_{12} - \varepsilon_{21})\,d\theta^1 d\theta^2\,\theta^1\theta^2 \quad \text{(using (8.5))}$$

$$= -\int d\theta^1 d\theta^2\,\theta^1\theta^2 \quad \text{(using (1.59))}$$

$$= 1 \quad \text{(using (8.6d))}$$

ii) Using the definition (8.11) we have

$$\int d^2\bar{\theta}\,(\bar{\theta}\bar{\theta}) = -\frac{1}{4}\int d\bar{\theta}_{\dot{A}}\,d\bar{\theta}_{\dot{B}}\,\varepsilon^{\dot{A}\dot{B}}(\bar{\theta}\bar{\theta})$$

$$= -\frac{1}{2}\int d\bar{\theta}_{\dot{A}}\,d\bar{\theta}_{\dot{B}}\,\varepsilon^{\dot{A}\dot{B}}\,\bar{\theta}_{\dot{1}}\bar{\theta}_{\dot{2}} \quad \text{(using (1.82))}$$

$$= -\frac{1}{2}\int [d\bar{\theta}_{\dot{1}}\,d\bar{\theta}_{\dot{2}} - d\bar{\theta}_{\dot{2}}\,d\bar{\theta}_{\dot{1}}]\,\bar{\theta}_{\dot{1}}\bar{\theta}_{\dot{2}}$$

$$= -\frac{1}{2}\int 2\,d\bar{\theta}_{\dot{1}}\,d\bar{\theta}_{\dot{2}}\,\bar{\theta}_{\dot{1}}\bar{\theta}_{\dot{2}} \quad \text{(using (8.5))}$$

$$= \int d\bar{\theta}_{\dot{1}}\left[\int d\bar{\theta}_{\dot{2}}\,\bar{\theta}_{\dot{2}}\right]\bar{\theta}_{\dot{1}} = \int d\bar{\theta}_{\dot{1}}\,\bar{\theta}_{\dot{1}}$$

$$= 1$$

Proposition: The delta function on G_2, defined by

$$\int d^2\theta \, f(\theta)\, \delta^2(\theta) = f(0) \tag{8.14}$$

is given by

$$\delta^2(\theta) = (\theta\theta) \tag{8.15}$$

Proof: Let $f(\theta) \in G_2$. Then $f(\theta)$ has the expansion

$$f(\theta) = f(0) + \theta^A f_A^{(1)} + (\theta\theta)\, f^{(2)}$$

and

$$\int d^2\theta \, f(\theta) = \int d^2\theta \left[f(0) + \theta f^{(1)} + (\theta\theta) f^{(2)} \right]$$

$$= f(0) \int d^2\theta + f^{(1)A} \int d^2\theta \, \theta_A + f^{(2)} \int d^2\theta \, (\theta\theta)$$

(using (1.80))

$$= f^{(2)} \qquad \text{(using (8.13)}$$

Since

$$(\theta\theta)\theta^A = 0$$

we also have

$$\int d^2\theta \, f(\theta)\, \theta^A$$

$$= \int d^2\theta \left[f(0) + f^{(1)}\theta + (\theta\theta) f^{(2)} \right] \theta^A$$

$$= \int d^2\theta \left[f(0)\, \theta^A + f^{(1)B}\, \theta_B \, \theta^A \right]$$

$$= \int d^2\theta \left[f(0)\, \theta^A - f^{(1)}_B \, \theta^B \theta^A \right] \qquad \text{(using (1.80))}$$

$$= \int d^2\theta \left[f(0)\, \theta^A + \tfrac{1}{2} f_B^{(1)} \varepsilon^{BA} (\theta\theta) \right]$$

(using (1.83a))

$$= f(0) \int d^2\theta \, \theta^A + \tfrac{1}{2} f_B^{(1)} \varepsilon^{BA} \int d^2\theta \, (\theta\theta)$$

$$= -\tfrac{1}{2} \varepsilon^{AB} f_B^{(1)} \qquad \text{(with (8.13))}$$

$$= -\tfrac{1}{2} f^{(1)A}$$

and

$$\int d^2\theta\, f(\theta)\,(\theta\theta)$$
$$= \int d^2\theta\,[f(0) + \theta f^{(1)} + (\theta\theta) f^{(2)}](\theta\theta)$$
$$= \int d^2\theta\, f(0)\,(\theta\theta) \quad \text{(using (8.13))}$$
$$= f(0) \int d^2\theta\,(\theta\theta) \quad \text{(using (8.13))}$$
$$= f(0)$$

Comparing this result with (8.14) we find

$$\theta\theta = \delta^2(\theta)$$

which had to be shown. Similarly we obtain the following result.

Proposition: The delta function $\delta^2(\bar\theta)$ is given by

$$\delta^2(\bar\theta) = \bar\theta\bar\theta \tag{8.16}$$

and

$$\int d^2\bar\theta\,(\bar\theta\bar\theta) = \int d^2\bar\theta\,\delta^2(\bar\theta) = 1$$

We then have, using (8.12),

$$\int d^4\theta\, f(\theta)\,\delta^2(\bar\theta)$$
$$= \int d^2\theta \int d^2\bar\theta\, f(\theta)\,\delta^2(\bar\theta)$$
$$= \int d^2\theta \int d^2\bar\theta\,[f^{(0)} + \theta f^{(1)} + (\theta\theta) f^{(2)}]\,\delta^2(\bar\theta)$$
$$= \int d^2\theta\,[f^{(0)} + \theta f^{(1)} + (\theta\theta) f^{(2)}] \int d^2\bar\theta\,\delta^2(\bar\theta)$$
$$= \int d^2\theta\,[f^{(0)} + \theta f^{(1)} + (\theta\theta) f^{(2)}] \int d^2\bar\theta\,(\bar\theta\bar\theta)$$

(with (8.16))

$$= \int d^2\theta\,[f^{(0)} + \theta f^{(1)} + (\theta\theta) f^{(2)}] \quad \text{(with (8.13b))}$$
$$= f^{(0)} \int d^2\theta + f^{(1)A} \int d^2\theta\,\theta_A + f^{(2)} \int d^2\theta\,(\theta\theta)$$

(using (1.80))

$$= f^{(2)} \qquad \text{(using (8.13a))}$$

Hence

$$\int d^4\theta \, f(\theta)\, \delta^2(\bar{\theta}) = f^{(2)} \tag{8.17}$$

We now consider a general superfield with component expansion given by (6.75), and we evaluate the following integral

$$\int d^4\theta \;\Phi(x,\theta,\bar{\theta})$$

$$= \int d^4\theta \, \{ f(x) + \theta\phi(x) + \bar{\theta}\bar{\chi}(x) + \theta\theta\, m(x)$$
$$+ \bar{\theta}\bar{\theta}\, n(x) + \theta\sigma^{\mu}\bar{\theta}\, V_{\mu}(x) + (\theta\theta)\,\bar{\theta}\bar{\lambda}(x)$$
$$+ (\bar{\theta}\bar{\theta})\,\theta\psi(x) + (\theta\theta)(\bar{\theta}\bar{\theta})\, d(x) \}$$

$$= f(x)\int d^4\theta . 1 + \phi^A(x)\int d^4\theta \; \theta_A$$
$$+ \bar{\chi}_{\dot{A}}(x)\int d^2\theta\, d^2\bar{\theta}\; \bar{\theta}^{\dot{A}} + m(x)\int d^2\theta\, d^2\bar{\theta}\; \theta\theta$$
$$+ n(x)\int d^2\theta\, d^2\bar{\theta}\; \bar{\theta}\bar{\theta} + V_{\mu}(x)\int d^2\theta\, d^2\bar{\theta}\; \theta\sigma^{\mu}\bar{\theta}$$
$$- \bar{\lambda}^{\dot{A}}(x)\int d^2\theta\, d^2\bar{\theta}\, (\theta\theta)\,\bar{\theta}_{\dot{A}}$$
$$- \psi_A(x)\int d^2\theta\, d^2\bar{\theta}\, (\bar{\theta}\bar{\theta})\,\theta^A$$
$$+ d(x)\int d^2\theta\, d^2\bar{\theta}\; (\bar{\theta}\bar{\theta})(\theta\theta)$$

$$= d(x) \tag{8.18}$$

using previous formulae and (1.80). Thus integrating any superfield with respect to the Grassmann supercoordinates always projects out the highest order component field.

8.2 Lagrangians and Actions

We have already mentioned before, that the highest order component of a superfield transforms under supersymmetry transformations into a total derivative. A space-time integral of such a quantity is therefore invariant under supersymmetry transformations. This fundamental property provides the criterion for constructing supersymmetric Lagrangian densities. In order to construct a supersymmetric coupling of several superfields, one considers simply the resulting superfield obtained by multiplication and selects the highest order component field. But this highest order component field is obtained, as explained in Chapter 7, by integration with respect to the Grassmann numbers θ and $\bar{\theta}$. Hence a supersymmetric action integral can be described by

$$\begin{aligned}\mathcal{A} &:= \int d^4x \int d^4\theta \, \mathcal{L} \\ &= \int d^4x \int d^2\theta \, d^2\bar{\theta} \, \mathcal{L} \qquad (8.19)\end{aligned}$$

(using (8.12)) where $\mathcal{L}$ is a supersymmetric Lagrangian density (i.e. supersymmetric up to a total divergence).

8.2.1 Construction of Lagrangians from Scalar Superfields

The most general supersymmetric and renormalizable Lagrangian is given by the following expression

$$\begin{aligned}\mathcal{L} = \Phi_i^{+}\Phi_i + (g_i \Phi_i + \tfrac{1}{2} m_{ij} \Phi_i \Phi_j \\ + \tfrac{1}{3} \lambda_{ijk} \Phi_i \Phi_j \Phi_k)\, \delta^2(\bar{\theta})\end{aligned}$$

$$+ (g_i^* \Phi_i^+ + \frac{1}{2} m_{ij}^* \Phi_i^+ \Phi_j^+ + \frac{1}{3} \lambda_{ijk}^* \Phi_i^+ \Phi_j^+ \Phi_k^+) \delta^2(\theta)$$

(summation over i, j, k understood). (8.20)

Here $i,j = 1,2,\ldots,N$, where N is the number of scalar superfields. It can be shown (see, for instance, Wess and Bagger[2]) that the condition of renormalizability forbids powers of Φ in (8.20) higher than the third; however, linear terms are permissible. To find a supersymmetric Lagrangian such as (8.20) for chiral superfields, one can use arguments from dimensional analysis. From previous considerations we know that the Grassmann number θ has dimension $+\frac{1}{2}$ in length. According to (7.5) the component fields of a chiral superfield are a scalar, a spinor and another scalar field. The spinor field has, as usual, the dimension $-\frac{3}{2}$, and the two complex scalar fields differ by one unit of dimension in view of the extra factor $\theta\theta$ in front of one of them. We therefore assign the chiral superfield Φ the dimension - 1. From

$$\int d\theta\, \theta = 1$$

we deduce that $\int d\theta$ has dimension $-\frac{1}{2}$; this implies, in particular, that $\int d^4\theta$ has dimension -2 in length. This therefore leads to a unique choice for a free (quadratic) massless action without dimensional parameters, i.e. to the expression

$$\mathcal{A}_{kin} = \int d^4x\, d^4\theta\, \Phi_i \Phi_i^+$$

where

dimension of d^4x = 4, dimension of $d^4\theta$ = - 2, and dimension of $\Phi_i \Phi_i^+$ = - 2, such that $\mathcal{A}_{kin}$ is dimensionless. These arguments also justify the mass and interaction terms of (8.20), i.e.

$$\mathcal{A}_{\text{mass}} = \frac{1}{2}\int d^4x\, d^2\theta\, m_{ij}\, \Phi_i \Phi_j$$

where dimension of m_{ij} = - 1, dimension of $\Phi_i\Phi_j$ = - 2, dimension of $d^2\theta$ = - 1, and dimension of d^4x = 4, such that $\mathcal{A}_{\text{mass}}$ is dimensionless, and

$$\mathcal{A}_{\text{int}} = \frac{1}{3}\int d^4x\, d^2\theta\, \lambda_{ijk}\, \Phi_i \Phi_j \Phi_k$$

where dimension of $\Phi_i\Phi_j\Phi_k$ = -3, dimension of $d^2\theta$ = - 1, dimension of d^4x = 4, and the coupling λ_{ijk} is dimensionless.

The couplings m_{ij} and λ_{ijk} are symmetric in their indices.

Now from (7.16) we know that the product $\Phi_i\Phi_k$ has the expansion

$$\begin{aligned}\Phi_i \Phi_k = {} & A_i(y)A_k(y) + 2^{\frac{1}{2}}\theta\{A_i(y)\psi_k(y) \\ & + \psi_i(y)A_k(y)\} + (\theta\theta)\{A_i(y)F_k(y) \\ & + F_i(y)A_k(y) - \psi_i(y)\psi_k(y)\}\end{aligned}$$

so that

$$\begin{aligned}\Phi_i^+ \Phi_k^+ = {} & A_i^*(z)A_k^*(z) + 2^{\frac{1}{2}}\bar{\theta}\{A_i^*(z)\bar{\psi}_k(z) \\ & + \bar{\psi}_i(z)A_k^*(z)\} + (\bar{\theta}\bar{\theta})\{A_i^*(z)F_k^*(z) \\ & + F_i^*(z)A_k^*(z) - \bar{\psi}_i(z)\bar{\psi}_k(z)\}\end{aligned}$$

The component expansion of the product of three fields Φ is given by (7.17); the expansion of $\Phi_i^+\Phi_j^+\Phi_k^+$ is then obtained by complex conjugation. The expansion of $\Phi_i^+\Phi_i$ is given by (7.27). We emphasize that according to (7.31) $\Phi_i^+\Phi_i$ is a vector superfield, the term of highest order in this multiplet being of the form

$$(\theta\theta)(\bar{\theta}\bar{\theta})\,D(x)$$

Inserting (8.20) into the action integral (8.19) we obtain

$$\begin{aligned}\mathcal{O}\!\iota = \int d^4x \int d^4\theta\, \{ & \Phi_i^+ \Phi_i + g_i \Phi_i\, \delta^2(\bar{\theta}) \\ & + \tfrac{1}{2} m_{ij} \Phi_i \Phi_j\, \delta^2(\bar{\theta}) + \tfrac{1}{3}\lambda_{ijk} \Phi_i \Phi_j \Phi_k\, \delta^2(\bar{\theta}) \\ & + g_i^* \Phi_i^+ \delta^2(\theta) + \tfrac{1}{2} m_{ij}^* \Phi_i^+ \Phi_j^+ \delta^2(\theta) \\ & + \tfrac{1}{3}\lambda_{ijk}^* \Phi_i^+ \Phi_j^+ \Phi_k^+ \delta^2(\theta) \}\end{aligned} \quad (8.21)$$

Inserting the component field expansions for the various terms and using the rules of integration developed in Section 8.1, we obtain from the coefficient of $(\theta\theta)(\bar{\theta}\bar{\theta})$ in (7.27) the contribution

$$\begin{aligned}&\int d^4x \int d^4\theta\, \Phi_i^+ \Phi_i \\ &= \int d^4x \{ \tfrac{1}{2}\partial_\mu A_i^*(x)\partial^\mu A_i(x) - \tfrac{1}{4} A_i^*(x)\Box A_i(x) \\ &\quad - \tfrac{1}{4} A_i(x) \Box A_i^*(x) + \tfrac{i}{2}\partial_\mu \psi_i(x)\sigma^\mu \bar{\psi}_i(x) \\ &\quad - \tfrac{i}{2}\psi_i(x)\sigma^\mu \partial_\mu \bar{\psi}_i(x) + F_i^*(x) F_i(x) \}\end{aligned} \quad (8.22)$$

From the coefficient of $(\theta\theta)$ in (7.16) we obtain

$$\begin{aligned}&\tfrac{1}{2} m_{ij} \int d^4x\, d^4\theta\, \Phi_i \Phi_j\, \delta^2(\bar{\theta}) \\ &= \tfrac{1}{2} m_{ij} \int d^4x\, d^2\theta\, \Phi_i \Phi_j \\ &= \tfrac{1}{2} m_{ij} \int d^4x [A_i(x) F_j(x) + F_j(x) A_i(x) \\ &\qquad\qquad - \psi_i(x)\psi_j(x)] \\ &= \int d^4x\ m_{ij} [A_i(x) F_j(x) - \tfrac{1}{2}\psi_i(x)\psi_j(x)]\end{aligned} \quad (8.23)$$

since $m_{ij} = - m_{ji}$. Similarly, or by complex conjugation of (8.23), we obtain

$$\frac{1}{2} m^*_{ij} \int d^4x\, d^4\theta\, \Phi_i^+ \Phi_j^+ \delta^2(\theta)$$

$$= \frac{1}{2} m^*_{ij} \int d^4x\, d^2\bar{\theta}\, \Phi_i^+ \Phi_j^+$$

$$= \int d^4x\, m^*_{ij} \left[A_i^*(x) F_j^*(x) - \frac{1}{2} \bar{\psi}_i(x) \bar{\psi}_j(x) \right] \tag{8.24}$$

Using (7.17) we obtain in a similar way

$$\frac{1}{3} \lambda_{ijk} \int d^4x \int d^4\theta\, \Phi_i \Phi_j \Phi_k \delta^2(\bar{\theta})$$

$$= \frac{1}{3} \lambda_{ijk} \int d^4x \int d^2\theta\, \Phi_i \Phi_j \Phi_k$$

$$= \int d^4x \Big\{ \frac{1}{3} \lambda_{ijk} \big[A_i(x) A_j(x) F_k(x)$$
$$+ A_i(x) F_j(x) A_k(x) + F_i(x) A_j(x) A_k(x)$$
$$- \psi_i(x) \psi_j(x) A_k(x) - \psi_i(x) \psi_k(x) A_j(x)$$
$$- A_i(x) \psi_j(x) \psi_k(x) \big] \Big\}$$

$$= \int d^4x\, \lambda_{ijk} \big[A_i(x) A_j(x) F_k(x)$$
$$- \psi_i(x) \psi_j(x) A_k(x) \big] \tag{8.25}$$

since λ_{ijk} is symmetric in all three indices. Analogously we have

$$\int d^4x \int d^4\theta\, \frac{1}{3} \lambda^*_{ijk} \Phi_i^+ \Phi_j^+ \Phi_k^+$$

$$= \int d^4x\, \lambda^*_{ijk} \big[A_i^*(x) A_j^*(x) F_k^*(x)$$
$$- \bar{\psi}_i(x) \bar{\psi}_j(x) A_k^*(x) \big] \tag{8.26}$$

In addition

$$\int d^4x\, d^4\theta\, g_i \Phi_i \delta^2(\bar\theta)$$
$$= \int d^4x\, d^4\theta\, g_i (A_i + 2^{\frac{1}{2}}\theta\psi_i + \theta\theta F_i)\,\delta^2(\bar\theta)$$
$$= \int d^4x\, g_i F_i(x)$$

and

$$\int d^4x\, d^4\theta\, g_i^* \Phi_i^+ \delta^2(\bar\theta)$$
$$= \int d^4x\, g_i^* F_i^*(x)$$

We now observe that we are free to shift the coordinates from y and z to x. Such a change of variables does not alter the Lagrangian density because the highest order component of any scalar superfield has, for example, the form $\theta\theta D(y)$ and

$$\begin{aligned}\theta\theta D(y) &= \theta\theta D(x + i\theta\sigma\bar\theta)\\ &= \theta\theta[D(x) + i(\theta\sigma^\mu\bar\theta)\partial_\mu D(x)\\ &\quad - (\theta\sigma^\mu\bar\theta)(\theta\sigma^\rho\bar\theta)\partial_\mu\partial_\rho D(x)]\\ &= \theta\theta D(x)\end{aligned}$$

since $(\theta\theta)\theta = 0.$

Hence, changing the variable from y and z to x does not change the Lagrangian. Then the action (8.21) becomes, on adding (8.22) to (8.26),

$$\begin{aligned}\mathcal{O}\!\mathcal{l} = \int d^4x \{&\tfrac{1}{2}\partial_\mu A_i^*(x)\partial^\mu A_i(x) - \tfrac{1}{4}A_i^*(x)\Box A_i(x)\\ &- \tfrac{1}{4}A_i(x)\Box A_i^*(x) + \tfrac{i}{2}\partial_\mu\psi_i(x)\sigma^\mu\bar\psi_i(x)\\ &- \tfrac{i}{2}\psi_i(x)\sigma^\mu\partial_\mu\bar\psi_i(x) + F_i^*(x)F_i(x)\\ &+ [m_{ij}(A_i(x)F_j(x) - \tfrac{1}{2}\psi_i(x)\psi_j(x))\end{aligned}$$

$$+ \lambda_{ijk} \left(A_i(x) A_j(x) F_k(x) - \psi_i(x) \psi_j(x) A_k(x) \right) + g_i F_i(x) \quad + \text{hermitian conjugate} \Big] \qquad (8.27)$$

We observe that $\mathfrak{A}$ does not contain a kinetic term for the field $F_i(x)$. Such fields are called "auxiliary fields". These fields can be eliminated from the theory with the help of their respective equations of motion; these fields thus provide potentials in terms of the dynamical fields. The reason for the nonoccurence of derivatives of the fields F_i is that these derivatives would be associated with θ^3 which is zero. We now rewrite a few terms of (8.27).

Proposition: We can show that

$$\frac{1}{2} \partial_\mu A_i^*(x) \partial^\mu A_i(x) - \frac{1}{4} A_i^*(x) \Box A_i(x) - \frac{1}{4} A_i(x) \Box A_i^*(x) = - A_i^*(x) \Box A_i(x) + \text{total derivatives} \qquad (8.28)$$

Verification: Consider

$$\frac{1}{2} \partial_\mu A_i^*(x) \partial^\mu A_i(x) = \frac{1}{2} \partial_\mu [A_i^*(x) \partial^\mu A_i(x)] - \frac{1}{2} A_i^*(x) \Box A_i(x)$$

and

$$-\frac{1}{4} A_i(x) \Box A_i^*(x) = -\frac{1}{4} A_i(x) \partial_\mu \partial^\mu A_i^*(x) = -\frac{1}{4} \partial_\mu [A_i(x) \partial^\mu A_i^*(x)] + \frac{1}{4} \partial_\mu A_i(x) \partial^\mu A_i^*(x)$$

$$= -\frac{1}{4}\partial_\mu\left[A_i(x)\partial^\mu A_i^*(x)\right] + \frac{1}{4}\partial_\mu\left[A_i(x)\partial^\mu A_i^*(x)\right]$$
$$-\frac{1}{4}A_i^*(x)\Box A_i(x)$$

Hence

$$\frac{1}{2}\partial_\mu A_i^*(x)\partial^\mu A_i(x) - \frac{1}{4}A_i(x)\Box A_i^*(x)$$
$$-\frac{1}{4}A_i^*(x)\Box A_i(x)$$
$$= -A_i^*(x)\Box A_i(x) \quad \text{+ total derivatives.}$$

Using (1.115) we have

$$\frac{i}{2}(\partial_\mu\psi_i(x))\sigma^\mu\bar{\psi}_i(x) - \frac{i}{2}\psi_i(x)\sigma^\mu(\partial_\mu\bar{\psi}_i(x))$$
$$= -\frac{i}{2}\bar{\psi}_i\bar{\sigma}^\mu(\partial_\mu\psi_i(x)) + \frac{i}{2}(\partial_\mu\bar{\psi}_i(x))\bar{\sigma}^\mu\psi_i(x)$$
$$= -\frac{i}{2}\partial_\mu(\bar{\psi}_i\bar{\sigma}^\mu\psi_i) + i(\partial_\mu\bar{\psi}_i(x))\bar{\sigma}^\mu\psi_i(x)$$
$$= i(\partial_\mu\bar{\psi}_i(x))\bar{\sigma}^\mu\psi_i(x) \quad \text{+ total derivatives} \tag{8.29}$$

Inserting (8.28) and (8.29) in (8.27) we obtain the final expression for the action (8.20) in terms of component fields, i.e.

$$\mathcal{A} = \int d^4x \Big\{ F_i^*(x)F_i(x) + i(\partial_\mu\bar{\psi}_i(x))\bar{\sigma}^\mu\psi_i(x)$$
$$- A_i^*(x)\Box A_i(x)$$
$$+ \Big[m_{ij}\big(A_i(x)F_j(x) - \frac{1}{2}\psi_i(x)\psi_j(x)\big)$$
$$+ \lambda_{ijk}\big(A_i(x)A_j(x)F_k(x)$$
$$- \psi_i(x)\psi_j(x)A_k(x)\big)$$
$$+ g_iF_i(x) \quad \text{+ hermitian conjugate}\Big]$$

$$= \int d^4x \{ i (\partial_\mu \bar{\psi}_i(x)) \bar{\sigma}^\mu \psi_i(x) - A_i^*(x) \Box A_i(x)$$
$$+ (m_{ik} A_i(x) + \lambda_{ijk} A_i(x) A_j(x)) F_k(x)$$
$$+ (m_{ik}^* A_i^*(x) + \lambda_{ijk}^* A_i^*(x) A_j^*(x)) F_k^*(x)$$
$$- \frac{1}{2} m_{ij} \psi_i(x) \psi_j(x)$$
$$- \lambda_{ijk} \psi_i(x) \psi_j(x) A_k(x)$$
$$- \frac{1}{2} m_{ij}^* \bar{\psi}_i(x) \bar{\psi}_j(x)$$
$$- \lambda_{ijk}^* \bar{\psi}_i(x) \bar{\psi}_j(x) A_k^*(x)$$
$$+ F_i^*(x) F_j^*(x) + g F_i(x) + g_i^* F_i^*(x) \}$$

(8.30)

This is the socalled "off-shell" form of the action which involves the auxiliary fields $F_i(x)$. The auxiliary fields $F_i(x)$ can be eliminated with the help of the equations of motion, thus giving the "on-shell" form of the action integral. Consider the Euler-Lagrange equations

$$\frac{\partial \mathcal{L}}{\partial F_i(x)} - \partial_\mu \frac{\partial \mathcal{L}}{\partial(\partial_\mu F_i(x))} = 0$$

for the fields $F_i(x)$. Since the Lagrangian density does not contain derivatives of the fields $F_i(x)$, we have

$$\frac{\partial \mathcal{L}}{\partial F_i(x)} = 0$$

Hence

$$\frac{\partial \mathcal{L}}{\partial F_i(x)} = F_i^*(x) + g_i + m_{ij} A_j(x)$$
$$+ \lambda_{ijk} A_j(x) A_k(x) = 0$$

$$\frac{\partial \mathcal{L}}{\partial F_i^*(x)} = F_i(x) + g_i^* + m_{ij}^* A_j^*(x) + \lambda_{ijk}^* A_j^*(x) A_k^*(x) = 0 \qquad (8.31)$$

From these equations which determine the auxiliary fields $F_i(x)$ we see that if the scalar fields $A_i(x)$ all have dimension $(\text{length})^{-1}$, $m \sim (\text{length})^{-1}$ and λ is dimensionless, then $F_i(x)$ must have dimension $(\text{length})^{-2}$. This is consistent with the arguments given earlier.

Replacing $F_i(x)$, $F_i^*(x)$ in (8.30) by their expressions in terms of $A_i(x)$, $A_i^*(x)$, we obtain

$$\begin{aligned}
\mathcal{A} = \int d^4x \{ & i\partial_\mu \bar{\psi}_i(x) \bar{\sigma}^\mu \psi_i(x) - A_i^*(x) \Box A_i(x) \\
& - \tfrac{1}{2} m_{ij} \psi_i(x) \psi_j(x) - \tfrac{1}{2} m_{ij}^* \bar{\psi}_i(x) \bar{\psi}_j(x) \\
& - \lambda_{ijk} \psi_i(x) \psi_j(x) A_k(x) - \lambda_{ijk}^* \bar{\psi}_i(x) \bar{\psi}_j(x) A_k^*(x) \\
& + [g_k + m_{ki} A_i(x) + \lambda_{kij} A_i(x) A_j(x)] F_k(x) \\
& + [g_k^* + m_{ki}^* A_i^*(x) + \lambda_{kij}^* A_i^*(x) A_j^*(x)] F_k^*(x) \\
& + F_k^*(x) F_k(x) \}
\end{aligned}$$

Replacing $F_i(x)$, $F_i^*(x)$ in this expression by their equivalents in terms of $A_j(x)$, $A_j^*(x)$ obtained from equations (8.31), we obtain the on-shell form of the action integral, i.e.

$$\begin{aligned}
\mathcal{A} = \int d^4x \{ & i\partial_\mu \bar{\psi}_i(x) \bar{\sigma}^\mu \psi_i(x) - A_i^*(x) \Box A_i(x) \\
& - \tfrac{1}{2} m_{ij} \psi_i(x) \psi_j(x) - \tfrac{1}{2} m_{ij}^* \bar{\psi}_i(x) \bar{\psi}_j(x) \\
& - \lambda_{ijk} \psi_i(x) \psi_j(x) A_k(x)
\end{aligned}$$

$$- \lambda^*_{ijk} \bar{\Psi}_i(x) \bar{\Psi}_j(x) A^*_k(x)$$

$$- V(A_i, A^*_j) \} \qquad (8.32)$$

where

$$V(A_i, A^*_j) = F_k^*(x) F_k(x)$$

is the potential of this supersymmetric field theory, and $F_k(x)$ and $F_k^*(x)$ are functions of $A_i(x)$, $A^*_i(x)$ given by (8.31). The potential $V = |F_k(x)|^2$ is always greater than or equal to zero (this being a consequence of supersymmetry). Configurations for which $F_k = 0$ are absolute minima of the potential.

To extract the particle content of the field theory described by the action (8.32), we choose the simplest case of only one scalar superfield Φ, so that all indices can be dropped in (8.32) and our starting point is the Lagrangian density

$$\begin{aligned} \mathcal{L} = &\, i\partial_\mu \bar{\Psi}(x) \bar{\sigma}^\mu \psi(x) - A^*(x) \Box A(x) \\ & - \tfrac{1}{2} m \psi^2(x) - \tfrac{1}{2} m^* \bar{\Psi}^2(x) - \lambda \psi^2(x) A(x) \\ & - \lambda^* \bar{\Psi}^2(x) A^*(x) - V(A, A^*) \end{aligned}$$

where

$$\begin{aligned} V(A, A^*) &= |F(x)|^2 = F(x) F^*(x) \\ &= (g + mA(x) + \lambda A^2(x))(g^* + m^* A^*(x) + \lambda^* A^{*2}(x)) \\ &= |g|^2 + g m^* A^*(x) + g \lambda^* A^{*2}(x) + m g^* A(x) \\ &\quad + |m|^2 |A(x)|^2 + m \lambda^* A^*(x) |A(x)|^2 \\ &\quad + \lambda g^* A^2(x) + m^* \lambda |A(x)|^2 A(x) \\ &\quad + |\lambda|^2 (|A(x)|^2)^2 \end{aligned}$$

Furthermore, assuming that g, m and λ are real, we obtain (using (1.115))

$$\begin{aligned}\mathcal{L} = &-i\psi(x)\sigma^\mu\partial_\mu\bar{\Psi}(x) - A^*(x)\Box A(x) - g^2\\ &-\frac{1}{2}m(\psi^2(x)+\bar{\Psi}^2(x))\\ &-mg(A(x)+A^*(x)) - \lambda^2(|A(x)|^2)^2\\ &-\lambda(\bar{\Psi}^2(x)A^*(x)+\psi^2(x)A(x))\\ &-g\lambda(A^{*2}(x)+A^2(x))\\ &-m\lambda|A(x)|^2(A(x)+A^*(x))\end{aligned} \tag{8.33}$$

Then the Euler-Lagrange equations for the various fields are:

i) for $\psi(x)$:

$$\frac{\partial\mathcal{L}}{\partial\psi(x)} - \partial_\mu\frac{\partial\mathcal{L}}{\partial(\partial_\mu\psi(x))} = 0$$

where

$$\frac{\partial\mathcal{L}}{\partial\psi(x)} = -i\sigma^\mu\partial_\mu\bar{\Psi}(x) - m\psi(x) - 2\lambda\psi(x)A(x)$$

$$\frac{\partial\mathcal{L}}{\partial(\partial_\mu\psi)} = 0$$

so that

$$-i\sigma^\mu\partial_\mu\bar{\Psi}(x) = m\psi(x) + 2\lambda\psi(x)A(x) \tag{8.34}$$

ii) for $\bar{\Psi}(x)$:

$$\frac{\partial\mathcal{L}}{\partial\bar{\Psi}(x)} - \partial_\mu\frac{\partial\mathcal{L}}{\partial(\partial_\mu\bar{\Psi}(x))} = 0$$

where

$$\frac{\partial \mathcal{L}}{\partial \overline{\Psi}(x)} = -m\overline{\Psi}(x) - 2\lambda\overline{\Psi}(x)A^*(x)$$

$$\frac{\partial \mathcal{L}}{\partial(\partial_\mu \overline{\Psi}(x))} = -i\,\Psi(x)\sigma^\mu$$

so that

$$i\,\partial_\mu \Psi(x)\sigma^\mu = m\overline{\Psi}(x) + 2\lambda\overline{\Psi}(x)A^*(x) \tag{8.35}$$

iii) for $A^*(x)$:

$$\frac{\partial \mathcal{L}}{\partial A^*(x)} - \partial_\mu \frac{\partial \mathcal{L}}{\partial(\partial_\mu A^*(x))} = 0$$

where

$$\begin{aligned}\frac{\partial \mathcal{L}}{\partial A^*(x)} = &-\Box A(x) - \lambda\overline{\Psi}^2(x) - m^2 A(x)\\ &- 2m\lambda|A(x)|^2 - m\lambda A^2(x)\\ &- 2\lambda^2|A(x)|^2 A(x) - mg - 2g\lambda A^*(x)\end{aligned}$$

$$\frac{\partial \mathcal{L}}{\partial(\partial_\mu A^*(x))} = 0$$

so that

$$\begin{aligned}(\Box + m^2)A(x) = -\lambda\{&\overline{\Psi}^2(x) + 2m|A(x)|^2\\ &+ mA^2(x) + 2\lambda|A(x)|^2 A(x)\\ &+ 2gA^*(x)\} + mg\end{aligned} \tag{8.36}$$

If we set $M := m$ and $\lambda = 0 = g$ (free theory) we obtain from(8.34) and (8.35)

$$i\,\sigma^\mu \partial_\mu \overline{\Psi}(x) + M\Psi(x) = 0$$

$$i\,\partial_\mu \Psi(x)\sigma^\mu - M\overline{\Psi}(x) = 0$$

which are equivalent to

$$(i\gamma^{\mu}_{W}\partial_{\mu} - M)\Psi_M(x) = 0 \qquad (8.35')$$

where γ^{μ}_{W} are the γ matrices in the Weyl representation and Ψ_M is a four-component Majorana spinor with left- and right-handed components ψ and $\bar{\psi}$ respectively. Furthermore, (8.36) becomes (also with g = 0)

$$(\Box + M^2)A(x) = 0 \qquad (8.36')$$

where A(x) is a complex scalar field. Hence the Lagrangian (8.33) describes the field theory (in the case g, λ = 0) of a spinor field and a complex scalar field, each with the same mass M. This is, of course, simply the action of the free Wess-Zumino model (discussed in Chapter 5), formulated here in terms of two-component Weyl spinors.

In order to see that the real and imaginary parts of the complex "scalar" field A(x) do, in fact, define the scalar and pseudoscalar components of the Wess-Zumino supermultiplet, we recall that the Weyl formulation employed here does not preserve a well defined parity property. Previously we saw that (in our present notation) for a chiral superfield representing the lowest supersymmetric multiplet (cf. (7.5)) the expansion of the corresponding superfield, i.e.

$$\Phi(y,\theta) = A(y) + 2^{\frac{1}{2}}\theta\psi(y) + \theta\theta F(y)$$

implies the following transformations of the component fields (cf. (7.19))

$$\delta_S' A(x) = 2^{\frac{1}{2}}\alpha\psi(x)$$

$$\delta_S'(2^{\frac{1}{2}}\psi_A(x)) = 2\alpha_A F(x) + 2i(\sigma^{\mu}\bar{\alpha})_A\partial_{\mu}A(x)$$

$$\delta_S' F(x) = -i\partial_{\mu}(2^{\frac{1}{2}}\psi(x))\sigma^{\mu}\bar{\alpha}$$

where $F^* = -mA$ from (8.31).

Comparing these relations with (5.11') we see that indeed

$$2^{\frac{1}{2}}\psi(x)\Big|_{\text{here}} = \psi(x)\Big|_{(5.11')}$$

$$A(x)\Big|_{\text{here}} = f^{*}(x)\Big|_{(5.11')}$$

$$\alpha\Big|_{\text{here}} = \varepsilon\Big|_{(5.11')}$$

Thus the real and imaginary parts of the complex field A(x) of (8.36') are the real scalar and pseudoscalar fields of the Wess-Zumino model, the combination (A + B) /2 arising as a result of the decomposition of 4x4 Dirac quantities of well defined parity into 2x2 Weyl matrices.

8.2.2 Construction of Lagrangians from Vector Superfields

In the previous section we discussed Lagrangians constructed entirely from scalar superfields which describe spin zero and spin-$\frac{1}{2}$ particles. Of course, one also wants to construct model theories which describe spin-1 particles, the ultimate aim being the construction of supersymmetric Yang-Mills gauge theories. In this section we restrict ourselves to the construction of the Lagrangian of a supersymmetric abelian gauge theory. As discussed in Section 7.3, the supersymmetric generalization of the electromagnetic field strength is given by the chiral superfields W_A and $\overline{W}_{\dot{A}}$ which have been constructed from vector superfields. Since W_A is chiral, as demonstrated by (7.45), $W^A W_A$ is a scalar field (i.e. a Lorentz scalar as in (1.77)) and therefore the $\theta\theta$ - component of $W^A W_A$ is of interest in the construction of supersymmetrically invariant Lagrangians. (We have seen that the $\theta\theta$ -component of a scalar superfield Φ always transforms into a space-time derivative). Hence our first task is to calculate the component expansion of the scalar superfield $W^A W_A$.

Proposition: The component field expansion of the scalar superfield $W^A W_A$ is given by

$$
\begin{aligned}
W^A W_A &= \lambda^2(y) + 4\,D(y)\lambda(y)\theta \\
&\quad + \lambda(y)\sigma^{\mu\nu}\theta F_{\mu\nu}(y) \\
&\quad + \theta\theta\{4D^2(y) - 2i\lambda(y)\sigma^\mu\partial_\mu\bar{\lambda}(y) \\
&\qquad - \tfrac{1}{2}F_{\mu\nu}(y)F^{\mu\nu}(y) \\
&\qquad - \tfrac{i}{2}F^*_{\mu\nu}(y)F^{\mu\nu}(y)\}
\end{aligned}
\tag{8.37}
$$

where $F^*_{\mu\nu}(y)$ is the dual field strength of $F_{\mu\nu}(y)$

Proof: Using (7.52) we have

$$
\begin{aligned}
&W^A W_A \\
&= \{\lambda^A(y) + 2D(y)\theta^A + \varepsilon^{AB}(\sigma^{\mu\nu})_B{}^C\theta_C F_{\mu\nu}(y) \\
&\quad - i(\theta\theta)\varepsilon^{AB}(\sigma^\mu)_{B\dot{B}}\partial_\mu\bar{\lambda}^{\dot{B}}(y)\}\cdot\{\lambda_A(y) \\
&\quad + 2D(y)\theta_A + (\sigma^{\rho\sigma})_A{}^B\theta_B F_{\rho\sigma}(y) \\
&\quad - i(\theta\theta)(\sigma^\mu)_{A\dot{C}}\partial_\mu\bar{\lambda}^{\dot{C}}(y)\}
\end{aligned}
$$

Using various results obtained earlier, properties such as (1.80), the Grassmann property $(\theta\theta)\theta_A = 0$ etc. this becomes

$$
\begin{aligned}
&W^A W_A \\
&= \lambda^2(y) + 2\,D(y)\,(\lambda(y)\theta) + \lambda(y)\sigma^{\rho\sigma}\theta F_{\rho\sigma}(y)
\end{aligned}
$$

$$-i(\theta\theta)\lambda(y)\sigma^{\mu}\partial_{\mu}\bar{\lambda}(y)+2D(y)\theta\lambda(y)$$
$$+4D^{2}(y)\theta\theta+2D(y)(\theta\sigma^{\rho\sigma}\theta)F_{\rho\sigma}(y)$$
$$+\varepsilon^{AB}(\sigma^{\mu\nu})_{B}{}^{C}\theta_{C}\lambda_{A}(y)F_{\mu\nu}(y)$$
$$+\varepsilon^{AB}(\sigma^{\mu\nu})_{B}{}^{C}\theta_{C}(\sigma^{\rho\sigma})_{A}{}^{D}\theta_{D}F_{\mu\nu}(y)F_{\rho\sigma}(y)$$
$$-i(\theta\theta)\varepsilon^{AB}(\sigma^{\mu})_{B\dot{B}}\partial_{\mu}\bar{\lambda}^{\dot{B}}(y)\lambda_{A}(y)$$
$$+2\varepsilon^{AB}(\sigma^{\mu\nu})_{B}{}^{C}\theta_{C}\theta_{A}D(y)F_{\mu\nu}(y)$$

Now

$$\varepsilon^{AB}(\sigma^{\mu\nu})_{B}{}^{C}\theta_{C}\lambda_{A}(y)F_{\mu\nu}(y)$$
$$=-\lambda_{A}(y)\varepsilon^{AB}(\sigma^{\mu\nu})_{B}{}^{C}\theta_{C}F_{\mu\nu}(y)$$
$$=\varepsilon^{BA}\lambda_{A}(y)(\sigma^{\mu\nu})_{B}{}^{C}\theta_{C}F_{\mu\nu}(y)$$
$$=(\lambda(y)\sigma^{\mu\nu}\theta)F_{\mu\nu}(y)$$

and

$$-i(\theta\theta)\varepsilon^{AB}\sigma^{\mu}_{B\dot{B}}\partial_{\mu}\bar{\lambda}^{\dot{B}}(y)\lambda_{A}(y)$$
$$=i(\theta\theta)\lambda_{A}(y)\varepsilon^{AB}(\sigma^{\mu}\partial_{\mu}\bar{\lambda}(y))_{B}$$
$$=-i(\theta\theta)\lambda(y)\sigma^{\mu}\partial_{\mu}\bar{\lambda}(y)$$

and

$$2\varepsilon^{AB}(\sigma^{\mu\nu})_{B}{}^{C}\theta_{C}\theta_{A}D(y)F_{\mu\nu}(y)$$
$$=2\theta\sigma^{\mu\nu}\theta\, D(y)F_{\mu\nu}(y)$$
$$=0$$

since

$$\theta\sigma^{\mu\nu}\theta=0 \quad \text{(cf. (1.128))}$$

Finally

$$\varepsilon^{AB}(\sigma^{\mu\nu})_B{}^C\theta_C(\sigma^{\rho\sigma})_A{}^D\theta_D F_{\mu\nu}(y)F_{\rho\sigma}(y)$$
$$= \varepsilon^{AB}(\sigma^{\mu\nu})_B{}^C\theta_C\theta_D(\sigma^{\rho\sigma})_A{}^D F_{\mu\nu}(y)F_{\rho\sigma}(y)$$
$$= \frac{1}{2}(\theta\theta)\varepsilon^{AB}(\sigma^{\mu\nu})_B{}^C\varepsilon_{CD}(\sigma^{\rho\sigma})_A{}^D F_{\mu\nu}(y)F_{\rho\sigma}(y)$$

(using (1.83b))

$$= \frac{1}{2}(\theta\theta)\varepsilon^{AB}(\sigma^{\mu\nu})_B{}^C\varepsilon_{AD}(\sigma^{\rho\sigma})_C{}^D F_{\mu\nu}(y)F_{\rho\sigma}(y)$$

(using (1.127))

$$= \frac{1}{2}(\theta\theta)\varepsilon^{AB}\varepsilon_{AD}(\sigma^{\mu\nu}\sigma^{\rho\sigma})_B{}^D F_{\mu\nu}(y)F_{\rho\sigma}(y)$$
$$= -\frac{1}{2}(\theta\theta)\delta^B{}_D(\sigma^{\mu\nu}\sigma^{\rho\sigma})_B{}^D F_{\mu\nu}(y)F_{\rho\sigma}(y)$$
$$= -\frac{1}{2}(\theta\theta)\mathrm{Tr}[\sigma^{\mu\nu}\sigma^{\rho\sigma}]F_{\mu\nu}(y)F_{\rho\sigma}(y)$$
$$= -\frac{1}{2}(\theta\theta)\left\{\frac{1}{2}[\eta^{\mu\rho}\eta^{\nu\sigma}-\eta^{\mu\sigma}\eta^{\nu\rho}]+\frac{i}{2}\varepsilon^{\mu\nu\rho\sigma}\right\}\cdot F_{\mu\nu}(y)F_{\rho\sigma}(y)$$

(using (1.125))

$$= (\theta\theta)\left[-\frac{1}{4}F_{\mu\nu}(y)F^{\mu\nu}(y)+\frac{1}{4}F_{\mu\nu}(y)F^{\nu\mu}(y) - \frac{i}{4}\varepsilon^{\mu\nu\rho\sigma}F_{\mu\nu}(y)F_{\rho\sigma}(y)\right]$$
$$= (\theta\theta)\left[-\frac{1}{2}F_{\mu\nu}(y)F^{\mu\nu}(y)-\frac{i}{4}F_{\mu\nu}(y)\cdot \varepsilon^{\mu\nu\rho\sigma}F_{\rho\sigma}(y)\right]$$
$$= (\theta\theta)\left[-\frac{1}{2}F_{\mu\nu}(y)F^{\mu\nu}(y) - \frac{i}{2}F_{\mu\nu}(y)F^{*\mu\nu}(y)\right]$$

where $F^{\mu\nu *}$ is the dual field strength defined by

$$F^{\mu\nu*}(y) = \frac{1}{2}\varepsilon^{\mu\nu\rho\sigma}F_{\rho\sigma}(y) \qquad (8.37a)$$

Collecting all terms we obtain

$$W^A W_A$$
$$= \lambda^2(y) + 4D(y)\lambda(y)\theta + 2\lambda(y)\sigma^{\mu\nu}\theta F_{\mu\nu}(y)$$
$$+ (\theta\theta)\{4D^2(y) - 2i\lambda(y)\sigma^{\mu}\partial_{\mu}\bar{\lambda}(y)$$
$$- \frac{1}{2}F_{\mu\nu}(y)F^{\mu\nu}(y) - \frac{i}{2}F_{\mu\nu}(y)F^{\mu\nu*}(y)\}$$

We have thus obtained expression (8.37). We observe that the superfield $W^A W_A$ depends only on y and θ which is characteristic of a left-handed chiral superfield. We now derive the corresponding expansion for $\overline{W}_{\dot{A}}\overline{W}^{\dot{A}}$.

Proposition: The component field expansion of the scalar superfield $\overline{W}_{\dot{A}}\overline{W}^{\dot{A}}$ is given by

$$\overline{W}_{\dot{A}}\overline{W}^{\dot{A}} = \bar{\lambda}^2(z) + 4D(z)\bar{\lambda}(z)\bar{\theta}$$
$$- 2\bar{\lambda}(z)\bar{\sigma}^{\mu\nu}\bar{\theta}F_{\mu\nu}(z)$$
$$+ (\bar{\theta}\bar{\theta})\{4D^2(z) + 2i(\partial_{\mu}\lambda(z))\sigma^{\mu}\bar{\lambda}(z)$$
$$- \frac{1}{2}F_{\mu\nu}(z)F^{\mu\nu}(z)$$
$$+ \frac{i}{2}F_{\mu\nu}(z)*F^{\mu\nu}(z)\}$$

$$(8.38)$$

Proof: Using (7.53) we have
$$\overline{W}_{\dot{A}}\overline{W}^{\dot{A}} = \{\bar{\lambda}_{\dot{A}}(z) + 2D(z)\bar{\theta}_{\dot{A}} - \varepsilon_{\dot{A}\dot{B}}(\bar{\sigma}^{\mu\nu}\bar{\theta})^{\dot{B}}F_{\mu\nu}(z)$$
$$+ i(\bar{\theta}\bar{\theta})(\partial_{\mu}\lambda(z)\sigma^{\mu})_{\dot{A}}\}\cdot$$
$$\cdot\{\bar{\lambda}^{\dot{A}}(z) + 2D(z)\bar{\theta}^{\dot{A}} - (\bar{\sigma}^{\rho\sigma})^{\dot{A}}_{\ \dot{B}}\bar{\theta}^{\dot{B}}F_{\rho\sigma}(z)$$

$$-i(\bar{\theta}\bar{\theta})(\partial_\mu \lambda(x)\sigma^\mu)_{\dot{B}}\,\varepsilon^{\dot{B}\dot{A}}\}$$

$$= \bar{\lambda}^2(x) + 2D(x)\bar{\lambda}(x)\bar{\theta} - (\bar{\lambda}(x)\bar{\sigma}^{\mu\nu}\bar{\theta})F_{\mu\nu}(x)$$

$$-i(\bar{\theta}\bar{\theta})\bar{\lambda}_{\dot{A}}(x)(\partial_\mu \lambda(x)\sigma^\mu)_{\dot{B}}\,\varepsilon^{\dot{B}\dot{A}}$$

$$+ 2D(x)\bar{\theta}\bar{\lambda}(x) + 4(\bar{\theta}\bar{\theta})D^2(x)$$

$$-2D(x)(\bar{\theta}\bar{\sigma}^{\rho\sigma}\bar{\theta})F_{\rho\sigma}(x)$$

$$-\varepsilon_{\dot{A}\dot{B}}(\bar{\sigma}^{\mu\nu})^{\dot{B}}{}_{\dot{C}}\,\bar{\theta}^{\dot{C}}\bar{\lambda}^{\dot{A}}(x)F_{\mu\nu}(x)$$

$$-2\varepsilon_{\dot{A}\dot{B}}(\bar{\sigma}^{\mu\nu})^{\dot{B}}{}_{\dot{C}}\,\bar{\theta}^{\dot{C}}\bar{\theta}^{\dot{A}}D(x)F_{\mu\nu}(x)$$

$$+\varepsilon_{\dot{A}\dot{B}}(\bar{\sigma}^{\mu\nu})^{\dot{B}}{}_{\dot{C}}\,\bar{\theta}^{\dot{C}}(\bar{\sigma}^{\rho\sigma})^{\dot{A}}{}_{\dot{D}}\,\bar{\theta}^{\dot{D}}F_{\mu\nu}(x)F_{\rho\sigma}(x)$$

$$+ i(\bar{\theta}\bar{\theta})\,\partial_\mu \lambda(x)\sigma^\mu \bar{\lambda}(x)$$

Now

$$-i(\bar{\theta}\bar{\theta})\bar{\lambda}_{\dot{A}}(x)(\partial_\mu \lambda(x)\sigma^\mu)_{\dot{B}}\,\varepsilon^{\dot{B}\dot{A}}$$

$$= i(\bar{\theta}\bar{\theta})(\partial_\mu \lambda(x)\sigma^\mu)_{\dot{B}}\,\varepsilon^{\dot{B}\dot{A}}\bar{\lambda}_{\dot{A}}(x)$$

$$= i(\bar{\theta}\bar{\theta})(\partial_\mu \lambda(x)\sigma^\mu \bar{\lambda}(x))$$

and

$$\bar{\theta}\bar{\sigma}^{\mu\nu}\bar{\theta}$$

$$= \bar{\theta}_{\dot{A}}(\bar{\sigma}^{\mu\nu})^{\dot{A}}{}_{\dot{B}}\,\bar{\theta}^{\dot{B}}$$

$$= \bar{\theta}_{\dot{A}}(\bar{\sigma}^{\mu\nu})^{\dot{A}}{}_{\dot{B}}\,\varepsilon^{\dot{B}\dot{C}}\bar{\theta}_{\dot{C}}$$

$$= \bar{\theta}_{\dot{A}}\bar{\theta}_{\dot{C}}(\bar{\sigma}^{\mu\nu})^{\dot{A}}{}_{\dot{B}}\,\varepsilon^{\dot{B}\dot{C}}$$

$$= -\frac{1}{2}(\bar{\theta}\bar{\theta})\varepsilon_{\dot{A}\dot{C}}(\bar{\sigma}^{\mu\nu})^{\dot{A}}{}_{\dot{B}}\,\varepsilon^{\dot{B}\dot{C}}$$

(using (1.83d))

$$= \frac{1}{2}(\bar{\theta}\bar{\theta})(\bar{\sigma}^{\mu\nu})^{\dot{A}}{}_{\dot{B}}\,\varepsilon^{\dot{B}\dot{C}}\varepsilon_{\dot{C}\dot{A}}$$

$$= \frac{1}{2}(\bar{\theta}\bar{\theta})(\bar{\sigma}^{\mu\nu})^{\dot{A}}{}_{\dot{A}}$$

$$= 0$$

since

$$Tr[\bar{\sigma}^{\mu\nu}] = 0$$

Also

$$-\varepsilon_{\dot{A}\dot{B}}(\bar{\sigma}^{\mu\nu})^{\dot{B}}{}_{\dot{C}}\bar{\theta}^{\dot{C}}\bar{\lambda}^{\dot{A}}(x)F_{\mu\nu}(x)$$

$$= \bar{\lambda}^{\dot{A}}(x)\varepsilon_{\dot{A}\dot{B}}(\bar{\sigma}^{\mu\nu})^{\dot{B}}{}_{\dot{C}}\bar{\theta}^{\dot{C}}F_{\mu\nu}(x)$$

$$= -\varepsilon_{\dot{B}\dot{A}}\bar{\lambda}^{\dot{A}}(x)(\bar{\sigma}^{\mu\nu})^{\dot{B}}{}_{\dot{C}}\bar{\theta}^{\dot{C}}F_{\mu\nu}(x)$$

$$= -\bar{\lambda}(x)\bar{\sigma}^{\mu\nu}\bar{\theta}F_{\mu\nu}(x)$$

and

$$\varepsilon_{\dot{A}\dot{B}}(\bar{\sigma}^{\mu\nu})^{\dot{B}}{}_{\dot{C}}\bar{\theta}^{\dot{C}}(\bar{\sigma}^{\rho\sigma})^{\dot{A}}{}_{\dot{D}}\bar{\theta}^{\dot{D}}F_{\mu\nu}(x)F_{\rho\sigma}(x)$$

$$= \tfrac{1}{2}(\bar{\theta}\bar{\theta})\varepsilon_{\dot{A}\dot{B}}(\bar{\sigma}^{\mu\nu})^{\dot{B}}{}_{\dot{C}}\varepsilon^{\dot{C}\dot{D}}(\bar{\sigma}^{\rho\sigma})^{\dot{A}}{}_{\dot{D}}F_{\mu\nu}(x)F_{\rho\sigma}(x)$$

(using (1.83c))

$$= \tfrac{1}{2}(\bar{\theta}\bar{\theta})\varepsilon_{\dot{C}\dot{B}}(\bar{\sigma}^{\mu\nu})^{\dot{B}}{}_{\dot{A}}(\bar{\sigma}^{\rho\sigma})^{\dot{A}}{}_{\dot{D}}\varepsilon^{\dot{C}\dot{D}} \cdot$$
$$\cdot F_{\mu\nu}(x)F_{\rho\sigma}(x)$$

(using (1.127'))

$$= -\tfrac{1}{2}(\bar{\theta}\bar{\theta})(\bar{\sigma}^{\mu\nu}\bar{\sigma}^{\rho\sigma})^{\dot{B}}{}_{\dot{D}}\varepsilon^{\dot{D}\dot{C}}\varepsilon_{\dot{C}\dot{B}}F_{\mu\nu}(x)F_{\rho\sigma}(x)$$

$$= -\tfrac{1}{2}(\bar{\theta}\bar{\theta})Tr[\bar{\sigma}^{\mu\nu}\bar{\sigma}^{\rho\sigma}]F_{\mu\nu}(x)F_{\rho\sigma}(x)$$

$$= -\tfrac{1}{2}(\bar{\theta}\bar{\theta})\Big[\tfrac{1}{2}(\eta^{\mu\rho}\eta^{\nu\sigma} - \eta^{\mu\sigma}\eta^{\nu\rho})$$
$$- \tfrac{i}{2}\varepsilon^{\mu\nu\rho\sigma}\Big]F_{\mu\nu}(x)F_{\rho\sigma}(x)$$

(using (1.125') and (8.39) below)

$$= -\tfrac{1}{4}(\bar{\theta}\bar{\theta})[F_{\mu\nu}(x)F^{\mu\nu}(x) - F_{\mu\nu}(x)F^{\nu\mu}(x)$$
$$-i\varepsilon^{\mu\nu\rho\sigma}F_{\mu\nu}(x)F_{\rho\sigma}(x)]$$

$$= -\frac{1}{2}(\bar\theta\bar\theta)\,[F_{\mu\nu}(x)F^{\mu\nu}(x) - iF_{\mu\nu}(x)F^{\mu\nu *}(x)]$$

(with (8.37a))

In deriving the last expression we made use of the following formula.

<u>Proposition</u>: The following relation holds

$$Tr[\bar\sigma^{\mu\nu}\bar\sigma^{\rho\sigma}] = \frac{1}{2}[\eta^{\mu\rho}\eta^{\nu\sigma} - \eta^{\mu\sigma}\eta^{\nu\rho}] - \frac{i}{2}\varepsilon^{\mu\nu\rho\sigma} \tag{8.39}$$

<u>Proof</u>: This proof is similar to that of formula (1.125). Consider, using (1.119b),

$$Tr[\bar\sigma^{\mu\nu}\bar\sigma^{\rho\sigma}]$$
$$= -\frac{1}{16}Tr[(\bar\sigma^{\mu}\sigma^{\nu} - \bar\sigma^{\nu}\sigma^{\mu})(\bar\sigma^{\rho}\sigma^{\sigma} - \bar\sigma^{\sigma}\sigma^{\rho})]$$
$$= -\frac{1}{16}Tr[\bar\sigma^{\mu}\sigma^{\nu}\bar\sigma^{\rho}\sigma^{\sigma}] + \frac{1}{16}Tr[\bar\sigma^{\mu}\sigma^{\nu}\bar\sigma^{\sigma}\sigma^{\rho}]$$
$$+ \frac{1}{16}Tr[\bar\sigma^{\nu}\sigma^{\mu}\bar\sigma^{\rho}\sigma^{\sigma}] - \frac{1}{16}Tr[\bar\sigma^{\nu}\sigma^{\mu}\bar\sigma^{\sigma}\sigma^{\rho}]$$
$$= -\frac{1}{16}Tr[\sigma^{\sigma}\bar\sigma^{\mu}\sigma^{\nu}\bar\sigma^{\rho}] + \frac{1}{16}Tr[\sigma^{\rho}\bar\sigma^{\mu}\sigma^{\nu}\bar\sigma^{\sigma}]$$
$$+ \frac{1}{16}Tr[\sigma^{\sigma}\bar\sigma^{\nu}\sigma^{\mu}\bar\sigma^{\rho}] - \frac{1}{16}Tr[\sigma^{\rho}\bar\sigma^{\nu}\sigma^{\mu}\bar\sigma^{\sigma}]$$
$$= -\frac{1}{8}\{\eta^{\sigma\mu}\eta^{\nu\rho} + \eta^{\mu\nu}\eta^{\sigma\rho} - \eta^{\sigma\nu}\eta^{\mu\rho} - i\varepsilon^{\sigma\mu\nu\rho}\}$$
$$+ \frac{1}{8}\{\eta^{\rho\mu}\eta^{\nu\sigma} + \eta^{\mu\nu}\eta^{\rho\sigma} - \eta^{\rho\nu}\eta^{\mu\sigma} - i\varepsilon^{\rho\mu\nu\sigma}\}$$
$$+ \frac{1}{8}\{\eta^{\sigma\nu}\eta^{\mu\rho} + \eta^{\mu\nu}\eta^{\sigma\rho} - \eta^{\sigma\mu}\eta^{\nu\rho} - i\varepsilon^{\sigma\nu\mu\rho}\}$$

$$-\frac{1}{8}\{\eta^{\rho\nu}\eta^{\mu\sigma}+\eta^{\nu\mu}\eta^{\sigma\rho}-\eta^{\mu\rho}\eta^{\nu\sigma}-i\varepsilon^{\rho\nu\mu\sigma}\}$$

(using (1.126))

$$=\frac{1}{2}\eta^{\mu\rho}\eta^{\nu\sigma}-\frac{1}{2}\eta^{\mu\sigma}\eta^{\nu\rho}-\frac{i}{2}\varepsilon^{\mu\nu\rho\sigma}$$

as claimed by (8.39).

With the rewritten versions of the above expressions we then obtain

$$\begin{aligned}
&\overline{W}_{\dot{A}}\overline{W}^{\dot{A}}\\
&=\bar{\lambda}^2(z)+4D(z)\bar{\lambda}(z)\bar{\theta}\\
&\quad -2(\bar{\lambda}(z)\bar{\sigma}^{\mu\nu}\bar{\theta})F_{\mu\nu}(z)\\
&\quad +2i(\bar{\theta}\bar{\theta})(\partial_\mu\lambda(z)\sigma^\mu\bar{\lambda}(z))\\
&\quad +4(\bar{\theta}\bar{\theta})D^2(z)-\frac{1}{2}(\bar{\theta}\bar{\theta})[F_{\mu\nu}(z)F^{\mu\nu}(z)\\
&\qquad\qquad -iF_{\mu\nu}(z)F^{\mu\nu*}(z)]\\
&=\bar{\lambda}^2(z)+4D(z)\bar{\lambda}(z)\bar{\theta}\\
&\quad -2(\bar{\lambda}(z)\bar{\sigma}^{\mu\nu}\bar{\theta})F_{\mu\nu}(z)\\
&\quad +(\bar{\theta}\bar{\theta})\{4D^2(z)+2i\partial_\mu\lambda(z)\sigma^\mu\bar{\lambda}(z)\\
&\qquad -\frac{1}{2}F_{\mu\nu}(z)F^{\mu\nu}(z)\\
&\qquad +\frac{i}{2}F_{\mu\nu}(z)F^{\mu\nu*}(z)\}
\end{aligned}$$

This is the expression claimed by (8.38).

For the construction of Lagrangians it is important to observe that the $\theta\theta$ -component of $W^A W_A$ is invariant under the shift of coordinates

$$y^\mu = x^\mu + i\theta\sigma^\mu\bar{\theta} \longrightarrow x^\mu$$

Thus

$$\theta\theta\, W^A W_A\Big|_{\theta\theta}$$

$$= (\theta\theta)\big[4D^2(y) - 2i\lambda(y)\sigma^\mu\partial_\mu\bar{\lambda}(y) - \tfrac{1}{2}F_{\mu\nu}(y)F^{\mu\nu}(y) - \tfrac{i}{2}F_{\mu\nu}(y)F^{\mu\nu *}(y)\big]$$

$$= (\theta\theta)\big[4D^2(x + i\theta\sigma\bar{\theta}) - 2i\,\lambda(x + i\theta\sigma\bar{\theta})\sigma^\mu\partial_\mu\bar{\lambda}(x + i\theta\sigma\bar{\theta}) - \tfrac{1}{2}F_{\mu\nu}(x + i\theta\sigma\bar{\theta})F^{\mu\nu}(x + i\theta\sigma\bar{\theta}) - \tfrac{i}{2}F_{\mu\nu}(x + i\theta\sigma\bar{\theta})F^{\mu\nu *}(x + i\theta\sigma\bar{\theta})\big]$$

$$= (\theta\theta)\big[4D^2(x) - 2i\lambda(x)\sigma^\mu\partial_\mu\bar{\lambda}(x) - \tfrac{1}{2}F_{\mu\nu}(x)F^{\mu\nu}(x) - \tfrac{i}{2}F_{\mu\nu}(x)F^{\mu\nu *}(x)\big]$$

since a typical term in the expression can be expanded using (1.118) like

$$4(\theta\theta)D^2(x + i\theta\sigma\bar{\theta})$$

$$= 4(\theta\theta)\big[D(x)D(x) + D(x)\,i(\theta\sigma\bar{\theta})\partial_\mu D(x) - \tfrac{1}{4}D(x)(\theta\theta)(\bar{\theta}\bar{\theta})\Box D(x) + i(\theta\sigma^\mu\bar{\theta})\partial_\mu D(x).D(x) - \tfrac{1}{4}(\theta\theta)(\bar{\theta}\bar{\theta})\Box D(x).D(x)\big]$$

$$= 4(\theta\theta)\big[D^2(x) + 2iD(x)(\theta\sigma^\mu\bar{\theta})\partial_\mu D(x) - \tfrac{1}{2}(\theta\theta)(\bar{\theta}\bar{\theta})\Box D(x).D(x)\big]$$

$$= 4(\theta\theta)D^2(x)$$

since all powers of θ higher than the second vanish. The same invariance applies to the $\bar{\theta}\bar{\theta}$ -component of $\bar{W}_{\dot{A}}\bar{W}^{\dot{A}}$, where we can shift the coordinates according to

$$z^\mu = x^\mu - i(\theta\sigma^\mu\bar{\theta}) \longrightarrow x^\mu$$

without altering the $\bar{\theta}\bar{\theta}$-component, i.e.

$$(\bar{\theta}\bar{\theta})\bar{W}_{\dot{A}}\bar{W}^{\dot{A}}\Big|_{\bar{\theta}\bar{\theta}}$$

$$= (\bar{\theta}\bar{\theta})\{4D^2(z) + 2i\,\partial_\mu\lambda(z)\sigma^\mu\bar{\lambda}(z) - \tfrac{1}{2}F_{\mu\nu}(z)F^{\mu\nu}(z) + \tfrac{i}{2}F_{\mu\nu}(z)F^{\mu\nu *}(z)\}$$

$$= (\bar{\theta}\bar{\theta})\{4D^2(x) + 2i\,\partial_\mu\lambda(x)\sigma^\mu\bar{\lambda}(x) - \tfrac{1}{2}F_{\mu\nu}(x)F^{\mu\nu}(x) + \tfrac{i}{2}F_{\mu\nu}(x)F^{\mu\nu *}(x)\}$$

We can now construct the supersymmetric and gauge-invariant generalization of the action of a pure abelian gauge theory. As demonstrated at the end of Section 8.1, integration with respect to θ or $\bar{\theta}$ corresponds to projecting out the highest order component of the corresponding superfield. As stated above, the highest order components of $W^A W_A$ and $\bar{W}_{\dot{A}}\bar{W}^{\dot{A}}$ are independent of any particular representation (i.e. invariant under shifts of coordinates). We may therefore shift both scalar superfields to the x-representation and add the highest order components. This procedure gives the desired action

$$\mathcal{A} = \int d^4x \int d^4\theta \{W^A W_A \delta^2(\bar{\theta}) + \bar{W}_{\dot{A}}\bar{W}^{\dot{A}}\delta^2(\theta)\} \qquad (8.40)$$

In terms of its component fields this integral becomes

$$\mathcal{O}\!\iota = \int d^4x \,\{8\,D^2(x) - F_{\mu\nu}(x)F^{\mu\nu}(x) - 2i\lambda(x)\sigma^\mu\partial_\mu\bar{\lambda}(x) + 2i(\partial_\mu\lambda(x))\sigma^\mu\bar{\lambda}(x)\}$$

Integrating the last term by parts we find

$$\begin{aligned} &2i\int d^4x\,(\partial_\mu\lambda(x))\sigma^\mu\bar{\lambda}(x) \\ &= 2i\int d^4x\,[\partial_\mu(\lambda(x)\sigma^\mu\cdot\bar{\lambda}(x)) - \lambda(x)\sigma^\mu(\partial_\mu\bar{\lambda}(x))] \\ &= 2i\int d\overset{3}{S}_\mu\,\lambda(x)\sigma^\mu\bar{\lambda}(x) - 2i\int d^4x\,\lambda(x)\sigma^\mu\partial_\mu\bar{\lambda}(x) \end{aligned}$$

Assuming the fields λ , $\bar{\lambda}$ fall off sufficiently fast at infinity so that the surface integral vanishes we obtain

$$\mathcal{O}\!\iota = \int d^4x\,[8\,D^2(x) - F_{\mu\nu}(x)F^{\mu\nu}(x) - 4i\lambda(x)\sigma^\mu\partial_\mu\bar{\lambda}(x)] \qquad (8.41)$$

This is the action for a pure supersymmetric abelian gauge theory with the following properties:

a) The field D(x) is an auxiliary field which can be eliminated with the help of the equations of motion.
b) Supersymmetry requires a massless fermionic partner $\lambda(x)$ of the massless gauge boson $V_\mu(x)$ which is called the "gauge fermion" or "gaugino" or "photino". Thus the fermion field $\lambda(x)$ (in nonsupersymmetric contexts a matter field) now becomes part of the gauge field.

c) The invariance of (8.41) under supersymmetry transformations is manifest, since the integrand is a $\theta\theta$- or correspondingly $\bar{\theta}\bar{\theta}$ -component of a scalar superfield, which transforms under supersymmetry transformations into a total space-time derivative.
d) The invariance of (8.41) under gauge transformations (7.35) is also manifest since the action is a functional of the fields D(x), $F_{\mu\nu}(x) = \partial_\mu V_\nu(x) - \partial_\nu V_\mu(x)$ and λ (x) which are gauge invariant fields (cf. (7.37)).

<u>Remarks</u>: Knowing the general procedure for constructing supersymmetric Lagrangians it is straight-forward to construct other supersymmetric theories like e.g. supersymmetric quantum mechanics or supersymmetric nonlinear σ -models, the latter being of considerable interest in connection with superstrings. The Lagrangian density of a nonsupersymmetric nonlinear σ -model is given by

$$\mathcal{L} = -\frac{1}{2} g_{ij}(\phi)\, \partial_\mu \phi^i(x)\, \partial^\mu \phi^j(x)$$

where $\phi^i(x)$, i = 1,..., N denotes a set of N scalar fields and g_{ij} is the metric of a manifold. To obtain the supersymmetric version one replaces the scalar fields $\phi^i(x)$ by scalar superfields $\Phi^i(x,\theta)$. In order to obtain the component field expansion of the resulting superfield Lagrangian, one performs the θ -integration to project out the highest order component. Alternatively one can use the projection technique of Section 9.2. The resulting Lagrangian density is [53]

$$\begin{aligned}\mathcal{L} &= h_{ij*}(A,\bar{A})\, \partial_\mu A^i(x)\, \partial^\mu \bar{A}^j(x) \\ &\quad - \frac{i}{2} h_{ij*}(A,\bar{A})\, \psi^i(x)\, \sigma^\mu \overleftrightarrow{D}_\mu \bar{\psi}^j(x) \\ &\quad + \frac{1}{4} R_{j*ilk*}(A,\bar{A}) (\bar{\psi}^j(x)\bar{\psi}^k(x))(\psi^i(x)\psi^l(x))\end{aligned}$$

Here the scalar fields A^i of the supermultiplets $\Phi^i(x,\theta)$ and their conjugates $\bar{A}^i$ parametrize a complex manifold, h_{ij^*} are the components of the hermitian metric and $R_{j^*ilk^*}$ are the components of the curvature tensor. D_μ denotes a covariant derivative given by

$$D_\mu \psi^i = \partial_\mu \psi^i + \Gamma^i{}_{mk}\, \partial_\mu A^m \psi^k(x)$$

where $\Gamma^i{}_{mk}$ are the Christoffel symbols. As was shown by Zumino[54], in order to be able to describe a supersymmetric nonlinear σ -model, the complex manifold which is parametrized by the scalar fields A^i, must have an additional structure, i.e. the hermitian metric must obey the relations

$$\frac{\partial h_{ij^*}(A,\bar{A})}{\partial A^k} = \frac{\partial h_{kj^*}(A,\bar{A})}{\partial A^i}$$

$$\frac{\partial h_{ij^*}(A,\bar{A})}{\partial \bar{A}^k} = \frac{\partial h_{ik^*}(A,\bar{A})}{\partial \bar{A}^j}$$

Such a complex manifold is called a Kähler manifold. It was shown by Zumino[54] that there is a one-to-one correspondence between supersymmetric nonlinear σ -models and Kähler manifolds. These properties all depend on the dimension of the underlying Minkowski space (here 3 + 1). If one is interested in a 1 + 1 -dimensional supersymmetric field theory, one obtains a similar result; in this case the manifold is real and Riemannian. For a readable introduction into complex manifolds and Kähler geometry we refer to work by Alvarez-Gaumé and Freedman[55] and to related literature[56-58].

CHAPTER 9

SPONTANEOUS BREAKING OF SUPERSYMMETRY

9.1 The Superpotential

We now discuss some aspects of the socalled "superpotential" in relation to the spontaneous breaking of supersymmetry. Our motivation for investigating properties of the superpotential first is the usefulness of the latter in the study of specific models of spontaneous breaking of supersymmetry such as the models of O'Raifeartaigh[59] and Fayet and Iliopoulos[60].

We remarked in connection with (8.20) that the most general supersymmetric and renormalizable Lagrangian density which involves only scalar superfields is given by

$$\mathcal{L} = \mathcal{L}_{\text{kin}} + \mathcal{L}_{\text{pot}}$$

where

$$\mathcal{L}_{\text{kin}} = \Phi_i^{+} \Phi_i$$

is the kinetic term and

$$\mathcal{L}_{\text{pot}} = (g_i \Phi_i + \frac{1}{2} m_{ij} \Phi_i \Phi_j$$

$$+\frac{1}{3}\lambda_{ijk}\Phi_i\Phi_j\Phi_k)\delta^2(\bar\theta) \quad + \text{h.c.}$$

the potential term. We consider first the case of a theory with a single scalar superfield (as in the Wess-Zumino model). Then

$$\mathcal{L} = \Phi^+\Phi\Big|_{\theta\theta\bar\theta\bar\theta} + \Big[(g\Phi + \frac{1}{2}m\Phi^2 + \frac{1}{3}\lambda\Phi^3)\Big|_{\theta\theta} + \text{h.c.}\Big] \tag{9.1}$$

where the subscripts indicate that the integration with respect to θ, $\bar\theta$ removes the delta function leaving only the coefficient of the appropriate term of the power series expansion. Denoting $W[\Phi]$ by

$$W[\Phi] := g\Phi + \frac{1}{2}m\Phi^2 + \frac{1}{3}\lambda\Phi^3 \tag{9.2}$$

we can write the action in the form

$$\mathcal{A} = \int d^4x \int d^4\theta \{\Phi^+\Phi + [W[\Phi]\delta^2(\bar\theta) + \text{h.c.}]\} \tag{9.3}$$

The functional $W[\Phi]$ which is a polynomial of Φ is called the superpotential. The superpotential contains mass terms and interactions. The integral

$$\int d^4\theta\, W[\Phi]\delta^2(\bar\theta) = \int d^2\theta\, W[\Phi] \tag{9.4}$$

appearing in the action (9.3) projects out the highest order component of the superpotential and is manifestly supersymmetric, since this component always transforms into a space-time derivative. We now show that the integral (9.4) is given by

$$\int d^2\theta\, W[\Phi] = \frac{\partial W}{\partial A} F - \frac{1}{2}\frac{\partial^2 W}{\partial A^2}\psi\psi \qquad (9.5)$$

where

$$\frac{\partial W}{\partial A} := \frac{\partial W(A)}{\partial A}$$

Here W(A) is the superpotential (9.2) with Φ replaced by the first component A, the quantities A, ψ, F being the component fields of the scalar superfield Φ, i.e.

$$\Phi(y,\theta) = A(y) + 2^{\frac{1}{2}}\theta\psi(y) + \theta\theta F(y) \qquad (9.6)$$

The second term on the right hand side of (9.5) is a mass term of the fermionic part of the theory. A systematic derivation of (9.5) will be given in Section 9.2. We first demonstrate (9.5) explicitly for the case in which the superpotential is given by (9.2), i.e.

$$W[\Phi] = g\Phi + \frac{1}{2}m\Phi^2 + \frac{1}{3}\lambda\Phi^3$$

Then

$$\begin{aligned}
&\int d^2\theta\, W[\Phi] \\
&= g\Phi\Big|_{\theta\theta} + \frac{1}{2}m\Phi^2\Big|_{\theta\theta} + \frac{1}{3}\lambda\Phi^3\Big|_{\theta\theta} \\
&= gF(y) + \frac{1}{2}m[2F(y)A(y) - \psi(y)\psi(y)] \\
&\qquad + \frac{1}{3}\lambda[3A^2(y)F(y) - 3A(y)\psi(y)\psi(y)]
\end{aligned}$$

(using (7.5), (7.16) and (7.17))

$$\begin{aligned}
&= gF(y) + m[F(y)A(y) - \frac{1}{2}\psi^2(y)] \\
&\qquad + \lambda[A^2(y)F(y) - A(y)\psi^2(y)]
\end{aligned}$$

On the other hand

$$W(A) = gA(y) + \frac{1}{2} m A^2(y) + \frac{1}{3}\lambda A^3(y) \tag{9.6a}$$

and

$$\frac{\partial W(A)}{\partial A} = g + m A(y) + \lambda A^2(y) \tag{9.6b}$$

$$\frac{\partial^2 W(A)}{\partial A^2} = m + 2\lambda A(y) \tag{9.6c}$$

Then

$$\frac{\partial W(A)}{\partial A} F(y) - \frac{1}{2} \frac{\partial^2 W(A)}{\partial A^2} \psi^2(y)$$

$$= gF(y) + mA(y)F(y) + \lambda A^2(y)F(y)$$

$$- \frac{1}{2} m \psi^2(y) - \lambda \psi^2(y) A(y)$$

$$= gF(y) + m\left[F(y)A(y) - \frac{1}{2}\psi^2(y)\right]$$

$$+ \lambda\left[A^2(y)F(y) - \psi^2(y)A(y)\right]$$

$$= \int d^2\theta \, W[\Phi]$$

In general, if we have a set of n scalar superfields Φ_i, $i = 1, 2, \ldots, n$, and

$$W[\Phi] = W[\Phi_1, \ldots, \Phi_n]$$

then

$$\int d^4\theta \, W[\Phi_1, \ldots, \Phi_n] \, \delta^2(\bar{\theta})$$

$$= \int d^2\theta \, W[\Phi_1, \ldots, \Phi_n]$$

$$= W_i F_i - \frac{1}{2} W_{ij} \psi_i \psi_j \tag{9.7}$$

where

$$W_i := \frac{\partial W(A_1, \ldots, A_n)}{\partial A_i} \tag{9.8a}$$

$$W_{ij} := \frac{\partial^2 W(A_1, \ldots, A_n)}{\partial A_i \partial A_j} \tag{9.8b}$$

The matrix with elements W_{ij} is called the fermionic mass matrix. Setting

$$W[\Phi_1, \ldots, \Phi_n] := \sum_{i=1}^{n} g_i \Phi_i + \frac{1}{2}\sum_{i,j=1}^{n} m_{ij} \Phi_i \Phi_j + \frac{1}{3}\sum_{i,j,k=1}^{n} \lambda_{ijk} \Phi_i \Phi_j \Phi_k$$

and

$$\overline{W}[\Phi_1^+, \ldots, \Phi_n^+] := \sum_{i=1}^{n} g_i^* \Phi_i^+ + \frac{1}{2}\sum_{i,j=1}^{n} m_{ij}^* \Phi_i^+ \Phi_j^+ + \frac{1}{3}\sum_{i,j,k=1}^{n} \lambda^*_{ijk} \Phi_i^+ \Phi_j^+ \Phi_k^+ \tag{9.8c}$$

we can rewrite (8.21) in terms of superpotentials

$$\begin{aligned}\mathcal{O}\!\iota &= \int d^4x \, \Big\{ \int d^4\theta \, \Phi_i^+ \Phi_i + \int d^2\theta \, W[\Phi_1, \ldots, \Phi_n] + \int d^2\bar\theta \, \overline{W}[\Phi_1^+, \ldots, \Phi_n^+] \Big\} \\ &= \int d^4x \, \Big\{ \int d^4\theta \, \Phi_i^+ \Phi_i + W_i F_i - \frac{1}{2} W_{ij} \psi_i \psi_j + \overline{W}_i F_i^* - \frac{1}{2} \overline{W}_{ij} \overline{\psi}_i \overline{\psi}_j \Big\} \end{aligned} \tag{9.9}$$

(using (9.7))

$$= \int d^4x \, \{ i\partial_\mu \overline{\psi}_i \bar\sigma^\mu \psi_i - A_i^*(x) \Box A_i(x)$$

$$+ |F_i(x)|^2 + W_i F_i(x) + W_i^* F_i^*(x)$$

$$- \frac{1}{2} W_{ij} \psi_i(x) \psi_j(x) - \frac{1}{2} \overline{W}_{ij} \overline{\psi}_i(x) \overline{\psi}_j(x)$$

$$+ \text{total derivatives} \}$$

(using (7.27b))

The auxiliary fields $F_i(x)$ and $F_i^*(x)$ can be eliminated with the help of their respective Euler-Lagrange equations. Thus, since the action $\mathcal{O}\!\mathcal{L}$ contains no derivatives of $F_i(x)$, $F_i^*(x)$, we have

$$\frac{\partial \mathcal{L}}{\partial F_i^*(x)} = F_i(x) + W_i^* = 0$$
$$\frac{\partial \mathcal{L}}{\partial F_i(x)} = F_i^*(x) + W_i = 0 \qquad (9.10)$$

and we obtain

$$\mathcal{O}\!\mathcal{L} = \int d^4x \, [i \partial_\mu \overline{\psi}_i(x) \bar{\sigma}^\mu \psi_i(x) - A_i^*(x) \Box A_i(x)$$
$$- \frac{1}{2} W_{ij} \psi_i(x) \psi_j(x) - \frac{1}{2} \overline{W}_{ij} \overline{\psi}_i(x) \overline{\psi}_j(x)$$
$$- |W_i|^2]$$

where

$$|W_i|^2 = \sum_{i=1}^{n} W_i \overline{W}_i = \sum_{i=1}^{n} \left| \frac{\partial W}{\partial A_i} \right|^2 \quad \text{(with (9.8))}$$
$$= \sum_{i=1}^{n} |F_i(x)|^2 \quad \text{(using (9.10))}$$
$$= V(A, A^*) \quad \text{(using (8.32)}$$

Hence the scalar potential, i.e. that part of the Lagrangian density which does not contain any derivative or fermionic field, is related to the superpotential W by

$$V(A,A^*) = \sum_{i=1}^{n} \left| \frac{\partial W}{\partial A_i} \right|^2 \tag{9.11}$$

In the particular case of a single scalar field Φ (i.e. in the case of the Wess-Zumino model) V (A, A^*) is obtained by inserting (9.6b) into (9.11), i.e.

$$V(A,A^*) = |g + mA(y) + \lambda A^2(y)|^2$$

We observe that the scalar potential is always greater than or equal to zero.

Now the invariance of the theory (i.e. the action) with respect to supersymmetry transformations implies that the Hamiltonian H commutes with the generators of the supersymmetry transformation, i.e.

$$[H, Q_A] = 0 = [H, \bar{Q}_{\dot{A}}]$$

The ground state $|0\rangle$ satisfies $H|0\rangle = 0$. (We see from (5.109) that this is trivially satisfied in the Wess-Zumino model). Thus besides $H|0\rangle = 0$, also $Q_A|0\rangle = 0 = \bar{Q}_{\dot{A}}|0\rangle$ and the ground state is invariant under supersymmetry transformations. The energy of the ground state, i.e. the eigenvalue of $H|0\rangle$, is zero, i.e. minimized, if the scalar potential (9.11) which is positive definite is zero. Thus in a theory constructed from chiral scalar superfields Φ_i , supersymmetry is unbroken if and only if the scalar potential (9.11) is zero and hence

$$\frac{\partial W}{\partial A_i} = 0 \tag{9.12}$$

for all i = 1,..., n at the minimum. Hence a nonvanishing derivative $\partial W/\partial A_i$ for some index i at the absolute minimum, signals the breaking of supersymmetry.

9.2 Projection Technique

Before we discuss supersymmetry breaking in more detail we develop a projection technique which is a useful tool for evaluating integrals over Grassmann variables. This technique is based on the fact that Grassmann integration is equivalent to Grassmann differentiation. We made already explicit use of this method in evaluating the Grassmann integral of the superpotential (see (9.5)).

We start by considering a chiral superfield

$$\Phi(y,\theta) = A(y) + 2^{\frac{1}{2}}\,\theta\psi(y) + (\theta\theta)F(y) \tag{9.13a}$$

which obeys the constraint

$$\bar{D}^{(1)}_{\dot{A}}(y)\,\Phi(y,\theta) = 0 \tag{9.13b}$$

We show that one can obtain the component fields of this supermultiplet by the following projection operations:

$$A(x) = \Phi(y,\theta)\Big|_{\theta=\bar{\theta}=0} = \Phi(x,\theta,\bar{\theta})\Big|_{\theta=\bar{\theta}=0} \tag{9.14a}$$

$$\begin{aligned}\psi_A(x) &= \frac{1}{2^{1/2}} D^{(1)}_A\,\Phi(y,\theta)\Big|_{\theta=\bar{\theta}=0} \\ &= \frac{1}{2^{1/2}} D_A\,\Phi(x,\theta,\bar{\theta})\Big|_{\theta=\bar{\theta}=0}\end{aligned} \tag{9.14b}$$

$$\begin{aligned}F(x) &= -\frac{1}{4} D^{(1)\,2}\,\Phi(y,\theta)\Big|_{\theta=\bar{\theta}=0} \\ &= -\frac{1}{4} D^2\,\Phi(x,\theta,\bar{\theta})\Big|_{\theta=\bar{\theta}=0}\end{aligned} \tag{9.14c}$$

where (cf. (7.2))

$$y_\mu = x_\mu + i\theta\sigma_\mu\bar{\theta}$$

The projections given by the relations (9.14a,b,c) are seen to be independent of the particular field representation $\Phi(y,\theta)$ or $\Phi(x,\theta,\bar{\theta})$ since the change of representation is caused by this shift of variables.

We now verify the relations (9.14a,b,c).

i) Setting $\theta=\bar{\theta}=0$ in (9.13a) yields A(x) trivially.

ii) Consider, using (6.47),

$$\frac{1}{2^{1/2}}\mathcal{D}^{(1)}_A\Phi(y,\theta)$$

$$=\frac{1}{2^{1/2}}\left(\partial_A+2i(\sigma^\mu\bar{\theta})_A\partial_\mu\right)\Phi(y,\theta)\Big|_{\theta=\theta=0}$$

$$=\frac{1}{2^{1/2}}\partial_A\left(A(y)+2^{\frac{1}{2}}\theta^B\psi_B(y)+\theta\theta F(y)\right)\Big|_{\theta=\bar{\theta}=0}$$

$$+2^{\frac{1}{2}}i(\sigma^\mu\bar{\theta})_A\left[\partial_\mu A(y)+2^{\frac{1}{2}}\theta^B\partial_\mu\psi_B(y)\right.$$

$$\left.+\theta\theta\partial_\mu F(y)\right]\Big|_{\theta=\bar{\theta}=0}$$

$$=\left(\psi_A(y)+2^{\frac{1}{2}}\theta_A F(y)\right)\Big|_{\theta=\bar{\theta}=0}$$

(using (6.4f) and (6.4p))

$$=\psi_A(x)$$

iii) From (6.47) we have

$$\mathcal{D}^{(1)A}=\varepsilon^{AB}\mathcal{D}^{(1)}_B$$

$$=-\partial^A+2i(\varepsilon\sigma^\mu\bar{\theta})^A\partial_\mu$$

(using (6.4g))

Then

$$\mathcal{D}^{(1)2}=\mathcal{D}^{(1)A}\mathcal{D}^{(1)}_A$$

$$=\left(-\partial^A+2i(\varepsilon\sigma^\mu\bar{\theta})^A\partial_\mu\right)\left(\partial_A+2i(\sigma^\nu\bar{\theta})_A\partial_\nu\right)$$

$$= -\partial^A\partial_A + 2i(\sigma^\nu\bar\theta)_A\partial_\nu\partial^A + 2i(\varepsilon\sigma^\mu\bar\theta)^A\partial_\mu\partial_A$$
$$-4(\varepsilon\sigma^\mu\bar\theta)^A(\sigma^\nu\bar\theta)_A\partial_\mu\partial_\nu$$

Now

$$(\varepsilon\sigma^\mu\bar\theta)^A(\sigma^\nu\bar\theta)_A\partial_\mu\partial_\nu$$
$$= \varepsilon^{AB}\sigma^\mu_{B\dot B}\bar\theta^{\dot B}\sigma^\nu_{A\dot C}\bar\theta^{\dot C}\partial_\mu\partial_\nu$$
$$= \tfrac{1}{2}(\bar\theta\bar\theta)\varepsilon^{\dot B\dot C}\varepsilon^{AB}\sigma^\mu_{B\dot B}\sigma^\nu_{A\dot C}\partial_\mu\partial_\nu$$

(using (1.83c))

$$= -\tfrac{1}{2}(\bar\theta\bar\theta)\,\mathrm{Tr}[\bar\sigma^\mu\sigma^\nu]\partial_\mu\partial_\nu$$

(using (1.88a))

$$= -(\bar\theta\bar\theta)\Box \quad \text{(using (1.91))}$$

and

$$2i(\sigma^\nu\bar\theta)_A\partial_\nu\partial^A + 2i(\varepsilon\sigma^\mu\bar\theta)^A\partial_\mu\partial_A$$
$$= 2i\sigma^\nu_{A\dot B}\bar\theta^{\dot B}\partial_\nu\partial^A + 2i(\varepsilon\sigma^\mu\bar\theta)^A\partial_\mu\partial_A$$
$$= 2i\delta_A{}^B\sigma^\nu_{B\dot B}\bar\theta^{\dot B}\partial_\nu\partial^A + 2i(\varepsilon\sigma^\mu\bar\theta)^A\partial_\mu\partial_A$$
$$= 2i\varepsilon_{AC}\varepsilon^{CB}\sigma^\nu_{B\dot B}\bar\theta^{\dot B}\partial_\nu\partial^A + 2i(\varepsilon\sigma^\mu\bar\theta)^A\partial_\mu\partial_A$$
$$= 2i\varepsilon^{CB}\sigma^\nu_{B\dot B}\bar\theta^{\dot B}\partial_\nu\partial_C + 2i(\varepsilon\sigma^\mu\bar\theta)^A\partial_\mu\partial_A$$
$$= 2i(\varepsilon\sigma^\mu\bar\theta)^A\partial_\mu\partial_A + 2i(\varepsilon\sigma^\mu\bar\theta)^A\partial_\mu\partial_A$$

(using (6.4g))

$$= 4i(\varepsilon\sigma^\mu\bar\theta)^A\partial_\mu\partial_A$$

Hence

$$D^{(1)2} = -\partial^A\partial_A + 4i(\varepsilon\sigma^\mu\bar\theta)^A\partial_\mu\partial_A + 4(\bar\theta\bar\theta)\Box\,.$$

Then the last component of the supermultiplet is obtained with

$$\mathcal{D}^{(1)2}\,\Phi(y,\theta)\Big|_{\theta=\bar\theta=0}$$

$$= -\partial^A\partial_A\left[A(y)+2^{\frac{1}{2}}\theta\psi(y)+\theta\theta F(y)\right]\Big|_{\theta=\bar\theta=0}$$

$$+4i(\varepsilon\sigma^\mu\bar\theta)^A\partial_\mu\partial_A\Phi(y,\theta)\Big|_{\theta=\bar\theta=0}$$

$$+4(\bar\theta\bar\theta)\Box\Phi(y,\theta)\Big|_{\theta=\bar\theta=0}$$

(with (9.13a) and (9.15))

$$= \varepsilon^{AB}\partial_B\partial_A\,\theta\theta F(y)\Big|_{\theta=0} \quad \text{(with (6.4g))}$$

$$= -\varepsilon^{AB}\partial_A\partial_B\,\theta\theta F(y)\Big|_{\theta=0} \quad \text{(with (6.4i))}$$

$$= -4F(x) \quad \text{(with (6.4q))}$$

Hence

$$F(x) = -\frac{1}{4}\mathcal{D}^{(1)2}\,\Phi(y,\theta)\Big|_{\theta=\bar\theta=0}$$

which had to be shown.

From Grassmann integration we know that the integral $\int d^2\theta$ projects out the $\theta\theta$ -component of any function integrated over; in particular for the chiral superfield $\Phi(y,\theta)$ we have

$$\int d^2\theta\,\Phi(y,\theta) = F(x) \tag{9.16}$$

Thus, comparing (9.14) and (9.16) we have formally

$$\int d^2\theta = -\frac{1}{4}\mathcal{D}^{(1)2}\Big|_{\theta=\bar\theta=0} = -\frac{1}{4}D^2\Big|_{\theta=\bar\theta=0} \tag{9.17}$$

We can now derive the expression (9.5) for the superpotential by using the above procedure. Thus

$$\int d^2\theta \, W[\Phi]$$

$$= -\frac{1}{4} D^{(1)2} W[\Phi]\Big|_{\theta=\bar{\theta}=0}$$

$$= -\frac{1}{4} D^{(1)A} D^{(1)}{}_A W[\Phi]\Big|_{\theta=\bar{\theta}=0}$$

$$= -\frac{1}{4} D^{(1)A}\left[\frac{\partial W[\Phi]}{\partial \Phi} D_A{}^{(1)} \Phi\right]\Big|_{\theta=\bar{\theta}=0}$$

$$= -\frac{1}{4}\left[\frac{\partial^2 W[\Phi]}{\partial\Phi\partial\Phi} D^{(1)A}\Phi D_A{}^{(1)}\Phi + \frac{\partial W[\Phi]}{\partial\Phi} D^{(1)2}\Phi\right]\Big|_{\theta=\bar{\theta}=0}$$

$$= -\frac{1}{4}\left[2\,\frac{\partial^2 W[A]}{\partial A^2}\psi^A\psi_A - 4\,\frac{\partial W[A]}{\partial A}F\right]$$

$$= \frac{\partial W[A]}{\partial A}F - \frac{1}{2}\frac{\partial^2 W[A]}{\partial A^2}\psi\psi \qquad (9.18)$$

It is now easy to extend these calculations to the case of n chiral superfields (cf. (9.7)), i.e.

$$\int d^2\theta \, W[\Phi_1, \ldots, \Phi_n]$$

$$= -\frac{1}{4} D^{(1)2} W[\Phi_1, \ldots, \Phi_n]\Big|_{\theta=\bar{\theta}=0}$$

$$= -\frac{1}{4} D^{(1)A}\left[\frac{\partial W[\Phi_1, \ldots, \Phi_n]}{\partial \Phi_i} D_A^{(1)} \Phi_i\right]\Big|_{\theta=\bar{\theta}=0}$$

$$= -\frac{1}{4}\Big[\sum_{i,j=1}^{n} \frac{\partial^2 W[\Phi_1,\ldots,\Phi_n]}{\partial\Phi_i\,\partial\Phi_j}(D^{(1)A}\Phi^i)\cdot$$
$$\cdot(D_A^{(1)}\Phi^j)$$
$$+\sum_{i=1}^{n}\frac{\partial W[\Phi_1,\ldots,\Phi_n]}{\partial\Phi_i}D^{(1)2}\Phi_i\Big]\Big|_{\theta=\bar\theta=0}$$

$$= \sum_{i=1}^{n}\frac{\partial W[A]}{\partial A_i}F_i(x) - \frac{1}{2}\sum_{i,j=1}^{n}\frac{\partial^2 W[A]}{\partial A_i\,\partial A_j}\psi_i(x)\psi_j(x) \tag{9.19}$$

For an antichiral (i.e. right-handed) superfield we have the expansion

$$\Phi^+(z,\bar\theta) = A^*(z) + 2^{\frac{1}{2}}\bar\theta\bar\psi(z) + \bar\theta\bar\theta F^*(z) \tag{9.20a}$$

obeying the defining constraint equation

$$D_A\Phi^+ = 0 \tag{9.20b}$$

In much the same way as (9.14) we obtain the component fields of the antichiral supermultiplet by the projections

$$A^*(x) = \Phi^+(z,\bar\theta)\Big|_{\theta=\bar\theta=0} = \Phi^+(x,\theta,\bar\theta)\Big|_{\theta=\bar\theta=0} \tag{9.21a}$$

$$\bar\psi_{\dot A}(x) = \frac{1}{2^{1/2}}\bar D^{(2)}_{\dot A}\Phi^+(z,\bar\theta)\Big|_{\theta=\bar\theta=0}$$
$$= \frac{1}{2^{1/2}}\bar D_{\dot A}\Phi^+(x,\theta,\bar\theta)\Big|_{\theta=\bar\theta=0} \tag{9.21b}$$

$$F^*(x) = -\frac{1}{4}(\bar D^{(2)})^2\Phi^+(z,\bar\theta)\Big|_{\theta=\bar\theta=0}$$
$$= -\frac{1}{4}\bar D^2\Phi^+(x,\theta,\bar\theta)\Big|_{\theta=\bar\theta=0} \tag{9.21c}$$

where from (7.10)

$$z^\mu = x^\mu - i\theta\sigma^\mu\bar{\theta}$$

We now verify the relations (9.21a,b,c).

i) Obviously, setting $\theta = \bar{\theta} = 0$ in the component field expansion of the antichiral superfield $\Phi^+(z,\bar{\theta})$ gives

$$A^*(x) = \Phi^+(z,\bar{\theta})\Big|_{\theta=\bar{\theta}=0}$$

ii) Consider, using (6.50),

$$\frac{1}{2^{1/2}}\bar{D}^{(2)}_{\dot{A}}\Phi^+(z,\bar{\theta})\Big|_{\theta=\bar{\theta}=0}$$

$$= \frac{1}{2^{1/2}}\left[-\bar{\partial}_{\dot{A}} - 2i(\theta\sigma^\mu)_{\dot{A}}\partial_\mu\right]\Phi^+(z,\bar{\theta})\Big|_{\theta=\bar{\theta}=0}$$

$$= -\frac{1}{2^{1/2}}\bar{\partial}_{\dot{A}}\left[A^*(z) + 2^{\frac{1}{2}}\bar{\theta}\bar{\Psi}(z) + \bar{\theta}\bar{\theta}F^*(x)\right]\Big|_{\theta=\bar{\theta}=0}$$

$$- 2^{\frac{1}{2}}i(\theta\sigma^\mu)_{\dot{A}}\partial_\mu\Phi^+(z,\bar{\theta})\Big|_{\theta=\bar{\theta}=0}$$

$$= \left[-\bar{\partial}_{\dot{A}}\bar{\theta}_{\dot{B}}\bar{\Psi}^{\dot{B}}(z) - \frac{1}{2^{1/2}}\bar{\partial}_{\dot{A}}(\bar{\theta}\bar{\theta})F^*(z)\right]\Big|_{\theta=\bar{\theta}=0}$$

$$= \left[\varepsilon_{\dot{A}\dot{B}}\bar{\Psi}^{\dot{B}}(z) + 2^{\frac{1}{2}}\bar{\theta}_{\dot{A}}F^*(z)\right]\Big|_{\theta=\bar{\theta}=0}$$

(using (6.4k))

$$= \bar{\Psi}_{\dot{A}}(x)$$

Before we demonstrate the last relation we verify the following expression

$$\bar{D}^{(2)2} = \bar{D}^{(2)}_{\dot{A}}\bar{D}^{(2)\dot{A}}$$

$$= -\bar{\partial}\bar{\partial} - 4i(\theta\sigma^{\mu}\bar{\partial})\partial_{\mu} + 4\theta\theta\Box \qquad (9.22)$$

In order to prove this we recall that (according to (6.41) rewritten with the help of (1.88))

$$\bar{D}^{(2)\dot{A}} = \bar{\partial}^{\dot{A}} + 2i(\theta\sigma^{\mu}\bar{\varepsilon})^{\dot{A}}\partial_{\mu}$$

Hence

$$\bar{D}^{(2)}_{\dot{A}}\bar{D}^{(2)\dot{A}}$$

$$= \left(-\bar{\partial}_{\dot{A}} - 2i(\theta\sigma^{\mu})_{\dot{A}}\partial_{\mu}\right)\left(\bar{\partial}^{\dot{A}} + 2i(\theta\sigma^{\nu}\varepsilon)^{\dot{A}}\partial_{\nu}\right)$$

(using (6.50))

$$= -\bar{\partial}_{\dot{A}}\bar{\partial}^{\dot{A}} - 2i(\theta\sigma^{\mu}\bar{\partial})\partial_{\mu} - 2i\,\bar{\partial}_{\dot{A}}(\theta\sigma^{\mu}\bar{\varepsilon})^{\dot{A}}\partial_{\mu} + 4(\theta\sigma^{\mu})_{\dot{A}}(\theta\sigma^{\nu}\bar{\varepsilon})^{\dot{A}}\partial_{\mu}\partial_{\nu}$$

$$= -\bar{\partial}\bar{\partial} - 2i(\theta\sigma^{\mu}\bar{\partial})\partial_{\mu} + 2i(\theta\sigma^{\mu})_{\dot{B}}\varepsilon^{\dot{B}\dot{A}}\bar{\partial}_{\dot{A}}\partial_{\mu} + 4\theta^{A}\theta^{B}\sigma^{\mu}_{A\dot{A}}\sigma^{\nu}_{B\dot{B}}\varepsilon^{\dot{B}\dot{A}}\partial_{\mu}\partial_{\nu}$$

$$= -\bar{\partial}\bar{\partial} - 2i(\theta\sigma^{\mu}\bar{\partial})\partial_{\mu} - 2i(\theta\sigma^{\mu}\bar{\partial})\partial_{\mu} + 2(\theta\theta)\varepsilon^{AB}\varepsilon^{\dot{A}\dot{B}}\sigma^{\mu}_{A\dot{A}}\sigma^{\nu}_{B\dot{B}}\partial_{\mu}\partial_{\nu}$$

(using (6.4j) and (1.83a))

$$= -\bar{\partial}\bar{\partial} - 4i(\theta\sigma^{\mu}\bar{\partial})\partial_{\mu} + 2(\theta\theta)\mathrm{Tr}(\sigma^{\mu}\bar{\sigma}^{\nu})\partial_{\mu}\partial_{\nu}$$

(using (1.88a))

$$= -\bar{\partial}\bar{\partial} - 4i(\theta\sigma^{\mu}\bar{\partial})\partial_{\mu} + 4(\theta\theta)\Box$$

(using (1.91))

Hence, with (9.22) we obtain

$$\bar{D}^{(2)2}\Phi^{+}(z,\bar{\theta})\Big|_{\theta=\bar{\theta}=0}$$

$$= (-\bar\partial\bar\partial - 4i(\theta\sigma^\nu\bar\partial)\partial_\mu + 4(\theta\theta)\Box)\Phi^+(z,\theta)\Big|_{\theta=\bar\theta=0}$$

$$= -\bar\partial\bar\partial\,\Phi^+(z,\bar\theta)\Big|_{\theta=\bar\theta=0}$$

$$= -\bar\partial\bar\partial\left[A^*(x) + 2^{\frac{1}{2}}\bar\theta\bar\Psi(x) + \bar\theta\bar\theta F^*(x)\right]\Big|_{\theta=\bar\theta=0}$$

$$= -\bar\partial\bar\partial(\bar\theta\bar\theta)F^*(x)\Big|_{\theta=\bar\theta=0}$$

$$= \varepsilon_{\dot A\dot B}\,\bar\partial^{\dot B}\bar\partial^{\dot A}(\bar\theta\bar\theta)F^*(x)\Big|_{\theta=\bar\theta=0}$$

(using (6.4j))

$$= -\varepsilon_{\dot A\dot B}\,\bar\partial^{\dot A}\bar\partial^{\dot B}(\bar\theta\bar\theta)F^*(x)\Big|_{\theta=\bar\theta=0}$$

$$= -4F^*(x) \quad \text{(with (6.4q))}$$

As in the case of chiral superfields we can project out the component field of the highest order term in two ways, either by evaluation of the Grassmann integral

$$\int d^2\bar\theta\,\Phi^+(z,\bar\theta) = F^*(x) \tag{9.23a}$$

or by differentiation as above, i.e.

$$-\frac{1}{4}\bar D^{(2)^2}\Phi^+(z,\theta) = F^*(x)$$

From these equivalent procedures we deduce by comparison of (9.21) and (9.23a) the formal relation

$$\int d^2\bar\theta = -\frac{1}{4}\bar D^{(2)^2}\Big|_{\theta=\bar\theta=0} = -\frac{1}{4}\bar D^2\Big|_{\theta=\bar\theta=0} \tag{9.23b}$$

In the following we write simply $\Big|_0$ for $\Big|_{\theta=\bar\theta=0}$.

The projection technique is not restricted to the evaluation of the Grassmann integral of the superpotential but can also be applied to other cases and is particularly useful for finding the component expansion of supersymmetric actions formulated in terms of chiral and antichiral superfields. Consider an arbitrary superfield with the power series expansion (6.6). From the general formalism of Grassmann integration we know that $\int d^4\theta$ projects out the $(\theta\theta)(\bar{\theta}\bar{\theta})$ component of this superfield, i.e. (cf. (6.6) and (8.18))

$$\int d^4\theta\, \Phi(x,\theta,\bar{\theta}) = d(x) \tag{9.24}$$

Alternatively this component can be obtained by the application of the operator $\frac{1}{16}\bar{D}_{\dot{A}} D^2 \bar{D}^{\dot{A}}$

to the superfield, i.e.

$$\frac{1}{16}\bar{D}_{\dot{A}} D^2 \bar{D}^{\dot{A}} \Phi(x,\theta,\bar{\theta})\Big|_0 = d(x) \tag{9.25}$$

which again yields d(x). In order to verify (9.25) we consider

$$\frac{1}{16}\bar{D}_{\dot{A}} D^2 \bar{D}^{\dot{A}} \Phi(x,\theta,\bar{\theta})\Big|_0$$

$$= \frac{1}{16}\bar{D}_{\dot{A}} D^A D_A \bar{D}^{\dot{A}} \Phi(x,\theta,\bar{\theta})\Big|_0$$

$$= \frac{1}{16}(-\bar{\partial}_{\dot{A}})(-\partial^A)(\partial_A)(\bar{\partial}^{\dot{A}}) \Phi(x,\theta,\bar{\theta})\Big|_0$$

(using (6.40) and (6.41) and ignoring other terms which vanish when evaluated at $\theta = \bar{\theta} = 0$)

$$= \frac{1}{16}\partial^A \partial_A \bar{\partial}_{\dot{A}} \bar{\partial}^{\dot{A}} \Phi(x,\theta,\bar{\theta})\Big|_0$$

$$= \frac{1}{16} \varepsilon^{AB} \partial_B \partial_A \varepsilon_{\dot{A}\dot{B}} \bar{\partial}^{\dot{B}} \bar{\partial}^{\dot{A}} \Phi(x,\theta,\bar{\theta})\Big|_0$$

(using (6.4g) and (6.4j))

$$= \frac{1}{16} \varepsilon^{AB} \partial_B \partial_A \varepsilon_{\dot{A}\dot{B}} \bar{\partial}^{\dot{B}} \bar{\partial}^{\dot{A}} (\theta\theta)(\bar{\theta}\bar{\theta}) d(x)\Big|_0$$

(using (6.6))

$$= d(x) \quad \text{(with (6.4q))}$$

Hence, comparing (9.24) and (9.25) we see that formally

$$\int d^4\theta = \frac{1}{16} \bar{D}_{\dot{A}} D^2 \bar{D}^{\dot{A}}\Big|_0 \tag{9.26}$$

With this correspondence we can rederive the kinetic part of the action (8.30) in a more elegant way. Consider (using (9.26))

$$\int d^4\theta\, \Phi^+ \Phi$$

$$= \frac{1}{16} \bar{D}_{\dot{A}} D^2 \bar{D}^{\dot{A}} \Phi^+ \Phi\Big|_0$$

$$= \frac{1}{16} \bar{D}_{\dot{A}} D^2 \{(\bar{D}^{\dot{A}} \Phi^+)\Phi + \Phi^+(\bar{D}^{\dot{A}} \Phi)\}\Big|_0$$

$$= \frac{1}{16} \bar{D}_{\dot{A}} D^A D_A \{(\bar{D}^{\dot{A}} \Phi^+)\Phi\}\Big|_0 \quad \text{(with (9.13b))}$$

$$= \frac{1}{16} \bar{D}_{\dot{A}} D^A \{(D_A \bar{D}^{\dot{A}} \Phi^+)\Phi - (\bar{D}^{\dot{A}} \Phi^+)(D_A \Phi)\}\Big|_0$$

$$= \frac{1}{16} \bar{D}_{\dot{A}} \Big[(D^2 \bar{D}^{\dot{A}} \Phi^+)\Phi + (D_A \bar{D}^{\dot{A}} \Phi^+)(D^A \Phi) - (D^A \bar{D}^{\dot{A}} \Phi^+)(D_A \Phi) + (\bar{D}^{\dot{A}} \Phi^+)(D^2 \Phi)\Big]\Big|_0$$

$$= \frac{1}{16} \{(\bar{D}_{\dot{A}} D^2 \bar{D}^{\dot{A}} \Phi^+)\Phi - (D^2 \bar{D}^{\dot{A}} \Phi^+)(\bar{D}_{\dot{A}} \Phi) + (\bar{D}_{\dot{A}} D_A \bar{D}^{\dot{A}} \Phi^+)(D^A \Phi) + (D_A \bar{D}^{\dot{A}} \Phi^+)(\bar{D}_{\dot{A}} D^A \Phi)$$

$$-(\bar{D}_{\dot{A}} D^A \bar{D}^{\dot{A}} \Phi^+)(D_A \Phi)$$
$$-(D^A \bar{D}^{\dot{A}} \Phi^+)(\bar{D}_{\dot{A}} D_A \Phi)$$
$$+(\bar{D}^2 \Phi^+)(D^2 \Phi) - (\bar{D}^{\dot{A}} \Phi^+)(\bar{D}_{\dot{A}} D^2 \Phi)\}\Big|_0$$

$$= \frac{1}{16}\{(D^A \bar{D}^2 D_A \Phi^+)\Phi$$
$$+ \bar{D}_{\dot{A}}(-\bar{D}^{\dot{A}} D_A + 2i(\sigma^\mu \bar{\varepsilon})_A{}^{\dot{A}} \partial_\mu)\Phi^+(D^A \Phi)$$
$$+(-\bar{D}^{\dot{A}} D_A + 2i(\sigma^\mu \bar{\varepsilon})_A{}^{\dot{A}} \partial_\mu)\Phi^+ \cdot$$
$$\cdot(-D^A \bar{D}_{\dot{A}} - 2i(\varepsilon\sigma^\nu)^A{}_{\dot{A}} \partial_\nu)\Phi$$
$$-[\bar{D}_{\dot{A}}(-\bar{D}^{\dot{A}} D^A - 2i(\bar{\sigma}^\mu)^{\dot{A}A} \partial_\mu)\Phi^+](D_A \Phi)$$
$$-[(-\bar{D}^{\dot{A}} D^A - 2i(\bar{\sigma}^\mu)^{\dot{A}A} \partial_\mu)\Phi^+] \cdot$$
$$\cdot[(-D_A \bar{D}_{\dot{A}} - 2i\sigma^\nu_{A\dot{A}} \partial_\nu)\Phi]$$
$$+(\bar{D}^2 \Phi^+)(D^2 \Phi)$$
$$-(\bar{D}^{\dot{A}} \Phi^+)(D^2 \bar{D}_{\dot{A}} + 4i D^A \sigma^\rho_{A\dot{A}} \partial_\rho)\Phi\}\Big|_0$$

(using (9.13b), (6.65), (7.1), (6.53), (6.58))

$$= \frac{1}{16}\{2i(\sigma^\mu \bar{\varepsilon})_A{}^{\dot{A}} \partial_\mu(\bar{D}_{\dot{A}} \Phi^+)(D^A \Phi)$$
$$+4(\sigma^\mu \bar{\varepsilon})_A{}^{\dot{A}}(\partial_\mu \Phi^+)(\varepsilon\sigma^\nu)^A{}_{\dot{A}}(\partial_\nu \Phi)$$
$$+2i\bar{D}_{\dot{A}}(\bar{\sigma}^\mu)^{\dot{A}A}(\partial_\mu \Phi^+)(D_A \Phi)$$
$$+4(\bar{\sigma}^\mu)^{\dot{A}A}(\partial_\mu \Phi^+)\sigma^\nu_{A\dot{A}}(\partial_\nu \Phi)$$
$$+(\bar{D}^2 \Phi^+)(D^2 \Phi)$$

$$- 4i (\bar{D}^{\dot{A}} \Phi^{+}) \sigma^{\mu}_{A\dot{A}} \partial_\mu (D^A \Phi) \} \Big|_0$$

(using (9.13b) and (9.20b))

$$= \frac{1}{16} \{ 4i \sigma^{\mu}_{A\dot{B}} \varepsilon^{\dot{B}\dot{A}} \partial_\mu \bar{\psi}_{\dot{A}}(x) \psi^A(x)$$

$$+ 4 \sigma^{\mu}_{A\dot{B}} \varepsilon^{\dot{B}\dot{A}} \varepsilon^{AC} \sigma^{\nu}_{C\dot{A}} \partial_\mu A^*(x) \partial_\nu A(x)$$

$$+ 4i \bar{\sigma}^{\mu\,\dot{A}A} \partial_\mu \bar{\psi}_{\dot{A}}(x) \psi_A(x)$$

$$+ 4 \, Tr(\bar{\sigma}^\mu \sigma^\nu) \partial_\mu A^*(x) \partial_\nu A(x)$$

$$+ 16 F^*(x) F(x)$$

$$- 8i \bar{\psi}^{\dot{A}}(x) \sigma^{\mu}_{A\dot{A}} \partial_\mu \psi^A(x) \}$$

(using (9.14) and (9.21))

$$= \frac{1}{16} \{ 16 F^*(x) F(x) - 4i \psi(x) \sigma^\mu (\partial_\mu \bar{\psi}(x))$$

$$+ 4 Tr(\bar{\sigma}^\mu \sigma^\nu) \partial_\mu A^*(x) \partial_\nu A(x)$$

$$+ 4i (\partial_\mu \bar{\psi}(x)) \bar{\sigma}^\mu \psi(x)$$

$$+ 4 Tr(\bar{\sigma}^\mu \sigma^\nu) (\partial_\mu A^*(x)) (\partial_\nu A(x))$$

$$+ 8i (\partial_\mu \psi(x)) \sigma^\mu \bar{\psi}(x) \}$$

(using (1.88))

$$= F^*(x) F(x) + \partial_\mu A^*(x) \partial^\mu A(x)$$

$$+ \frac{i}{2} (\partial_\mu \bar{\psi}(x)) \bar{\sigma}^\mu \psi(x)$$

$$- \frac{i}{2} \bar{\psi}(x) \bar{\sigma}^\mu (\partial_\mu \psi(x))$$

(using (1.115) and (1.91))

$$= F^*(x) F(x) - A^*(x) \Box A(x)$$

$$+ i (\partial_\mu \bar{\psi}(x)) \bar{\sigma}^\mu \psi(x) + \text{total derivatives}$$

where we used (8.28) and (8.29) in the last step. We see that the expression obtained agrees with the previous result (7.27b).

The projection technique to find component field expansions of supersymmetric actions can also be used to obtain the component field expansions of spinor superfields such as W_A, which are used in the construction of action integrals of supersymmetric gauge theories. Since the spinor superfield carries an external Lorentz index, i.e. W_A transforms according to the $(\frac{1}{2}, 0)$ representation of SL(2,C), the separation into components by projection requires reduction with respect to the group SL(2,C). Consider the component field expansion of the spinor superfield W_A, i.e. (cf. (7.52))

$$W_A(y,\theta) = \lambda_A(y) + 2D(y)\theta_A + (\sigma^{\mu\nu}\theta)_A F_{\mu\nu}(y) - i(\theta\theta)\sigma^{\mu}_{A\dot{B}}\partial_\mu \bar{\lambda}^{\dot{B}}(y) \tag{9.27}$$

We demonstrate below that as for scalar fields we can obtain the component fields $\lambda_A(x)$, $D(x)$ and $F_{\mu\nu}(x)$ by projection, i.e.

$$\lambda_A(x) = W_A(y,\theta)\Big|_0 \tag{9.28a}$$

$$\begin{aligned} D(x) &= -\frac{1}{4}D^{(1)A}W_A(y,\theta)\Big|_0 \\ &= -\frac{1}{4}D^A W_A(x,\theta,\bar{\theta})\Big|_0 \end{aligned} \tag{9.28b}$$

$$\begin{aligned} (\sigma_{\mu\nu}\varepsilon^T)_{AB}F^{\mu\nu}(x) &= -\frac{1}{2}(D^{(1)}_A W_B + D^{(1)}_B W_A)\Big|_0 \\ &\equiv -\frac{1}{2}D^{(1)}_{(A}W_{B)}\Big|_0 \end{aligned} \tag{9.28c}$$

Projecting out the symmetric combination $D_{(A}W_{B)}$ (equation (9.28c) defines this notation) and the contraction $D^A W_A$ correspond to reduction with respect to the Lorentz group in the following sense. The spinor superfield W_A describes an irreducible supermultiplet (of mass zero) with components $\lambda_A(x)$, D(x) and $F_{\mu\nu}(x)$. The Clifford vacuum of this multiplet corresponds to the field

$$W_A|_0 = \lambda_A(x)$$

i.e. the Clifford vacuum carries a spinor index (see the discussion following (4.34)). According to the general procedure of Chapter 4 the other components of this multiplet are obtained by application of the supercharge $\bar{Q}_{\dot{A}}$ to the Clifford vacuum. In the superfield formalism this operation is represented by the projection $D_A W_B|_0$. With respect to the Lorentz group the projection operation $D_A W_B|_0$ corresponds to the Kronecker product of two $(\frac{1}{2}, 0)$ representations with sum decomposition given by (1.129b), i.e.

$$(\tfrac{1}{2}, 0) \otimes (\tfrac{1}{2}, 0) = (0, 0) \oplus (1, 0)$$

where D(x) transforms according to (0,0) and $F_{\mu\nu}(x)$ transforms according to the (1, 0) representation.

We now verify the projections (9.28a,b,c).

i) Equation (9.28a) is obvious.

ii) Consider, using (6.47) and then (6.4g) and (9.27),

$$D^{(1)A} W_A(y,\theta)|_0$$

$$= \varepsilon^{AB}(\partial_B + 2i\sigma^{\mu}_{B\dot{B}}\,\bar{\theta}^{\dot{B}}\partial_\mu) W_A|_0$$

$$= -\partial^A[\lambda_A(y) + 2D(y)\theta_A + (\sigma^{\mu\nu}\theta)_A F_{\mu\nu}(y) - i(\theta\theta)\sigma^{\mu}_{A\dot{B}}\,\partial_\mu\bar{\lambda}^{\dot{B}}(y)]|_0$$

$$+2i\varepsilon^{AB}\sigma^{\mu}_{B\dot{B}}\,\bar{\theta}^{\dot{B}}\,\partial_{\mu}W_{A}(y,\theta)\Big|_{0}$$

$$=\left[-2D(y)\,\partial^{A}\theta_{A}-(\sigma^{\mu\nu})_{A}{}^{B}\,\partial^{A}\theta_{B}\,F_{\mu\nu}(y)\right]\Big|_{0}$$

$$=\left[-2D(y)\,\delta^{A}{}_{A}-(\sigma^{\mu\nu})_{A}{}^{B}\,\delta^{A}{}_{B}\,F_{\mu\nu}(y)\right]\Big|_{0}$$

$$=\left[-4D(y)-\mathrm{Tr}(\sigma^{\mu\nu})F_{\mu\nu}(y)\right]\Big|_{0}$$

$$=-4D(x)$$

since, using (1.119a) and (1.91),

$$\mathrm{Tr}[\sigma^{\mu\nu}]$$

$$=\frac{i}{4}\left[\mathrm{Tr}(\sigma^{\mu}\bar{\sigma}^{\nu})-\mathrm{Tr}(\sigma^{\nu}\bar{\sigma}^{\mu})\right]$$

$$=\frac{i}{2}(\eta^{\mu\nu}-\eta^{\nu\mu})$$

$$=0$$

iii) Consider, using (6.47),

$$\frac{1}{2}(D^{(1)}_{A}W_{B}+D^{(1)}_{B}W_{A})$$

$$=\frac{1}{2}\Big\{(\partial_{A}+2i\sigma^{\mu}_{A\dot{B}}\,\bar{\theta}^{\dot{B}}\,\partial_{\mu})W_{B}$$

$$+(\partial_{B}+2i\sigma^{\mu}_{B\dot{B}}\,\bar{\theta}^{\dot{B}}\,\partial_{\mu})W_{A}\Big\}\Big|_{0}$$

$$=\frac{1}{2}(\partial_{A}W_{B}+\partial_{B}W_{A})\Big|_{0}$$ + terms which vanish at $\theta=\bar{\theta}=0$

$$=\frac{1}{2}\Big\{2D(y)\,\partial_{A}\theta_{B}+(\sigma^{\mu\nu})_{B}{}^{C}\,\partial_{A}\theta_{C}\,F_{\mu\nu}(y)$$

$$+2D(y)\,\partial_{B}\theta_{A}+(\sigma^{\mu\nu})_{A}{}^{C}\,\partial_{B}\theta_{C}\,F_{\mu\nu}(y)\Big\}\Big|_{0}$$

(using (9.27))

$$= \{ - D(y)\varepsilon_{AB} - \tfrac{1}{2}(\sigma^{\mu\nu})_B{}^C \varepsilon_{AC} F_{\mu\nu}(y)$$

$$- D(y)\varepsilon_{BA} - \tfrac{1}{2}(\sigma^{\mu\nu})_A{}^C \varepsilon_{BC} F_{\mu\nu}(y)\}\Big|_0$$

(using (6.4f))

$$= \tfrac{1}{2}\{(\sigma^{\mu\nu})_B{}^C \varepsilon_{CA} + (\sigma^{\mu\nu})_A{}^C \varepsilon_{CB}\} F_{\mu\nu}(y)\Big|_0$$

$$= -\tfrac{1}{2}[(\sigma^{\mu\nu}\varepsilon^T)_{BA} + (\sigma^{\mu\nu}\varepsilon^T)_{AB}] F_{\mu\nu}(x)$$

$$= -(\sigma^{\mu\nu}\varepsilon^T)_{AB} F_{\mu\nu}(x) \quad \text{(with (1.127))}$$

Hence

$$-\tfrac{1}{2} D^{(1)}_{(A} W_{B)}\Big|_0 = (\sigma^{\mu\nu}\varepsilon^T)_{AB} F_{\mu\nu}(x)$$

It is convenient to calculate another expression here, which will be needed later. Thus consider

$$D^{(1)2} W_A\Big|_0 = D^{(1)B} D^{(1)}_B W_A\Big|_0$$

$$= (-\partial^B + 2i\varepsilon^{BC}\sigma^\mu_{C\dot{B}}\bar{\theta}^{\dot{B}}\partial_\mu)\cdot$$

$$\cdot(\partial_B + 2i\sigma^\nu_{B\dot{B}}\bar{\theta}^{\dot{B}}\partial_\nu) W_A\Big|_0$$

(with (6.47))

$$= -\partial^B\partial_B W_A\Big|_0$$

$$= -\partial^B\partial_B(\lambda_A(y) + 2D(y)\theta_A + \sigma^{\mu\nu}{}_A{}^C\theta_C F_{\mu\nu}(y)$$

$$- i(\theta\theta)\sigma^\mu_{A\dot{B}}\partial_\mu\bar{\lambda}^{\dot{B}}(y))\Big|_0$$

(using (9.27))

$$= i\,\partial^{B}\partial_{B}(\theta\theta)\,\sigma^{\mu}_{A\dot{B}}\,\partial_{\mu}\bar{\lambda}^{\dot{B}}(y)\Big|_{0}$$

$$= 4i\,\sigma^{\mu}_{A\dot{B}}\,\partial_{\mu}\bar{\lambda}^{\dot{B}}(x) \quad \text{(with (6.4q))} \tag{9.29}$$

Raising the index we have

$$D^{(1)2}W^{A}\Big|_{0} = 4i\,\varepsilon^{AB}\sigma^{\mu}_{B\dot{B}}\,\partial_{\mu}\bar{\lambda}^{\dot{B}}(x) \tag{9.30}$$

The same type of analysis can be applied to the anti-chiral spinor superfield $\overline{W}_{\dot{A}}$ which transforms according to the $(0, \frac{1}{2})$ representation of SL(2,C). This superfield has the component field expansion (cf. (7.53))

$$\begin{aligned}\overline{W}_{\dot{A}}(z,\bar{\theta}) = {} & \bar{\lambda}_{\dot{A}}(z) + 2D(x)\bar{\theta}_{\dot{A}} \\ & - \varepsilon_{\dot{A}\dot{B}}(\bar{\sigma}^{\mu\nu})^{\dot{B}}{}_{\dot{C}}\,\bar{\theta}^{\dot{C}}F_{\mu\nu}(z) \\ & + i(\bar{\theta}\bar{\theta})(\partial_{\mu}\lambda(z)\sigma^{\mu})_{\dot{A}}\end{aligned} \tag{9.31}$$

and the component fields are given by the projections

$$\bar{\lambda}_{\dot{A}}(x) = \overline{W}_{\dot{A}}(z,\bar{\theta})\Big|_{0} \tag{9.32a}$$

$$D(x) = -\frac{1}{4}\bar{D}^{(2)}_{\dot{A}}\,\overline{W}^{\dot{A}}(z,\bar{\theta})\Big|_{0} \tag{9.32b}$$

(cf. (7.57)) and

$$\begin{aligned}&(\bar{\varepsilon}\bar{\sigma}^{\mu\nu})_{\dot{A}\dot{B}}\,F_{\mu\nu}(x) \\ &= \frac{1}{2}\left(\bar{D}^{(2)}_{\dot{A}}\,\overline{W}_{\dot{B}}(z,\bar{\theta}) + \bar{D}^{(2)}_{\dot{B}}\,\overline{W}_{\dot{A}}(z,\bar{\theta})\right)\Big|_{0} \\ &=: \frac{1}{2}\bar{D}^{(2)}_{(\dot{A}}\,\overline{W}_{\dot{B})}(z,\bar{\theta})\Big|_{0}\end{aligned} \tag{9.32c}$$

Furthermore

$$\bar{D}^{(2)2} W_{\dot{A}}(z,\bar{\theta})\Big|_0 = \bar{D}^2 \bar{W}_{\dot{A}}(x,\theta,\bar{\theta})\Big|_0$$

$$= -4i\,(\partial_\mu \lambda(x) \sigma^\mu)_{\dot{A}} \qquad (9.32d)$$

We verify (9.32c). Consider, using (6.50),

$$\frac{1}{2}\bar{D}^{(2)}_{(\dot{A}} \bar{W}_{\dot{B})}(z,\bar{\theta})\Big|_0$$

$$= -\frac{1}{2}\bar{\partial}_{(\dot{A}} \bar{W}_{\dot{B})}(z,\bar{\theta})\Big|_0$$ + terms which vanish for $\theta = \bar{\theta} = 0$

$$= -\frac{1}{2}\Big[2D(z)\bar{\partial}_{(\dot{A}} \bar{\theta}_{\dot{B})} - \varepsilon_{(\dot{B}\dot{D}} \bar{\sigma}^{\mu\nu\dot{D}}{}_{\dot{C}} \bar{\partial}_{\dot{A})} \bar{\theta}^{\dot{C}} F_{\mu\nu}\Big]\Big|_0$$

(using (9.31))

$$= -\frac{1}{2}\Big[2D(x)(\varepsilon_{\dot{B}\dot{A}} + \varepsilon_{\dot{A}\dot{B}}) - \varepsilon_{\dot{B}\dot{D}}(\bar{\sigma}^{\mu\nu})^{\dot{D}}{}_{\dot{C}}\, \delta_{\dot{A}}{}^{\dot{C}} F_{\mu\nu}(x) - \varepsilon_{\dot{A}\dot{D}}(\bar{\sigma}^{\mu\nu})^{\dot{D}}{}_{\dot{C}}\, \delta_{\dot{B}}{}^{\dot{C}} F_{\mu\nu}(x)\Big]$$

(using (6.4f), (6.41))

$$= \frac{1}{2}\Big[\varepsilon_{\dot{B}\dot{D}}\bar{\sigma}^{\mu\nu\dot{D}}{}_{\dot{A}} + \varepsilon_{\dot{A}\dot{D}}\bar{\sigma}^{\mu\nu\dot{D}}{}_{\dot{B}}\Big]F_{\mu\nu}(x)$$

$$= \varepsilon_{\dot{A}\dot{D}}(\bar{\sigma}^{\mu\nu})^{\dot{D}}{}_{\dot{B}} F_{\mu\nu}(x)$$ (with (1.127'))

$$= (\bar{\varepsilon}\,\bar{\sigma}^{\mu\nu})_{\dot{A}\dot{B}} F_{\mu\nu}(x)$$

This verifies (9.32c).

With the above projections it is now possible to compute the Grassmann integrated action of a pure supersymmetric abelian gauge theory. From (8.40) we know that the supersymmetric action of this theory is (expressed in terms of the spinor superfields W_A, $\overline{W}_{\dot{A}}$)

$$\mathcal{A} = \int d^4x\, \tilde{\mathcal{A}}$$

where

$$\begin{aligned}
\tilde{\mathcal{A}} &= \int d^2\theta\, W^A W_A + \int d^2\bar{\theta}\, \overline{W}_{\dot{A}} \overline{W}^{\dot{A}} \\
&= -\tfrac{1}{4} D^2 (W^A W_A)\Big|_0 - \tfrac{1}{4} \bar{D}^2 (\overline{W}_{\dot{A}} \overline{W}^{\dot{A}})\Big|_0
\end{aligned}$$

(using (9.17) and (9.23b))

$$\begin{aligned}
&= -\tfrac{1}{4} D^B [(D_B W^A) W_A - W^A (D_B W_A)]\Big|_0 \\
&\quad - \tfrac{1}{4} \bar{D}_{\dot{B}} [(\bar{D}^{\dot{B}} \overline{W}_{\dot{A}}) \overline{W}^{\dot{A}} - \overline{W}_{\dot{A}} (\bar{D}^{\dot{B}} \overline{W}^{\dot{A}})]\Big|_0 \\
&= -\tfrac{1}{4} [(D^2 W^A) W_A + (D_B W^A)(D^B W_A) \\
&\qquad - (D^B W^A)(D_B W_A) + W^A (D^2 W_A)]\Big|_0 \\
&\quad - \tfrac{1}{4} [(\bar{D}^2 \overline{W}_{\dot{A}}) \overline{W}^{\dot{A}} + (\bar{D}^{\dot{B}} \overline{W}_{\dot{A}})(\bar{D}_{\dot{B}} \overline{W}^{\dot{A}}) \\
&\qquad - (\bar{D}_{\dot{B}} \overline{W}_{\dot{A}})(\bar{D}^{\dot{B}} \overline{W}^{\dot{A}}) + \overline{W}_{\dot{A}} (\bar{D}^2 \overline{W}^{\dot{A}})]\Big|_0 \\
&= -\tfrac{1}{4} [(D^2 W^A) W_A + W^A (D^2 W_A)]\Big|_0 \\
&\quad + \tfrac{1}{2} (D^B W^A)(D_B W_A)\Big|_0 \\
&\quad - \tfrac{1}{4} [(\bar{D}^2 \overline{W}_{\dot{A}}) \overline{W}^{\dot{A}} + \overline{W}_{\dot{A}} (\bar{D}^2 \overline{W}^{\dot{A}})]\Big|_0 \\
&\quad + \tfrac{1}{2} (\bar{D}_{\dot{B}} \overline{W}_{\dot{A}})(\bar{D}^{\dot{B}} \overline{W}^{\dot{A}})\Big|_0
\end{aligned} \tag{9.33}$$

Using the projections of the spinor superfields W_A and $\overline{W}_{\dot{A}}$ calculated above, we obtain

i) $(\mathcal{D}^2 W^A) W_A|_0$

$$= 4i\varepsilon^{AB}\sigma^\mu_{B\dot{B}}\,\partial_\mu\bar{\lambda}^{\dot{B}}(x)\,\lambda_A(x)$$

(using (9.28a) and (9.30))

$$= -4i\lambda_A(x)\varepsilon^{AB}\sigma^\mu_{B\dot{B}}\,\partial_\mu\bar{\lambda}^{\dot{B}}(x)$$

$$= 4i\,\varepsilon^{BA}\lambda_A(x)\sigma^\mu_{B\dot{B}}\,\partial_\mu\bar{\lambda}^{\dot{B}}(x)$$

$$= 4i\lambda(x)\sigma^\mu\partial_\mu\bar{\lambda}(x)$$

and

$$W^A(\mathcal{D}^2 W_A)|_0$$

$$= \lambda^A(x).\,4i\sigma^\mu_{A\dot{B}}\,\partial_\mu\bar{\lambda}^{\dot{B}}(x) \quad \text{(with (9.29))}$$

$$= 4i\lambda(x)\sigma^\mu\bar{\lambda}(x)$$

so that

$$-\frac{1}{4}[(\mathcal{D}^2 W^A)W_A + W^A(\mathcal{D}^2 W_A)]|_0$$

$$= -2i\lambda(x)\sigma^\mu\partial_\mu\bar{\lambda}(x) \tag{9.34}$$

ii) The next contribution to be evaluated is

$$\frac{1}{2}(\mathcal{D}^B W^A)(\mathcal{D}_B W_A)$$

$$= \frac{1}{4}(\mathcal{D}^B W^A + \mathcal{D}^A W^B)(\mathcal{D}_B W_A)$$
$$\quad + \frac{1}{4}(\mathcal{D}^B W^A - \mathcal{D}^A W^B)(\mathcal{D}_B W_A)$$

$$= \frac{1}{8}(\mathcal{D}^B W^A + \mathcal{D}^A W^B)(\mathcal{D}_B W_A + \mathcal{D}_A W_B)$$
$$+ \frac{1}{8}(\mathcal{D}^B W^A + \mathcal{D}^A W^B)(\mathcal{D}_B W_A - \mathcal{D}_A W_B)$$
$$+ \frac{1}{8}(\mathcal{D}^B W^A - \mathcal{D}^A W^B)(\mathcal{D}_B W_A + \mathcal{D}_A W_B)$$

$$-\frac{1}{8}(D^B W^A - D^A W^B)(D_B W_A - D_A W_B)$$

$$= \frac{1}{8} D^{(B} W^{A)} D_{(B} W_{A)} + \frac{1}{8} D^{(B} W^{A)} D_{[B} W_{A]}$$

$$+ \frac{1}{8} D^{[B} W^{A]} D_{(B} W_{A)} + \frac{1}{8} D^{[B} W^{A]} D_{[B} W_{A]}$$

where

$$D_{(B} W_{A)} := D_B W_A + D_A W_B$$

$$D_{[B} W_{A]} := D_B W_A - D_A W_B$$

Now, contracting a symmetric second rank spinor with an antisymmetric one always gives zero. Furthermore, using (1.129c) and (1.129f) we have

$$\frac{1}{8} D^{[B} W^{A]} D_{[B} W_{A]}\Big|_0$$

$$= -\frac{1}{8} (DW)(DW)\, \varepsilon^{BA} \varepsilon_{BA}\Big|_0$$

$$= \frac{1}{8} (DW)(DW) \varepsilon^{AB} \varepsilon_{BA}\Big|_0$$

$$= \frac{1}{8} (DW)(DW)\, \delta^A{}_A\Big|_0 = \frac{1}{4} (DW)(DW)\Big|_0$$

$$= 4\, D^2(x) \quad \text{(with (9.28b))} \qquad (9.35)$$

Finally the product of two symmetric expressions gives

$$\frac{1}{8} D^{(B} W^{A)} D_{(B} W_{A)}$$

$$= \frac{1}{8} \varepsilon^{BC} \varepsilon^{AD} D_{(C} W_{D)} D_{(B} W_{A)}$$

$$= \frac{1}{8} \varepsilon^{BC} \varepsilon^{AD} (-2\sigma_{\mu\nu} \varepsilon^T)_{CD} F^{\mu\nu}(x).$$

$$\cdot\left(-2\sigma_{\rho\sigma}\varepsilon^T\right)_{BA} F^{\rho\sigma}(x)$$

(using (9.28c))

$$= \frac{1}{2}\varepsilon^{BC}(\sigma_{\mu\nu}\varepsilon^T)_{CD}(-\varepsilon^{DA})(\sigma_{\rho\sigma}\varepsilon^T)_{AB}\cdot$$
$$\cdot F^{\mu\nu}(x)F^{\rho\sigma}(x)$$

(using (1.127))

$$= -\frac{1}{2}\varepsilon^{BC}(\sigma_{\mu\nu})_C{}^E\varepsilon^T_{ED}\varepsilon^{DA}(\sigma_{\rho\sigma})_A{}^L\varepsilon^T_{LB}\cdot$$
$$\cdot F^{\mu\nu}(x)F^{\rho\sigma}(x)$$
$$= -\frac{1}{2}(\varepsilon^T_{LB}\varepsilon^{BC})(\varepsilon^T_{ED}\varepsilon^{DA})(\sigma_{\mu\nu})_C{}^E(\sigma_{\rho\sigma})_A{}^L\cdot$$
$$\cdot F^{\mu\nu}(x)F^{\rho\sigma}(x)$$
$$= -\frac{1}{2}(\varepsilon^{CB}\varepsilon^T_{BL})(\varepsilon^{AD}\varepsilon^T_{DE})(\sigma_{\mu\nu})_C{}^E(\sigma_{\rho\sigma})_A{}^L\cdot$$
$$\cdot F^{\mu\nu}(x)F^{\rho\sigma}(x)$$
$$= -\frac{1}{2}\delta^C{}_L\delta^A{}_E(\sigma_{\mu\nu})_C{}^E(\sigma_{\rho\sigma})_A{}^L F^{\mu\nu}(x)F^{\rho\sigma}(x)$$
$$= -\frac{1}{2}(\sigma_{\mu\nu})_L{}^E(\sigma_{\rho\sigma})_E{}^L F^{\mu\nu}(x)F^{\rho\sigma}(x)$$
$$= -\frac{1}{2}(\sigma_{\mu\nu}\sigma_{\rho\sigma})_L{}^L F^{\mu\nu}(x)F^{\rho\sigma}(x)$$
$$= -\frac{1}{2}\mathrm{Tr}[\sigma_{\mu\nu}\sigma_{\rho\sigma}]F^{\mu\nu}(x)F^{\rho\sigma}(x)$$
$$= -\left\{\frac{1}{4}(\eta^{\mu\rho}\eta^{\nu\sigma}-\eta^{\mu\sigma}\eta^{\nu\rho})+\frac{i}{4}\varepsilon^{\mu\nu\rho\sigma}\right\}\cdot$$
$$\cdot F_{\mu\nu}(x)F_{\rho\sigma}(x)$$

(using (1.125))

$$= -\frac{1}{2}F_{\mu\nu}(x)F^{\mu\nu}(x) - \frac{i}{2}F_{\mu\nu}(x)F^{\mu\nu *}(x)$$

(using (8.37a)) (9.36)

Similarly, from the $\overline{W}_{\dot{A}}\overline{W}^{\dot{A}}$-term in the action integral one obtains the followong contributions to the component field expansion

i) $$-\frac{1}{4}(\bar{D}^2\overline{W}_{\dot{A}})\overline{W}^{\dot{A}}\Big|_0$$

$$= i(\partial_\mu\lambda(x)\sigma^\mu)_{\dot{A}}\bar{\lambda}^{\dot{A}}(x) \quad \text{(using (9.32a,d))}$$

$$= i(\partial_\mu\lambda(x))\sigma^\mu\bar{\lambda}(x) \tag{9.37}$$

and

$$\frac{1}{2}(\bar{D}_{\dot{B}}\overline{W}_{\dot{A}})(\bar{D}^{\dot{B}}\overline{W}^{\dot{A}})\Big|_0$$

$$= \frac{1}{8}(\bar{D}_{(\dot{B}}\overline{W}_{\dot{A})})(\bar{D}^{(\dot{B}}\overline{W}^{\dot{A})})\Big|_0 + \frac{1}{8}(\bar{D}_{[\dot{B}}\overline{W}_{\dot{A}]})(\bar{D}^{[\dot{B}}\overline{W}^{\dot{A}]})\Big|_0$$

$$+ \frac{1}{8}\bar{D}_{(\dot{B}}\overline{W}_{\dot{A})}\varepsilon^{\dot{B}\dot{C}}\varepsilon^{\dot{A}\dot{D}}\bar{D}_{(\dot{C}}\overline{W}_{\dot{D})}\Big|_0$$

$$+ \frac{1}{8}(\bar{D}_{[\dot{B}}\overline{W}_{\dot{A}]})(\bar{D}^{[\dot{B}}\overline{W}^{\dot{A}]})\Big|_0$$

$$= \frac{1}{2}(\bar{\varepsilon}\bar{\sigma}^{\mu\nu})_{\dot{B}\dot{A}}\varepsilon^{\dot{B}\dot{C}}\varepsilon^{\dot{A}\dot{D}}(\bar{\varepsilon}\bar{\sigma}^{\rho\sigma})_{\dot{C}\dot{D}}\cdot$$

$$\cdot F_{\mu\nu}(x)F_{\rho\sigma}(x) - 2D^2(x)\varepsilon_{\dot{B}\dot{A}}\varepsilon^{\dot{B}\dot{A}}$$

(using (9.32b,c) and (1.129g))

$$= \frac{1}{2}(\bar{\varepsilon}\bar{\sigma}^{\mu\nu})_{\dot{B}\dot{A}}\varepsilon^{\dot{B}\dot{C}}\varepsilon^{\dot{A}\dot{D}}(\bar{\varepsilon}\bar{\sigma}^{\rho\sigma})_{\dot{D}\dot{C}}\cdot$$

$$\cdot F_{\mu\nu}(x)F_{\rho\sigma}(x) + 2D^2(x)\delta_{\dot{A}}{}^{\dot{A}}$$

(using (1.127'))

$$= \frac{1}{2}\varepsilon_{\dot{B}\dot{E}}(\bar{\sigma}^{\mu\nu})^{\dot{E}}{}_{\dot{A}}\varepsilon^{\dot{B}\dot{C}}\varepsilon^{\dot{A}\dot{D}}\varepsilon_{\dot{D}\dot{F}}(\bar{\sigma}^{\rho\sigma})^{\dot{F}}{}_{\dot{C}}\cdot$$

$$\cdot F_{\mu\nu}(x)F_{\rho\sigma}(x) + 4D^2(x)$$

$$= -\frac{1}{2}\varepsilon_{\dot{E}\dot{B}}\,\varepsilon^{\dot{B}\dot{C}}\,\delta^{\dot{A}}_{\dot{F}}\,(\bar{\sigma}^{\mu\nu})^{\dot{E}}_{\dot{A}}\,(\bar{\sigma}^{\rho\sigma})^{\dot{F}}_{\dot{C}}\,\cdot$$

$$\cdot F_{\mu\nu}(x)F_{\rho\sigma}(x) + 4D^2(x)$$

$$= -\frac{1}{2}\delta_{\dot{E}}{}^{\dot{C}}(\bar{\sigma}^{\mu\nu}\bar{\sigma}^{\rho\sigma})^{\dot{E}}_{\dot{C}}\,F_{\mu\nu}(x)F_{\rho\sigma}(x)$$

$$+ 4D^2(x)$$

$$= -\frac{1}{2}\mathrm{Tr}[\bar{\sigma}^{\mu\nu}\bar{\sigma}^{\rho\sigma}]F_{\mu\nu}(x)F_{\rho\sigma}(x) + 4D^2(x)$$

$$= -\left[\frac{1}{4}(\eta^{\mu\rho}\eta^{\nu\sigma} - \eta^{\mu\sigma}\eta^{\nu\rho}) - \frac{i}{4}\varepsilon^{\mu\nu\rho\sigma}\right]\cdot$$

$$\cdot F_{\mu\nu}(x)F_{\rho\sigma}(x) + 4D^2(x)$$

(using (1.125'))

$$= -\frac{1}{2}F_{\mu\nu}(x)F^{\mu\nu}(x) + \frac{i}{2}F_{\mu\nu}(x)F^{\mu\nu *}(x)$$

$$+ 4D^2(x) \tag{9.39}$$

(using (8.37a))

Hence, inserting (9.34) to (9.39) into (9.33) we obtain

$$\int d^2\theta\, W^A W_A + \int d^2\bar{\theta}\, \bar{W}_{\dot{A}}\bar{W}^{\dot{A}}$$

$$= 2i\lambda(x)\sigma^\mu\partial_\mu\bar{\lambda}(x) + 4D^2(x) - \frac{1}{2}F_{\mu\nu}(x)F^{\mu\nu}(x)$$

$$- \frac{i}{2}F_{\mu\nu}(x)F^{\mu\nu *}(x) + 2i(\partial_\mu\lambda(x))\sigma^\mu\bar{\lambda}(x)$$

$$- \frac{1}{2}F_{\mu\nu}(x)F^{\mu\nu}(x) + \frac{i}{2}F_{\mu\nu}(x)F^{\mu\nu *}(x)$$

$$+ 4D^2(x)$$

$$= -F_{\mu\nu}(x)F^{\mu\nu}(x) + 8D^2(x) - 4i\lambda(x)\sigma^\mu\partial_\mu\bar{\lambda}(x)$$

$$+ \text{total derivatives} \tag{9.40}$$

This expression is seen to agree with (8.41) which was obtained by Grassmann integration.

9.3 Spontaneous Symmetry Breaking

If supersymmetric gauge theories are to find realistic applications in high energy physics, both supersymmetry and the gauge symmetry must be broken spontaneously . Supersymmetry must be broken because in experiments one does not observe degenerate Bose-Fermi multiplets.

The spontaneous breaking of gauge symmetry is well understood[61-64], but supersymmetry imposes additional conditions which require further discussion. In this section we restrict ourselves to the case of supersymmetric theories constructed from scalar superfields (the socalled O'Raifeartaigh mechanism of supersymmetry breaking). The breaking of supersymmetry in supersymmetric gauge theories (the Fayet-Iliopoulos mechanism of supersymmetry breaking) will be considered later.

The peculiarity of the spontaneous breakdown of supersymmetry is due to the fact that the Hamiltonian of any supersymmetric theory is related to the supercharges Q_A and $\bar{Q}_{\dot{A}}$ by (see below)

$$H = \frac{1}{4}\{Q_1\bar{Q}_{\dot{1}} + \bar{Q}_{\dot{1}}Q_1 + Q_2\bar{Q}_{\dot{2}} + \bar{Q}_{\dot{2}}Q_2\} \tag{9.40}$$

In order to verify this relation we recall that (cf.(6.2))

$$\{Q_A, \bar{Q}_{\dot{B}}\} = 2\sigma^{\mu}_{A\dot{B}}P_{\mu}$$

Multiplying from the right by $(\bar{\sigma}^{\nu})^{\dot{B}A}$ and summing over A and $\dot{B}$ we obtain

$$\{Q_A, \bar{Q}_{\dot{B}}\}(\bar{\sigma}^{\nu})^{\dot{B}A} = 2(\sigma^{\mu})_{A\dot{B}}(\bar{\sigma}^{\nu})^{\dot{B}A}P_{\mu}$$

$$= 2\,\mathrm{Tr}[\sigma^{\mu}\bar{\sigma}^{\nu}]P_{\mu}$$

$$= 4\eta^{\mu\nu}P_{\mu} \qquad \text{(using (1.91))}$$

$$= 4P_{\nu}$$

Taking $\nu = 0$, we obtain

$$P^{0} = H = \frac{1}{4}\{Q_A, \bar{Q}_{\dot{B}}\}(\bar{\sigma}^{0})^{\dot{B}A}$$

$$= \frac{1}{4}(Q_A\bar{Q}_{\dot{B}} + \bar{Q}_{\dot{B}}Q_A)(\bar{\sigma}^{0})^{\dot{B}A}$$

$$= \frac{1}{4}(Q_1\bar{Q}_{\dot{1}} + \bar{Q}_{\dot{1}}Q_1 + Q_2\bar{Q}_{\dot{2}} + \bar{Q}_{\dot{2}}Q_2)$$

where

$$\bar{\sigma}^{0} = 1_{2\times 2}$$

Equation (9.13) tells us that the spectrum of the Hamiltonian is semipositive definite, i.e.

$$\langle\psi|H|\psi\rangle \geqslant 0 \qquad (9.41)$$

for every state $|\psi\rangle$. Since H is semipositive definite, those states with vanishing energy density are the lowest lying eigenstates and therefore the supersymmetric ground states of the theory. Such states, denoted by $|0\rangle$, are supersymmetric, because (as pointed out above)

$$\langle 0|H|0\rangle = 0$$

implies in view of (9.40)

$$Q_A|0\rangle = 0 \qquad \text{for } A = 1,2$$

$$\bar{Q}_{\dot{A}}|0\rangle = 0 \qquad \text{for } \dot{A} = \dot{1},\dot{2}$$

This fact, however, has the consequence that according to (6.22)

$$\delta_S|0\rangle = i(\alpha^{A}Q_A + \bar{\alpha}_{\dot{A}}\bar{Q}^{\dot{A}})|0\rangle = 0 \qquad (9.42)$$

Hence a supersymmetric vacuum state $|0\rangle$ implies that the supercharges Q_A and $\bar{Q}_{\dot{A}}$ annihilate the vacuum, as expected on general grounds. Therefore the vacuum energy and consequently $V(A, A^*)$ are not only well defined, but are also bound to vanish, i.e.

$$E_{vac} = \langle 0|H|0\rangle = 0 \tag{9.43}$$

This situation is illustrated in Fig.3 for a theory with scalar superfields.

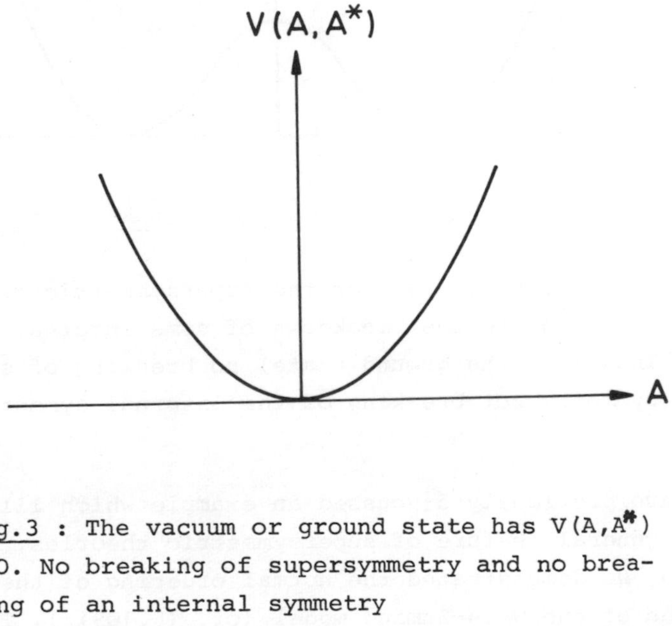

Fig.3 : The vacuum or ground state has $V(A,A^*) = 0$. No breaking of supersymmetry and no breaking of an internal symmetry

Supersymmetric ground states are always at $E_{vac} = 0$ and may still be degenerate with other states which have $E_{vac} = 0$ as illustrated in Fig. 4, indicating the possible breakdown of some internal symmetry. However, in the case illustrated supersymmetry is unbroken because the ground state energy, the minimum of the potential, is zero.

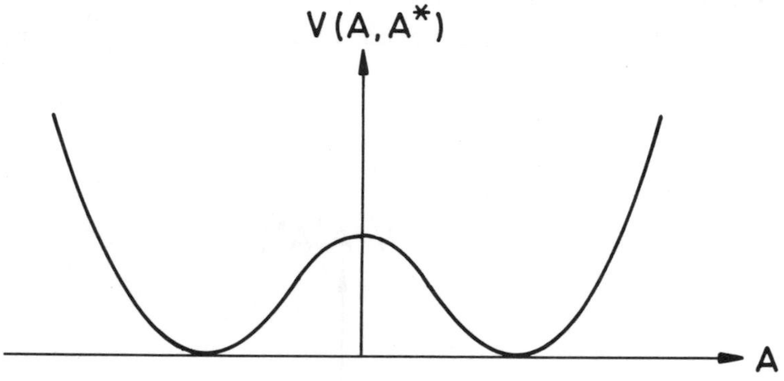

Fig. 4 : Degeneracy of the supersymmetric ground state due to the breakdown of some internal symmetry. Thus (for the ground state) no breaking of supersymmetry but breaking of the internal symmetry

We have previously discussed an example which illustrates this general feature of supersymmetric theories. In Chapter 5 we demonstrated the normal ordering of the Hamiltonian of the Wess-Zumino model (cf. (5.109)). This, of course, is equivalent[65] to fixing the energy of the vacuum to be zero.

We now consider the case in which the vacuum state is not annihilated by the supercharges, i.e.

$$Q_A|0\rangle \neq 0 \tag{9.44}$$

which implies in view of (9.40)

$$E_{vac} = \langle 0|H|0\rangle \neq 0 \tag{9.45}$$

We therefore arrive at the following conclusion: If supersymmetry is not spontaneously broken, i.e. if the vacuum is invariant under supersymmetry transformations (cf. (9.42)) the energy of the vacuum is zero. Conversely, if there exists a state for which the expectation value of the Hamiltonian is zero, supersymmetry is not spontaneously broken. Furthermore, if supersymmetry is spontaneously broken, the vacuum energy is positive (cf. (9.45)). The case of broken supersymmetry is illustrated in Fig. 5, where the expectation value of the scalar field is zero, but supersymmetry is spontaneously broken because the energy of the ground state is greater than zero.

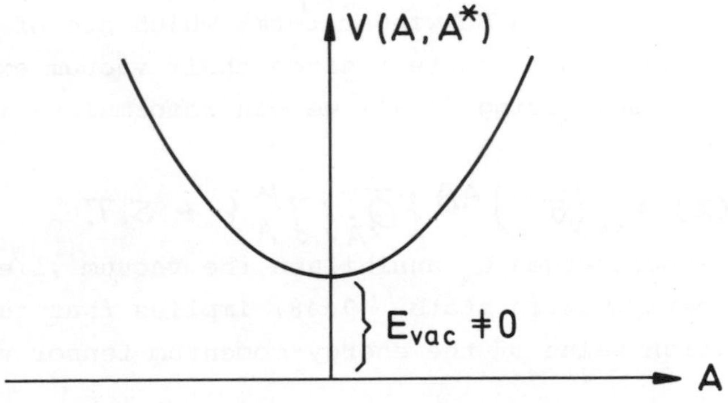

Fig.5 : The case of broken supersymmetry but no breaking of internal symmetry

9.3.1 The Goldstone Theorem

The fact that a positive nonzero vacuum energy indicates supersymmetry breaking is a special case of the more general case that supersymmetry is spontaneously broken if any anticommutator $\{Q, X\}$ (X: some operator) has a nonvanishing vacuum expectation value. The basic anticommutator of a supersymmetric theory (cf. (6.2))

$$\{Q_A, \bar{Q}_{\dot{B}}\} = 2\sigma^{\mu}_{A\dot{B}} P_{\mu}$$

can be considered as arising from integration of the local relation

$$\{\bar{Q}_{\dot{A}}, J^{\mu}_{A}(x)\} = 2\sigma_{\nu A\dot{A}} T^{\mu\nu}(x) + S.T. \tag{9.46}$$

where J^{μ}_{A} is the conserved supercurrent introduced in Section 5.2 and related to the supercharges Q_A by

$$Q_A := \int d^3x \, J^0_A(x), \quad \partial_\nu J^\nu_A = 0 \tag{9.47}$$

(cf. (5.19)). In (9.46) $T^{\mu\nu}(x)$ is the energy-momentum tensor of the corresponding theory and S.T. denotes additional socalled Schwinger terms which are of no interest in the present context since their vacuum expectation values vanish. Using (1.91) we can reformulate (9.46) in the form

$$T^{\mu\nu}(x) = \frac{1}{4}(\bar{\sigma}^{\nu})^{\dot{A}A}\{\bar{Q}_{\dot{A}}, J^{\mu}_{A}\} + S.T. \tag{9.48}$$

If the supercharges Q_A annihilate the vacuum ,i.e. if $|0\rangle$ is a supersymmetric state, (9.48) implies that the vacuum expectation value of the energy-momentum tensor vanishes, i.e.

$$\langle 0|T^{\mu\nu}(x)|0\rangle = 0 \tag{9.49}$$

However, when supersymmetry is spontaneously broken, (9.49) ceases to be true, even in the socalled tree approximation, i.e. at the classical level; instead of (9.49) we write in this case

$$\langle 0|T^{\mu\nu}(x)|0\rangle = E_{vac}\,\eta^{\mu\nu} \tag{9.50}$$

Inserting (9.48) we obtain

$$E_{vac}\,\eta^{\mu\nu} = \frac{1}{4}\bar{\sigma}^{\nu\dot{A}A}\langle 0|\{\bar{Q}_{\dot{A}}, J^{\mu}_{A}(x)\}|0\rangle \tag{9.51}$$

and a nonvanishing value of this expression means that supersymmetry is spontaneously broken; in other words, the vacuum is not invariant,

$$\bar{Q}_{\dot{A}}|0\rangle \neq 0$$

where $\bar{Q}_{\dot{A}}$, or more generally simply Q, is the generator of the appropriate continuous symmetry transformation. Since for a scalar field A(x) (cf. (5.114a))

$$\delta A(x) = -i\bar{\varepsilon}_a[Q_a, A(x)]$$

we have

$$\delta\langle 0|A(x)|0\rangle = -i\bar{\varepsilon}_a\langle 0|[Q_a, A(x)]|0\rangle$$

and hence $Q|0\rangle \neq 0$ implies

$$\delta\langle 0|A(x)|0\rangle \neq 0$$

We now show that this implies the existence of a massless fermion which is called the Goldstone fermion or "goldstino"[66].

If j^μ is a conserved current, such as the supercurrent (5.16) which implies a conserved Majorana spinor current k^μ_a , i.e.

$$\partial_\mu j^\mu = 0, \quad \partial_\mu k^\mu_a = 0$$

with Majorana index a = 1,2,3,4 , or equivalently

$$\partial_\mu J^\mu{}_A = 0, \quad \partial_\mu \bar{J}^\mu{}_{\dot{A}} = 0$$

in the Weyl formulation, then for any local operator A(y)

$$\begin{aligned} 0 &= \int_V d^3x\,[\partial_\mu j^\mu(x), A(y)] \\ &= \frac{d}{dt}\int_V d^3x\,[j^0(\vec{x},t), A(y)] \\ &\quad + \int_S d\vec{s}\cdot[\vec{j}(\vec{x},t), A(y)] \end{aligned}$$

$$= \frac{d}{dt}[Q(t), A(y)]$$

if the surface S enclosing the volume V is made large enough so that on it

$$[\vec{j}(\vec{x},t), A(y)] \sim 0$$

It follows that

$$0 = \frac{d}{dt}\langle 0|[Q(t), A(y)]|0\rangle \qquad (9.52)$$

Inserting a complete set of four-momentum eigenstates $|p_n\rangle$ and using translation invariance, i.e.

$$\langle 0|j_0(\vec{x},t)|p_n\rangle = \langle 0|j_0(0)|p_n\rangle e^{-ip_n x}$$

(which follows from

$$e^{icP}F(x)e^{-icP} = F(x+c)$$

i.e.

$$\partial_\mu F(x) = i[P_\mu, F(x)]$$

and so

$$\langle a|\partial_\mu F(x)|b\rangle = i(p_\mu^a - p_\mu^b)\langle a|F(x)|b\rangle$$

$$= i(p_\mu^a - p_\mu^b)\langle a|F(0)|b\rangle \cdot$$

$$\cdot e^{i(p_\mu^a - p_\mu^b)x^\mu}$$

for an arbitrary function F(x) in selfevident terminology) we have

$$\langle 0|[Q(t), A(x')]|0\rangle$$

$$= \int d^3x \langle 0|[j_0(\vec{x},t), A(x')]|0\rangle$$

$$= \int d^3x \sum_n [\langle 0|j_0(\vec{x},t)|p_n\rangle\langle p_n|A(x')|0\rangle$$

$$- \langle 0|A(x')|p_n\rangle\langle p_n|j_0(\vec{x},t)|0\rangle]$$

$$= \int d^3x \sum_n [\langle 0|j_0(0)|p_n\rangle\langle p_n|A(0)|0\rangle . \cdot e^{-ip_n(x-x')}$$

$$- \langle 0|A(0)|p_n\rangle\langle p_n|j_0(0)|0\rangle . e^{ip_n(x-x')}]$$

$$= \sum_n (2\pi)^3 \delta(\vec{p}_n) .$$

$$\cdot [\langle 0|j_0(0)|p_n\rangle\langle p_n|A(0)|0\rangle e^{-iE_n(t-t')}$$

$$- \langle 0|A(0)|p_n\rangle\langle p_n|j_0(0)|0\rangle e^{iE_n(t-t')}]$$

Hence (9.52) implies

$$0 = \sum_n (2\pi)^3 \delta(\vec{p}_n)(-iE_n) .$$

$$\cdot [\langle 0|j_0(0)|p_n\rangle\langle p_n|A(0)|0\rangle e^{-iE_n(t-t')}$$

$$+ \langle 0|A(0)|p_n\rangle\langle p_n|j_0(0)|0\rangle e^{iE_n(t-t')}]$$

Thus if a state $|p_n\rangle$ exists such that

$$\langle 0|j_0(0)|p_n\rangle\langle p_n|A(0)|0\rangle \neq 0$$

it must have

$$E_n \delta(\vec{p}_n) = 0, \quad E_n = (m_n^2 + \vec{p}_n^{\,2})^{1/2}$$

i.e. the mass m_n of the appropriate particle state must be zero. Since

$$\langle 0|j_0(0)|p_n\rangle$$

must be invariant under Lorentz transformations U, i.e.

$$\langle 0|j_0(0)|p_n\rangle$$

$$= \langle 0|\mathcal{U}^+(\mathcal{U} j_0(0)\mathcal{U}^+)\mathcal{U}|p_n\rangle$$

the operator $\dot{j}_0(0)$ must transform with respect to the same representation of the Lorentz group as the state $|p_n\rangle$. Hence if $\dot{j}_0(0)$ is a spinor current, $|p_n\rangle$ must be a spinor state. This therefore proves the Goldstone theorem for either of the supersymmetric or nonsupersymmetric cases.

9.3.2 Remarks on the Wess-Zumino Model

We now consider the Wess-Zumino model with respect to the possible breakdown of supersymmetry. To this end we have to evaluate the scalar potential V and then calculate $\partial W/\partial A_i$; supersymmetry is unbroken if $\partial W/\partial A_i = 0$ as stated by (9.12). The action of the Wess-Zumino model is given by (cf. (9.3))

$$\mathcal{A}_{WZ} = \int d^4x \int d^4\theta \; \Phi^+\Phi + \int d^4x \int d^2\theta \, W[\Phi] + \int d^4x \int d^2\bar{\theta} \, \bar{W}[\Phi]$$

where the superpotential $W[\Phi]$ takes the form

$$W[\Phi] := g\Phi + \tfrac{1}{2} m\Phi^2 + \tfrac{1}{3}\lambda\Phi^3$$

Then

$$W(A) = gA(x) + \tfrac{1}{2} m A^2(x) + \tfrac{1}{3}\lambda A^3(x)$$

where A is the scalar field of the supermultiplet Φ, i.e.

$$\Phi = A(x) + 2^{\frac{1}{2}}\theta\psi(x) + (\theta\theta)F(x)$$

Differentiating W(A) with respect to A gives

$$\frac{dW(A)}{dA} = g + mA(x) + \lambda A^2(x)$$

Now, it is always possible to find solutions $A_{\pm}$ (i.e. degenerate bosonic ground states) to the equation

$$\frac{dW}{dA} = 0$$

i.e.

$$A^2(x) + \frac{m}{\lambda} A(x) + \frac{g}{\lambda} = 0 \tag{9.53}$$

with

$$A_{\pm} = -\frac{m}{2\lambda} \pm \left[\frac{m^2}{4\lambda^2} - \frac{g}{\lambda}\right]^{\frac{1}{2}}$$

no matter how we adjust the parameters m, g and λ of the theory (remembering that A(x) is a complex scalar field). Hence according to the above discussions the vacuum energy is zero, if we choose A as solution of (9.53) and there is no supersymmetry breaking. This case is illustrated in Fig. 3 and Fig. 4 respectively, depending on the choice of the parameters, i.e. for g = O, m = O Fig. 3 applies, whereas for m = O, g $\neq$ O, $\lambda \neq$ O Fig. 4 applies. In the latter case the "internal symmetry" of

$$V(A, A^*) = |dW/dA|^2$$

which is violated is the phase symmetry

$$A \longrightarrow e^{i\alpha} A$$

9.4 The O'Raifeartaigh Model

9.4.1 Spontaneous Breaking of Supersymmetry in the O'Raifeartaigh Model

The model of O'Raifeartaigh is a supersymmetric field theory constructed from chiral scalar superfields. In his paper O'Raifeartaigh demonstrated that at least three different scalar superfields are required to yield a model which exhibits the spontaneous breakdown of supersymmetry.

We denote the three chiral scalar superfields of the O'Raifeartaigh model by A,X and Y with component field expansions

$$\left.\begin{aligned} A &= a + 2^{1/2}\,\theta\psi_A + \theta\theta F_A \\ X &= x + 2^{1/2}\,\theta\psi_X + \theta\theta F_X \\ Y &= y + 2^{1/2}\,\theta\psi_Y + \theta\theta F_Y \end{aligned}\right\} \qquad (9.54)$$

and corresponding expansions can be written down for the conjugate fields A^+, X^+ and Y^+. The kinetic part of the Lagrangian of the O'Raifeartaigh model is given by

$$\mathcal{L}_{kin} = (A^+A + X^+X + Y^+Y)\Big|_{\theta\theta\,\bar{\theta}\bar{\theta}\text{-component}} \qquad (9.55)$$

and the superpotential of the theory is taken to be

$$W[A,X,Y] := \lambda AY + gX(A^2 - M^2) \qquad (9.56)$$

where λ , g and M are real and nonzero. The auxiliary fields of (9.54), i.e. F_A, F_X and F_Y, are (according to (9.8), (9.10)) given by the partial derivatives of the superpotential, i.e.

$$\begin{aligned} F_A^* &= -\,\partial W(a,x,y)/\partial a \\ F_X^* &= -\,\partial W(a,x,y)/\partial x \\ F_Y^* &= -\,\partial W(a,x,y)/\partial y \end{aligned}$$

Hence

$$\frac{\partial W(a,x,y)}{\partial a} = \lambda y + 2gax$$

$$\frac{\partial W(a,x,y)}{\partial x} = g(a^2 - M^2)$$

$$\frac{\partial W(a,x,y)}{\partial y} = \lambda a \tag{9.57}$$

According to our general discussion (cf. (9.12)) supersymmetry is unbroken if and only if we can find a simultaneous solution to the set of equations

$$\frac{\partial W}{\partial a} = \frac{\partial W}{\partial x} = \frac{\partial W}{\partial y} = 0 \tag{9.58}$$

However, the model has been constructed in such a way that this solution does not exist. According to (9.57)

$$\frac{\partial W}{\partial a} = 0 \tag{9.59}$$

is solved by

$$y = -\frac{2g}{\lambda} a x \tag{9.60}$$

But

$$\frac{\partial W}{\partial x} = 0 \qquad \text{and} \qquad \frac{\partial W}{\partial y} = 0$$

are inconsistent with this since

$\partial W / \partial y = 0$ implies $a = 0$ $(\lambda \neq 0)$

and inserted into $\partial W / \partial x = 0$ gives $g M^2 = 0$ which is impossible since g and M are assumed to be nonzero. Hence supersymmetry must be broken spontaneously.

Since we cannot find a simultaneous solution to the set of equations (9.58) at least one of the auxiliary fields F_A, F_X, F_Y acquires a nonvanishing vacuum expectation value. This is the general feature of spontaneously broken supersymmetry.

To obtain the ground state of the model, we have to minimize the scalar potential $V(a,x,y; a^*, x^*, y^*)$. According to (9.11) the scalar potential is related to the superpotential by

$$V(a,x,y;a^*,x^*,y^*)$$

$$= \left|\frac{\partial W}{\partial a}\right|^2 + \left|\frac{\partial W}{\partial x}\right|^2 + \left|\frac{\partial W}{\partial y}\right|^2$$

$$= |\lambda y + 2gax|^2 + g^2|a^2 - M^2|^2 + \lambda^2|a|^2$$

(using (9.57))

Differentiating the scalar potential with respect to x and y we obtain (setting these expressions equal to zero)

$$\frac{\partial V}{\partial x} = (\lambda y^* + 2g x^* a^*) 2ga \overset{!}{=} 0$$

$$\frac{\partial V}{\partial y} = (\lambda y^* + 2g x^* a^*) \lambda \overset{!}{=} 0$$

The second equation is satisfied if (9.60) holds. Then also the derivative $\partial V/\partial x$ vanishes and (9.60) gives the minimum value of y, but x remains undetermined; such a "degeneracy" is a general feature of spontaneously broken supersymmetry. Hence at the minimum we have

$$V(a,a^*) = g^2|a^2 - M^2|^2 + \lambda^2|a|^2 \tag{9.62}$$

and

$$\frac{\partial V(a,a^*)}{\partial a} = a^*\lambda^2 + 2g^2(a^{*2} - M^2)\,a = 0 \tag{9.63}$$

Equation (9.62) implies that a must be real, for setting

$$a = k + ih$$

then

$$\lambda^2(k - ih) - 2g^2M^2(k + ih) + 2g^2(k^2 + h^2)(k - ih) = 0$$

Equating real and imaginary parts of the left hand side of this equation to zero, we obtain

$$k\left[\lambda^2 - 2g^2M^2 + 2g^2(k^2+h^2)\right] = 0$$

$$-h\left[\lambda^2 + 2g^2M^2 + 2g^2(k^2+h^2)\right] = 0$$

The latter of these equations is satisfied if and only if h is zero, since the expression in the bracket is a sum of positive terms which cannot vanish. Thus (9.63) becomes

$$\frac{\partial V}{\partial a} = a\left\{\lambda^2 + 2g^2(a^2 - M^2)\right\} = 0 \qquad (9.64)$$

One solution to this equation is $a = a_1 = 0$. Inserting this into the scalar potential (9.62) we obtain the value of V at the extremum $a = a_1 = 0$, i.e.

$$V(a_1 = 0) = g^2 M^4 \qquad (9.65)$$

But $a = a_1 = 0$ is the position of a local maximum of V since

$$\frac{\partial^2 V}{\partial a^2} = \lambda^2 + 6g^2a^2 - 2g^2M^2$$

and

$$\left(\frac{\partial^2 V}{\partial a^2}\right)_{a=0} = 2g^2\left(\frac{\lambda^2}{2g^2} - M^2\right) < 0$$

for $M^2 > \lambda^2/2g^2$

The other solutions to (9.64) are obtained from

$$\lambda^2 + 2g^2a^2 - 2g^2M^2 = 0$$

i.e. the roots are

$$a_2, a_3 = \pm\left(M^2 - \frac{\lambda^2}{2g^2}\right)^{\frac{1}{2}}$$

provided $M^2 > \lambda^2/2g^2$ (in which case a is real). The value of the scalar potential at the extrema a_2, a_3 is

$$V[a_2, a_3]$$

$$= g^2\left|M^2 - \frac{\lambda^2}{2g^2} - M^2\right|^2 + \lambda^2\left|M^2 - \frac{\lambda^2}{2g^2}\right|$$

$$= \frac{\lambda^4}{4g^2} + \lambda^2\left(M^2 - \frac{\lambda^2}{2g^2}\right)$$

$$= \lambda^2\left(M^2 - \frac{\lambda^2}{4g^2}\right) \tag{9.66}$$

and

$$\left(\frac{\partial^2 V}{\partial a^2}\right)_{a_2, a_3} = \lambda^2 + 6g^2\left(M^2 - \frac{\lambda^2}{2g^2}\right) - 2g^2 M^2$$

$$= 4g^2\left(M^2 - \frac{\lambda^2}{2g^2}\right)$$

$$> 0 \quad \text{for} \quad M^2 > \lambda^2/2g^2$$

Hence a_2, a_3 are the positions of minima of the scalar potential (9.61) provided we adjust the parameters M, λ and g such that

$$M^2 > \frac{\lambda^2}{2g^2} \tag{9.67}$$

Furthermore, the difference between V(a_1) and V(a_2 or a_3) is

$$V(a_1) - V(a_2, a_3)$$

$$= g^2 M^4 - \lambda^2\left(M^2 - \frac{\lambda^2}{4g^2}\right)$$

$$= g^2\left(M^2 - \frac{\lambda^2}{2g^2}\right)^2 > 0$$

Hence with our choice of parameters (9.67), the absolute minimum is degenerate and at

$$a_2, a_3 = \pm\left(M^2 - \frac{\lambda^2}{2g^2}\right)^{\frac{1}{2}}$$

where

$$V(a_2, a_3) > 0$$

Hence the scalar potential has the shape shown in Fig. 6.

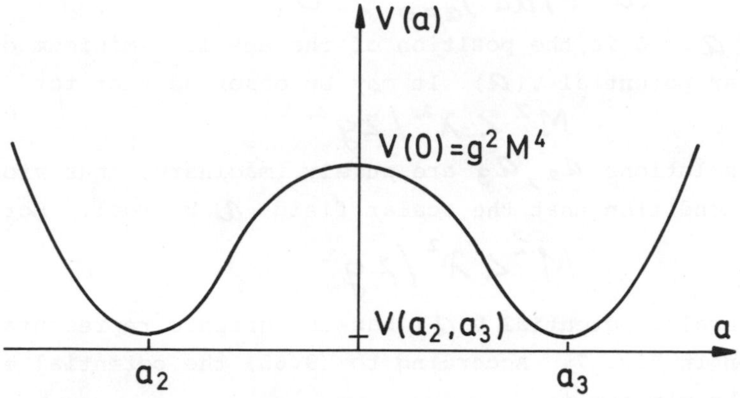

Fig.6 : The scalar potential V(a) of the O'Raifeartaigh model in the range $M^2 > \lambda^2/2g^2$. In the case shown both supersymmetry and an internal symmetry are broken

The scalar potential (9.62) has the internal U(1) phase symmetry

$$a \longrightarrow e^{i\alpha}a \tag{9.68}$$

Now from (9.62)

$$\begin{aligned} V(a) &= g^2(a^2-M^2)^2+\lambda^2a^2 \\ &= g^2M^4+a^2(\lambda^2-2M^2g^2)+g^2a^4 \end{aligned}$$

The spontaneous breaking of the symmetry (9.68) occurs if and only if the factor $(\lambda^2-2M^2g^2)$ is negative (this corresponds to the well known case of a negative $(\text{mass})^2$-term in the simple Higgs model). However, this corresponds exactly to the choice (9.66) for the parameters M, g and λ.

In the case $M^2 < \lambda^2/2g^2$ or $\lambda^2 - 2M^2g^2 > 0$ no breaking of the internal symmetry appears[67] and the only solution to (9.64) is $a = 0$. In this case

$$(d^2V/da^2)_{a=0} > 0$$

and $a = 0$ is the position of the absolute minimum of the scalar potential $V(a)$. It may be observed that for

$$M^2 < \lambda^2/2g^2$$

the solutions a_2, a_3 are purely imaginary, thus violating the condition that the scalar field a be real. For

$$M^2 < \lambda^2/2g^2$$

the scalar potential $V(a)$ has the graphic representation shown in Fig. 7. According to (9.65) the potential energy of the minimum is

$$V(a=0) = g^2 M^4 > 0$$

indicating the breakdown of supersymmetry.

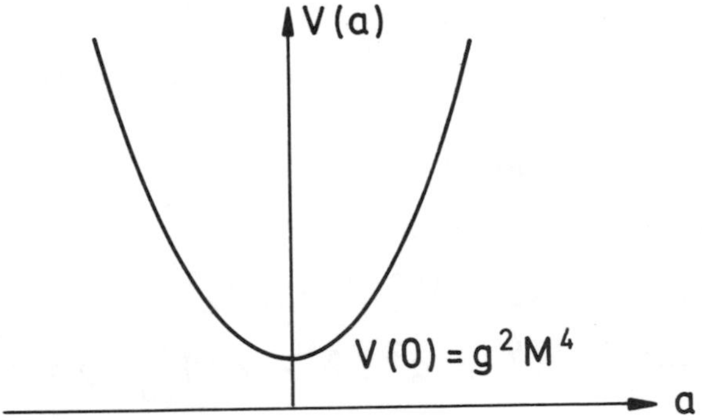

Fig.7 : The scalar potential $V(a)$ of the O'Raifeartaigh model in the range $M^2 < \lambda^2/2g^2$. Supersymmetry is broken but no internal symmetry

9.4.2 The Mass Spectrum of the O'Raifeartaigh Model

In general if supersymmetry is broken, then according to (9.12) $\partial W / \partial A_i = W_i \neq 0$ for some values of i. On the other hand at the minimum of the scalar potential we have

$$\frac{\partial V}{\partial A_i} = 0$$

Using (9.11) this implies

$$\frac{\partial V}{\partial A_i} = \frac{\partial}{\partial A_i} \sum_j \left| \frac{\partial W}{\partial A_j} \right|^2 = \frac{\partial}{\partial A_i} \sum_j W_j^*(A) W_j(A)$$

$$= \sum_j \left(W_{ij}^* W_j + W_j^* W_{ij} \right) = 0$$

where (cf. (9.8))

$$W_{ij} := \frac{\partial^2 W}{\partial A_i \partial A_j}$$

Hence at the minimum

$$W_{ij} W_j^* = 0 \qquad (9.69)$$

Now since W_{ij} is the fermionic mass matrix (see (9.7)), the diagonalized matrix and thus the eigenvalues of W_{ij} determine the masses of the fermions. If supersymmetry is spontaneously broken, $W_j \neq 0$ for some j, and hence some of these eigenvalues must be zero according to (9.69). Thus in order to determine the fermionic mass spectrum of the O'Raifeartaigh model we have to diagonalize the matrix W_{ij}. The detailed analysis given below leads to the following results for the case $M^2 < \lambda^2/2g^2$. One eigenvalue of the fermionic mass matrix vanishes, and the second and third eigenvalues are λ. As we shall see ψ_X, the supersymmetric partner of F_X, turns out to be the Goldstone fermion; furthermore there are two massive fermions of mass λ which are linear combinations of $\psi_A(x)$ and $\psi_Y(x)$. The associated bosonic mass spectrum is the following: the scalar X(x) is massless, Y(x) has mass λ and a mass splitting occurs in the case of the scalars a (x), i.e. setting as before

$$a(x) = k(x) + i h(x)$$

where k(x) and h(x) are real scalar fields, we find

$$m_k^2 = \lambda^2 - 4g^2 M^2$$
$$m_h^2 = \lambda^2 + 4g^2 M^2$$

In order to obtain these results it is advantageous to write the Lagrangian of the O'Raifeartaigh model out explicitly in its component field expansion. The Lagrangian is given by (cf. (9.55), (9.56))

$$\begin{aligned} \mathcal{L} &= (A^+A + X^+X + Y^+Y)\big|_{\theta\theta\bar{\theta}\bar{\theta}} \\ &\quad + W[A,X,Y]\big|_{\theta\theta} \\ &\quad + \overline{W}[A^+,X^+,Y^+]\big|_{\bar{\theta}\bar{\theta}} \\ &= (A^+A + X^+X + Y^+Y)\big|_{\theta\theta\bar{\theta}\bar{\theta}} \\ &\quad + (\lambda AY + gXA^2 - gM^2X)\big|_{\theta\theta} \\ &\quad + (\lambda A^+Y^+ + gX^+A^{+2} - gM^2X^+)\big|_{\bar{\theta}\bar{\theta}} \end{aligned} \tag{9.70}$$

where $|_{\cdots}$ indicates that the appropriate components are to be taken. Inserting the component expansions (9.54) of the superfields A, X and Y and using (7.27b), (8.27) and (8.29) we obtain for the kinetic part

$$(A^+A + X^+X + Y^+Y)|_{\theta\theta\bar\theta\bar\theta}$$
$$= -[a^*(x)\Box a(x) + x^*(x)\Box x(x) + y^*(x)\Box y(x)]$$
$$+ F_A^*(x)F_A(x) + F_X^*(x)F_X(x) + F_Y^*(x)F_Y(x)$$
$$+ i[\partial_\mu \bar\psi_A(x)\bar\sigma^\mu \psi_A(x) + \partial_\mu \bar\psi_X(x)\bar\sigma^\mu \psi_X(x)$$
$$+ \partial_\mu \bar\psi_Y(x)\bar\sigma^\mu \psi_Y(x)] \tag{9.71}$$

Now the $\theta\theta$ -component of a product of two left-handed chiral scalar superfields Φ_i and Φ_k is (as we know from (7.16))

$$\Phi_i \Phi_k|_{\theta\theta} = A_i(x)F_k(x) + F_i(x)A_k(x) - \psi_i(x)\psi_k(x)$$

which implies in our particular case

$$\lambda AY|_{\theta\theta}$$
$$= \lambda\left(a(x)F_Y(x) + F_A(x)y(x) - \psi_A(x)\psi_Y(x)\right) \tag{9.72}$$

The $\theta\theta$ -term of a product of three scalar superfields has been obtained in (7.17) and is given by

$$\Phi_l \Phi_i \Phi_k|_{\theta\theta}$$
$$= A_l(x)A_i(x)F_k(x) + A_l(x)F_i(x)A_k(x)$$
$$+ F_l(x)A_i(x)A_k(x) - \psi_l(x)\psi_k(x)A_i(x)$$
$$- \psi_l(x)\psi_i(x)A_k(x) - A_l(x)\psi_i(x)\psi_k(x)$$

so that in the O'Raifeartaigh model

$$gXA^2|_{\theta\theta} = g[2x(x)a(x)F_A(x)$$

$$+ F_X(x)\,a^2(x) - 2\,a(x)\,\psi_X(x)\,\psi_A(x) - x(x)\,\psi_A^2(x)\Big] \tag{9.73}$$

and finally, using (9.54),

$$gM^2 X\Big|_{\theta\theta} = g\,M^2 F_X(x) \tag{9.74}$$

Similar expressions can be obtained for $\overline{W}[A^+, X^+, Y^+]$. Hence with (9.71) to (9.74) the component field expansion of the Lagrangian density of the O'Raifeartaigh model is (rewriting $a^*\Box a$ as $-(\partial_\mu a^*)(\partial^\mu a)$ as we can since

$$\int d^4x\; a^*\Box a = \int_{S_\infty} dS_\mu\, a^*\partial_\mu a - \int(\partial_\mu a^*)(\partial^\mu a)\,d^4x$$

where the first integral vanishes when integrated over an infinitely large surface)

$$\begin{aligned}
\mathcal{L} = {} & |\partial_\mu a(x)|^2 + |\partial_\mu x(x)|^2 + |\partial_\mu y(x)|^2 \\
& + |F_A(x)|^2 + |F_X(x)|^2 + |F_Y(x)|^2 \\
& + i\{(\partial_\mu \overline{\psi}_A(x))\,\overline{\sigma}^\mu \psi_A(x) + (\partial_\mu \overline{\psi}_X(x))\,\overline{\sigma}^\mu \psi_X(x) \\
& \quad + (\partial_\mu \overline{\psi}_Y(x))\,\overline{\sigma}^\mu \psi_Y(x)\} \\
& + \lambda\, a(x) F_Y(x) + \lambda F_A(x)\, y(x) - \lambda \psi_A(x)\psi_Y(x) \\
& + 2g\, x(x)\, a(x) F_A(x) + g F_X(x)\, a^2(x) \\
& - 2g\, a(x)\psi_X(x)\psi_A(x) - \tfrac{1}{2} x(x)\psi_A^2(x) \\
& - g M^2 F_X(x) \\
& + \lambda a^*(x) F_Y^*(x) + \lambda F_A^*(x)\, y^*(x)
\end{aligned}$$

$$- \lambda \bar{\Psi}_A(x) \bar{\Psi}_Y(x) + 2g\, x^*(x) a^*(x) F_A^*(x)$$
$$+ g F_X^*(x) a^{*2}(x) - 2g\, a^*(x) \bar{\Psi}_X(x) \bar{\Psi}_A(x)$$
$$- g\, x^*(x) \bar{\Psi}_A^2(x) - g M^2 F_X^*(x) \qquad (9.75)$$

The auxiliary fields $F_A(x)$, $F_X(x)$ and $F_Y(x)$ as well as their complex conjugates can be eliminated with the help of their respective Euler-Lagrange equations. We have

$$\frac{\partial \mathcal{L}}{\partial F_A(x)} = F_A^*(x) + 2g\, x(x) a(x) + \lambda y(x) = 0$$

so that

$$F_A^*(x) = - \frac{\partial W}{\partial a} = -2g\, x(x) a(x) - \lambda y(x)$$

Also

$$\frac{\partial \mathcal{L}}{\partial F_A^*(x)} = F_A(x) + 2g\, x^*(x) a^*(x) + \lambda y^*(x) = 0$$

so that

$$F_A(x) = - \frac{\partial \bar{W}}{\partial a^*} = -2g\, x^*(x) a^*(x) - \lambda y^*(x)$$

and

$$\frac{\partial \mathcal{L}}{\partial F_X(x)} = F_X^*(x) + g\,(a^2(x) - M^2) = 0$$

so that

$$F_X^*(x) = - \frac{\partial W}{\partial x(x)} = -g\,[a^2(x) - M^2]$$

and

$$F_X(x) = - \frac{\partial \bar{W}}{\partial x^*(x)} = -g\,[a^{*2}(x) - M^2]$$

Finally

$$\frac{\partial \mathcal{L}}{\partial F_Y(x)} = F_Y^*(x) + \lambda a(x) = 0$$

so that
$$F_Y^*(x) = -\frac{\partial W}{\partial y} = -\lambda a(x)$$
and
$$F_Y(x) = -\frac{\partial \overline{W}}{\partial y^*} = -\lambda a^*(x)$$

We can therefore rewrite the scalar potential (9.32) in the form

$$\begin{aligned} V &= |F_A(x)|^2 + |F_X(x)|^2 + |F_Y(x)|^2 \\ &= (2g\,x(x)a(x) + \lambda y(x))(2g\,x^*(x)a^*(x) + \lambda y^*(x)) \\ &\quad + g^2(a^2(x) - M^2)(a^{*2} - M^2) + \lambda^2 a(x)a^*(x) \\ &= |\lambda y(x) + 2g\,x(x)a(x)|^2 + g^2|a^2(x) - M^2|^2 \\ &\qquad + \lambda^2 |a(x)|^2 \\ &= V(a, a^*, x, x^*, y, y^*) \end{aligned} \tag{9.75'}$$

(using (9.61)). Inserting this into the Lagrangian (9.75) we obtain a Lagrangian which depends only on the dynamical fields $a(x)$, $x(x)$, $y(x)$, $\psi_A(x)$, $\psi_X(x)$ and $\psi_Y(x)$, i.e.

$$\begin{aligned} \mathcal{L} &= |\partial_\mu a(x)|^2 + |\partial_\mu x(x)|^2 + |\partial_\mu y(x)|^2 \\ &\quad - |\lambda y(x) + 2g\,x(x)a(x)|^2 - g^2|a^2(x) - M^2|^2 \\ &\quad - \lambda^2 |a(x)|^2 \end{aligned}$$

$$+ i \left[(\partial_\mu \overline{\Psi}_A(x)) \bar{\sigma}^\mu \psi_A(x) + (\partial_\mu \overline{\Psi}_X(x)) \bar{\sigma}^\mu \psi_X(x) + (\partial_\mu \overline{\Psi}_Y(x) \bar{\sigma}^\mu \psi_Y(x) \right]$$

$$- \lambda \psi_A(x) \psi_Y(x) - 2g a(x) \psi_X(x) \psi_A(x)$$

$$- g x(x) \psi_A^2(x) - \lambda \overline{\Psi}_A(x) \overline{\Psi}_Y(x)$$

$$- 2g a^*(x) \overline{\Psi}_X(x) \overline{\Psi}_A(x) - g x^*(x) \overline{\Psi}_A^2(x)$$

$$+ g^2 M^2 (a^2 + a^{*2} - 2M^2) \tag{9.76}$$

Next we derive the Euler-Lagrange equations of the dynamical fields:

i) $a^*(x)$:

$$\frac{\partial \mathcal{L}}{\partial a^*(x)} - \partial_\mu \frac{\partial \mathcal{L}}{\partial(\partial_\mu a^*(x))} = 0$$

gives

$$- \Box a(x) - \lambda^2 a(x) + 4g^2 M^2 a^*(x)$$
$$= 2g \overline{\Psi}_X(x) \overline{\Psi}_A(x) + 2\lambda g y(x) x^*(x)$$
$$+ 4g^2 |x(x)|^2 a(x) + 2g^2 |a|^2 a(x)$$

i.e. (9.77)

$$(\Box + \lambda^2) a(x) - 4g^2 M^2 a^*(x)$$
$$= - 2\lambda g y(x) x^*(x) - 2g \overline{\Psi}_X(x) \overline{\Psi}_A(x)$$
$$- 4g^2 |x(x)|^2 a(x) - 2g^2 |a(x)|^2 a(x)$$

ii) $a(x)$:

In a similar way we obtain

$$(\Box+\lambda^2)\,a^*(x)-4g^2M^2a(x)$$
$$=-2\lambda g\,y^*(x)\,x(x)-2g\,\psi_X(x)\,\psi_A(x)$$
$$-4g^2|x(x)|^2a^*(x)-2g^2|a(x)|^2a^*(x) \tag{9.78}$$

iii) $x^*(x)$:

The Euler-Lagrange equation

$$\frac{\partial\mathcal{L}}{\partial x^*(x)}-\partial_\mu\frac{\partial\mathcal{L}}{\partial(\partial_\mu x^*(x))}=0$$

leads to

$$\Box x(x)=-2\lambda g\,y(x)\,a^*(x)-4g^2|a(x)|^2x(x)$$
$$-g\,\Psi_A^2(x) \tag{9.79}$$

iv) $y^*(x)$:

Here

$$\frac{\partial\mathcal{L}}{\partial y^*(x)}-\partial_\mu\frac{\partial\mathcal{L}}{\partial(\partial_\mu y^*(x))}=0$$

leads to

$$(\Box+\lambda^2)\,y(x)=-2\lambda g\,a(x)\,x(x) \tag{9.80}$$

v) $\Psi_A(x)$:

Here

$$\frac{\partial\mathcal{L}}{\partial\Psi_A(x)}-\partial_\mu\frac{\partial\mathcal{L}}{\partial(\partial_\mu\Psi_A(x))}=0$$

where

$$\frac{\partial \mathcal{L}}{\partial(\partial_\mu \overline{\Psi}_A(x))} = i\bar{\sigma}^\mu \psi_A(x)$$

$$\frac{\partial \mathcal{L}}{\partial \overline{\Psi}_A(x)} = -\lambda \overline{\Psi}_Y(x) - 2g\, a^*(x)\, \overline{\Psi}_X(x) - 2g\, x^*(x)\, \overline{\Psi}_A(x)$$

so that the equation becomes

$$i\bar{\sigma}^\mu \partial_\mu \psi_A(x) = -\lambda \overline{\Psi}_Y(x) - 2g\, a^*(x)\, \overline{\Psi}_X(x) - 2g\, x^*(x)\, \overline{\Psi}_A(x) \tag{9.81}$$

vi) $\overline{\Psi}_Y(x)$:

The equation of motion for $\overline{\Psi}_Y(x)$ is

$$i\bar{\sigma}^\mu \partial_\mu \psi_Y(x) = -\lambda\, \overline{\Psi}_A(x) \tag{9.82}$$

vii) $\overline{\Psi}_X(x)$:

In this case we obtain the equation

$$i\bar{\sigma}^\mu \partial_\mu \psi_X(x) = -2g\, a^*(x)\, \overline{\Psi}_A(x) \tag{9.83}$$

In order to obtain the mass spectrum of the O'Raifeartaigh model, we have to consider small oscillations of the component fields around the absolute minimum of the potential V (a, a^*, x, x^*, y, y^*) which we choose to be at $a \equiv \langle a \rangle = 0$, $y \equiv \langle y \rangle = 0$ and x arbitrary. This choice corresponds to the case illustrated in Fig. 7, where min $V = g^2M^4$ and $M^2 < \lambda^2/2g^2$, and the internal U(1) symmetry is not broken.

Then from (9.79) we see that the scalar field $x(x)$ is massless, and from (9.80) we deduce that the scalar

$y(x)$ has mass λ . From (9.81) and (9.82) we see (by comparison with the Weyl equations (1.144b) for Weyl components of a Majorana spinor) that there are two massive fermions of mass $-\lambda$ for λ negative, which are linear combinations of $\psi_A(x)$ and $\psi_Y(x)$, and (9.83) demonstrates that $\psi_X(x)$ is a massless fermion, the goldstino. Decomposing the complex scalar field $a(x)$ into real and imaginary parts

$$a(x) = k(x) + ih(x)$$
$$a^*(x) = k(x) - ih(x) \qquad (9.84)$$

where k(x) and h(x) are real scalar fields, and inserting (9.84) into (9.77) or (9.78) we obtain the remarkable result that the mass-squared of k(x) is

$$m_k^2 = \lambda^2 - 4g^2M^2 = \lambda^2 - 4\frac{V(0)}{M^2} \qquad (9.85)$$

and that of h(x) is

$$m_h^2 = \lambda^2 + 4g^2M^2 = \lambda^2 + 4\frac{V(0)}{M^2} \qquad (9.86)$$

where $V(o) = g^2M^4$. We now see that the boson-fermion mass degeneracy, a characteristic property of supersymmetry, is destroyed. The quantity V(o) is a measure of the magnitude of the breaking of supersymmetry.

We conclude these considerations with the following remarks:

i) At the tree graph level only the bosonic part of the mass spectrum is affected by broken supersymmetry. Here it is the scalar field $a(x)$ which shows a mass splitting. The reason for this is clear, since $a(x)$ is the only scalar field which couples to the Goldstone fermion field $\psi_X(x)$ via the term gXA^2 in the superpotential (9.56).

ii) An interesting property of the mass spectrum at the classical, i.e. tree graph level, is the fact that

$$m_k^2 + m_h^2 = 2\lambda^2$$

irrespective of the supersymmetry breaking. At the tree graph level the fermion masses remain unchanged, whereas the scalars are shifted by the same amount in opposite directions. This is a particular case of a more general result[68] which states that the supertrace (cf. (2.33)) of the squared mass matrix, i.e.

$$STr\, M^2 = \sum_J (-1)^{2J}(2J+1) M_J^2 = 0 \tag{9.87}$$

remains zero in the presence of spontaneous symmetry breaking (J = spin of the particle). It can be shown, however, that this result does not hold if one includes radiative corrections[69].

iii) In the above discussion we considered the case $M^2 < \lambda^2 / 2g^2$. This condition leads to the correct sign of the mass-term m_k^2 (cf. (9.85)). On the other hand, for $M^2 > \lambda^2 / 2g^2$ it is the scalar field h(x) = Im a(x) which is responsible for the spontaneous breakdown of the internal U(1) symmetry.

iv) We state without further calculation that the same fermion mass spectrum is obtained by finding the eigenvalues of the fermionic mass matrix W_{ij} (see (9.8b) and (9.69)).

CHAPTER 10

SUPERSYMMETRIC GAUGE THEORIES

10.1 Minimal Coupling

We now discuss the supersymmetric generalization of various gauge theories and in addition an alternative to the O'Raifeartaigh model as a method to obtain spontaneous breaking of supersymmetry. This alternative theory is known as the Fayet-Iliopoulos model[60].

In Section 7.2 we investigated vector superfields and the supersymmetric extension of abelian gauge transformations. In addition, in Section 8.2.2 we calculated the action integral of a free abelian gauge field, using the supersymmetric extension of the field strength

$$W_A = -\frac{1}{4}\bar{D}\bar{D}D_A V_{WZ}(x,\theta,\bar{\theta}) \tag{10.1}$$

where $V_{WZ}(x, \theta, \bar{\theta})$ is a vector superfield in the Wess-Zumino gauge (see (7.40)). The gauge invariant and supersymmetry invariant action is, as we have seen (see (8.40)),

$$\mathcal{O}\!\mathcal{l} = \int d^4x \int d^4\theta \, \{ W^A W_A \delta^2(\bar\theta) + \bar W_{\dot A} \bar W^{\dot A} \delta^2(\theta) \} \qquad (10.2)$$

In the following we write this and similar expressions in the Grassmann integrated form

$$\mathcal{O}\!\mathcal{l} = \int d^4x \, \{ W^A W_A |_{\theta\theta} + \bar W_{\dot A} \bar W^{\dot A} |_{\bar\theta\bar\theta} \}$$

The component field expansion yields (cf. (8.41))

$$\mathcal{O}\!\mathcal{l} = \int d^4x \, \{ 8 D^2(x) - F_{\mu\nu}(x) F^{\mu\nu}(x) - i \lambda(x) \sigma^\mu \partial_\mu \bar\lambda(x) \} \qquad (10.3)$$

where

$$F_{\mu\nu}(x) = \partial_\mu V_\nu(x) - \partial_\nu V_\mu(x)$$

is the usual field strength tensor, D(x) is an auxiliary scalar field and λ (x) is the supersymmetric partner (spin-$\frac{1}{2}$ field) of V_μ (x), now part of the gauge field.

The question now arises as to how one can include matter fields. This implies, of course, that we must search for gauge invariant interactions of scalar and vector superfields. We begin with the simplest case, the U(1) gauge group. We define the transformation of scalar superfields under global U(1) gauge transformations by the relation

$$\Phi_i' = \exp\{-i q_i \lambda\} \Phi_i \qquad (10.4)$$

Here the q_i are the U(1) charges of the superfields Φ_i and λ is the global U(1) rotation angle. Both q_i and λ are real and constant. Since constants are particular cases of superfields satisfying the constraints $D_A \lambda = 0, \bar D_{\dot A} \lambda = 0$,it follows that the transformed scalar superfield Φ_i' (being a product of such fields) is again a scalar superfield satisfying the constraint

$$\bar{D}_{\dot{A}} \Phi_i' = 0 \tag{10.5}$$

It is now not difficult to construct a Lagrangian density which is invariant under (10.4) for constant λ. We write

$$\begin{aligned}
\mathcal{L} &= \mathcal{L}_{\text{kin}} + W[\Phi]\big|_{\theta\theta} + \bar{W}[\Phi^+]\big|_{\bar{\theta}\bar{\theta}} \\
&= \Phi_i^+ \Phi_i\big|_{\theta\theta\bar{\theta}\bar{\theta}} \\
&\quad + \left[\tfrac{1}{2} m_{ij} \Phi_i \Phi_j + \tfrac{1}{3}\lambda_{ijk} \Phi_i \Phi_j \Phi_k \right. \\
&\qquad \left. + g_i \Phi_i \right]\Big|_{\theta\theta} + \text{h.c.}
\end{aligned} \tag{10.6}$$

where h.c. implies the hermitian conjugate of the term in square brackets. We then have (using (10.4))

$$\begin{aligned}
\mathcal{L}' &= \Phi_i'^+ \Phi_i'\big|_{\theta\theta\bar{\theta}\bar{\theta}} \\
&\quad + \left[\tfrac{1}{2} m_{ij} \Phi_i' \Phi_j' + \tfrac{1}{3}\lambda_{ijk} \Phi_i' \Phi_j' \Phi_k' \right. \\
&\qquad \left. + g_i \Phi_i' \right]\Big|_{\theta\theta} + \text{h.c.} \\
&= (\Phi_i^+ e^{iq_i\lambda})(e^{-iq_i\lambda}\Phi_i)\big|_{\theta\theta\bar{\theta}\bar{\theta}} \\
&\quad + \left[\tfrac{1}{2} m_{ij} e^{-i(q_i+q_j)\lambda} \Phi_i \Phi_j + g_i e^{-i\lambda q_i} \Phi_i \right. \\
&\qquad \left. + \tfrac{1}{3}\lambda_{ijk} e^{-i(q_i+q_j+q_k)\lambda} \Phi_i \Phi_j \Phi_k \right]\Big|_{\theta\theta} \\
&\qquad + \text{h. c.} \\
&= \Phi_i^+ \Phi_i\big|_{\theta\theta\bar{\theta}\bar{\theta}} + \left\{ \tfrac{1}{2} m_{ij} e^{-i(q_i+q_j)\lambda} \Phi_i \Phi_j \right.
\end{aligned}$$

$$+ g_i e^{-i\lambda q_i} \Phi_i$$
$$+ \frac{1}{3}\lambda_{ijk} e^{-i(q_i+q_j+q_k)} \Phi_i \Phi_j \Phi_k \Big\}\Big|_{\theta\theta}$$
$$+ \text{h. c.} \tag{10.7}$$

To obtain global U(1) phase invariance of the superpotential we have to demand

$$\left.\begin{aligned} g_i &= 0 \quad \text{if} \quad q_i \neq 0 \\ m_{ij} &= 0 \quad \text{if} \quad q_i + q_j \neq 0 \\ \lambda_{ijk} &= 0 \quad \text{if} \quad q_i + q_j + q_k \neq 0 \end{aligned}\right\} \tag{10.8}$$

The kinetic term of the Lagrangian is automatically U(1) invariant without any restrictions.

As demonstrated above, a global U(1) phase transformation maps scalar superfields onto scalar superfields, this map depending crucially on the fact that the rotation angle λ may be looked at as a scalar superfield which satisfies the constraints $\mathcal{D}\lambda = 0$, $\overline{\mathcal{D}}\lambda = 0$.

If we want to introduce local U(1) phase transformations the construction must be altered in the following way. Assume that in a local theory the parameter λ is a function of the space-time variable x, i.e. $\lambda = \lambda(x)$, and let

$$\Phi_i' = \exp\{-i\lambda(x) q_i\}\Phi_i$$

Then Φ_i' is no longer a scalar superfield since

$$\overline{D}_{\dot{A}} \Phi_i' = \overline{D}_{\dot{A}} \{\exp[-iq_i \lambda(x)]\Phi_i\}$$
$$= \exp[-iq_i\lambda(x)]\overline{D}_{\dot{A}}\Phi_i$$
$$- iq_i \overline{D}_{\dot{A}}(\lambda(x)) e^{-iq_i\lambda(x)}\Phi_i$$

$$= -i q_i \bar{D}_{\dot{A}} (\lambda(x)) e^{-i\lambda(x) q_i} \Phi_i$$

In order to obtain a transformed scalar superfield Φ_i' satisfying the constraint equation $\bar{D}_{\dot{A}} \Phi_i' = 0$ we must introduce a quantity

$$\Lambda = \Lambda(x, \theta, \bar{\theta})$$

which obeys the same constraint equation, i.e.

$$\bar{D}_{\dot{A}} \Lambda(x, \theta, \bar{\theta}) = 0 \qquad (10.9)$$

However, the quantity $\Lambda(x, \theta, \bar{\theta})$ satisfying (10.9) is nothing but a full scalar multiplet. Hence, in order to replace a global U(1) phase transformation by a local phase transformation one is forced to introduce a new scalar superfield $\Lambda(x, \theta, \bar{\theta})$ such that the transformed scalar superfields are again scalar superfields. Thus the local U(1) phase transformation is given by the relations

$$\begin{aligned}
&\Phi_i' = \exp\{-i q_i \cdot \Lambda(x, \theta, \bar{\theta})\} \Phi_i \\
&\bar{D}_{\dot{A}} \Lambda(x, \theta, \bar{\theta}) = 0 \\
&\Phi_i'^{+} = \Phi_i^{+} \exp\{i q_i \cdot \Lambda^{+}(x, \theta, \bar{\theta})\} \\
&D_A \Lambda^{+}(x, \theta, \bar{\theta}) = 0
\end{aligned} \qquad (10.10)$$

As in the case of ordinary local phase transformations, the Lagrangian (10.6) is not invariant under (10.10). The superpotential $W[\Phi]$ remains invariant with respect to the restrictions (10.8) , but the kinetic part does not since

$$\mathcal{L}'_{\text{kin}} = \Phi_i'^{+} \Phi_i' \Big|_{\theta\theta\bar{\theta}\bar{\theta}}$$

$$= \exp\{iq_i \cdot (\Lambda^+ - \Lambda)\}\, \Phi_i^+ \Phi_i \big|_{\theta\theta\bar\theta\bar\theta}$$

6using (10.10))

$$\neq \Phi_i^+ \Phi_i$$

To obtain an invariant kinetic part of the Lagrangian we have to introduce a compensating field as in ordinary gauge theories. In the supersymmetric case we must choose a vector superfield V = V (x, θ , $\bar\theta$) which has the transformation law under supersymmetric gauge transformations (as we know from (7.35) with $\Phi = i\Lambda$)

$$V'(x,\theta,\bar\theta)$$
$$= V(x,\theta,\bar\theta) + i\,[\Lambda(x,\theta,\bar\theta) - \Lambda^+(x,\theta,\bar\theta)] \tag{10.11}$$

Then the invariant expression is given by

$$\mathcal{L}_{\text{kin}} = \Phi_i^+ e^{q_i V} \Phi_i \big|_{\theta\theta\bar\theta\bar\theta} \tag{10.12}$$

In order to check the invariance we consider

$$\mathcal{L}'_{\text{kin}} = \Phi_i'^+ e^{q_i \cdot V'} \Phi_i' \big|_{\theta\theta\bar\theta\bar\theta}$$
$$= \Phi_i^+ e^{iq_i \cdot \Lambda^+} \exp\{q_i \cdot (V + i\Lambda - i\Lambda^+)\} \cdot$$
$$\cdot\, e^{-iq_i \cdot \Lambda}\, \Phi_i \big|_{\theta\theta\bar\theta\bar\theta}$$

(with (10.10) and (10.11))

$$= \Phi_i^+ e^{iq_i \cdot \Lambda^+} e^{-iq_i \cdot \Lambda^+} e^{q_i \cdot V} e^{iq_i \cdot \Lambda} e^{-iq_i \cdot \Lambda} \cdot$$
$$\cdot\, \Phi_i \big|_{\theta\theta\bar\theta\bar\theta}$$

$$= \Phi_i^+ e^{q_i V} \Phi_i \big|_{\theta\theta\bar{\theta}\bar{\theta}}$$

$$= \mathcal{L}_{\text{kin}}$$

as had to be shown. Hence, as we shall demonstrate below, (10.12) is the supersymmetric generalization of the principle of minimal coupling in ordinary gauge theories. With (10.12) and the action (8.40) for pure abelian supersymmetric gauge theories the most general U(1) gauge invariant supersymmetric action integral is

$$\mathcal{A} = \int d^4x \int d^4\theta \, \{ W^A W_A \, \delta^2(\bar{\theta}) + \bar{W}_{\dot{A}} \bar{W}^{\dot{A}} \, \delta^2(\theta)$$
$$+ \Phi_i^+ e^{q_i V} \Phi_i$$
$$+ W[\Phi] \, \delta^2(\bar{\theta}) + \bar{W}[\Phi^+] \, \delta^2(\theta) \}$$

(10.13)

where the superpotential $W[\Phi]$ is a polynomial of the superfield Φ . Renormalizability allows $W[\Phi]$ to be at most of order three in Φ as we remarked in Section 8.2.1.

In order to demonstrate that the formal expression (10.12) is indeed the supersymmetric minimal coupling, we must reexpress the kinetic term (10.12) in terms of component fields and show that the component field $V_\mu(x)$ of V appears in (10.12) only in the covariant derivative. This is most easily done in the Wess-Zumino gauge (7.40), in which case we can use (7.41) and (7.42). For simplicity we consider only the case of a single scalar superfield Φ with component expansion

$$\Phi = A(y) + 2^{\frac{1}{2}} \, \theta\psi(y) + \theta\theta F(y)$$
$$\Phi^+ = A^*(y) + 2^{\frac{1}{2}} \, \bar{\theta}\bar{\psi}(y) + \bar{\theta}\bar{\theta} F^*(y)$$

Then

$$\Phi^+ e^{qV} \Phi \Big|_{\theta\theta\bar{\theta}\bar{\theta}}$$

$$= \Phi^+ \Big\{ 1 + q\big[(\theta\sigma^\mu\bar{\theta}) V_\mu(x) + (\theta\theta)(\bar{\theta}\bar{\lambda}(x)) + (\bar{\theta}\bar{\theta})(\theta\lambda(x)) + (\theta\theta)(\bar{\theta}\bar{\theta}) D(x) + (\theta\theta)(\bar{\theta}\bar{\theta}) \frac{q}{4} V_\mu(x) V^\mu(x)\big]\Big\} \Phi \Big|_{\theta\theta\bar{\theta}\bar{\theta}}$$

(using (7.42))

$$= \Phi^+\Phi \Big|_{\theta\theta\bar{\theta}\bar{\theta}} + q(\theta\sigma^\mu\bar{\theta} V_\mu(x) \Phi^+\Phi)\Big|_{\theta\theta\bar{\theta}\bar{\theta}} + q(\theta\theta\,\bar{\theta}\bar{\lambda}(x)\Phi^+\Phi)\Big|_{\theta\theta\bar{\theta}\bar{\theta}} + q(\bar{\theta}\bar{\theta}\,\theta\lambda(x)\Phi^+\Phi)\Big|_{\theta\theta\bar{\theta}\bar{\theta}} + q\Big(\theta\theta\bar{\theta}\bar{\theta}\Big\{D(x) + \frac{1}{4} q V_\mu(x) V^\mu(x)\Big\}\Phi^+\Phi\Big)\Big|_{\theta\theta\bar{\theta}\bar{\theta}}$$

The field Φ^+ can be shifted to the right because this involves only an even number of anticommutations of Grassmann variables. Using (7.27b), (8.28) and (8.29) for the product $\Phi^+\Phi$ we obtain

$$\Phi^+ e^{qV} \Phi \Big|_{\theta\theta\bar{\theta}\bar{\theta}}$$

$$= |F(x)|^2 + i\partial_\mu \bar{\psi}(x)\bar{\sigma}^\mu\psi(x) - A^*(x)\Box A(x) + iq(\theta\sigma^\mu\bar{\theta})(\theta\sigma^\nu\bar{\theta}) V_\mu(x) \cdot \{(\partial_\nu A(x))A^*(x) - (\partial_\nu A^*(x))A(x)\}\Big|_{\theta\theta\bar{\theta}\bar{\theta}} + 2q(\theta\sigma^\mu\bar{\theta}) V_\mu(x)\, \bar{\theta}\bar{\psi}(x)\, \theta\psi(x)\Big|_{\theta\theta\bar{\theta}\bar{\theta}}$$

$$+ 2^{\frac{1}{2}} q\, (\theta\theta)\, \bar\theta\bar\lambda(x)\ \bar\theta\bar\Psi(x)\, A(x)\big|_{\theta\theta\,\bar\theta\bar\theta}$$

$$+ 2^{\frac{1}{2}} q\, (\bar\theta\bar\theta)\, \theta\lambda(x)\ \theta\psi(x)\, A^{*}(x)\big|_{\theta\theta\,\bar\theta\bar\theta}$$

$$+ q\,[D(x) + \tfrac{1}{4} q\, V_\mu(x) V^\mu(x)]\, |A(x)|^2$$

+ total derivatives (10.14)

We now use (1.118), i.e.

$$(\theta\sigma^\mu\bar\theta)(\theta\sigma^\nu\bar\theta) = \tfrac{1}{2}\eta^{\mu\nu}(\theta\theta)(\bar\theta\bar\theta) \qquad (10.15)$$

and

$$(\theta\sigma^\mu\bar\theta)(\bar\theta\bar\Psi(x))(\theta\psi(x))$$
$$= \theta^A \sigma^\mu_{A\dot A}\bar\theta^{\dot A}\,\bar\theta_{\dot B}\,\bar\Psi^{\dot B}\,\theta^B\psi_B(x)$$
$$= -\theta^A\theta^B \sigma^\mu_{A\dot A}\,\bar\theta^{\dot A}\varepsilon_{\dot B\dot C}\bar\theta^{\dot C}\bar\Psi^{\dot B}(x)\,\psi_B(x)$$
$$= -\tfrac{1}{2}(\theta\theta)\varepsilon^{AB}\sigma^\mu_{A\dot A}\,\bar\theta^{\dot A}\bar\theta^{\dot C}\varepsilon_{\dot B\dot C}\,\bar\Psi^{\dot B}(x)\,\psi_B(x)$$

(using (1.83a))

$$= \tfrac{1}{4}(\theta\theta)(\bar\theta\bar\theta)\varepsilon^{AB}\sigma^\mu_{A\dot A}\varepsilon^{\dot A\dot C}\varepsilon_{\dot B\dot C}\,\bar\Psi^{\dot B}(x)\,\psi_B(x)$$

(using (1.83c))

$$= \tfrac{1}{4}(\theta\theta)(\bar\theta\bar\theta)\varepsilon^{BA}\varepsilon^{\dot C\dot A}\sigma^\mu_{A\dot A}\varepsilon_{\dot B\dot C}\,\bar\Psi^{\dot B}(x)\,\psi_B(x)$$
$$= -\tfrac{1}{4}(\theta\theta)(\bar\theta\bar\theta)\,\bar\sigma^{\mu\,\dot C B}\,\bar\Psi_{\dot C}(x)\,\psi_B(x)$$

(using (1.88))

$$= -\tfrac{1}{4}(\theta\theta)(\bar\theta\bar\theta)\,\bar\Psi(x)\bar\sigma^\mu\psi(x) \qquad (10.16)$$

Furthermore

$$(\theta\theta)\,(\bar\theta\bar\lambda(x))(\bar\theta\bar\Psi(x))$$
$$= (\theta\theta)\,\bar\theta_{\dot A}\bar\lambda^{\dot A}(x)\ \bar\theta_{\dot B}\bar\Psi^{\dot B}(x)$$

$$= -(\theta\theta)\bar\theta_{\dot A}\bar\theta_{\dot B}\bar\lambda^{\dot A}(x)\bar\psi^{\dot B}(x)$$

$$= \frac{1}{2}(\theta\theta)(\bar\theta\bar\theta)\varepsilon_{\dot A\dot B}\bar\lambda^{\dot A}(x)\bar\psi^{\dot B}(x)$$

(using (1.83d))

$$= -\frac{1}{2}(\theta\theta)(\bar\theta\bar\theta)\bar\lambda(x)\bar\psi(x) \qquad (10.17)$$

and similarly

$$\bar\theta\bar\theta\ \theta\lambda(x)\ \theta\psi(x) = -\frac{1}{2}\bar\theta\bar\theta\ \theta\theta\ \lambda(x)\psi(x) \qquad (10.18)$$

Inserting (10.15) to (10.18) into (10.14) we obtain

$$\begin{aligned}
&\Phi^+ e^{qV}\Phi\Big|_{\theta\theta\bar\theta\bar\theta} \\
&= |F(x)|^2 + i(\partial_\mu\bar\psi(x))\bar\sigma^\mu\psi(x) - A^*(x)\Box A(x) \\
&\quad + \frac{1}{2}iq\,V^\mu(x)\left[(\partial_\mu A(x))A^*(x) - (\partial_\mu A^*(x))A(x)\right] \\
&\quad - \frac{1}{2}q\,V_\mu(x)\bar\psi(x)\bar\sigma^\mu\psi(x) \\
&\quad - \frac{1}{2^{1/2}}q\left[\bar\lambda(x)\bar\psi(x)A(x) + \lambda(x)\psi(x)A^*(x)\right] \\
&\quad + q\left[D(x) + \frac{1}{4}q\,V_\mu(x)V^\mu(x)\right]|A(x)|^2 \\
&\quad + \text{total derivatives} \qquad (10.19)
\end{aligned}$$

Now consider the expression (defining D_μ)

$$\begin{aligned}
&|D_\mu A(x)|^2 \equiv \left|(\partial_\mu - \tfrac{1}{2}iq\,V_\mu(x))A(x)\right|^2 \equiv \\
&= \left\{(\partial_\mu - \tfrac{1}{2}iq\,V_\mu(x))A(x)\right\}\left\{(\partial^\mu + \tfrac{1}{2}iqV^\mu(x))A^*(x)\right\} \\
&= \left\{\partial_\mu A(x) - \tfrac{1}{2}iqV_\mu(x)A(x)\right\}\left\{\partial^\mu A^*(x) + \tfrac{1}{2}iqV^\mu(x)A^*(x)\right\} \\
&= |\partial_\mu A(x)|^2 + \frac{1}{4}q^2 V_\mu(x)V^\mu(x)|A(x)|^2
\end{aligned}$$

$$+ \tfrac{1}{2} i q V^\mu(x) \{ (\partial_\mu A(x)) A^*(x) - A(x) (\partial_\mu A^*(x)) \}$$

$$= - A^*(x) \Box A(x) + \tfrac{1}{4} q^2 V_\mu(x) V^\mu(x) \, |A(x)|^2$$

$$+ \tfrac{1}{2} i q V^\mu(x) [(\partial_\mu A(x)) A^*(x) - A(x) (\partial_\mu A^*(x))]$$

+ total derivatives (10.20)

We define

$$\mathcal{D}_\mu := \partial_\mu - i q \left(\frac{V_\mu(x)}{2} \right) \tag{10.21}$$

as the gauge covariant derivative. The factor $\frac{1}{2}$, which at first seems unusual, has its origin in the choice of coefficients of the component field expansion (7.34) and can be removed by redefining V_μ .Furthermore

$$i (\partial_\mu \bar{\psi}(x)) \bar{\sigma}^\mu \psi(x) - \tfrac{1}{2} q V_\mu(x) \bar{\psi}(x) \bar{\sigma}^\mu \psi(x)$$

$$= [i (\partial_\mu + \tfrac{1}{2} i q V_\mu(x)) \bar{\psi}(x)] \bar{\sigma}^\mu \psi(x)$$

$$= i (\mathcal{D}_\mu^* \bar{\psi}(x)) \bar{\sigma}^\mu \psi(x) \tag{10.22}$$

Hence with (10.20) and (10.22) we obtain

$$\Phi^+ e^{qV} \Phi \Big|_{\theta\theta\bar{\theta}\bar{\theta}}$$

$$= |F(x)|^2 + i (\mathcal{D}_\mu^* \bar{\psi}(x)) \bar{\sigma}^\mu \psi(x) + |\mathcal{D}_\mu A(x)|^2$$

$$- \frac{1}{2^{1/2}} q [\bar{\lambda}(x) \bar{\psi}(x) A(x) + \lambda(x) \psi(x) A^*(x)]$$

$$+ q D(x) |A(x)|^2 \tag{10.23}$$

This component field expansion demonstrates that (10.12) is the correct supersymmetric extension of the principle of minimal coupling, i.e. in that term of $\mathcal{L}$ which couples V to Φ, the vector field V_μ contained in V is completely absorbed in the gauge covariant derivative (10.21).

We mention in passing that the $\theta\theta\bar{\theta}\bar{\theta}$-component of

$$\Phi^+ e^{qV} \Phi$$

contains no terms of dimension higher than the fourth which ensures that the Lagrangian (10.13) remains renormalizable. However, this depends crucially on the fact that we expanded exp (qV) in the Wess-Zumino gauge, in which $V^3_{WZ} = 0$ (cf. (7.41)).

10.2 Super Quantum Electrodynamics

As an application of the ideas of Section 10.1 we extend ordinary electrodynamics (including interactions with matter fields) to a supersymmetric theory. We have to introduce two scalar superfields Φ_1 and Φ_2 which transform under local U(1) phase transformations according to (10.10), i.e.

$$\Phi_1 \rightarrow \Phi_1' = exp\{-i q_1 \Lambda(x,\theta,\bar{\theta})\}\Phi_1 \qquad (10.24)$$

$$\Phi_2 \rightarrow \Phi_2' = exp\{-i q_2 \Lambda(x,\theta,\bar{\theta})\}\Phi_2$$

where $q_1 = +e$ and $q_2 = -e$. Then according to the restrictions (10.8) the couplings and masses of the superpotential of the superaction (10.13) are subject to the following constraints

$$g_1 = g_2 = 0 \quad \text{since} \quad q_1 = e \neq 0$$
$$\text{and} \quad q_2 = -e \neq 0$$

$$m_{12} =: m \neq 0 \quad \text{since} \quad q_1 + q_2 = e - e = 0$$

and all couplings λ_{ijk} vanish since

$$2q_1 + q_2 \neq 0, \quad 3q_1 \neq 0$$
$$2q_2 + q_1 \neq 0, \quad 3q_2 \neq 0 \tag{10.25}$$

Inserting these restrictions into the superpotential $W[\Phi]$ we obtain a U(1) gauge invariant action, i.e.

$$\begin{aligned} \mathcal{A}_{\text{SQED}} = \int d^4x \int d^4\theta \{ & W^A W_A \delta^2(\bar{\theta}) \\ & + \overline{W}_{\dot{A}} \overline{W}^{\dot{A}} \delta^2(\theta) + \Phi_1^+ e^{eV} \Phi_1 \\ & + \Phi_2^+ e^{-eV} \Phi_2 - m (\Phi_1 \Phi_2 \delta^2(\bar{\theta}) \\ & + \Phi_1^+ \Phi_2^+ \delta^2(\theta)) \} \end{aligned} \tag{10.26}$$

(using (10.6) and (10.12))

We now reexpress $\mathcal{A}_{\text{SQED}}$ in terms of its component fields. According to (8.41) we have

$$\begin{aligned} & \int d^4x \int d^4\theta \{ W^A W_A \delta^2(\bar{\theta}) + \overline{W}_{\dot{A}} \overline{W}^{\dot{A}} \delta^2(\theta) \} \\ & = \int d^4x \{ 8 D^2(x) - F_{\mu\nu}(x) F^{\mu\nu}(x) \\ & \qquad - 4i \lambda(x) \sigma^\mu \partial_\mu \bar{\lambda}(x) \} \end{aligned} \tag{10.27}$$

Using (10.23) we obtain for the kinetic terms of the matter fields Φ_1 and Φ_2

$$\int d^4x \int d^4\theta\, \Phi_1^+ e^{eV(x,\theta,\bar\theta)} \Phi_1$$

$$= |F_1(x)|^2 + i\,(D_\mu^* \bar\Psi_1(x))\,\bar\sigma^\mu \psi_1(x) + |D_\mu A_1(x)|^2$$

$$- \frac{1}{2^{1/2}} e\left[\bar\lambda(x)\bar\Psi_1(x)A_1(x) + \lambda(x)\psi_1(x)A_1^*(x)\right]$$

$$+ e\,D(x)\,|A_2(x)|^2 \qquad (10.28)$$

where the covariant derivatives are from (10.21)

$$D_\mu = \partial_\mu - \tfrac{1}{2} ie\, V_\mu(x)\,, \quad D_\mu^* = \partial_\mu + \tfrac{1}{2} ie\, V_\mu(x)$$

and

$$\int d^4x \int d^4\theta\, \Phi_2^+ e^{-eV(x,\theta,\bar\theta)} \Phi_2$$

$$= |F_2(x)|^2 + i\,(D_\mu \bar\Psi_2(x))\,\bar\sigma^\mu \psi_2(x)$$

$$+ |D_\mu^* A_2(x)|^2$$

$$+ \frac{1}{2^{1/2}} e\left[\bar\lambda(x)\bar\Psi_2(x)A_2(x) + \lambda(x)\psi_2(x)A_2^*(x)\right]$$

$$- e\,D(x)\,|A_2(x)|^2 \qquad (10.29)$$

Finally the mass-terms give, using (7.16),

$$\int d^4x \int d^4\theta\, m\, \Phi_1 \Phi_2\, \delta^2(\bar\theta)$$

$$= \int d^4x\, m\, \Phi_1\Phi_2\big|_{\theta\theta}$$

$$= \int d^4x\, m\left[A_1(x)F_2(x) + A_2(x)F_1(x) - \psi_1(x)\psi_2(x)\right] \qquad (10.30)$$

and

$$\int d^4x \int d^4\theta\, m\, \Phi_1^+ \Phi_2^+\, \delta^2(\theta)$$

$$= \int d^4x \, m \, \Phi_1^+ \Phi_2^+ \Big|_{\bar\theta\bar\theta}$$

$$= \int d^4x \, m \left[A_1^*(x) F_2^*(x) + A_2^*(x) F_1^*(x) - \bar\psi_1(x) \bar\psi_2(x) \right] \qquad (10.31)$$

Inserting (10.27) to (10.31) into the action (10.26) we obtain

$$\begin{aligned} \mathcal{A}_{SQED} = \int d^4x \Big\{ & 8 D^2(x) - F_{\mu\nu}(x) F^{\mu\nu}(x) \\ & - 4i \lambda(x) \sigma^\mu \partial_\mu \bar\lambda(x) + |F_1(x)|^2 \\ & + |F_2(x)|^2 + i (D_\mu^* \bar\psi_1(x)) \bar\sigma^\mu \psi_1(x) \\ & + i D_\mu \bar\psi_2(x) \bar\sigma^\mu \psi_2(x) + |D_\mu A_1(x)|^2 \\ & + |D_\mu A_2(x)|^2 + e D(x) \left(|A_1(x)|^2 - |A_2(x)|^2 \right) \\ & - \frac{1}{2^{1/2}} e \Big[\bar\lambda(x) \left(\bar\psi_1(x) A_1(x) - \bar\psi_2(x) A_2(x) \right) \\ & \quad + \lambda(x) \left(\psi_1(x) A_1^*(x) - \psi_2(x) A_2^*(x) \right) \Big] \\ & - m \Big[A_1(x) F_2(x) + A_1^*(x) F_2^*(x) \\ & \quad + A_2(x) F_1(x) + A_2^*(x) F_1^*(x) \\ & \quad - \psi_1(x) \psi_2(x) - \bar\psi_1(x) \bar\psi_2(x) \Big] \Big\} \end{aligned} \qquad (10.32)$$

We now eliminate the auxiliary fields $F_1(x)$, $F_2(x)$ and $D(x)$ using their respective Euler-Lagrange equations, i.e.

$$\frac{\partial \mathcal{L}}{\partial F_1(x)} = F_1^*(x) - m A_2(x) = 0$$

so that

$$F_1^*(x) = mA_2(x), \quad F_1(x) = mA_2^*(x)$$

and

$$\frac{\partial \mathcal{L}}{\partial F_2(x)} = F_2^*(x) - mA_1(x) = 0$$

so that

$$F_2^*(x) = mA_1(x), \quad F_2(x) = mA_1^*(x)$$

Hence

$$|F_1(x)|^2 + |F_2(x)|^2 = m^2\left(|A_1(x)|^2 + |A_2(x)|^2\right)$$

and

$$-m\left(A_1(x)F_2(x) + A_1^*(x)F_2^*(x) + A_2(x)F_1(x) + A_2^*(x)F_1^*(x)\right)$$
$$= -2m^2\left(|A_1(x)|^2 + |A_2(x)|^2\right)$$

Furthermore

$$\frac{\partial \mathcal{L}}{\partial D(x)} = 16D(x) + e\left(|A_1(x)|^2 - |A_2(x)|^2\right) = 0$$

so that

$$D(x) = -\frac{e}{16}\left(|A_1(x)|^2 - |A_2(x)|^2\right)$$

Then

$$8D^2(x) + eD(x)\left(|A_1(x)|^2 - |A_2(x)|^2\right)$$
$$= -\frac{1}{32}e^2\left(|A_1(x)|^2 - |A_2(x)|^2\right)$$

In terms of the dynamical fields the action (10.32) then becomes

$$\mathcal{A}_{\text{SQED}} = \int d^4x \Big\{ -F_{\mu\nu}(x)F^{\mu\nu}(x) - 4i\lambda(x)\sigma^\mu\partial_\mu\bar{\lambda}(x) + i(D_\mu^*\bar{\psi}_1(x))\bar{\sigma}^\mu\psi_1(x) + i(D_\mu\bar{\psi}_2(x))\bar{\sigma}^\mu\psi_2(x) + |D_\mu A_1(x)|^2 + |D_\mu^* A_2(x)|^2 - \frac{e}{2^{1/2}}\big[\bar{\lambda}(x)(\bar{\psi}_1(x)A_1(x) - \bar{\psi}_2(x)A_2(x)) + \lambda(x)(\psi_1(x)A_1^*(x) - \psi_2(x)A_2^*(x))\big] + m[\psi_1(x)\psi_2(x) + \bar{\psi}_1(x)\bar{\psi}_2(x)] - m^2(|A_1(x)|^2 + |A_2(x)|^2) - \frac{e^2}{32}(|A_1(x)|^2 - |A_2(x)|^2) \Big\} \quad (10.33)$$

In order to see that the action (10.33) describes the correct field content, i.e. in particular a massive Dirac field, we define the four-component Dirac spinor by (cf. (1.130))

$$\Psi(x) := \begin{pmatrix} \psi_{1A}(x) \\ \bar{\psi}_2{}^{\dot{A}}(x) \end{pmatrix} \quad (10.34)$$

and the Dirac adjoint of (10.34) is (cf. (1.178a)) with a suitable change of nomenclature

$$\bar{\Psi}(x) = \left(\psi_2{}^B(x),\ \bar{\psi}_{1\dot{B}}(x)\right) \quad (10.35)$$

We now consider the following terms of (10.33):

$$i(D_\mu^* \bar{\Psi}_1(x))\bar{\sigma}^\mu \psi_1(x) + i(D_\mu \bar{\Psi}_2(x))\bar{\sigma}^\mu \psi_2(x)$$

$$= i[\partial_\mu + \tfrac{1}{2} ie V_\mu(x)]\bar{\Psi}_1(x)\bar{\sigma}^\mu \psi_1(x)$$
$$+ i[\partial_\mu - \tfrac{1}{2} ie V_\mu(x)]\bar{\Psi}_2(x)\bar{\sigma}^\mu \psi_2(x)$$

$$= i(\partial_\mu \bar{\Psi}_1(x))\bar{\sigma}^\mu \psi_1(x) - \tfrac{1}{2} e\bar{\Psi}_1(x)\bar{\sigma}^\mu V_\mu(x)\psi_1(x)$$
$$+ i(\partial_\mu \bar{\Psi}_2(x))\bar{\sigma}^\mu \psi_2(x) + \tfrac{1}{2} e\bar{\Psi}_2(x)\bar{\sigma}^\mu V_\mu(x)\psi_2(x)$$

$$= -i\bar{\Psi}_1(x)\bar{\sigma}^\mu(\partial_\mu \psi_1(x)) - \tfrac{1}{2} e\bar{\Psi}_1(x)\bar{\sigma}^\mu V_\mu(x)\psi_1(x)$$
$$+ i(\partial_\mu \bar{\Psi}_2(x))\bar{\sigma}^\mu \psi_2(x) + \tfrac{1}{2} e\bar{\Psi}_2(x)\bar{\sigma}^\mu V_\mu(x)\psi_2(x)$$

+ total derivatives

$$= -i\bar{\Psi}_1(x)\bar{\sigma}^\mu(\partial_\mu - \tfrac{1}{2} ie V_\mu(x))\psi_1(x)$$
$$- i\psi_2(x)\sigma^\mu \partial_\mu \bar{\Psi}_2(x) - \tfrac{1}{2} e\psi_2(x)\sigma^\mu V_\mu(x)\bar{\Psi}_2(x)$$

+ total derivatives

(using (1.115))

$$= -i\bar{\Psi}_1(x)\bar{\sigma}^\mu[\partial_\mu - \tfrac{1}{2} ie V_\mu(x)]\psi_1(x)$$
$$- i\bar{\Psi}_2(x)\sigma^\mu[\partial_\mu - \tfrac{1}{2} ie V_\mu(x)]\psi_2(x)$$

+ total derivatives

$$= -i\bar{\Psi}_1(x)\bar{\sigma}^\mu D_\mu \psi_1(x) - i\psi_2(x)\sigma^\mu D_\mu \bar{\Psi}_2(x)$$

+ total derivatives

$$= -i(\psi_2(x), \bar{\Psi}_1(x))\begin{pmatrix} 0 & \sigma^\mu D_\mu \\ \bar{\sigma}^\mu D_\mu & 0 \end{pmatrix}\begin{pmatrix} \psi_1(x) \\ \bar{\Psi}_2(x) \end{pmatrix}$$

+ total derivatives

$$= -i\,\overline{\Psi}(x)\gamma_W^{\mu} D_\mu \Psi(x) \text{ + total derivatives}$$

(using (10.35), (10.34), (1.136)) (10.36)

where the gauge covariant derivative is given by

$$D_\mu = \partial_\mu - \tfrac{1}{2} i e\, V_\mu(x) \qquad (10.37)$$

Hence in (10.33) we can rewrite the following terms

$$i(D_\mu^* \overline{\psi}_1(x))\bar{\sigma}^\mu \psi_1(x) + i(D_\mu \overline{\psi}_2(x))\bar{\sigma}^\mu \psi_2(x)$$
$$+ m[\psi_1(x)\psi_2(x) + \overline{\psi}_1(x)\overline{\psi}_2(x)]$$
$$= -i\,\overline{\Psi}(x)\gamma_W^\mu D_\mu \Psi(x) + m(\psi_2(x), \overline{\psi}_1(x))\begin{pmatrix}\psi_1(x)\\ \overline{\psi}_2(x)\end{pmatrix}$$
$$= -\{i\,\overline{\Psi}(x)\gamma_W^\mu D_\mu \Psi(x) - m\,\overline{\Psi}(x)\Psi(x)\}$$
$$\equiv -\,\overline{\Psi}(x)(i\not{D} - m)\Psi(x) \qquad (10.38)$$

This part of the action (10.33) is the Lagrangian for a massive electron field (of mass m and charge q = - e) which is minimally coupled to a U(1) gauge boson field $V_\mu(x)$.[70]

10.3 The Fayet-Iliopoulos Model

We now consider the Fayet-Iliopoulos mechanism of spontaneous breaking of supersymmetry.[60] This mechanism is an alternative to the O'Raifeartaigh model discussed in Chapter 9 as a mechanism to obtain spontaneous breaking of supersymmetry. In the Fayet-Iliopoulos model one considers the breaking of supersymmetry in gauge theories with abelian gauge groups. The basic elements of this theory are the Lagrangian of the action (10.26) of supersymmetric electrodynamics and the $\theta\theta\bar{\theta}\bar{\theta}$ -component of a vector superfield. The Lagrangian is

$$\begin{aligned}\mathcal{L}_{FI} = &\; W^A W_A \big|_{\theta\theta} + \bar{W}_{\dot{A}} \bar{W}^{\dot{A}} \big|_{\bar{\theta}\bar{\theta}} \\ &+ \Phi_1^+ e^{eV} \Phi_1 \big|_{\theta\theta\bar{\theta}\bar{\theta}} + \Phi_2^+ e^{-eV} \Phi_2 \big|_{\theta\theta\bar{\theta}\bar{\theta}} \\ &- m \left(\Phi_1 \Phi_2 \big|_{\theta\theta} + \Phi_1^+ \Phi_2^+ \big|_{\bar{\theta}\bar{\theta}} \right) \\ &+ 2kV \big|_{\theta\theta\bar{\theta}\bar{\theta}}\end{aligned} \tag{10.39}$$

Expanding this expression in terms of its component fields, using (10.32) and (7.29), we obtain

$$\begin{aligned}\mathcal{L}_{FI} = &- F_{\mu\nu}(x) F^{\mu\nu}(x) - 4i\lambda(x)\sigma^\mu \partial_\mu \bar{\lambda}(x) \\ &+ i D_\mu^* \bar{\psi}_1(x) \bar{\sigma}^\mu \psi_1(x) + |D_\mu A_1(x)|^2 \\ &+ i D_\mu \bar{\psi}_2(x) \bar{\sigma}^\mu \psi_2(x) + |D_\mu^* A_2(x)|^2 \\ &- \frac{e}{2^{1/2}} \{ \bar{\lambda}(x)(\bar{\psi}_1(x) A_1(x) - \bar{\psi}_2(x) A_2(x)) \\ &\qquad + \lambda(x)(\psi_1(x) A_1^*(x) - \psi_2(x) A_2^*(x)) \} \\ &+ e D(x) \left(|A_1(x)|^2 - |A_2(x)|^2 \right) \\ &- m [A_1(x) F_2(x) + A_1^*(x) F_2^*(x) \\ &\qquad + A_2(x) F_1(x) + A_2^*(x) F_1^*(x)] \\ &+ m \left(\psi_1(x) \psi_2(x) + \bar{\psi}_1(x) \bar{\psi}_2(x) \right) \\ &+ 2k D(x) + |F_1(x)|^2 + |F_2(x)|^2 \\ &+ 8 D^2(x)\end{aligned} \tag{10.40}$$

For later convenience we write out explicitly the expressions of the gauge covariant derivatives of the scalar fields $A_1(x)$ and $A_2(x)$ in (10.40) (all expressions are complete except for total derivatives which are deleted)

$$
\begin{aligned}
|\mathcal{D}_\mu A_1(x)|^2 \\
&= -A_1^*(x)\,\Box A_1(x) + \tfrac{1}{4}e^2|A_1(x)|^2 V_\mu(x)V^\mu(x) \\
&\quad + \tfrac{1}{2}ie\,V^\mu(x)\big[(\partial_\mu A_1(x))A_1^*(x) \\
&\qquad\qquad - A_1(x)(\partial_\mu A_1^*(x))\big]
\end{aligned}
$$

(using (10.10) and (10.21))

and

$$
\begin{aligned}
|\mathcal{D}_\mu^* A_2(x)|^2 \\
&= -A_2^*(x)\,\Box A_2(x) + \tfrac{1}{4}e^2|A_2(x)|^2 V_\mu(x)V^\mu(x) \\
&\quad - \tfrac{1}{2}ie\,V^\mu(x)\big[(\partial_\mu A_2(x))A_2^*(x) \\
&\qquad\qquad - A_2(x)(\partial_\mu A_2^*(x))\big]
\end{aligned}
$$

Hence the Lagrangian density becomes

$$
\begin{aligned}
\mathcal{L}_{FI} &= -F_{\mu\nu}(x)F^{\mu\nu}(x) - 4i\,\lambda(x)\sigma^\mu\partial_\mu\bar{\lambda}(x) \\
&\quad + i\,(\mathcal{D}_\mu^*\bar{\psi}_1(x))\bar{\sigma}^\mu\psi_1(x) \\
&\quad + i\,(\mathcal{D}_\mu\bar{\psi}_2(x))\bar{\sigma}^\mu\psi_2(x) \\
&\quad - A_1^*(x)\,\Box A_1(x) + \tfrac{1}{4}e^2|A_1(x)|^2 V_\mu(x)V^\mu(x) \\
&\quad + \tfrac{1}{2}ie\,V^\mu(x)\big[(\partial_\mu A_1(x))A_1^*(x) \\
&\qquad\qquad - A_1(x)(\partial_\mu A_1^*(x))\big] \\
&\quad - A_2^*(x)\,\Box A_2(x) + \tfrac{1}{4}e^2|A_2(x)|^2 V_\mu(x)V^\mu(x) \\
&\quad - \tfrac{1}{2}ie\,V^\mu(x)\big[(\partial_\mu A_2(x))A_2^*(x)
\end{aligned}
$$

$$- A_2(x)\left(\partial_\mu A_2^*(x)\right)]$$
$$- \frac{e}{2^{1/2}}\{\bar{\lambda}(x)[\bar{\psi}_1(x)A_1(x) - \bar{\psi}_2(x)A_2(x)]$$
$$+ \lambda(x)[\psi_1(x)A_1^*(x) - \psi_2(x)A_2^*(x)]\}$$
$$+ m[\psi_1(x)\psi_2(x) + \bar{\psi}_1(x)\bar{\psi}_2(x)]$$
$$+ 8D^2(x) + |F_1(x)|^2 + |F_2(x)|^2$$
$$- m\left(A_1(x)F_2(x) + A_1^*(x)F_2^*(x)\right.$$
$$\left. + A_2(x)F_1(x) + A_2^*(x)F_1^*(x)\right)$$
$$+ D(x)\left(2k + e\{|A_1(x)|^2 - |A_2(x)|^2\}\right) \tag{10.41}$$

Next we eliminate the auxiliary fields $F_1(x)$, $F_2(x)$ and $D(x)$ via their equations of motion. Thus

$$\frac{\partial \mathcal{L}_{FI}}{\partial F_1^*(x)} = F_1(x) - mA_2^*(x) = 0$$

so that

$$F_1(x) = mA_2^*(x), \quad F_1^*(x) = mA_2(x) \tag{10.42}$$

and

$$\frac{\partial \mathcal{L}_{FI}}{\partial F_2^*(x)} = F_2(x) - mA_1^*(x) = 0$$

so that

$$F_2(x) = mA_1^*(x), \quad F_2^*(x) = mA_1(x) \tag{10.43}$$

and finally

$$\frac{\partial \mathcal{L}_{FI}}{\partial D(x)} = 16\, D(x) + 2k + e\left(|A_1(x)|^2 - |A_2(x)|^2\right)$$
$$= 0$$

such that

$$D(x) = -\frac{1}{16}\left[2k + e\left(|A_1(x)|^2 - |A_2(x)|^2\right)\right] \tag{10.44}$$

Then we have (with (10.44))

$$8\, D^2(x) + D(x)\left[2k + e\left(|A_1(x)|^2 - |A_2(x)|^2\right)\right]$$
$$= 8\, D^2(x) - 16\, D^2(x) = -\, 8 D^2(x)$$

and (with (10.42), (10.43))

$$|F_1(x)|^2 + |F_2(x)|^2$$
$$-m\left\{A_1(x) F_2(x) + A_1^*(x) F_2^*(x) + A_2(x) F_1(x) + A_2^*(x) F_1^*(x)\right\}$$
$$= |F_1(x)|^2 + |F_2(x)|^2 - 2\,|F_1(x)|^2 - 2\,|F_2(x)|^2$$
$$= -|F_1(x)|^2 - |F_2(x)|^2$$

With these substitutions the Lagrangian (10.41) takes the form

$$\begin{aligned}\mathcal{L}_{FI} = &- F_{\mu\nu}(x) F^{\mu\nu}(x) - 4i\,\lambda(x)\sigma^{\mu}\partial_{\mu}\bar{\lambda}(x)\\ &+ i\,(D_{\mu}^{*}\bar{\psi}_1(x))\,\bar{\sigma}^{\mu}\psi_1(x)\\ &+ i\,(D_{\mu}\bar{\psi}_2(x))\,\bar{\sigma}^{\mu}\psi_2(x)\\ &- A_1^*(x)\,\square\, A_1(x) + \frac{e^2}{4}|A_1(x)|^2 V_{\mu}(x) V^{\mu}(x)\end{aligned}$$

$$+\frac{1}{2}ie\,V^{\mu}(x)\big[(\partial_\mu A_1(x))A_1^*(x) - A_1(x)(\partial_\mu A_1^*(x))\big]$$

$$-A_2^*(x)\,\Box A_2(x)+\frac{e^2}{4}|A_2(x)|^2 V_\mu(x)V^\mu(x)$$

$$-\frac{1}{2}ie\,V^{\mu}(x)\big[(\partial_\mu A_2(x))A_2^*(x) - A_2(x)(\partial_\mu A_2^*(x))\big]$$

$$-\frac{e}{2^{1/2}}\{\bar\lambda(x)[\bar\psi_1(x)A_1(x)-\bar\psi_2(x)A_2(x)] + \lambda(x)[\psi_1(x)A_1^*(x)-\psi_2(x)A_2^*(x)]\}$$

$$+m[\psi_1(x)\psi_2(x)+\bar\psi_1(x)\bar\psi_2(x)]$$

$$-\mathcal{V}(A_1,A_2)$$

where

$$\mathcal{V}(A_1,A_2)=8D^2(x)+|F_1(x)|^2+|F_2(x)|^2 \tag{10.46}$$

is the scalar potential of the Fayet-Iliopoulos model where $D(x)$, $F_1(x)$ and $F_2(x)$ are solutions of equations (10.42), (10.43) and (1o.44). Using these Euler-Lagrange equations, the explicit form of the scalar potential is

$$\mathcal{V}(A_1,A_2)$$

$$=\frac{1}{32}\big[2k+e\,(|A_1(x)|^2-|A_2(x)|^2)\big]^2 + m^2|A_1(x)|^2+m^2|A_2(x)|^2$$

Hence

$$\mathcal{V}(A_1,A_2)=\frac{1}{8}k^2+(m^2+\frac{ek}{8})|A_1(x)|^2$$

$$+ (m^2 - \frac{ek}{8}) |A_2(x)|^2$$
$$+ \frac{1}{32} e^2 \{|A_1(x)|^2 - |A_2(x)|^2\}^2$$

(10.47)

We observe that there is no solution to the Euler-Lagrange equations of the auxiliary fields, i.e. (10.42) to (10.44), such that the scalar potential (10.46) or (10.47) vanishes. Thus according to the general considerations of Section 9.3 supersymmetry must be broken spontaneously.

We now distinguish between the cases $m^2 > ek/8$ and $m^2 < ek/8$ and deal with each case separately.

i) <u>The case $4\,m^2 > ek/2$</u>

The absolute minimum of the scalar potential is determined by the vanishing of the partial derivatives

$$\frac{\partial \mathcal{V}}{\partial A_1^*(x)} = \left[(m^2 + \frac{ek}{8}) + \frac{e^2}{16}(|A_1(x)|^2 - |A_2(x)|^2)\right] A_1(x) \tag{10.48a}$$

$$\frac{\partial \mathcal{V}}{\partial A_2^*(x)} = \left[(m^2 - \frac{ek}{8}) - \frac{e^2}{16}(|A_1(x)|^2 - |A_2(x)|^2)\right] A_2(x) \tag{10.48b}$$

i.e. by the equations

$$\frac{\partial \mathcal{V}(A_1, A_2)}{\partial A_1^*(x)} = 0\,, \quad \frac{\partial \mathcal{V}(A_1, A_2)}{\partial A_2^*(x)} = 0 \tag{10.49}$$

Considering solutions with $A_1^{min}(x) = 0$ (the value of A_1 at the minimum), we see that $A_2^{min}(x)$, determined by (10.48b), is given by

$$\left[(m^2 - \frac{ek}{8}) + \frac{e^2}{16} |A_2^{min}(x)|^2\right] A_2^{min}(x) = 0$$

Thus if $A_2^{min}(x) \neq 0$, we must have $m^2 < ek/8$ and in the domain $m^2 > ek/8$ we can only have the solution $A_2^{min}(x) = 0$. Furthermore, at the absolute minimum $A_1^{min}(x) = 0$, $A_2^{min}(x) = 0$ the scalar potential $\vartheta(A_1, A_2)$ has the value

$$\vartheta(0,0) = \frac{k^2}{8} > 0 \tag{10.50}$$

indicating that supersymmetry is spontaneously broken. In this case both scalar fields $A_1(x)$ and $A_2(x)$ have real masses and vanishing expectation values, indicating that the internal U(1) phase symmetry is unbroken. Within this range of parameters (i.e. $4m^2 > ek/2$) the Fayet-Iliopoulos model with Lagrangian (10.45) describes two complex scalar fields $A_1(x)$ and $A_2(x)$, one of mass m_1 with

$$m_1^2 = m^2 + \frac{ek}{8}$$

and the other of mass m_2 with

$$m_2^2 = m^2 - \frac{ek}{8}$$

as well as three spinor fields $\psi_1(x)$, $\psi_2(x)$ and $\lambda(x)$, and one real vector field $V_\mu(x)$. The masses of the spinor fields and the vector field remain unchanged by the breaking of supersymmetry. In particular, as demonstrated in Section 10.2, $\psi_1(x)$ and $\psi_2(x)$ combine to a massive Dirac spinor field with mass m, whereas the spinor field $\lambda(x)$ and the gauge vector field $V_\mu(x)$ remain massless. We observe again that, just as in the case of the O'Raifeartaigh model (see remark ii) at the end of Chapter 9), the bosonic mass spectrum gives

$$m_1^2 + m_2^2 = 2\,m^2 \tag{10.51}$$

and it is only the bosonic mass spectrum which is affected by the breaking of supersymmetry.

The vector field $V_\mu(x)$ plays the role of a gauge field for the unbroken internal U(1) phase symmetry. The spinor field λ (x) is the Goldstone fermion which arises from spontaneously broken supersymmetry. From the transformation law of λ(x) under supersymmetry transformations (see (6.93) with the identifications $\psi_A(x) \equiv \lambda_A(x)$ and $d(x) \equiv D(x)$), i.e.

$$\delta_S \lambda_A(x) = 2\alpha_A D(x) + \cdots$$

we see that λ(x) transforms inhomogeneously as soon as the auxiliary field D(x) acquires a nonvanishing vacuum expectation value

$$D(x) = -\frac{k}{8}$$

(from (10.44) with $A_1(x) = A_2(x) = 0$), i.e.

$$\delta_S \lambda_A(x) = -\alpha_A \frac{k}{4} + \cdots \tag{10.52}$$

This is a typical feature of the Goldstone fermion. In general we can say: nonzero vacuum expectation values of auxiliary fields induce the breakdown of supersymmetry.

ii) <u>The case $4m^2 < ek/2$</u>

When $4m^2 < ek/2$, $A_1^{min}(x) = A_2^{min}(x) = 0$ does not represent the position of the absolute minimum of the potential (10.47). Instead, as we saw above, choosing $A_1^{min}(x) = 0$, $A_2^{min}(x)$ is determined by the equation

$$4m^2 - \frac{ek}{2} + \frac{e^2}{4}|A_2^{min}(x)|^2 = 0$$

or

$$\begin{aligned} \vartheta^2 := |A_2^{min}(x)|^2 &= \frac{4}{e^2}\left(\frac{ek}{2} - 4m^2\right) \\ &= \frac{16}{e^2}\left(\frac{ek}{8} - m^2\right) > 0 \end{aligned} \tag{10.53}$$

Expanding the potential (10.47) around the minimum at $A_1^{min}(x) = 0$ and $A_2^{min}(x) = \vartheta$ breaks the internal U(1) gauge symmetry spontaneously. We consider small oscillations around the absolute minimum and set

$$
\begin{aligned}
A_1(x) &= A(x) \\
A_2(x) &= v + \tilde{A}(x)
\end{aligned}
\tag{10.54}
$$

where v is real and constant. To find the particle spectrum, we have to insert (10.54) into the Lagrangian (10.45). Then

$$
\begin{aligned}
\mathcal{L}_{FI} = & -F_{\mu\nu}(x)F^{\mu\nu}(x) - 4i\lambda(x)\sigma^\mu\partial_\mu\bar{\lambda}(x) \\
& + i(D^*_\mu\bar{\psi}_1(x))\bar{\sigma}^\mu\psi_1(x) \\
& + i(D_\mu\bar{\psi}_2(x))\bar{\sigma}^\mu\psi_2(x) \\
& - A^*(x)\Box A(x) + \frac{e^2}{4}|A(x)|^2 V_\mu(x)V^\mu(x) \\
& + \frac{1}{2}ie V^\mu(x)\big[(\partial_\mu A(x))A^*(x) \\
& \qquad - A(x)(\partial_\mu A^*(x))\big] \\
& - \big[(v+\tilde{A}^*(x))\Box(v+\tilde{A}(x))\big] \\
& - \frac{1}{2}ie V^\mu(x)\big[\{\partial_\mu(v+\tilde{A}(x))\}(v+\tilde{A}(x)) \\
& \qquad - (v+\tilde{A}(x))\partial_\mu(v+\tilde{A}^*(x))\big] \\
& + \frac{1}{4}e^2|v+\tilde{A}(x)|^2 V_\mu(x)V^\mu(x) \\
& - \frac{e}{2^{1/2}}\big[\bar{\lambda}(x)(\bar{\psi}_1(x)A(x) - \bar{\psi}_2(x)(v+\tilde{A}(x))) \\
& \qquad + \lambda(x)(\psi_1(x)A^*(x) - \psi_2(x)(v+\tilde{A}^*(x)))\big] \\
& + m\big[\psi_1(x)\psi_2(x) + \bar{\psi}_1(x)\bar{\psi}_2(x)\big]
\end{aligned}
$$

$$
\begin{aligned}
&-\frac{1}{8}k^2-\left(m^2+\frac{ek}{8}\right)|A(x)|^2 \\
&\quad -\left(m^2-\frac{ek}{8}\right)|v+\tilde{A}(x)|^2 \\
&\quad -\frac{e^2}{32}\left\{|A(x)|^2-|v+\tilde{A}(x)|^2\right\}^2 \\
= &-F_{\mu\nu}(x)F^{\mu\nu}(x)-4i\lambda(x)\sigma^\mu\partial_\mu\bar{\lambda}(x) \\
&+i\left(D_\mu^*\bar{\psi}_1(x)\right)\bar{\sigma}^\mu\psi_1(x) \\
&+i\left(D_\mu\bar{\psi}_2(x)\right)\bar{\sigma}^\mu\psi_2(x) \\
&-A^*(x)\Box A(x)-\tilde{A}^*(x)\Box\tilde{A}(x) \\
&+\frac{1}{2}ie\,V^\mu(x)\left[(\partial_\mu A(x))A^*(x)-A(x)(\partial_\mu A^*(x))\right] \\
&+\frac{e^2}{4}|A(x)|^2V_\mu(x)V^\mu(x)-v\Box\tilde{A}(x) \\
&-\frac{1}{2}ie\,V^\mu(x)\left[(\partial_\mu\tilde{A}(x))\tilde{A}^*(x)-\tilde{A}(x)(\partial_\mu\tilde{A}^*(x))\right. \\
&\qquad\qquad \left.+v\partial_\mu(\tilde{A}(x)-\tilde{A}^*(x))\right] \\
&+\frac{e^2}{4}|\tilde{A}(x)|^2V_\mu(x)V^\mu(x) \\
&+\frac{e^2}{4}v\left[\tilde{A}(x)+\tilde{A}^*(x)\right]V_\mu(x)V^\mu(x) \\
&-\frac{e}{2^{1/2}}\left[\bar{\lambda}(x)\left(\bar{\psi}_1(x)A(x)-\bar{\psi}_2(x)\tilde{A}(x)\right)\right. \\
&\qquad\qquad \left.+\lambda(x)\left(\psi_1(x)A^*(x)-\psi_2(x)\tilde{A}^*(x)\right)\right] \\
&+\frac{e}{2^{1/2}}v\bar{\lambda}(x)\bar{\psi}_2(x)+\frac{e}{2^{1/2}}v\lambda(x)\psi_2(x) \\
&+m\left[\psi_1(x)\psi_2(x)+\bar{\psi}_1(x)\bar{\psi}_2(x)\right] \\
&-\frac{1}{8}k^2-\left(m^2+\frac{ek}{8}\right)|A(x)|^2-\left(m^2-\frac{ek}{8}\right)|\tilde{A}(x)|^2
\end{aligned}
$$

$$-(m^2-\frac{ek}{8})v(\tilde{A}(x)+\tilde{A}^*(x))$$
$$-(m^2-\frac{ek}{8})v^2+\frac{1}{4}e^2v^2V_\mu(x)V^\mu(x)$$
$$-\frac{e^2}{32}\{|A(x)|^2-|\tilde{A}(x)|^2\}^2$$
$$-\frac{e^2}{32}v^2(\tilde{A}(x)+\tilde{A}^*(x))^2-\frac{e^2}{32}v^4$$
$$+\frac{e^2}{16}v\{|A(x)|^2-|\tilde{A}(x)|^2\}(\tilde{A}(x)+\tilde{A}^*(x))$$
$$+\frac{e^2}{16}v^2\{|A(x)|^2-|\tilde{A}(x)|^2\}$$
$$-\frac{e^2}{16}v^3(\tilde{A}(x)+\tilde{A}^*(x)) \qquad (10.55)$$

We can simplify the sums of several terms. Thus

$$\frac{1}{8}k^2+(m^2-\frac{ek}{8})v^2+\frac{e^2}{32}v^4$$
$$=\frac{1}{8}k^2+\frac{4}{e^2}(m^2-\frac{ek}{8})(\frac{ek}{2}-4m^2)$$
$$+\frac{1}{2e^2}(\frac{ek}{2}-4m^2)^2 \quad \text{(using (10.53))}$$
$$=2\frac{m^2}{e^2}(ek-4m^2) \qquad (10.56)$$

and

$$-(m^2+\frac{ek}{8})|A(x)|^2+\frac{e^2}{16}v^2|A(x)|^2$$
$$=-2m^2|A(x)|^2 \qquad (10.57)$$

(using (10.53)) and similarly

$$-(m^2-\frac{ek}{8})|\tilde{A}(x)|^2-\frac{e^2}{16}v^2|\tilde{A}(x)|^2=0 \qquad (10.58)$$

(again using (10.53)). Hence with (10.56) to (10.58)

we obtain

$$\begin{aligned}
\mathcal{L}_{FI} = &- F_{\mu\nu}(x) F^{\mu\nu}(x) - 4i\lambda(x)\sigma^{\mu}\partial_{\mu}\bar{\lambda}(x) \\
&+ i(D^{*}_{\mu}\bar{\psi}_1(x))\bar{\sigma}^{\mu}\psi_1(x) \\
&+ i(D_{\mu}\bar{\psi}_2(x))\bar{\sigma}^{\mu}\psi_2(x) \\
&- v\Box\tilde{A}(x) - \tilde{A}^{*}(x)\Box\tilde{A}(x) \\
&- \frac{1}{2}ieV^{\mu}(x)[(\partial_{\mu}\tilde{A}(x))\tilde{A}^{*}(x) \\
&\qquad - \tilde{A}(x)(\partial_{\mu}\tilde{A}^{*}(x)) \\
&\qquad + v\partial_{\mu}(\tilde{A}(x) - \tilde{A}^{*}(x))] \\
&+ \frac{e^2}{4}|\tilde{A}(x)|^2 V_{\mu}(x)V^{\mu}(x) \\
&+ \frac{e^2}{4}v(\tilde{A}(x) + \tilde{A}^{*}(x))V_{\mu}(x)V^{\mu}(x) \\
&- \frac{e^2}{2^{1/2}}[\bar{\lambda}(x)(\bar{\psi}_1(x)A(x) - \bar{\psi}_2(x)\tilde{A}(x)) \\
&\qquad + \lambda(x)(\psi_1(x)A^{*}(x) - \psi_2(x)\tilde{A}^{*}(x))] \\
&+ \frac{e}{2^{1/2}}v\bar{\lambda}(x)\bar{\psi}_2(x) + \frac{e}{2^{1/2}}v\lambda(x)\psi_2(x) \\
&+ m[\psi_1(x)\psi_2(x) + \bar{\psi}_1(x)\bar{\psi}_2(x)] \\
&- \tilde{\mathcal{V}}
\end{aligned} \tag{10.59}$$

where the potential $\tilde{\mathcal{V}}$ is given by

$$\begin{aligned}
\tilde{\mathcal{V}} = &\, 2\frac{m^2}{e^2}(ek - 4m^2) + 2m^2|A(x)|^2 \\
&- \frac{e^2}{4}v^2 V_{\mu}(x)V^{\mu}(x) \\
&+ \frac{e^2}{32}v^2[\tilde{A}(x) + \tilde{A}^{*}(x)]^2
\end{aligned}$$

$$- \frac{e^2}{16} v^3 (\tilde{A}(x) + \tilde{A}^*(x))$$

$$+ \frac{e^2}{32} \{ |A(x)|^2 - |\tilde{A}(x)|^2 \}^2$$

$$- \frac{e}{16} v \{ |A(x)|^2 - |\tilde{A}(x)|^2 \}^2 (\tilde{A}(x) + \tilde{A}^*(x))$$

$$+ \frac{e^2}{16} v^3 (\tilde{A}(x) + \tilde{A}^*(x))$$

$$+ (m^2 - \frac{ek}{8}) v (\tilde{A}(x) + \tilde{A}^*(x)) \qquad (10.60)$$

We now discuss what one can deduce from these results.

i) Supersymmetry is spontaneously broken as in the case $4m^2 > ek/2$, since the constant $2m^2 (ek - 4 m^2) /e^2$ appearing in the scalar potential (10.60) is strictly positive for $4 m^2 < ek/2$.

ii) The internal U(1) gauge symmetry is spontaneously broken because the complex scalar field $A_2(x)$ acquires a nonvanishing vacuum expectation value (cf. (10.53)).

iii) The usual Higgs-Kibble mechanism takes place, i.e. the vector field V_μ (x) acquires a mass, as we can see from the term

$$\frac{e^2 v^2}{4} V_\mu(x) V^\mu(x)$$

in the potential (10.60), by "eating up" the Goldstone boson field

$$\tilde{A}(x) - \tilde{A}^*(x)$$

which denotes the imaginary part of the complex scalar field $A_2(x)$. The real scalar field described by

$$\tilde{A}(x) + \tilde{A}^*(x) = 2 \, Re \, \tilde{A}(x)$$

has a mass term $(e^2 v^2/8)^{1/2}$.

iv) In addition the theory contains a complex scalar field A(x) with mass squared $8 m^2$.

v) We can completely eliminate the field

$$\tilde{A} - \tilde{A}^*$$

from the Lagrangian (10.59) by fixing the gauge (unitary gauge) and transforming the field $\tilde{A} - \tilde{A}^*$ into a longitudinal degree of freedom of the Vector field $V_\mu(x)$ (Higgs mechanism)[71].

vi) The breaking of the internal U(1) gauge symmetry also modifies the spinor mass terms, i.e.

$$m[\psi_1(x)\psi_2(x) + \bar{\psi}_1(x)\bar{\psi}_2(x)]$$
$$+\frac{ev}{2^{1/2}}[\bar{\lambda}(x)\bar{\psi}_2(x) + \lambda(x)\psi_2(x)]$$

Diagonalization of the mass terms leads to the following qualitative picture. The theory describes one massless spinor field which is a linear combination of $\lambda(x)$ and $\psi_1(x)$, and two massive spinor fields, i.e. $\psi(x) = \psi_2(x)$ and $\tilde{\psi}(x)$, the latter being a linear combination of $\psi_1(x)$ and $\lambda(x)$.

vii) From the above discussion we see that nonvanishing vacuum expectation values of auxiliary fields induce supersymmetry breaking, whereas nonzero vacuum expectation values of dynamical scalar fields lead to the breakdown of gauge symmetry. These features are illustrated by Fig. 8 and Fig. 9.

viii) There is a fundamental difference between supersymmetry breaking in the O'Raifeartaigh model and that in the Fayet-Iliopoulos model. In the case of the former, supersymmetry breaking is the consequence of a nonvanishing vacuum expectation value of an auxiliary field, whereas in the case of the latter it is the consequence of the nonvanishing vacuum expectation value of the component field D(x) of a vector superfield.

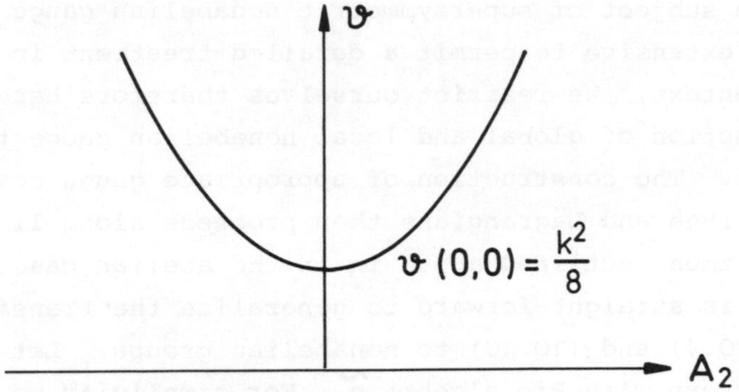

Fig.8 : Supersymmetry breaking in the Fayet-Iliopoulos model for $8\ m^2 > ek$

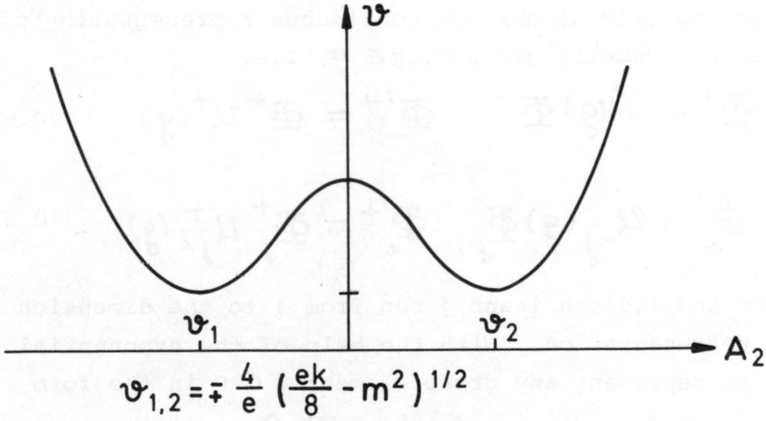

Fig.9 : Breaking of supersymmetry and gauge symmetry in the Fayet-Iliopoulos model for $8\ m^2 < ek$

10.4 Supersymmetric Nonabelian Gauge Theory[72,73]

The subject of supersymmetric nonabelian gauge theory is too extensive to permit a detailed treatment in the present context. We restrict ourselves therefore here to the introduction of global and local nonabelian gauge transformations. The construction of appropriate gauge covariant derivatives and Lagrangians then proceeds along lines similar to those outlined above in in the abelian case.

It is straight-forward to generalize the transformation laws (10.4) and (10.10) to nonabelian groups. Let G be the gauge group with Lie algebra $\tilde{g}$. For simplicity we shall assume $\tilde{g}$ to be **semi-simple**[35]. The basis elements of the Lie algebra $\tilde{g}$ are the hermitian operators T_a, $a = 1,\ldots,N$, N being the dimension of $\tilde{g}$. The superfields Φ and Φ^+ transform tensorially with respect to U(g), where U(g) is any nontrivial unitary continuous representation of the internal symmetry group G, $g \in G$, i.e.

$$\Phi' = \mathcal{U}(g)\Phi, \quad \Phi'^+ = \Phi^+\mathcal{U}^+(g) \tag{10.61}$$

or

$$\Phi_i' = \mathcal{U}_{ij}(g)\Phi_j, \quad \Phi_i'^+ = \Phi_j^+\mathcal{U}_{ji}^+(g) \tag{10.62}$$

where the indices i and j run from 1 to the dimension of the representation. With the help of the exponential map we represent any group element $g \in G$ in the form

$$g = \exp\{-i\Lambda^a T_a\}$$

where T_a are the generators (basis elements of the Lie algebra) and Λ^a are N parameters; hence the exponential map, i.e.

$$\exp : \tilde{g} \longrightarrow G$$

maps the Lie algebra $\tilde{g}$ onto the group G. Then the representation matrix U(g) of (10.61) can be written

$$\mathcal{U}(g) = \mathcal{U}(\exp\{-i\Lambda^a T_a\})$$
$$\equiv \exp[-i\Lambda] \qquad (10.63)$$

where Λ is a square matrix (dimensionality that of the representation) given by

$$(\Lambda_{ij}) = (\Lambda^a (T_a)_{ij}) \qquad (10.64)$$

The matrices $((T_a)_{ij})$ are the generators T_a of the Lie algebra $\tilde{g}$ in the particular representation.

As an example consider G = SU(2). If we choose the superfields Φ, Φ^+ to lie in the fundamental representation, the matrices $((T_a)_{ij})$ of (10.64) are given by the three Pauli matrices. If we choose the adjoint representation, then the matrices $((T_a)_{ij})$ are given by the structure constants ε_{kij} of the group SU(2).

Inserting (10.63) into (10.61) we obtain the generalization of the transformation law (10.4) to nonabelian global phase transformations, i.e.

$$\Phi' = e^{-i\Lambda}\Phi, \quad \Phi'^+ = \Phi^+ e^{i\Lambda^+} \qquad (10.65a)$$

where (cf.(10.5))

$$\bar{D}_{\dot{A}}\Lambda = 0, \quad D_A \Lambda^+ = 0 \qquad (10.65b)$$

so that the chiral field remains a chiral field under the transformation. Now let X be any element of the Lie algebra $\tilde{g}$, and $\tilde{g}l$ the algebra associated with the the general linear group. Then we define the adjoint representation by the map[35]

$$\begin{aligned} \mathrm{ad}\,X : \tilde{g} &\longrightarrow \tilde{g}l(\tilde{g}) \\ Y &\longrightarrow (\mathrm{ad}\,X)(Y) := [X, Y] \end{aligned} \qquad (10.66)$$

Furthermore, the Killing form B is a bilinear map defined on the algebra $\tilde{g}$, i.e. (see e.g. ref. 35)

$$B : \tilde{g} \times \tilde{g} \longrightarrow R$$
$$(X, Y) \longrightarrow B(X,Y) := Tr(ad\,X\,ad\,Y) \tag{10.67}$$

To obtain a coordinate representation of ad X and the Killing form B(X,Y) we calculate

$$(ad\,T_a)(T_b) = [T_a, T_b] \quad \text{(with (10.66))}$$
$$= i t_{ab}{}^{c} T_c \tag{10.68}$$

where $t_{ab}{}^{c}$ are the structure constants of the internal symmetry group G. Hence according to (10.68) the matrix corresponding to $(ad\,T_a) \in \tilde{g}l\,(\tilde{g})$ is given by the structure constants $(it_a)_b{}^c$ of the group G. Then the coordinate representation of the Killing form (10.67) becomes

$$b_{ab} = B(T_a, T_b)$$
$$= Tr((ad\,T_a)(ad\,T_b)) \quad \text{(with (10.67))}$$
$$= Tr(i\,(t_a)_c{}^d\, i\,(t_b)_d{}^c) \quad \text{(with (10.68))}$$
$$= -(t_a)_c{}^d (t_b)_d{}^c \tag{10.69}$$

Since we assumed G to be semi-simple, the Killing form B is nondegenerate[74] and therefore we can always normalize the generators T_a such that

$$B(T_a, T_b) = Tr(ad\,T_a\;ad\,T_b)$$
$$= k\,\delta_{ab}\,,\; k > 0 \tag{10.70}$$

In the literature[75] (10.70) is written

$$Tr(T_a T_b) = k\,\delta_{ab} \tag{10.71}$$

where T_a, T_b are the generators in the adjoint representation.

Proposition: With the normalization (10.70) of the generators in the adjoint representation the structure constants of the gauge group G, i.e. t_{abc}, are completely antisymmetric in their indices since

$$t_{bca} = -\frac{i}{k} Tr\left(T_a [T_b, T_c]\right) \tag{10.72}$$

Proof: Consider

$$-\frac{i}{k} Tr\left(T_a [T_b, T_c]\right)$$

$$= -\frac{i}{k} Tr\left(T_a\, i\, t_{bc}{}^{d}\, T_d\right)$$

$$= \frac{1}{k} t_{bc}{}^{d}\, k\, \delta_{ad}$$

$$= t_{bca}$$

In terms of the Killing form B (cf. (10.67)) this result reads

$$B(T_a, [T_b, T_c]) = i k\, t_{bca}$$

and from Tr (AB) = Tr (BA) one derives that t_{abc} is completely antisymmetric.

The Lagrangian of (10.13) is invariant under local nonabelian gauge transformations provided we extend the transformation law of the vector superfield (10.11) in the following representation independent way, i.e.

$$e^{V'} = e^{-i\Lambda^{+}} e^{V} e^{i\Lambda}, \quad V_{ij} = V^{a} (T_a)_{ij} \tag{10.73}$$

Then the expression

$$Tr\left[\Phi^{+} e^{V} \Phi\right]$$

is invariant under local nonabelian gauge transformations, i.e.

$$\mathrm{Tr}[\Phi'^{+} e^{V'} \Phi']$$
$$= \mathrm{Tr}[\Phi^{+} e^{i\Lambda^{+}} e^{-i\Lambda^{+}} e^{V} e^{i\Lambda} e^{-i\Lambda} \Phi]$$

(with (10.65) and (10.73))

$$= \mathrm{Tr}[\Phi^{+} e^{V} \Phi] \tag{10.74}$$

In order to show that (10.73) is independent of a particular representation we use the Baker-Campbell-Hausdorff formula (cf. (6.9)) to evaluate the product of exponentials with matrix arguments. Thus

$$\exp[-i\Lambda^{+}]\exp[V]\exp[i\Lambda]$$
$$= \exp[-i\Lambda^{+}]\exp[V + i\Lambda + \frac{i}{2}[V,\Lambda] + \frac{1}{12}[\Lambda,[\Lambda,V]] + \frac{i}{12}[V,[V,\Lambda]] + \cdots]$$
$$= \exp\{V + i\Lambda - i\Lambda^{+} + \frac{i}{2}[V,\Lambda] + \frac{1}{12}[\Lambda,[\Lambda,V]] + \frac{i}{12}[V,[V,\Lambda]]$$
$$- \frac{i}{2}\Big[\Lambda^{+}, V + i\Lambda + \frac{i}{2}[V,\Lambda] + \frac{1}{12}[\Lambda,[\Lambda,V]] + \frac{i}{12}[V,[V,\Lambda]]\Big] + \cdots\}$$
$$= \exp\{V + i(\Lambda - \Lambda^{+}) + \frac{i}{2}[V,\Lambda] - \frac{i}{2}[\Lambda^{+},V] + \frac{1}{12}[\Lambda^{+},\Lambda] + \frac{1}{4}[\Lambda^{+},[V,\Lambda]]$$
$$+ \frac{1}{12}[\Lambda,[\Lambda,V]] + \frac{i}{12}[V,[V,\Lambda]] + \cdots\}$$

$$
\begin{aligned}
&= \exp\{V^a T_a + i(\Lambda^a - \Lambda^{a+})T_a \\
&\qquad + \frac{i}{2} V^c \Lambda^b [T_c, T_b] \\
&\qquad - \frac{i}{2} \Lambda^{c+} V^b [T_c, T_b] \\
&\qquad + \frac{1}{2} \Lambda^{+c} \Lambda^b [T_c, T_b] \\
&\qquad + \frac{1}{4} \Lambda^{+c} V^b \Lambda^d [T_c, [T_b, T_d]] \\
&\qquad + \frac{1}{12} \Lambda^c \Lambda^b V^d [T_c, [T_b, T_d]] \\
&\qquad + \frac{i}{12} V^c V^b \Lambda^d [T_c, [T_b, T_d]] \\
&\qquad + \cdots \}
\end{aligned}
$$

(using $T_a = T_a^+$)

$$
\begin{aligned}
&= \exp\{[V^a + i(\Lambda^a - \Lambda^{+a})]T_a \\
&\qquad - \frac{1}{2} V^c \Lambda^b t_{cb}{}^a T_a + \frac{1}{2} \Lambda^{+c} V^b t_{cb}{}^a T_a \\
&\qquad + \frac{i}{2} \Lambda^{+c} \Lambda^b t_{cb}{}^a T_a \\
&\qquad + [-\frac{1}{4} \Lambda^{+c} V^b \Lambda^d - \frac{1}{12} \Lambda^c \Lambda^b V^d \\
&\qquad\qquad - \frac{i}{12} V^c V^b V^d] t_{bd}{}^e t_{ce}{}^a T_a \\
&\qquad + \cdots \}
\end{aligned}
$$

with (10.68))

$$
\begin{aligned}
&= \exp\{[V^a + i(\Lambda^a - \Lambda^{a+}) + \frac{1}{2} t_{cb}{}^a (\Lambda^{+c} V^b \\
&\qquad + i\Lambda^{+c} \Lambda^b - V^c \Lambda^b) + \frac{1}{4} t_{bd}{}^e t_{ce}{}^a \cdot \\
&\qquad \cdot (-\Lambda^{+c} V^b \Lambda^d - \frac{1}{3} \Lambda^c \Lambda^b V^d \\
&\qquad - \frac{i}{3} V^c V^b \Lambda^d) + \cdots] T_a \}
\end{aligned} \qquad (10.75)
$$

Hence we obtain only commutators of group generators in computing the product of exponentials of (10.73). This shows that we can express the transformed vector field V' as

$$V' = V'^a T_a \tag{10.76}$$

Equation (10.76) demonstrates that the transformation law (10.73) is independent of the particular representation of the group since the information about the chosen representation is contained in the generators T_a.

We now calculate the infinitesimal form of the transformation (10.73), i.e.

$$\delta V = V' - V$$

or

$$\begin{aligned}\delta(e^V) &= e^{V'} - e^V \\ &= e^{-i\Lambda^+} e^V e^{i\Lambda} - e^V \\ &\approx -i\Lambda^+ e^V + i e^V \Lambda \end{aligned} \tag{10.73'}$$

We begin by defining the Lie derivative.

Definition: If V and X are elements of the algebra $\tilde{g}$, the Lie derivative of X with respect to V is defined by

$$L_V X := [V, X] \tag{10.77}$$

Proposition: With the Lie derivative defined by (10.77) we have

$$e^{L_V} X = e^V X e^{-V} \tag{10.78}$$

Proof: We restrict ourselves here to a formal verification. A rigorous proof of (10.78) can be found in the literature[76]. The left hand side of (10.78) is defined by the power series expansion of the exponential; thus

$$e^{L_V} X = \left[\sum_{n=0}^{\infty} \frac{1}{n!} (L_V)^n \right] X$$

$$= X + [V, X] + \frac{1}{2} [V, [V, X]] + \cdots$$

$$\cdots + \frac{1}{k!} [V, [V, [\cdots [V, X] \cdots]]] + \cdots$$

$$= X + VX - XV + \frac{1}{2} V^2 X + \frac{1}{2} X V^2 - VXV + \cdots$$

$$= (1 + V + \frac{1}{2} V^2 + \cdots) X (1 - V + \frac{1}{2} V^2 + \cdots)$$

$$= e^V X e^{-V}$$

which had to be demonstrated.

Now obviously

$$[V, e^V] = 0 \tag{10.79}$$

and variation of this expression yields

$$\delta [V, e^V] = 0$$

i.e.

$$\delta (V e^V - e^V V) = 0$$

and so

$$(\delta V) e^V + V \delta (e^V) - \delta (e^V) V - e^V \delta V = 0$$

or

$$(\delta V) e^V - e^V (\delta V) + [V, \delta (e^V)] = 0$$

Multiplying this expression from the left and the right by $e^{-V/2}$ we obtain

$$e^{-V/2} (\delta V) e^{V/2} - e^{V/2} (\delta V) e^{-V/2}$$
$$+ e^{-V/2} [V, \delta (e^V)] e^{-V/2} = 0$$

Hence

$$e^{-V/2}(\delta V)e^{V/2} - e^{V/2}(\delta V)e^{-V/2}$$
$$= -e^{-V/2}(L_V \delta e^V)e^{-V/2}$$

(with (10.77)

Using (10.78) we obtain

$$e^{-L_{V/2}}(\delta V) - e^{L_{V/2}}(\delta V)$$
$$= -e^{-V/2}(L_V \delta e^V)e^{-V/2}$$

so that

$$2\sinh(L_{V/2})(\delta V)$$
$$= e^{-V/2}(L_V \delta e^V)e^{-V/2}$$
$$= e^{-V/2}(L_V[e^{V'} - e^V])e^{-V/2}$$
$$= e^{-V/2}(L_V[-i\Lambda^+ e^V + ie^V\Lambda])e^{-V/2}$$

(using (10.73')

$$= -iL_V[e^{-V/2}\Lambda^+ e^{V/2} - e^{V/2}\Lambda e^{-V/2}]$$

(using (10.79) and (10.77))

$$= -iL_V(e^{-L_{V/2}}\Lambda^+ - e^{L_{V/2}}\Lambda)$$

(using (10.78))

$$= -\frac{i}{2}L_V\{e^{L_{V/2}}\Lambda^+ - e^{L_{V/2}}\Lambda + e^{-L_{V/2}}\Lambda^+ - e^{-L_{V/2}}\Lambda - e^{L_{V/2}}\Lambda^+ - e^{L_{V/2}}\Lambda + e^{-L_{V/2}}\Lambda^+ + e^{-L_{V/2}}\Lambda\}$$

$$= -iL_V\{\tfrac{1}{2}(e^{L_{V/2}} + e^{-L_{V/2}})(\Lambda^+ - \Lambda)$$

$$-\tfrac{1}{2}(e^{L_{V/2}} - e^{-L_{V/2}})(\Lambda^+ + \Lambda)\}$$

$$= -i L_V \{\cosh L_{V/2} (\Lambda^+ - \Lambda) - \sinh L_{V/2} (\Lambda^+ + \Lambda)\}$$

Hence

$$2 \sinh (L_{V/2})\, \delta V = i L_V \{\sinh L_{V/2} (\Lambda^+ + \Lambda) - \cosh L_{V/2} (\Lambda^+ - \Lambda)\}$$

Now

$$\tfrac{1}{2} L_V X = \tfrac{1}{2}[V, X] \quad \text{(from (10.77)}$$

$$= [\tfrac{1}{2} V, X]$$

$$= L_{V/2} X \quad \text{(using (10.77))}$$

so that formally

$$\tfrac{1}{2} L_V = L_{V/2}$$

Then formally

$$\delta V = \frac{i}{2} L_V \left[\frac{1}{\sinh L_{V/2}} \sinh L_{V/2} (\Lambda^+ + \Lambda) - \frac{1}{\sinh L_{V/2}} \cosh L_{V/2} (\Lambda^+ - \Lambda) \right]$$

$$\delta V = V' - V$$

$$= i L_{V/2} [(\Lambda^+ + \Lambda) + \coth L_{V/2} (\Lambda - \Lambda^+)] \tag{10.80}$$

From (10.80) we obtain the infinitesimal transformation by expanding coth in its power series, i.e.

$$\begin{aligned} \delta V &= V' - V \\ &= i(\Lambda - \Lambda^+) + \frac{i}{2}[V, \Lambda^+ + \Lambda] + o(\Lambda^2) \end{aligned} \tag{10.81}$$

This is the desired infinitesimal transformation.

The supersymmmetric extension of the field strength is now defined in analogy to (7.43), (7.44) by

$$\begin{aligned} W_A &:= -\frac{1}{4}\bar{D}\bar{D}\, e^{-V} D_A e^{V} \\ \bar{W}_{\dot{A}} &:= -\frac{1}{4} DD\, e^{-V} \bar{D}_{\dot{A}} e^{V} \end{aligned} \tag{10.82}$$

where according to (10.73) the vector superfields V are matrix quantities

$$V_{ij} = V^a (T_a)_{ij}$$

and the generators T_a are given in the adjoint representation.

<u>Proposition</u>: The field strength W_A transforms according to

$$W_A \to W_A' = e^{-i\Lambda(x,\theta,\bar{\theta})} W_A e^{i\Lambda(x,\theta,\bar{\theta})} \tag{10.83}$$

and a similar relation holds for $\bar{W}_{\dot{A}}$.

<u>Proof</u>: We observe first that if e^V transforms according to (10.73), then e^{-V} transforms as

$$e^{-V'} = e^{-i\Lambda} e^{-V} e^{i\Lambda^+} \tag{10.84}$$

such that

$$\begin{aligned} e^{V'} e^{-V'} &= e^{-i\Lambda^+} e^{V} e^{i\Lambda} e^{-i\Lambda} e^{-V} e^{i\Lambda^+} \\ &= 1 = e^{-V'} e^{V'} \end{aligned}$$

Then we have

$$W_A' = -\frac{1}{4}\bar{D}\bar{D}e^{-V'}D_A e^{V'}$$

$$= -\frac{1}{4}\bar{D}\bar{D}(e^{-i\Lambda}e^{V}e^{i\Lambda^+})D_A(e^{-i\Lambda^+}e^{V}e^{i\Lambda})$$

(using (10.73), (10.84))

$$= -\frac{1}{4}e^{-i\Lambda}\bar{D}\bar{D}(e^{-V}e^{i\Lambda^+})D_A(e^{-i\Lambda^+}e^{V}e^{i\Lambda})$$

(using (10.65b))

$$= -\frac{1}{4}e^{-i\Lambda}\bar{D}\bar{D}e^{-V}(D_A e^{V})e^{i\Lambda}$$

$$-\frac{1}{4}e^{-i\Lambda}\bar{D}\bar{D}D_A e^{i\Lambda}$$

$$= e^{-i\Lambda}W_A e^{i\Lambda} - \frac{1}{4}e^{-i\Lambda}\bar{D}\bar{D}(D_A e^{i\Lambda})$$

(using (10.82))

$$= e^{-i\Lambda}W_A e^{i\Lambda} - \frac{1}{4}e^{-i\Lambda}\bar{D}\bar{D}D_A e^{i\Lambda}$$

$$-\frac{1}{4}e^{-i\Lambda}\bar{D}D_A\bar{D}e^{i\Lambda}$$

(using (10.65b))

$$= e^{-i\Lambda}W_A e^{i\Lambda} - \frac{1}{4}e^{-i\Lambda}\bar{D}\{\bar{D}, D_A\}e^{i\Lambda}$$

(since $\bar{D}_{\dot{A}}\bar{D}^{\dot{A}} = -\bar{D}^{\dot{A}}\bar{D}_{\dot{A}}$)

$$= e^{-i\Lambda}W_A e^{i\Lambda} - \frac{1}{2}e^{-i\Lambda}\bar{D}^{\dot{A}}\sigma^{\mu}_{A\dot{A}}P_{\mu}e^{i\Lambda}$$

(using (6.53))

$$= e^{-i\Lambda}W_A e^{i\Lambda} + \frac{1}{2}e^{-i\Lambda}P_{\mu}\sigma^{\mu}_{A\dot{A}}\bar{D}^{\dot{A}}e^{i\Lambda}$$

(since $[\bar{D}, P_{\mu}] = 0$)

$$= e^{-i\Lambda}W_A e^{i\Lambda} \qquad \text{(using (10.65b))}$$

The transformation of the second of relations (10.82) is demonstrated in a similar way.

For a treatment of further properties of supersymmetric nonabelian gauge theories (in particular for the introduction of gauge covariant derivatives and Lagrangians) we refer to other literature.[2,3] The above introduction provides the basic tools with which the study of these theories can be pursued along the lines of the previous discussion of abelian gauge theories.

REFERENCES

1. F. Legovini: Supersymmetry, International School for Advanced Studies, Trieste, Report No. 24/83/E.P., 1983, unpublished.
2. J. Wess and J. Bagger: Supersymmetry and Supergravity, Princeton University Press, Princeton, 1983.
3. P. Fayet and S. Ferrara: Supersymmetry, Physics Reports 32, 249 (1977).
4. M. de Roo: Supersymmetry, Lectures given at the University of Groningen, Internal Report No. 168, 1980/81, unpublished.
5. H.E. Haber and G.L. Kane: Is Nature Supersymmetric ? Scientific American 254, 42 (1986).
6. H.E. Haber and G.L. Kane: The Search for Supersymmetry: Probing Physics beyond the Standard Model, Physics Reports 117, 75 (1985).
7. Proceedings of the Thirteenth SLAC Summer Institute on Particle Physics, Supersymmetry, SLAC Report No. 296, Ed. E.C. Brennan, SLAC, Stanford, 1985.
8. H.P. Nilles: Supersymmetry, Supergravity and Particle Physics, Physics Reports 110, 1 (1984).
9. N. Dragon, U. Ellwanger and M.G. Schmidt: Supersymmetry and Supergravity, Reports on Progress in Physics, to be published (1987).
10. J. Wess: Symmetrie, Supersymmetrie, Supergravitation, Physikalische Blätter 43, 1 (1987).

11. B.A. Campbell and G. Fogleman: Supersymmetry and Supergravity: A Short Review, TRIUMF Report No. TRI-PP-126, TRIUMF, Vancouver, 1983.
12. J. Wess: Supersymmetry, Lectures given at XV. Int. Universitätswochen, Schladming, Austria, Acta Physica Austriaca, Suppl. XV., 475 (1976).
13. S. Ferrara: Supersymmetry and Supergravity for Nonpractitioners, Proceedings of International School of Physics "E. Fermi", Course 81, Ed. G. Costa and R.R. Gatto, North-Holland Publ. Co., Amsterdam , 237 (1982).
14. E. Witten: Introduction to Supersymmetry, in The Unity of the Fundamental Interactions, Ed. A. Zichichi, Plenum Press, New York, 305 (1983).
15. S.J. Gates, M.T. Grisaru, M. Rocek and W. Siegel: Superspace, Benjamin/Cummings,Reading, Massachusetts, 1983.
16. M.F. Sohnius: Introducing Supersymmetry, Physics Reports 128, 39 (1985).
17. P. West: Introduction to Supersymmetry and Supergravity, World Scientific , Singapore, 1986.
18. P.P. Srivastava: Supersymmetry, Superfield and Supergravitation, An Introduction, Adam-Hilger, London, 1986.
19. P.G.O. Freund: Introduction to Supersymmetry, Cambridge University Press, Cambridge, 1986.
20. A. Salam and J. Strathdee: Supersymmetry and Superfields, Fortschritte der Physik 26, 57 (1978).
21. Proceedings of the 28th Scottish Universities Summer School in Physics, Superstrings and Supergravity, Ed. A.T. Davies and D.G. Sutherland, SUSSP Publication, Edinburgh, 1986.
22. Proceedings of the NATO Advanced Study Institute on Supersymmetry, Ed. K. Dietz, R. Flume, G.v. Gehlen and V. Rittenberg, Plenum Press, New York, 1985.

23. F. Cooper and B. Freedman: Aspects of Supersymmetric Quantum Mechanics, Annals of Physics 146, 262 (1983).
24. H.J.W. Müller-Kirsten and A. Wiedemann: Dirac Quantization of a Supersymmetric Field Theory, Zeitschrift für Physik C, 35, 471 (1987).
25. S. Coleman and J. Mandula: All Possible Symmetries of the S-Matrix, Phys. Rev. 159, 1251 (1967).
26. J. Wess and B. Zumino: A Lagrangian Model Invariant under Supergauge Transformations, Phys. Lett. 49B, 52 (1974).
27. J. Wess and B. Zumino: Supergauge Transformations in Four Dimensions, Nucl. Phys. B70, 39 (1974).
28. R. Haag, J.T. Lopuszanski and M.F. Sohnius: All Possible Generators of Supersymmetries of the S-Matrix, Nucl. Phys. B88, 257 (1975).
29. M.D. Scadron: Advanced Quantum Theory, Springer, New York, 1979, Chapter 3.B.
30. T. Bröcker and T. tom Dieck: Representations of Compact Lie Groups, Springer, New York, 1985, GTM 98, Chapter III.8 .
31. W. Miller: Symmetry Groups and Their Applications, Academic Press, New York, 1972, Chapter 8.
32. W. Miller, ref. 31, p. 395.
33. R.U. Sexl and H.K. Urbantke: Relativität, Gruppen, Teilchen, Springer, New York, 1979.
34. M.D. Scadron, ref. 29.
35. A.S. Sciarrino and P. Sorba: Group Theory in Particle Physics, LAPP Report No. TH-79, to be published in Methods and Formulae in High Energy Physics Data Analysis, LAPP, Annecy-le-Vieux, 1979.
36. See for instance H.P. Nilles, ref. 8.
37. R.U. Sexl and H.K. Urbantke, ref. 33, Chapter 8.3 .

38. P. Roman: Theory of Elementary Particles, North-Holland Publ. Co. , Amsterdam, 1960, p. 75.
39. P. Roman, ref. 38, p. 72.
40. See e.g. L.I. Schiff: Quantum Mechanics, McGraw-Hill Publ. Co. , New York, 1955, Chapter 52.
41. A.O. Barut and R. Raczka:Theory of Group Representations and Applications, Second revised edition, Polish Scientific Publishers, Warsaw, 1980, p. 43, also give considerations concerning the unification of the Poincaré algebra with internal symmetry algebras. Stronger group theoretical lemmas are given on p. 629, and it is pointed out (p.630) that the infinite-parameter Lie algebra associated with noncompact dynamical groups (which lead to infinite particle multiplets (cf. pp. 411, 609) does not contradict these theorems.
42. R. Slanski: Symmetries of Theories in Higher Dimensions, Los Alamos Report No. LA-UR-85-2768, Los Alamos, 1985, to be published in Proceedings of Lectures at 1985 Les Houches Summer School.
43. W. Miller, ref. 31, p. 394.
44. See e.g. W. Miller, ref. 31, p. 161.
45. E. Witten: Constraints on Supersymmetry Breaking, Nucl. Phys. B202, 253 (1982).
46. D. Friedan and P. Windey: Supersymmetric Derivation of Atiyah-Singer Index and the Chiral Anomaly, Nucl. Phys. B235, 395 (1984).
47. L. Alvarez-Gaumé: Supersymmetry and the Atiyah-Singer Index Theorem, Commun. Math. Phys. 90, 161 (1983).
48. H. Goldstein: Classical Mechanics, sixth edition, Addison-Wesley Publ. Co., Reading, Mass., 1959, Chapter 8.
49. A. Salam and J. Strathdee: Supergauge Transformations, Nucl. Phys. B76, 477 (1974).

50. S. Ferrara, J. Wess and B. Zumino: Supergauge Multiplets and Superfields, Phys. Lett. 51B, 239 (1974).
51. F.A. Berezin: The Method of Second Quantization, Translation by M. Mugibayashi and A. Jeffrey, Academic Press, New York, 1966, p. 49.
52. G. Feldman and P.T. Matthews: Super Particles, Superfields and Yang-Mills Lagrangians, Cambridge University Report No. DAMTP 85-11 , Cambridge, 1985.
53. J.A. Bagger: Supersymmetric σ-Models, SLAC Report No. SLAC-PUB-3461, SLAC, Stanford, 1984.
54. B. Zumino: Supersymmetry and Kähler Manifolds, Phys. Lett. 87B, 203 (1979).
55. L. Alvarez-Gaumé and D.Z. Freedman: A Simple Introduction to Complex Manifolds, published in Unification of the Fundamental Particle Interactions, Erice 1980, ed. S. Ferrara, J. Ellis and P. van Nieuwenhuizen, Plenum Press, New York, 1980.
56. S. Goldberg: Curvature and Homology, Dover, New York, 1982.
57. R.O. Wells: Differential Analysis on Complex Manifolds, GTM 65, Springer, New York, 1980.
58. K. Yano: Differential Geometry on Complex and Almost Complex Spaces, Pergamon Press, New York, 1965.
59. L. O'Raifeartaigh: Spontaneous Symmetry Breaking for Chiral Scalar Superfields, Nucl. Phys. B96, 331 (1975).
60. P. Fayet and J. Iliopoulos: Spontaneously Broken Supergauge Symmetries and Goldstone Spinors, Phys. Lett. 51B, 461 (1974).
61. C. Quigg: Gauge Theories of the Strong, Weak and Electromagnetic Interactions, Benjamin/Cummings, Reading, Massachusetts, 1983.
62. L. Okun: Leptons and Quarks, North-Holland Publ. Co., Amsterdam, 1982.

63. P. Ramond: Field Theory, A Modern Primer, Benjamin/ Cummings, Reading, Massachusetts, 1981.
64. C. Itzykson and J.B. Zuber: Quantum Field Theory, McGraw-Hill, New York, 1980.
65. J.D. Bjorken and S.D. Drell: Relativistic Quantum Fields, McGraw-Hill, New York, 1965, Chapter 1.
66. See e.g. C. Itzykson and J.B. Zuber, ref. 64.
67. See e.g. T.-P. Cheng and L.-F. Li: Gauge Theory of Elementary Particle Physics, The Clarendon Press, Oxford, 1984, Chapter 5.3 .
68. S. Ferrara, L. Giradello and F. Palumbo: General Mass Formula in Broken Supersymmetry, Phys. Rev. D20, 403 (1979).
69. L. Giradello and J. Iliopoulos: Quantum Corrections to a Mass Formula in Broken Supersymmetry, Phys. Lett. B88, 85 (1975).
70. See e.g.T.-P. Cheng and L.-F. Li, ref. 67, Chapter 8.
71. See e.g.T.-P. Cheng and L.-F- Li, ref. 67, Section 8.3.
72. S. Ferrara and B. Zumino: Supergauge Invariant Yang-Mills Theories, Nucl. Phys. B79, 413 (1974).
73. A. Salam and J. Strathdee: Supersymmetry and Non-abelian Gauges, Phys. Lett. 51B, 353 (1974).
74. T. Bröcker and T. tom Dieck, ref. 30, Proposition (5.13).
75. J. Wess and J. Bagger, ref. 2, Chapter VII, Equation (7.13).
76. W. Miller, ref. 31, Lemma 5.3 .

INDEX